The Quantum Theory of Nonlinear Optics

Playing a prominent role in communications, quantum science, and laser physics, quantum nonlinear optics is an increasingly important field. This book presents a self-contained treatment of field quantization, and covers topics such as the canonical formalism for fields, phase-space representations, and the encompassing problem of quantization of electrodynamics in linear and nonlinear media. Starting with a summary of classical nonlinear optics, it then explains in detail the calculation techniques for quantum nonlinear optical systems and their applications, quantum and classical noise sources in optical fibers, and applications of nonlinear optics to quantum information science. Supplemented by end-of-chapter exercises and detailed examples of applications to different systems, this book is a valuable resource for graduate students and researchers in nonlinear optics, condensed matter physics, quantum information, and atomic physics. A solid foundation in quantum mechanics and classical electrodynamics is assumed, but no prior knowledge of nonlinear optics is required.

Peter D. Drummond is a Distinguished Professor in the Faculty of Science, Engineering and Technology, Swinburne University of Technology, Melbourne. His current research focuses on ultra-cold atomic physics, quantum information, and bio-informatics.

Mark Hillery is a Professor at the Department of Physics and Astronomy, Hunter College, and in the Physics Graduate Program, Graduate Center, City University of New York. His research focuses on the field of quantum information.

This book forms part of an informal series of books, all of which originated as review articles published in *Acta Physica Slovaca*. The journal can be accessed for free at www.physics.sk/aps.

Vladimir Buzek, editor of the journal

"This book is a valuable contribution to the scientific literature by addressing issues that fall at the boundary between quantum optics and nonlinear optics. It is exactly at this boundary where one might expect exciting advances to develop in the coming years. The authors have done a good job in selecting the topics for inclusion in their very fine text."

Robert W. Boyd
Canada Excellence Research Chair in Quantum Nonlinear Optics, University of Ottawa
Professor of Optics and of Physics, University of Rochester

"Two of the pioneers of quantum optics have produced a clear introduction to the quantum theory of nonlinear optical processes with applications ranging from laser physics to quantum information. The powerful method of P representations to treat quantum stochastic processes is introduced with exemplary clarity and many examples. This is an essential introduction for graduate students, who will appreciate the carefully chosen problem sets, and a valuable reference for experienced researchers in the field."

Gerard J. Milburn
Centre Director, Centre for Engineered Quantum Systems, University of Queensland

The Quantum Theory of Nonlinear Optics

PETER D. DRUMMOND
Swinburne University of Technology, Melbourne

MARK HILLERY
Hunter College, City University of New York

CAMBRIDGE
UNIVERSITY PRESS

University Printing House, Cambridge CB2 8BS, United Kingdom

One Liberty Plaza, 20th Floor, New York, NY 10006, USA

477 Williamstown Road, Port Melbourne, VIC 3207, Australia

314-321, 3rd Floor, Plot 3, Splendor Forum, Jasola District Centre, New Delhi - 110025, India

79 Anson Road, #06-04/06, Singapore 079906

Cambridge University Press is part of the University of Cambridge.

It furthers the University's mission by disseminating knowledge in the pursuit of education, learning and research at the highest international levels of excellence.

www.cambridge.org
Information on this title: www.cambridge.org/9781107004214

Cover artwork by Igor Minarik.

First published 2014
Reprinted 2016

A catalogue record for this publication is available from the British Library

Library of Congress Cataloging in Publication data
Drummond, P. D. (Peter D.), 1950–
The quantum theory of nonlinear optics / Peter D. Drummond, Swinburne University of Technology, Melbourne, Mark Hillery, Hunter College, City University of New York.
pages cm
Includes bibliographical references and index.
ISBN 978-1-107-00421-4 (hardback)
1. Nonlinear optics. 2. Quantum optics. 3. Quantum theory. I. Hillery, Mark Stephen. II. Title.
QC446.2.D78 2014
535′.2 – dc23 2013030430

ISBN 978-1-107-00421-4 Hardback

Contents

Preface

This book grew out of our work in the field of the quantum theory of nonlinear optics. Some of this work we have done together and some following our own paths. One major emphasis of this work has been the quantization of electrodynamics in the presence of dielectric media. This is a subject that is often given short shrift in many treatments, and we felt that a book in which it receives a more extensive discussion was warranted.

M.H. would like to thank his thesis advisor, Eyvind Wichmann, for an excellent education in quantum mechanics and quantum field theory, and M. Suhail Zubairy for introducing him to the field of nonlinear optics with quantized fields. Others who played a major role are Leonard Mlodinow, with whom the initial work on quantization in nonlinear media was done, and Janos Bergou and Vladimir Buzek, long-time collaborators with whom it has been a pleasure to work. He also thanks Carol Hutchins for many things.

P.D.D. wishes to acknowledge his parents and family for their invaluable support. The many colleagues who helped form his approach include Crispin Gardiner and the late Dan Walls, who pioneered quantum optics in New Zealand. Subhash Chaturvedi, Howard Carmichael, Steve Carter, Paul Kinsler, Joel Corney, Piotr Deuar and Kaled Dechoum have contributed greatly to this field. He also thanks Margaret Reid, who has played a leading role in some of the developments outlined in the quantum information section, and Qiongyi He, who provided illustrations.

Introduction

Nonlinear optics is the study of the response of dielectric media to strong optical fields. The fields are sufficiently strong that the response of the medium is, as its name implies, nonlinear. That is, the polarization, which is the dipole moment per unit volume in the medium, is not a linear function of the applied electric field. In the equation for the polarization, there is a linear term, but, in addition, there are terms containing higher powers of the electric field. This leads to significant new types of behavior, one of the most notable being that frequencies different from that of the incident electromagnetic wave, such as harmonics or subharmonics, can be generated. Linear media do not change the frequency of light incident upon them. The first observation of a nonlinear optical effect was, in fact, second-harmonic generation – a laser beam entering a nonlinear medium produced a second beam at twice the frequency of the original. Another type of behavior that becomes possible in nonlinear media is that the index of refraction, rather than being a constant, is a function of the intensity of the light. For a light beam with a nonuniform intensity profile, this can lead to self-focusing of the beam.

Most nonlinear optical effects can be described using classical electromagnetic fields, and, in fact, the initial theory of nonlinear optics was formulated assuming the fields were classical. When the fields are quantized, however, a number of new effects emerge. Quantized fields are necessary if we want to describe fields that originate from spontaneous emission. For example, in a process known as spontaneous parametric down-conversion, a beam of light at one frequency, the pump, produces a beam at half the original frequency, the signal. This second beam is a result of spontaneous emission. The quantum properties of the down-converted beam are novel, a result of the fact that its photons are produced in pairs, one pump photon disappearing to produce, simultaneously, two signal photons. This leads to strong correlations between pairs of photons in the signal beam. In particular, the photons produced in this way can be quantum mechanically entangled. In addition, the signal beam can have smaller phase fluctuations than is possible with classical light. Both of these properties have made light produced by parametric down-conversion useful for applications in the field of quantum information.

The quantization of electrodynamics in nonlinear media is not straightforward, and so we will treat the canonical quantization of fields in some detail. Field quantization is a subject that is treated in just a few pages in most books on quantum optics, but here we will be much more thorough. We will discuss two approaches to this problem. The first is the quantization of the macroscopic Maxwell equations. The goal here is to obtain a quantized theory that has the macroscopic Maxwell equations as its Heisenberg equations of motion. The second approach is to make a model for the medium and quantize the entire matter–field system. Once this is done, an effective Hamiltonian describing the behavior of the fields in the medium can be found.

Many nonlinear optical systems can be discussed by employing only a few modes, and we shall employ this approach in discussing a number of simple systems. This will allow us to discuss some of the quantum mechanical correlations of the fields that these systems produce. Attributes such as squeezing and entanglement are properties that quantum fields can have but that classical fields cannot. As we shall see, quantum fields with these unusual features are produced by nonlinear optical systems. We initially explore the properties of these systems in free space, but we then move on to see what happens when they are placed in an optical cavity.

In order to discuss systems in cavities, we need to present the theory of open quantum systems and some of the mathematical techniques that have been developed to treat them. The nonlinear interaction couples a small number of cavity modes either to themselves or to each other, but these modes are coupled to modes outside the cavity through an output mirror. The external modes can be treated as a reservoir. Thus, our discussion of nonlinear devices in cavities will entail the introduction of reservoirs, operator Langevin and master equations, and techniques for turning these operator equations into c-number equations that can be more easily solved. It will also entail an input–output theory to relate the properties of the field inside the cavity to those of the field outside the cavity. It is the field outside the cavity, of course, that is usually measured.

It is also possible to treat more complicated systems, such as a field propagating in a nonlinear fiber. As well being the backbone of modern communications systems, optical fibers can generate strong nonlinear and quantum effects. They support the existence of quantum solitons, and their output fields can demonstrate squeezing and polarization squeezing. All of these effects have been demonstrated experimentally. In comparing theory to experiment, it is necessary to take into account the quantum noise sources in fibers, and we show how this can be done. This allows a detailed and quantitative test of the theoretical techniques explained here. It is also a demonstration of the quantum dynamics of a many-body system, since, as we shall see, fiber optics is equivalent to a system of interacting bosons in one dimension.

We conclude with a short chapter on the applications of nonlinear optics to the field of quantum information. We show how a degenerate parametric amplifier can be used to approximately clone quantum states, and how the squeezed states that are produced by such a device can be used to teleport them. We explain how these quantum states can be used to demonstrate the Einstein–Podolsky–Rosen paradox, and to generate a violation of the Bell inequality, issues that are important for fundamental physics. Understanding these issues and how they can be experimentally tested requires an understanding of both quantum mechanics and nonlinear optics.

The reader for whom this book is intended is a graduate student who has taken one-year graduate courses in electromagnetic theory and quantum theory. Essential results from these areas are summarized where needed. We do not assume any knowledge of quantum optics or quantum field theory. We also hope that physicists working in other fields will find the book useful.

We would like to emphasize that this textbook is not a review article, so we have made no attempt to provide a comprehensive bibliography of the field. We provide limited lists of additional reading at the ends of the chapters. This text is a result of our having worked in this field, and it emphasizes the points of view we have developed in doing so. There are certainly many other ways in which the quantum theory of nonlinear optics can be approached, and, in some cases, these are presented in textbooks by other authors.

Our presentation begins with a very brief survey of some topics in the classical theory on nonlinear optics. It is useful to see some of the basic ideas in a simpler classical context before jumping into the more complicated quantum case. The first chapter will provide a rather quick overview that will nonetheless serve as a foundation for what follows.

1 Classical nonlinear optics

Before discussing nonlinear optics with quantized fields, it is useful to have a look at what happens with classical electromagnetic fields in nonlinear dielectric media. The theory of nonlinear optics was originally developed using classical fields by Armstrong, Bloembergen, Ducuing and Pershan in 1962, stimulated by an experiment by Franken, Hill, Peters and Weinreich in which a second harmonic of a laser field was produced by shining the laser into a crystal. This classical theory is sufficient for many applications. For the most part, quantized fields were introduced later, although a quantum theory for the parametric amplifier, a nonlinear device in which three modes are coupled, was developed by Louisell, Yariv and Siegman as early as 1961. In any case, a study of the classical theory will give us an idea of some of the effects to look for when we formulate the more complicated quantum theory.

What we will present here is a very short introduction to the subject. Our intent is to use the classical theory to present some of the basic concepts and methods of nonlinear optics. Further information can be found in the list of additional reading at the end of the chapter. The discussion here is based primarily on the presentations in the books by N. Bloembergen and by R. W. Boyd.

1.1 Linear polarizability

We wish to survey some of the effects caused by the linear polarizability of a dielectric medium. When an electric field $\mathbf{E}(\mathbf{r}, t)$ is applied to a dielectric medium, a polarization, that is, a dipole moment per unit volume, is created in the medium. Maxwell's equations for a nonmagnetic material, but with the polarization included, are

$$\begin{aligned} \nabla \cdot \mathbf{D} &= 0, & \nabla \cdot \mathbf{B} &= 0, \\ \nabla \times \mathbf{E} &= -\frac{\partial \mathbf{B}}{\partial t}, & \nabla \times \mathbf{H} &= \frac{\partial \mathbf{D}}{\partial t}. \end{aligned} \tag{1.1}$$

Here $\mathbf{D} = \epsilon_0 \mathbf{E} + \mathbf{P}$ is the displacement field, and $\mathbf{B} = \mu \mathbf{H}$ is the magnetic field. We use the bold notation $\mathbf{D} = (D_1, D_2, D_3)$ to indicate a 3-vector field at position $\mathbf{r} = (x, y, z)$ and time t, and generally omit the space-time arguments of fields for brevity. We use SI units, so ϵ_0 is the vacuum permittivity, and μ is the magnetic permeability. Here we separate the polarizability term so that we can more readily analyze nonlinear effects.

Differentiating the equation for $\nabla \times \mathbf{H}$ with respect to time and making use of the equation for $\nabla \times \mathbf{E}$ gives

$$\nabla \times \nabla \times \mathbf{E} = -\mu \frac{\partial^2 \mathbf{D}}{\partial t^2}. \tag{1.2}$$

An alternative form, in terms of the polarization field, is

$$\nabla \times \nabla \times \mathbf{E} + \frac{1}{c^2}\frac{\partial^2 \mathbf{E}}{\partial t^2} = -\mu \frac{\partial^2 \mathbf{P}}{\partial t^2}. \tag{1.3}$$

Here $c = 1/\sqrt{\mu\epsilon_0}$. Examining Eq. (1.3), we first note that $\nabla \times \nabla \times \mathbf{E} = \nabla(\nabla \cdot \mathbf{E}) - \nabla^2 \mathbf{E}$. In free space, $\nabla \cdot \mathbf{E} = 0$, but this is no longer true in a medium. In many cases it is, however, small and can be neglected. For example, if the field is close to a plane wave, this term will be small. We shall assume that it can be neglected in most situations we consider, which leads to the form of the wave equation most commonly used in nonlinear optics:

$$\nabla^2 \mathbf{E} - \frac{1}{c^2}\frac{\partial^2 \mathbf{E}}{\partial t^2} = \mu \frac{\partial^2 \mathbf{P}}{\partial t^2}. \tag{1.4}$$

In typical cases of interest, $\mu \approx \mu_0$, where μ_0 is the vacuum permeability, so that c is the vacuum light velocity to a good approximation. We retain the full permeability in the Maxwell equations for generality, as there are small contributions to the magnetic permeability – typically of $O(10^{-6})$ – in dielectric materials.

1.1.1 Linear polarizability

If the field is not too strong, the response of the medium is linear. This means that the polarization, $\mathbf{P}$, is linear in the applied field, and, in general,

$$P_j = \epsilon_0[\boldsymbol{\chi}^{(1)} \cdot \mathbf{E}]_j = \epsilon_0 \sum_{k=1}^{3} \chi_{jk}^{(1)} E_k. \tag{1.5}$$

In this equation, $[\boldsymbol{\chi}^{(1)}]_{ij} = \chi_{ij}^{(1)}$ is the linear susceptibility tensor of the medium. This tells us that the polarization acts as a source for the field, and, in particular, if the polarization has terms oscillating at a particular frequency, then those terms will give rise to components of the field oscillating at the same frequency. For a linear dielectric medium with no significant magnetization, the permeability equals the vacuum permeability. In a linear, isotropic medium, we can omit the tensor subscripts, writing $\mathbf{P} = \epsilon_0 \chi^{(1)} \mathbf{E}$. Since

$$\mathbf{D} = \boldsymbol{\epsilon} \cdot \mathbf{E} = \epsilon_0 \mathbf{E} + \mathbf{P}, \tag{1.6}$$

it follows that the electric permittivity in a dielectric is given by

$$\boldsymbol{\epsilon} = \epsilon_0[1 + \boldsymbol{\chi}^{(1)}], \tag{1.7}$$

and the exact linear wave equation, Eq. (1.2), reduces to

$$\nabla \times \nabla \times (\boldsymbol{\epsilon}^{-1} \cdot \mathbf{D}) = -\mu \frac{\partial^2 \mathbf{D}}{\partial t^2}. \tag{1.8}$$

An alternative form, valid for homogeneous, isotropic dielectrics, is obtained from Eq. (1.4), which simplifies to

$$\nabla^2 \mathbf{E} = \frac{1}{v_p^2}\frac{\partial^2 \mathbf{E}}{\partial t^2}, \tag{1.9}$$

where $v_p = 1/\sqrt{\mu\epsilon}$ is the phase velocity of electromagnetic waves in the medium.

The results in this chapter will generally assume an input of a plane-wave, nearly monochromatic, laser beam with polarization $\hat{\mathbf{e}}$, angular frequency $\omega = 2\pi f$, wavevector $k = 2\pi/\lambda$, and slowly varying envelope $\mathcal{E}$ in time,

$$\mathbf{E}(\mathbf{r}, t) = \mathcal{E}(\mathbf{r}, t)\hat{\mathbf{e}}\, e^{-i\omega t} + c.c. \tag{1.10}$$

(where $c.c.$ denotes complex conjugate). For an envelope that is slowly varying in time *and* space, we introduce

$$\mathbf{E}(\mathbf{r}, t) = \mathcal{A}(\mathbf{r}, t)\hat{\mathbf{e}}\, e^{-i(\omega t - kx)} + c.c. \tag{1.11}$$

If $\mathbf{P} = 0$ and $\mu = \mu_0$, the resulting speed of light in the vacuum is $c = 1/\sqrt{\epsilon_0\mu_0}$. More generally, the phase velocity of electromagnetic radiation in the dielectric medium is v_p, and is given by

$$v_p = \frac{1}{\sqrt{\mu\epsilon}} = \frac{c}{n_r} = \frac{\omega}{k}. \tag{1.12}$$

Here n_r is the refractive index, which is given in terms of the linear susceptibility by

$$n_r = \sqrt{1 + \chi^{(1)}}. \tag{1.13}$$

Inserting Eq. (1.11) into Eq. (1.9), and dropping second-derivative terms in time and the propagation, that is, the x, direction (that these terms are small follows from the assumption that the envelope varies slowly compared to the wavelength and optical frequency), we find that

$$\left[\frac{\partial}{\partial x} + \frac{1}{v_p}\frac{\partial}{\partial t} - \frac{i}{2k}\nabla_\perp^2\right]\mathcal{A}(\mathbf{r}, t) = 0. \tag{1.14}$$

Here, $\nabla_\perp^2 \equiv \partial^2/\partial y^2 + \partial^2/\partial z^2$. This equation is called the paraxial wave equation. It has characteristic traveling plane-wave solutions of the form $f(x - v_p t)$, consisting of waveforms traveling at the phase velocity, v_p. When a transverse variation is included, this equation leads to focusing and diffraction effects. Here $\lambda f = v_p = c/n_r$, in terms of the frequency f and wavelength λ. Since $\nabla \times \mathbf{E} = -\partial \mathbf{B}/\partial t$, it follows that the corresponding magnetic field is

$$\mathbf{B}(\mathbf{r}, t) = (\mathcal{A}/v_p)\hat{\mathbf{k}} \times \hat{\mathbf{e}}\, e^{-i(\omega t - kx)} + c.c. \tag{1.15}$$

Let us note that our definition of complex amplitudes follows the convention of Glauber. In some texts, complex amplitudes are defined as $\mathbf{E}(\mathbf{r}, t) = \Re[\mathcal{E}_C(\mathbf{r}, t)\hat{\mathbf{e}}\, e^{i\omega t}]$. These are related to ours by $\mathcal{E}_C = 2\mathcal{E}$, and give rise to differences of powers of 2^{n-1} in the nonlinear equations for an nth-order nonlinearity in later sections.

1.1.2 Energy density, intensity and power

The dispersionless energy density $\mathcal{H}$ has the usual classical form of

$$\mathcal{H} = \tfrac{1}{2}[\epsilon|\mathbf{E}|^2(t) + \mu|\mathbf{H}|^2(t)]. \tag{1.16}$$

Defining $\mathcal{H}_{av}$ as the time-averaged energy density, the magnetic and electric field contributions to the energy of a plane-wave solution to Maxwell's equations are equal, and the total intensity for a dispersionless medium is

$$I_0 = v_p \mathcal{H}_{av} = 2v_p\epsilon|\mathcal{A}(\mathbf{r}, t)|^2. \tag{1.17}$$

More rigorously, we demonstrate later that there are dispersive corrections to the above results for energy, intensity and power, due to the frequency dependence of the dielectric response. These are assumed negligible for simplicity here.

Lasers have a transverse envelope function $u(\mathbf{r})$ that is typically Gaussian, with a beam radius or 'waist' of W_0 that varies in the x direction. At a beam focus, this depends primarily on the transverse coordinate $\mathbf{r}_\perp = (y, z)$, so that

$$\mathcal{E}(\mathbf{r}_\perp) = \mathcal{E}(0)e^{-|\mathbf{r}_\perp|^2/W_0^2}. \tag{1.18}$$

Integrating over the beam waist, the total laser power is

$$P = \tfrac{1}{2}\pi I_0 W_0^2 = \pi v_p \epsilon W_0^2 |\mathcal{E}|^2. \tag{1.19}$$

We will generally ignore transverse effects. However, these are important in understanding how beam intensities, which cause nonlinear effects, are related to laser powers. We will treat the more general case of dispersive energy later in this chapter.

1.2 Nonlinear polarizability

If the field is sufficiently strong, the linear relation breaks down and nonlinear terms must be taken into account. In Bloembergen's approach, we expand the polarization in a Taylor expansion in $\mathbf{E}$ to give

$$P_j = \epsilon_0 \left[\sum_k \chi_{jk}^{(1)} E_k + \sum_{k,l} \chi_{jkl}^{(2)} E_k E_l + \sum_{k,l,m} \chi_{jklm}^{(3)} E_k E_l E_m + \cdots \right]. \tag{1.20}$$

Here, we have kept the first three terms in the power series expansion of the polarization in terms of the field. The quantities $\chi^{(2)}$ and $\chi^{(3)}$ are the second- and third-order nonlinear susceptibilities, respectively. We should also note that Eq. (1.20) is often written as a matrix or generalized tensor multiplication, in the form

$$\begin{aligned} \mathbf{P}(\mathbf{E}) &= \mathbf{P}^L + \mathbf{P}^{NL} \\ &= \epsilon_0 \sum_{n>0} \chi^{(n)} : \mathbf{E}^{\otimes n}, \end{aligned} \tag{1.21}$$

where $\mathbf{P}^L \equiv \epsilon_0 \boldsymbol{\chi}^{(1)} \cdot \mathbf{E} = (\boldsymbol{\epsilon} - 1) \cdot \mathbf{E}$ is the linear response, while $\mathbf{P}^{NL}$ is the nonlinear polarization. We use the notation $\mathbf{E}^{\otimes n}$ to indicate a vector Kronecker product, mapping a vector into an nth-order tensor, so

$$[\mathbf{E}^{\otimes n}]_{i_1 \ldots i_n} \equiv E_{i_1} \cdots E_{i_n}. \tag{1.22}$$

We can use the $\boldsymbol{\chi}$ coefficients to expand the displacement field $\mathbf{D}$ directly in terms of $\mathbf{E}$, which simplifies results in later chapters. We define $\boldsymbol{\epsilon}^{(1)} = \boldsymbol{\epsilon}$ and $\boldsymbol{\epsilon}^{(n)} = \epsilon_0 \boldsymbol{\chi}^{(n)}$, so that $\boldsymbol{\epsilon}^{(n)}$ becomes an $(n+1)$th-order tensor. Then

$$\mathbf{D}(\mathbf{E}) = \sum_{n>0} \boldsymbol{\epsilon}^{(n)} : \mathbf{E}^{\otimes n}. \tag{1.23}$$

Not all the terms in this series are necessarily present. The even terms like $\chi^{(2)}$ are only present if the medium is not invariant under spatial inversion ($\mathbf{r} \to -\mathbf{r}$). This follows from the fact that, if the medium is invariant under spatial inversion, the $\chi^{(2)}$ for the inverted medium will be the same as that for the original medium, i.e. under spatial inversion, we will have $\chi^{(2)} \to \chi^{(2)}$. However, under spatial inversion, we also have that $\mathbf{P} \to -\mathbf{P}$ and $\mathbf{E} \to -\mathbf{E}$. Consequently, while $\mathbf{P} \to -\mathbf{P}$ implies that we should have

$$\chi^{(2)} : \mathbf{E} \otimes \mathbf{E} \to -\chi^{(2)} : \mathbf{E} \otimes \mathbf{E}, \tag{1.24}$$

the relations $\chi^{(2)} \to \chi^{(2)}$ and $\mathbf{E} \to -\mathbf{E}$ show us that instead we have

$$\chi^{(2)} : \mathbf{E} \otimes \mathbf{E} \to \chi^{(2)} : \mathbf{E} \otimes \mathbf{E}. \tag{1.25}$$

The only way these conditions can be consistent is if $\chi^{(2)} = 0$. Therefore, for many materials, we do, in fact, have $\chi^{(2)} = 0$, and all even terms vanish for the same reason. Then, the first nonzero nonlinear susceptibility is $\chi^{(3)}$. The other symmetry properties of the medium also directly affect the susceptibilities. For example, in the common case of an amorphous solid with complete spherical symmetry, the susceptibilities are diagonal matrices and tensors. In this case, we refer to them as scalars, and one can simply drop the indices. Similarly, for plane-polarized radiation, under conditions where the polarization is always in the same plane, it is also possible to ignore the indices, though for a different reason.

We can also define an inverse permittivity tensor, $\boldsymbol{\eta}^{(n)}$, as a coefficient of a power series expansion of the macroscopic *electric* field in terms of the macroscopic displacement field. This greatly simplifies the treatment of quantized fields. In the subsequent chapters on quantization, we will make use of this expansion, which has the form

$$\mathbf{E}(\mathbf{D}) = \sum_{n>0} \boldsymbol{\eta}^{(n)} : \mathbf{D}^{\otimes n}. \tag{1.26}$$

This is an equally valid approach to nonlinear response, as these are simply two alternative power series expansions.

It is possible to express the inverse permittivity tensors $\eta^{(j)}$ in terms of the permittivities $\epsilon^{(j)}$. Let us assume that we know the $\chi^{(n)}$, and hence the $\epsilon^{(n)}$ coefficients already. Combining the two expansions, we have

$$\mathbf{E} = \sum_{n>0} \boldsymbol{\eta}^{(n)} : \left[\sum_{m>0} \boldsymbol{\epsilon}^{(m)} : \mathbf{E}^{\otimes m} \right]^{\otimes n} \tag{1.27}$$

We now simply equate equal powers of **E**, so that formally:

$$\begin{aligned} \mathbf{1} &= \boldsymbol{\eta}^{(1)} : \boldsymbol{\epsilon}^{(1)}, \\ \mathbf{0} &= \boldsymbol{\eta}^{(2)} : \boldsymbol{\epsilon}^{(1)}\boldsymbol{\epsilon}^{(1)} + \boldsymbol{\eta}^{(1)} : \boldsymbol{\epsilon}^{(2)}, \\ \mathbf{0} &= \boldsymbol{\eta}^{(3)} : \boldsymbol{\epsilon}^{(1)}\boldsymbol{\epsilon}^{(1)}\boldsymbol{\epsilon}^{(1)} + \boldsymbol{\eta}^{(2)} : \boldsymbol{\epsilon}^{(2)}\boldsymbol{\epsilon}^{(1)} + \boldsymbol{\eta}^{(2)} : \boldsymbol{\epsilon}^{(1)}\boldsymbol{\epsilon}^{(2)} + \boldsymbol{\eta}^{(1)} : \boldsymbol{\epsilon}^{(3)}. \end{aligned} \tag{1.28}$$

Writing this out in detail (and recalling that we define $\eta_{ij} = \eta_{ij}^{(1)}$), we see that, for the lowest-order terms,

$$\eta_{ij} = [\epsilon^{-1}]_{ij}, \quad \eta_{jnp}^{(2)} = -\eta_{jk}\epsilon_{klm}^{(2)}\eta_{ln}\eta_{mp}, \tag{1.29}$$

where we introduce the Einstein summation convention, in which repeated indices are summed over, and we note that there will be contributions from all the terms $\epsilon^{(1)}, \ldots, \epsilon^{(n)}$ to the inverse nonlinear coefficient $\eta^{(n)}$.

1.2.1 Second-order nonlinearity

We can calculate the effects of nonlinearities by substituting the response functions, i.e. the expansions for the polarization of the medium, into Maxwell's equations. In our initial survey of nonlinear optical effects, we shall ignore all indices, and treat all quantities as scalars; this corresponds to an assumption of plane polarization in a single direction. Let us first look at second-order nonlinearities. If the applied field oscillates at frequency ω,

$$E(t) = \mathcal{E}_0[e^{i\omega t} + e^{-i\omega t}] = 2\mathcal{E}_0 \cos \omega t, \tag{1.30}$$

then the nonlinear part of the polarization, P^{NL}, will be

$$P^{NL}(t) = \epsilon_0 \chi^{(2)} E(t)^2 = 2\epsilon_0 \chi^{(2)} \mathcal{E}_0^2 (1 + \cos 2\omega t). \tag{1.31}$$

The polarization has a term oscillating at twice the applied frequency, and this will give rise to a field whose frequency is also 2ω. This process is known as second-harmonic generation. It can be, and is, used to double the frequency of the output of a laser by sending the beam through an appropriate material, that is, one with a nonzero value of $\chi^{(2)}$. As was mentioned earlier, this was the first nonlinear optical effect that was observed.

Now suppose our applied field oscillates at two frequencies:

$$E(t) = 2[\mathcal{E}_1 \cos \omega_1 t + \mathcal{E}_2 \cos \omega_2 t]. \tag{1.32}$$

The nonlinear polarization is then

$$\begin{aligned} P^{NL}(t) = 2\epsilon_0 \chi^{(2)} \{ &\mathcal{E}_1^2 (1 + \cos 2\omega_1 t) + \mathcal{E}_2^2 (1 + \cos 2\omega_2 t) \\ &+ 2\mathcal{E}_1\mathcal{E}_2 [\cos(\omega_1 + \omega_2)t + \cos(\omega_1 - \omega_2)t] \}. \end{aligned} \tag{1.33}$$

In this case, not only do we have terms oscillating at twice the frequencies of the components of the applied field, but we also have terms oscillating at the sum and difference of their frequencies. These processes are called sum- and difference-frequency generation, respectively.

1.2.2 Third-order nonlinearity

Let us move on to a third-order nonlinearity. For an applied field oscillating at a single frequency, as before, we find that (assuming that $\chi^{(2)} = 0$)

$$P^{NL}(t) = 2\epsilon_0 \chi^{(3)} \mathcal{E}_0^3 [\cos 3\omega t + 3\cos \omega t]. \quad (1.34)$$

The first term will clearly cause a field at 3ω, the third harmonic of the applied field, to be generated. In order to see the effect of the second term, it is useful to combine the linear and nonlinear parts of the polarization to get the total polarization,

$$P(t) = 2\epsilon_0(\chi^{(1)} + 3\chi^{(3)}\mathcal{E}_0^2)\mathcal{E}_0 \cos \omega t + 2\epsilon_0 \chi^{(3)} \mathcal{E}_0^3 \cos 3\omega t. \quad (1.35)$$

When there is no nonlinear polarization, the polarization is proportional to the field, and the constant of proportionality, $\chi^{(1)}$, is directly related to the refractive index of the material. When there is a nonlinearity, we see that the component of the polarization at the same frequency as the applied field is similar to what it is in the linear case, except that

$$\chi^{(1)} \to \chi^{(1)} + 3\chi^{(3)}\mathcal{E}_0^2. \quad (1.36)$$

This results in a refractive index that depends on the intensity of the applied field according to Eqs (1.13) and (1.17). The refractive index can therefore be written as

$$n_r(I) = \sqrt{1 + \chi^{(1)} + 3\chi^{(3)} I/(2v_p\epsilon)}. \quad (1.37)$$

This can be expanded as a power series in the intensity, so that, to lowest order,

$$n_r(I) = n_1 + n_2 I + \cdots, \quad (1.38)$$

where the nonlinear refractive index, n_2, is given by

$$n_2 = \frac{3\chi^{(3)}}{4\epsilon c}. \quad (1.39)$$

This is often called the Kerr effect, after its original discoverer.

1.3 Frequency dependence and dispersion

So far we have assumed that the response of the medium to an applied field, that is, the polarization at time t, depends only on the electric field at time t. This is, of course, an idealization. The response of a medium is not instantaneous, so that the polarization at time t depends on the field at previous times, not just the field at time t. The most general expression of this type has the form

$$P_j(t) = \epsilon_0 \Bigg[\sum_{k=1}^{3} \int_0^\infty d\tau\, \tilde{\chi}_{jk}^{(1)}(\tau) E_k(t-\tau) + \sum_{k,l=1}^{3} \int_0^\infty d^2\boldsymbol{\tau}\, \tilde{\chi}_{jkl}^{(2)}(\boldsymbol{\tau}) E_k(t-\tau_1) E_l(t-\tau_2) + \sum_{k,l,m=1}^{3} \int_0^\infty d^3\boldsymbol{\tau}\, \tilde{\chi}_{jklm}^{(3)}(\boldsymbol{\tau}) E_k(t-\tau_1) E_l(t-\tau_2) E_m(t-\tau_3) + \cdots \Bigg]. \tag{1.40}$$

In the case of a linear medium, this is expressed as

$$\mathbf{P}(t) = \epsilon_0 \int_0^\infty d\tau\, \tilde{\boldsymbol{\chi}}^{(1)}(\tau) \cdot \mathbf{E}(t-\tau). \tag{1.41}$$

1.3.1 Frequency-dependent susceptibility

Taking the Fourier transform of both sides of Eq. (1.41), and defining

$$\mathbf{P}(\omega) = \frac{1}{2\pi} \int_{-\infty}^{\infty} dt\, e^{-i\omega t} \mathbf{P}(t),$$
$$\mathbf{E}(\omega) = \frac{1}{2\pi} \int_{-\infty}^{\infty} dt\, e^{-i\omega t} \mathbf{E}(t), \tag{1.42}$$

we find that

$$\mathbf{P}(\omega) = \epsilon_0 \boldsymbol{\chi}^{(1)}(\omega) \cdot \mathbf{E}(\omega),$$
$$\mathbf{D}(\omega) = \boldsymbol{\epsilon}(\omega) \cdot \mathbf{E}(\omega), \tag{1.43}$$

where the Fourier transforms of the linear susceptibility and permittivity tensors are defined as

$$\boldsymbol{\chi}^{(1)}(\omega) = \int_0^\infty d\tau\, e^{-i\omega\tau} \tilde{\boldsymbol{\chi}}^{(1)}(\tau),$$
$$\boldsymbol{\epsilon}(\omega) = \epsilon_0 [1 + \boldsymbol{\chi}^{(1)}(\omega)]. \tag{1.44}$$

Therefore, we see that a medium response that is not instantaneous causes the linear susceptibility to become frequency-dependent. This phenomenon is known as dispersion. We note that the medium response is causal, that is, the polarization at time t depends only on the field at earlier times. This has implications for the polarizability, in particular that its Fourier transform should obey the Kramers–Kronig relation, which follows from the fact that $\boldsymbol{\epsilon}(\omega)$ can be extended to a function of a complex variable that is analytic in the lower half-plane and goes to ϵ_0 as $\omega \to \infty$ in the lower half-plane. This implies that there is a relation between the real and imaginary parts of $\boldsymbol{\epsilon}(\omega)$ on the real axis. Setting

$\epsilon(\omega) = \Re[\epsilon(\omega)] + i\Im[\epsilon(\omega)]$, we have that

$$\Re[\epsilon(\omega)]/\epsilon_0 - 1 = \frac{-2}{\pi} \mathrm{P} \int_0^\infty d\omega' \frac{\omega' \Im[\epsilon(\omega')]/\epsilon_0}{\omega'^2 - \omega^2},$$

$$\Im[\epsilon(\omega)]/\epsilon_0 = \frac{2\omega}{\pi} \mathrm{P} \int_0^\infty d\omega' \frac{\Re[\epsilon(\omega)]/\epsilon_0 - 1}{\omega'^2 - \omega^2}, \tag{1.45}$$

where P indicates the principal value part of the integral.

Dispersion also causes the nonlinear susceptibilities to become frequency-dependent. For the second- and third-order nonlinearities, again assuming isotropy, we have

$$\mathbf{P}^{(2)}(t) = \epsilon_0 \int_0^\infty d\tau_1 \int_0^\infty d\tau_2 \, \tilde{\boldsymbol{\chi}}^{(2)}(\tau_1, \tau_2) : \mathbf{E}(t - \tau_1) \otimes \mathbf{E}(t - \tau_2),$$

$$\mathbf{P}^{(3)}(t) = \epsilon_0 \int_0^\infty d\tau_1 \int_0^\infty d\tau_2 \int_0^\infty d\tau_3$$

$$\times \tilde{\boldsymbol{\chi}}^{(3)}(\tau_1, \tau_2, \tau_3) : \mathbf{E}(t - \tau_1) \otimes \mathbf{E}(t - \tau_2) \otimes \mathbf{E}(t - \tau_3), \tag{1.46}$$

where $\mathbf{P}^{(2)}(t)$ and $\mathbf{P}^{(3)}(t)$ are the contributions of the second- and third-order nonlinearities to the polarization, respectively. If we now take the Fourier transforms of these equations, we find that

$$\mathbf{P}^{(2)}(\omega) = \epsilon_0 \int_{-\infty}^\infty d\omega_1 \int_{-\infty}^\infty d\omega_2 \, \delta(\omega - \omega_1 - \omega_2) \boldsymbol{\chi}^{(2)}(\omega_1, \omega_2) : \mathbf{E}(\omega_1) \otimes \mathbf{E}(\omega_2),$$

$$\mathbf{P}^{(3)}(\omega) = \epsilon_0 \int_{-\infty}^\infty d\omega_1 \cdots \int_{-\infty}^\infty d\omega_3 \, \delta(\omega - \omega_1 - \omega_2 - \omega_3)$$

$$\times \boldsymbol{\chi}^{(3)}(\omega_1, \omega_2, \omega_3) : \mathbf{E}(\omega_1) \otimes \mathbf{E}(\omega_2) \otimes \mathbf{E}(\omega_3), \tag{1.47}$$

where the Fourier transforms of the nonlinear susceptibility tensors are defined as

$$\boldsymbol{\chi}^{(2)}(\omega_1, \omega_2) = \int_0^\infty d\tau_1 \int_0^\infty d\tau_2 \, e^{-i(\omega_1\tau_1 + \omega_2\tau_2)} \tilde{\boldsymbol{\chi}}^{(2)}(\tau_1, \tau_2),$$

$$\boldsymbol{\chi}^{(3)}(\omega_1, \omega_2, \omega_3) = \int_0^\infty d\tau_1 \cdots \int_0^\infty d\tau_3 \, e^{-i(\omega_1\tau_1 + \omega_2\tau_2 + \omega_3\tau_3)} \tilde{\boldsymbol{\chi}}^{(3)}(\tau_1, \tau_2, \tau_3). \tag{1.48}$$

Note that, because it does not matter which frequency we call ω_1, which we call ω_2, and so on, we can define the frequency-dependent susceptibilities to be invariant under permutations of their arguments, and we shall assume that this is the case. Just as with $\boldsymbol{\chi}^{(1)}$, the frequency-dependent (not time-dependent) susceptibilities have the dimensions and approximate magnitudes of the approximate instantaneous susceptibilities, though they now depend on the frequency of the radiation.

1.4 Power and energy

Maxwell's equations allow a number of conservation laws to be obtained, depending on the symmetries of the medium. The most fundamental is energy conservation, which is determined by the use of Poynting's theorem. This is obtained from a fundamental vector identity for gradient and curl, which states that

$$\nabla \cdot (\mathbf{E} \times \mathbf{H}) = \mathbf{H} \cdot (\nabla \times \mathbf{E}) - \mathbf{E} \cdot (\nabla \times \mathbf{H}). \tag{1.49}$$

Using Maxwell's equations to replace the two curl operations by time derivatives then leads to

$$\begin{aligned} \nabla \cdot \mathbf{I} &= -\mathbf{H} \cdot \frac{\partial \mathbf{B}}{\partial t} - \mathbf{E} \cdot \frac{\partial \mathbf{D}}{\partial t} \\ &= -\frac{\partial \mathcal{H}}{\partial t}. \end{aligned} \tag{1.50}$$

This is known as Poynting's theorem, and it holds rigorously whenever Maxwell's equations are valid. Here, $\mathbf{I} = \mathbf{E} \times \mathbf{H}$ is usually interpreted as the directional energy flux density or intensity (in W m^{-2}) of an electromagnetic field, while $\mathcal{H}$ is the local energy density (in J m^{-3}), including the energy stored in the medium. There are some caveats associated with this interpretation, especially for static fields in media where the Poynting vector may be nonzero in a situation with no apparent energy flow.

1.4.1 Energy density

Assuming all the local fields and stored energies are initially zero at $t = -\infty$, we can write the energy density from Eq. (1.50) as

$$\mathcal{H} = \int_{-\infty}^{t} \left[\mathbf{H} \cdot \frac{\partial \mathbf{B}}{\partial t} + \mathbf{E} \cdot \frac{\partial \mathbf{D}}{\partial t} \right] dt. \tag{1.51}$$

The energy density clearly depends on the dielectric and magnetic response functions. If we suppose that there is a deterministic functional relationship between **H** and **B** and between **E** and **D**, we can write this as

$$\mathcal{H} = \mathcal{H}_M + \mathcal{H}_E = \int_0^{\mathbf{B}} \mathbf{H}(\mathbf{B}') \cdot d\mathbf{B}' + \int_0^{\mathbf{D}} \mathbf{E}(\mathbf{D}') \cdot d\mathbf{D}'. \tag{1.52}$$

For a linear magnetic response, and a nonlinear dielectric response according to Eq. (1.26), we immediately obtain

$$\mathcal{H} = \sum_{n>0} \frac{1}{n+1} \mathbf{D} \cdot \eta^{(n)} : \mathbf{D}^{\otimes n} + \tfrac{1}{2}\mu |\mathbf{H}|^2. \tag{1.53}$$

Alternatively, if we integrate by parts so that

$$\int_0^{\mathbf{D}} \mathbf{E}(\mathbf{D}') \cdot d\mathbf{D}' = \mathbf{E} \cdot \mathbf{D} - \int_0^{\mathbf{E}} \mathbf{D}(\mathbf{E}') \cdot d\mathbf{E}', \tag{1.54}$$

this can be written equivalently as

$$\mathcal{H} = \sum_{n>0} \frac{n}{n+1} \mathbf{E} \cdot \epsilon^{(n)} : \mathbf{E}^{\otimes n} + \tfrac{1}{2}\mu |\mathbf{H}|^2. \tag{1.55}$$

This agrees with the usual linear expression given in Eq. (1.16), in the appropriate limit. We will generally assume that these instantaneous response expressions give the nonlinear energy for *nonlinear* response functions that do not have a strong frequency dependence. However, we need to examine this result more carefully for the linear susceptibility, where dispersive corrections are significant.

1.4.2 Dispersive energy

The stored energy in a dispersive medium depends on the temporal response, and is not given exactly by Eq. (1.53), due to dispersive corrections. We will treat this in detail in the special case of a linear dispersive medium. Dispersive corrections are significant in this case, as the linear energy is usually the largest part of the energy. Inverting the Fourier transform, for the fields,

$$\mathbf{E}(t) = \int_{-\infty}^{\infty} d\omega\, e^{i\omega t}\mathbf{E}(\omega),$$
$$\mathbf{D}(t) = \int_{-\infty}^{\infty} d\omega\, e^{i\omega t}\mathbf{D}(\omega). \tag{1.56}$$

Noting that $\mathbf{D}(\omega) = \boldsymbol{\epsilon}(\omega)\mathbf{E}(\omega)$, and taking a time derivative, one obtains

$$\frac{\partial \mathbf{D}}{\partial t} = i\int_{-\infty}^{\infty} d\omega\, e^{i\omega t}\omega\boldsymbol{\epsilon}(\omega)\cdot\mathbf{E}(\omega). \tag{1.57}$$

The reality of the fields implies that $\mathbf{D}(-\omega) = \mathbf{D}^*(\omega)$, $\mathbf{E}(-\omega) = \mathbf{E}^*(\omega)$ and $\boldsymbol{\epsilon}(-\omega) = \boldsymbol{\epsilon}^*(\omega)$. Noting that $\mathbf{E}(t) = \mathbf{E}^*(t)$, we can now rewrite $\mathbf{E}\cdot\partial\mathbf{D}/\partial t$ as a product of the two Fourier integrals, so that

$$\mathbf{E}\cdot\frac{\partial \mathbf{D}}{\partial t} = i\int_{-\infty}^{\infty} d\omega \int_{-\infty}^{\infty} d\omega'\, e^{i(\omega-\omega')t}\omega\mathbf{E}^*(\omega')\cdot\boldsymbol{\epsilon}(\omega)\cdot\mathbf{E}(\omega). \tag{1.58}$$

Next, we split the frequency integral into equal parts. Making the substitutions $\omega \to -\omega'$ and $\omega' \to -\omega$ in one of the resulting integrals, and integrating in time to obtain the electric energy density, we obtain

$$\mathcal{H}_E = \frac{i}{2}\int_{-\infty}^{t} dt' \int_{-\infty}^{\infty} d\omega \int_{-\infty}^{\infty} d\omega'\, e^{i(\omega-\omega')t'}\mathbf{E}^*(\omega')\cdot[\omega\boldsymbol{\epsilon}(\omega) - \omega'\boldsymbol{\epsilon}^\dagger(\omega')]\cdot\mathbf{E}(\omega). \tag{1.59}$$

Assuming all the local fields and stored energies are initially zero at $t = -\infty$, we can carry out the time integral with a causal factor $\delta > 0$ inserted, and write the electric energy density in an exact form, as

$$\mathcal{H}_E = \lim_{\delta\to 0}\int_{-\infty}^{\infty} d\omega \int_{-\infty}^{\infty} d\omega'\, \mathbf{E}^*(\omega')\cdot\frac{[\omega'\boldsymbol{\epsilon}^\dagger(\omega') - \omega\boldsymbol{\epsilon}(\omega)]}{2(\omega'-\omega+i\delta)}\cdot\mathbf{E}(\omega)e^{i(\omega-\omega')t}. \tag{1.60}$$

This has an equivalent form, useful when quantizing displacement fields:

$$\mathcal{H}_E = \lim_{\delta\to 0}\int_{-\infty}^{\infty} d\omega \int_{-\infty}^{\infty} d\omega'\, \mathbf{D}^*(\omega')\cdot\frac{[\omega'\boldsymbol{\eta}(\omega) - \omega\boldsymbol{\eta}^\dagger(\omega')]}{2(\omega'-\omega+i\delta)}\cdot\mathbf{D}(\omega)e^{i(\omega-\omega')t}. \tag{1.61}$$

1.4.3 Group velocity

We now wish to relate this dispersive energy result to the well-known group velocity, which defines the average velocity of a waveform traveling in a dispersive medium. For a sufficiently narrow-band, nearly monochromatic pulse, it is known that the characteristic group velocity of a propagating pulse is

$$v_g = \frac{\partial \omega(k)}{\partial k}. \tag{1.62}$$

This is to be contrasted with Eq. (1.12) for the phase velocity. The two quantities are only equal for a homogeneous, nondispersive medium. In the case of a uniform, isotropic medium with dielectric dispersion, one can use the relation

$$k = \omega\sqrt{\mu\epsilon(\omega)} = \omega\sqrt{\frac{\mu}{\eta(\omega)}}. \tag{1.63}$$

Differentiating this dispersion relation gives

$$v_g = \left[\frac{\partial k}{\partial \omega}\right]^{-1}. \tag{1.64}$$

On neglecting magnetic dispersion, which is generally extremely small, we obtain the group velocity in terms of the dielectric properties:

$$\begin{aligned} v_g &= \left[\frac{k}{\omega} + \frac{k}{2\epsilon}\frac{\partial \epsilon}{\partial \omega}\right]^{-1} \\ &= \left[\frac{k}{\omega} - \frac{k}{2\eta}\frac{\partial \eta}{\partial \omega}\right]^{-1}. \end{aligned} \tag{1.65}$$

Next, we wish to relate this to the dispersive energy density given above. For a narrow-band field at ω_0, with a real (i.e. lossless) symmetric permittivity (these conditions imply that $\epsilon = \epsilon^\dagger$) that is slowly varying over the field bandwidth near ω_0, there are significant contributions to the integral in Eq. (1.60), at $\omega \approx \omega' \approx \pm\omega_0$. Noting that, near ω_0,

$$\frac{[\omega\epsilon(\omega) - \omega'\epsilon(\omega')]}{(\omega - \omega' - i\delta)} \approx \left.\frac{\partial}{\partial \omega}(\omega\epsilon(\omega))\right|_{\omega_0}. \tag{1.66}$$

In addition, because $\epsilon(\omega) = \epsilon(-\omega)$ if $\epsilon(\omega)$ is real (see Eq. (1.44) and its complex conjugate), then the derivative of $\omega\epsilon(\omega)$ is the same at both ω_0 and $-\omega_0$. We then have that

$$\begin{aligned} \mathcal{H}_E &\approx \frac{1}{2}\int_{-\infty}^{\infty} d\omega \int_{-\infty}^{\infty} d\omega'\, \mathbf{E}^*(\omega') \cdot \frac{\partial}{\partial \omega}(\omega\epsilon(\omega))|_{\omega_0} \cdot \mathbf{E}(\omega)e^{i(\omega-\omega')t} \\ &\approx \frac{1}{2}\mathbf{E}(t) \cdot \left[\epsilon(\omega_0) + \omega_0 \left.\frac{\partial \epsilon(\omega)}{\partial \omega}\right|_{\omega_0}\right] \cdot \mathbf{E}(t). \end{aligned} \tag{1.67}$$

Using the inverse permittivity, $\eta = \epsilon^{-1}$, this can also be written in an equivalent form in terms of displacement fields as

$$\mathcal{H}_E \approx \frac{1}{2}\mathbf{D}(t) \cdot \left[\eta(\omega_0) - \omega_0 \left.\frac{\partial \eta(\omega)}{\partial \omega}\right|_{\omega_0}\right] \cdot \mathbf{D}(t). \tag{1.68}$$

In summary, owing to dispersion, Eq. (1.53) is not exact, since the dielectric response is not strictly instantaneous. We must therefore include dispersive corrections, given in the linear case by Eq. (1.68). There are nonlinear dispersive corrections as well, obtained in a similar way. These are generally small, unless the nonlinearity is caused by a narrow resonance.

1.5 Order-of-magnitude estimates

We wish to obtain some order-of-magnitude estimates of the susceptibilities, both linear and nonlinear. Typically the optical refractive index, n_r, is of order one. For example, $n_r \approx 1.45$ for fused silica near the commonly used telecommunications wavelengths around $\lambda = 1.5\ \mu$m. We see that $\chi^{(1)}$ is therefore also generally of order unity.

This can be used to estimate the size of the nonlinear susceptibilities. If the nonlinearities in the medium are electronic in origin, then the nonlinear effects should be important when the applied field is of the same order as the electric field in an atom. The field in an atom is of order $E_{atom} \sim e/(4\pi\epsilon_0 a_0^2)$, where $e = 1.602\,176 \times 10^{-19}$ C is the elementary charge, and $a_0 = 5.291\,77 \times 10^{-11}$ m is the Bohr radius. The resulting internal electric field strength is approximately 10^{11} V m^{-1}.

For incident electromagnetic fields of this magnitude, the terms in the expansion for the polarization will be of roughly the same size. Using the fact that $\chi^{(1)} \sim 1$, we then find that

$$\chi^{(2)} \sim \frac{1}{E_{atom}} \simeq 10^{-12}\ \mathrm{m\,V^{-1}},$$
$$\chi^{(3)} \sim \frac{1}{E_{atom}^2} \simeq 10^{-24}\ \mathrm{m^2\,V^{-2}}. \tag{1.69}$$

These estimates are, in fact, quite good, although we have shown that there is a special symmetry requirement to obtain a nonzero $\chi^{(2)}$. From Eq. (1.15), the corresponding magnetic fields for plane waves of this intensity are around 10^3 T. This is stronger than the largest macroscopic magnetic field of current electromagnets. For example, the CERN Large Hadron Collider (LHC) currently uses superconducting magnets that have a magnetic field strength of 8.36 T. This shows that, as well as electric field effects, the magnetic forces like ponderomotive forces on free electrons can become very important at large field strengths.

1.5.1 Intensity, power and field strength

What powers and intensities are needed to observe nonlinear effects? From Eq. (1.17), an electric field strength of 10^{11} V m^{-1} corresponds to a threshold intensity of

$$I_{thr} \sim 10^{21}\ \mathrm{W\,m^{-2}}. \tag{1.70}$$

At these electric field strengths, ionization occurs very rapidly, and the Bloembergen nonlinear optics expansion breaks down. Such fields are obtainable with high-power lasers. Even at $I \sim 10^{18}$ W m^{-2}, strong multiphoton effects are observed. This leads to phase-matched high-harmonic generation (HHG). Such field strengths give rise to coherent kiloelectron-volt X-ray production from table-top lasers in the mid-infrared, via generation of harmonics with orders greater than 5000. More typical laser powers used in table-top nonlinear optics experiments have a power of

$$I \sim 10^9 \text{ W m}^{-2} = 1 \text{ GW m}^{-2}. \tag{1.71}$$

We see that there is a very large range of intensities where nonlinear effects are possible.

At the other extreme, a hand-held class II visible laser pointer with a 1 mW continuous-wave (CW) power output and a beam radius of 1 mm has an intensity of $I = P/A \sim$ 1 kW m^{-2}, a million times less. Here P is the average power and A is the effective cross-sectional area, as in Eq. (1.19). Surprisingly, even a 1 mW laser can have a high enough peak intensity for nonlinear effects, provided the output is a stream of short pulses.

To understand this, suppose the same 1 mW low-power laser is pulsed, with an output pulse train repetition rate of 100 MHz and a pulse duration of 10 fs. The result is an amplification of the peak power by a factor of 10^6. The use of pulsed lasers of this type means that, with 1 mW average laser power, one can easily reach peak powers of 1 kW, and focused intensities of $\sim$ 1 GW m^{-2}, or higher if the area is smaller. This is enough for nonlinear effects.

1.6 The two-level atom

As an illustration, let us compute the frequency-dependent susceptibilities for a medium consisting of two-level atoms. While such exotic atoms do not really exist, an approximate two-level behavior is obtained for near-resonant radiation in atomic systems with special preparation that only allows two energy eigenstates to be relatively strongly coupled. These systems display behavior typical of resonant transitions in more commonly found multi-level atomic systems. To treat this case in the simplest way, we introduce a semiclassical approximation, in which the atoms are treated quantum mechanically and the fields classically, with the classical polarization being calculated as a quantum mechanical average. Let the upper level of the relevant transition be $|e\rangle$, with an energy of $\hbar\nu_0$, and let the lower level be $|g\rangle$ with an energy of 0 (Figure 1.1).

1.6.1 Semiclassical Hamiltonian

The semiclassical Hamiltonian describing the interaction of a quantum atom at location $\mathbf{R}$ with an incident (classical) electromagnetic wave in the electric dipole approximation is

$$H = H^0 - \mathbf{E}(\mathbf{R}, t) \cdot [\mathbf{d}\sigma^{(+)} + \mathbf{d}^*\sigma^{(-)}]. \tag{1.72}$$

Here H^0 is the usual atomic Hamiltonian describing the Coulomb interactions between the electrons and the nucleus, $\sigma^{(+)} = |e\rangle\langle g|$ and $\sigma^{(-)} = |g\rangle\langle e|$ are the atomic raising and lowering operators, and $\mathbf{d}$ is the dipole matrix element of the transition that is being driven

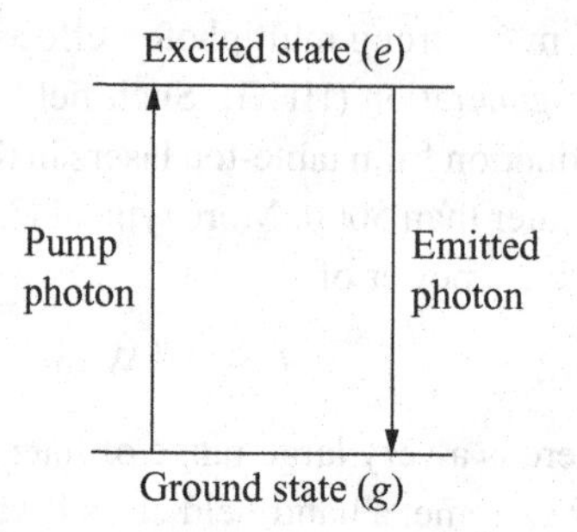

Fig. 1.1 Schematic diagram of a two-level atom. Pumping the atom from the ground state (g) to an excited state (e) with energy $\hbar\nu_0$ allows photons to be emitted, some in other directions and/or with other frequencies.

by the incident field. We give a more detailed treatment of quantum mechanical methods, and the derivation of this Hamiltonian, in later chapters. For now, we wish to use this as just a simple model of how nonlinear effects can arise in practice. At this stage, we still want to treat the field classically, for simplicity.

We shall assume that there is one optically active electron, and we suppose that the phases of the atomic wavefunctions can be chosen so that the dipole matrix element of the transition,

$$\mathbf{d} = -e\langle e|\mathbf{r} - \mathbf{R}|g\rangle, \tag{1.73}$$

is real. In the above equation, $-e = q_e = -1.602\,17 \times 10^{-19}$ C is the charge of the electron, $\mathbf{r}$ is the position operator of the optically active electron and $\mathbf{R}$ is the nuclear position, so that $\mathbf{r} - \mathbf{R}$ is the electron coordinate relative to the nucleus. For single-atom calculations, we take $\mathbf{R} = 0$. We suppose the electric field is nearly monochromatic and plane-polarized in the direction of the (real) unit vector $\hat{\mathbf{e}}$, so that we can introduce a complex function $\mathcal{E}(t)$, called a slowly varying envelope function:

$$\mathbf{E}(t) = \hat{\mathbf{e}}\mathcal{E}(t)e^{-i\omega_k t} + c.c. \tag{1.74}$$

The Hamiltonian describing the interaction of the two relevant levels with the incident electromagnetic wave is given in detail in Chapter 4. This uses the dipole approximation, which assumes that $|\langle a|\mathbf{r} - \mathbf{R}|b\rangle| \ll \lambda$, where $\lambda = 2\pi c/\omega_k$ is the wavelength of the incident field and ω_k is its angular frequency. This approximation is well satisfied at optical wavelengths. The resulting Hamiltonian is

$$H = \hbar\nu_0|a\rangle\langle a| - \tfrac{1}{2}\hbar[\Omega(t)e^{-i\omega_k t} + \Omega^*(t)e^{i\omega_k t}][\sigma^{(+)} + \sigma^{(-)}]. \tag{1.75}$$

Here, we have introduced an important semiclassical quantity, the Rabi frequency Ω, where

$$\Omega = \frac{2\,\hat{\mathbf{e}}\cdot\mathbf{d}}{\hbar}\mathcal{E}. \tag{1.76}$$

We will show that this gives the characteristic frequency of oscillation between the two levels under the action of an external field. We can obtain an estimate for the dipole moment by noting that $d = \mathbf{d}\cdot\hat{\mathbf{e}}$ is approximately the charge of the electron multiplied by the Bohr radius, which is about 10^{-29} C m.

While this estimate is reasonable in many cases, it needs to be taken with a grain of salt. There are well-known effects that can change the dipole moment enormously. For example, in a Rydberg atom, where one or more of the electrons are excited to very high principal quantum number, n, the mean radius of the outer electron scales with n^2. If $n = 100$, which has been achieved experimentally, the dipole moment will be 10^4 higher than our estimate. The large radius also means that the dipole approximation will break down at optical wavelengths of $\lambda \approx 0.5$ μm. At the other extreme, there are forbidden transitions for which $d = 0$ due to the symmetry properties of the atomic wavefunctions. We shall assume the transition we are considering is dipole-allowed, with d typically given by the value estimated above.

1.6.2 Rabi oscillation near resonance

We cannot solve the equations of motion resulting from this Hamiltonian in closed form, but this situation changes if we make what is called the rotating-wave approximation, which is valid near an atomic resonance. In the limit of no interaction, we have in the Heisenberg picture that $\sigma^{(+)} \sim e^{i\nu_0 t}$ and $\sigma^{(-)} \sim e^{-i\nu_0 t}$. That means that, if we are close to resonance, i.e. ω_k is close to ν_0, then the terms $\mathcal{E}(t)e^{-i\omega_k t}\sigma^{(+)}$ and $\mathcal{E}^*(t)e^{i\omega_k t}\sigma^{(-)}$ are slowly varying, while the terms $\mathcal{E}(t)e^{-i\omega_k t}\sigma^{(-)}$ and $\mathcal{E}^*(t)e^{i\omega_k t}\sigma^{(+)}$ are rapidly varying. Slowly varying terms will have a much larger effect on the dynamics than rapidly varying ones, so we drop the rapidly varying terms to give

$$H = \hbar\nu_0|e\rangle\langle e| - \tfrac{1}{2}\hbar[\Omega(t)e^{-i\omega_k t}\sigma^{(+)} + \Omega^*(t)e^{i\omega_k t}\sigma^{(-)}]. \tag{1.77}$$

This is the Hamiltonian in the rotating-wave approximation. This approximation is good near resonance, but less good when the detuning between the field frequency and atomic frequency is large, i.e. of the order of the optical frequency itself. It will suffice for our purposes here.

We now proceed to derive and solve the equations of motion for the atom in the Schrödinger picture, using the rotating-wave approximation. The quantum state of the atom can always be written as

$$|\psi(t)\rangle = c_e(t)e^{-i\nu_0 t}|e\rangle + c_g(t)|g\rangle, \tag{1.78}$$

and substituting this into the Schrödinger equation we find (with $\Omega(t) = \Omega$)

$$i\frac{d}{dt}\begin{pmatrix} c_e \\ c_g \end{pmatrix} = -\frac{1}{2}\begin{pmatrix} 0 & \Omega e^{i\Delta t} \\ \Omega^* e^{-i\Delta t} & 0 \end{pmatrix}\begin{pmatrix} c_e \\ c_g \end{pmatrix}, \tag{1.79}$$

where $\Delta = \nu_0 - \omega_k$.

If $\Delta = 0$ and Ω is real and constant, this can be rewritten in a form that uses a Pauli spin matrix,

$$\sigma^x = \begin{pmatrix} 0 & 1 \\ 1 & 0 \end{pmatrix}, \tag{1.80}$$

which we simply regard as a constant matrix. Therefore,

$$\frac{d\mathbf{c}}{dt} = \frac{i\Omega}{2}\sigma^x \mathbf{c}, \tag{1.81}$$

where $\mathbf{c} = (c_e, c_g)^{\mathrm{T}}$. This has the solution

$$\mathbf{c}(t) = \exp\left[i\frac{\Omega t}{2}\sigma^x\right]\mathbf{c}(0). \tag{1.82}$$

Using the algebraic property of the Pauli matrices that $(\sigma^x)^2 = 1$, it follows by expanding the exponential that

$$\begin{pmatrix} c_e(t) \\ c_g(t) \end{pmatrix} = \begin{pmatrix} \cos(\Omega t/2) & i\sin(\Omega t/2) \\ i\sin(\Omega t/2) & \cos(\Omega t/2) \end{pmatrix} \begin{pmatrix} c_e(0) \\ c_g(0) \end{pmatrix}. \tag{1.83}$$

This characteristic oscillatory response of a two-level system at resonance is called Rabi oscillation. If the atom starts in its ground state and is excited by a field that turns on rapidly on time-scales much *faster* than Ω^{-1}, then the excited-state population oscillates at the Rabi frequency:

$$|c_e(t)|^2 = \tfrac{1}{2}[1 - \cos(\Omega t)]. \tag{1.84}$$

When the detuning is not zero, the characteristic oscillation frequency is $\widetilde{\Omega} = \sqrt{\Delta^2 + \Omega^2}$, which is called the generalized Rabi frequency.

1.6.3 Adiabatic response

We now consider a different case, where the slowly varying electric field amplitude in the Hamiltonian, $\mathcal{E}(t)$, is taken to be time-dependent. We are going to start the atom in its ground state at $t = -\infty$ and adiabatically turn on the field on time-scales much *slower* than $\widetilde{\Omega}^{-1}$, until it reaches a value of $\mathcal{E}(t) = \mathcal{E}_0$ at $t = 0$, after which it remains steady. The reason for doing this is to eliminate transient effects, because now we are interested in the steady-state response of the system. Assuming that $c_g = \exp(i\omega t)$, and substituting this into Eq. (1.79), we find that

$$c_e = \frac{2\omega}{\Omega_0^*} e^{i(\omega+\Delta)t}, \tag{1.85}$$

where $\Omega_0 = 2\hat{\mathbf{e}} \cdot \mathbf{d}\mathcal{E}_0/\hbar$, and that ω must be a solution of the quadratic equation

$$\omega^2 + \Delta\omega - |\Omega_0|^2/4 = 0, \tag{1.86}$$

so that we have the following two possible values for ω:

$$\omega_\pm = \tfrac{1}{2}\left[-\Delta \pm (\Delta^2 + |\Omega_0|^2)^{1/2}\right]. \tag{1.87}$$

The existence of multiple levels in the adiabatic solutions, with splittings given by $\pm\widetilde{\Omega} = \pm\sqrt{\Delta^2 + \Omega_0{}^2}$, is called the Autler–Townes effect, and results in a doublet in the atomic spectrum.

The normalized solution that corresponds to ω_+ is

$$\begin{pmatrix} c_e \\ c_g \end{pmatrix} = \frac{1}{N_+} \begin{pmatrix} 2\omega_+ e^{-i\omega_- t} / \Omega_0^* \\ e^{i\omega_+ t} \end{pmatrix}, \tag{1.88}$$

and the one corresponding to ω_- is

$$\begin{pmatrix} c_e \\ c_g \end{pmatrix} = \frac{1}{N_-} \begin{pmatrix} 2\omega_- e^{-i\omega_+ t} / \Omega_0^* \\ e^{i\omega_- t} \end{pmatrix}. \tag{1.89}$$

The normalization constants in the above equations are

$$N_\pm = \left(1 + \left|\frac{2\omega_\pm}{\Omega_0}\right|^2\right)^{1/2}. \tag{1.90}$$

Examining these solutions in the limit $\mathcal{E}_0 \to 0$, we find that it is the solution corresponding to ω_+ that goes to the lower state $|g\rangle$, so this is the solution we need. In terms of the raising and lowering atomic operators, this means that from Eq. (1.88), we find that

$$\begin{aligned} \langle \sigma^{(+)}(t) \rangle &= c_e^* c_g e^{i\nu_0 t} \\ &= \frac{2\omega_+}{N_+^2 \Omega_0} e^{i(\nu_0 + \omega_- + \omega_+)t}. \end{aligned} \tag{1.91}$$

Next, by inserting the frequency solutions given in Eq. (1.87), and the solutions for the normalization constants from Eq. (1.90), we find that

$$\langle \sigma^{(+)}(t) \rangle = e^{i\omega_k t} \frac{2\Omega_0^* \omega_+}{|\Omega_0|^2 + 4\omega_+^2}. \tag{1.92}$$

We see that, far off-resonance, the atomic polarization is entrained to the laser frequency ω_k. This atomic polarization, oscillating at the incident laser frequency, gives rise to both the linear and nonlinear contributions to the refractive index, which we treat in the next section.

1.6.4 Microscopic polarization

Assuming that the atomic dipole is parallel to the incident field, the polarization of the atom can be computed by taking the expectation value of the dipole moment operator in the direction of the applied field:

$$\mathbf{p} = -e\langle \mathbf{r} \rangle = \mathbf{d}\langle \sigma^{(+)} + \sigma^{(-)} \rangle.$$

Note that because we are assuming that the atomic states have well-defined parity, $\langle a|\mathbf{r}|a\rangle = \langle b|\mathbf{r}|b\rangle = 0$.

Therefore, if $\mathbf{p} = p(t)\hat{\mathbf{e}}$, we find that

$$\begin{aligned} p(t) &= d\langle \sigma^{(+)} + \sigma^{(-)} \rangle \\ &= \frac{2d\omega_+}{|\Omega_0|^2 + 4\omega_+^2} [\Omega_0 e^{-i\omega_k t} + \Omega_0^* e^{i\omega_k t}]. \end{aligned} \tag{1.93}$$

In order to find the different susceptibilities, we expand this result in $\mathcal{E}_0$. First, on expanding the solution for ω_+, we note that

$$\omega_+ = \frac{|\Omega_0|^2}{4\Delta} - \frac{|\Omega_0|^4}{16\Delta^3}. \tag{1.94}$$

Inserting this result into the atomic polarization given in Eq. (1.93), we find that

$$p(t) = \frac{d}{2\Delta}\left(1 - \frac{|\Omega_0|^2}{2\Delta^2}\right)[\Omega_0 e^{-i\omega_k t} + \Omega_0^* e^{i\omega_k t}]. \tag{1.95}$$

Introducing the Fourier transforms as defined in Eq. (1.42), we obtain

$$\begin{aligned} p^{(1)}(\omega) &= \frac{d}{2\Delta}[\Omega_0^*\delta(\omega+\omega_k) + \Omega_0\delta(\omega-\omega_k)], \\ p^{(3)}(\omega) &= \frac{-d|\Omega_0|^2}{4\Delta^3}[\Omega_0^*\delta(\omega+\omega_k) + \Omega_0\delta(\omega-\omega_k)]. \end{aligned} \tag{1.96}$$

This means that the atomic polarizations can be expanded in a series in the incident field Fourier coefficients. In general this leads to the fundamental nonlinear response function in a power series form, as

$$\mathbf{p}\left(\sum_n \omega_n\right) = \epsilon_0 \sum_{n>0} \boldsymbol{\alpha}^{(n)}(\omega_1,\ldots,\omega_n) : \boldsymbol{E}_{loc}(\omega_1)\otimes\cdots\otimes\boldsymbol{E}_{loc}(\omega_n), \tag{1.97}$$

where $\boldsymbol{E}_{loc}(\omega_n)$ is the local field acting on each atom, and $\boldsymbol{\alpha}^{(n)}$ is the nth-order nonlinear local-field polarizability. Here we define the expansion in terms of the local field at the atomic position. As we see in the next section, this can differ from the macroscopic average.

1.6.5 Low-density susceptibility

If we have a medium consisting of ρ_n two-level atoms per unit volume, and we ignore local-field corrections (which is valid if the density of atoms is not too high, see Section 1.7), the macroscopic polarization of the medium, $\mathbf{P}(t) = P(t)\hat{\mathbf{e}}$, is just ρ_n times the polarization of each atom, so that $\chi^{(p)} = \rho_n\alpha^{(p)}$. We can now compare these equations with our equations for the susceptibilities. Recalling that $\Delta = \nu_0 - \omega$, where ν_0 is the atomic transition frequency and ω is the input radiation frequency, this gives for the low-density linear susceptibility that

$$\chi_0^{(1)}(\omega) = \chi_0^{(1)}(-\omega) = \rho_n\alpha^{(1)} = \frac{d^2\rho_n}{\epsilon_0\hbar(\nu_0-\omega)}. \tag{1.98}$$

This result is the dispersive equivalent to the familiar Lorentzian absorption near a resonance. We note that for excitations *below* the resonant frequency, i.e. for $\omega < \nu_0$, the susceptibility is positive. This means that the refractive index is larger than unity $n_r > 1$, so that $v_p < c$, i.e. the electromagnetic phase velocity is less than the speed of light, as it is in most dielectrics. Accordingly, this is called the normal dispersion regime.

For excitations *above* the atomic resonant frequency, i.e. $\omega > \nu_0$, the phase velocity is above the speed of light, which is called the anomalous dispersion regime. We note that a careful analysis in this regime shows that superluminal communication or energy transmission is still prohibited, since a transmitted signal must have modulation to allow communication, and so cannot involve just one frequency component!

The nonlinear susceptibility, also in the low-density regime, is then given by

$$\chi_0^{(3)}(-\omega,\omega,\omega) = n_\rho \alpha^{(1)}(-\omega,\omega,\omega)$$
$$= \frac{-2d^4\rho_n}{3\epsilon_0\hbar^3(\nu_0-\omega)^3}. \tag{1.99}$$

This last equation also holds if we permute the arguments of the third-order nonlinear susceptibilities. The factor of 3 in the denominator is a combinatoric factor that takes account of the fact that, in Eq. (1.97), there are three equivalent terms with frequency arguments $(-\omega,\omega,\omega)$, $(\omega,-\omega,\omega)$ and $(\omega,\omega,-\omega)$. Each of these cause a polarization at angular frequency ω.

We see that, in the normal dispersion regime, below a resonance, the characteristic values of the nonlinear susceptibility are negative. This means, from Eq. (1.39), that the effective refractive index tends to decrease with increasing intensity. The value of the intensity-dependent refractive index, n_2, is therefore

$$n_2 = \frac{d^4\rho_n}{2\epsilon\epsilon_0 c\hbar^3(\nu_0-\omega)^3}. \tag{1.100}$$

This implies that, in an intense, red-detuned CW laser beam *below* resonance, with $\nu_0 > \omega$ so that $n_2 < 0$, the refractive index is lower at the center of the beam than at the edges. This is equivalent to a concave lens, and results in self-defocusing. The reverse behavior of self-focusing is observed for blue-detuned lasers *above* resonance.

1.6.6 Spontaneous emission

At first sight, these results look very different from the earlier estimates appearing in Eq. (1.69). This is due to some approximations we have made in the above derivation. The susceptibilities calculated for two-level atoms appear to become infinite as $\Delta \to 0$, which is an unlikely result. The root of this problem lies in our assumption that the radiation field is classical. While this may be a good approximation for highly populated field modes, there are always an infinite number of modes that are nearly in the vacuum state that couple to an atom in free space, and these modes cannot be treated classically. Using the famous 'golden rule' of quantum mechanics, the vacuum modes interact with the atom to cause spontaneous emission with a rate

$$\Gamma = \frac{8\pi^2 d^2}{3\epsilon_0\hbar\lambda^3}. \tag{1.101}$$

For large detunings, $|\Delta| \gg \Gamma$, the effects due to the vacuum-state modes can be neglected, so this limits the applicability of our results derived from a two-level atom to the case of large detunings. For very large detunings, such that $|\Delta|/\nu_0 \sim 1$, we need to include counter-rotating terms as well, since the near-resonant approximation fails in this regime.

The size of Γ can be easily estimated. Assuming that $\lambda \sim 1\ \mu\text{m}$, for a near-infrared transition, and $d \sim e \times 10^{-10}\ \text{m} = 1.6 \times 10^{-29}$ C m for a typical atomic transition dipole moment in an atom of dimension 0.1 nm, one obtains an estimate of $\Gamma \sim 10$ MHz. This corresponds to a spontaneous lifetime of $\tau \sim 1/\Gamma \sim 100$ ns. Ultraviolet transitions, with shorter wavelengths, have shorter lifetimes unless they are forbidden for symmetry reasons. For example, the 2P to 1S transition (Lyman alpha line) in hydrogen has a wavelength of $\lambda = 121.6$ nm, and a correspondingly short lifetime of 1.6 ns.

We also mentioned that we ignored local-field effects, which are treated in the next section. This assumes that the atoms are noninteracting, which limits our results to low densities satisfying $n \ll 1/\lambda^3$. Setting $\Delta = \Gamma$ and $n = 1/\lambda^3$ in the above expressions gives us the limits that our approximations impose on our expressions for the first- and third-order susceptibilities:

$$|\chi^{(1)}(\omega_k)| < 1, \qquad |\chi^{(3)}(\omega_k)| < \frac{\lambda^6\epsilon_0^2}{d^2}. \tag{1.102}$$

A more general method of calculating susceptibilities is by the use of perturbation theory. One starts with the density matrix equations describing the interaction of the electromagnetic field with a medium. These equations include damping effects. The equations are then solved perturbatively, where the perturbation is the field–matter interaction. The first-order term yields the linear susceptibility, the second-order term yields $\chi^{(2)}$, and so on. This type of calculation allows us to take into account an arbitrary number of atomic or molecular energy levels, and the effects of the terms that were dropped when we made the rotating-wave approximation. Detailed accounts of these methods can be found in textbooks on nonlinear optics.

1.7 Local-field corrections

An additional subtlety in calculating susceptibilities from first principles is that the atomic structure of matter means that the polarization is extremely inhomogeneous on atomic scales. As a result, all the susceptibilities, which are usually quoted as macroscopic averages, need quantitative corrections called local-field corrections, to convert from atomic to bulk values.

This is caused in part by an important separation of these scales that allows us to carry out macroscopic averaging. While atomic inhomogeneous structures have a length-scale of ~ 0.1 nm $= 10^{-10}$ m, typical optical and infrared wavelengths are around $\sim 1\ \mu\text{m} = 10^{-6}$ m. It is this difference in length-scales of around 10^4 in magnitude that permits the use of relatively uniform dielectric response functions at optical wavelengths. When matter has structure on length-scales that approach a wavelength, considerable modifications occur, leading to such phenomena as waveguides and metamaterials.

The atomic structure of a dielectric material means that electric fields in nonlinear optics are inhomogeneous, like a scrambled egg. In the macroscopic theory we take an average over the atomic structure. This is like averaging over microscopic differences between yolk and egg white. More precisely, the electric field appearing in the equations is a field averaged over a volume large compared to the atomic spacing, but small compared to an

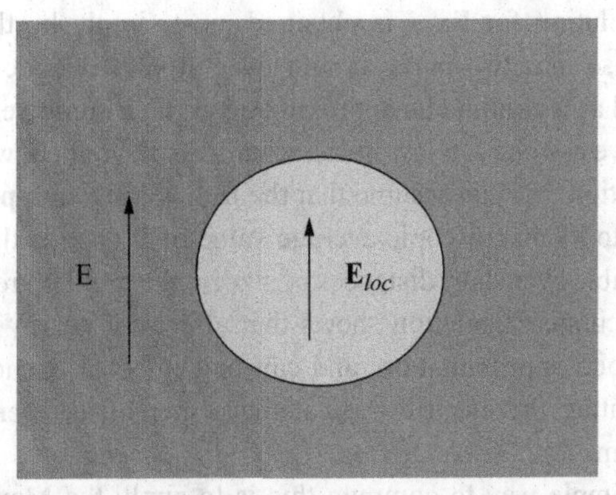

Fig. 1.2 Local-field corrections. The local electric field $\mathbf{E}_{loc}$ inside a spherical cavity is different from the macroscopic average field $\mathbf{E}$ outside.

optical wavelength. So far, we have implicitly assumed that the electric field acting on each dipole is just the same as the macroscopic average electric field. However, just as the polarization field clearly has variations on an atomic scale, so does the electric field.

1.7.1 Local and macroscopic fields

For an accurate atomic calculation of the susceptibility, we need to 'unscramble the egg' to find the true microscopic or local electric field $\mathbf{E}_{loc}$ acting on each atom. We must ask: Is $\mathbf{E}_{loc}$ at each atomic location really equal to the macroscopic average field $\mathbf{E}$? The answer is that they are *not* the same, and a correction called the local-field correction should be included (Figure 1.2). The size of this correction increases with the refractive index and density.

In general, this is highly nontrivial to calculate. Fortunately, in the case of crystalline structures, where there is a regular lattice, the local-field or 'unscrambling' problem was historically solved by Clausius and Mosotti, and by Lorentz. To understand their solution, consider a regular lattice of dipoles, in which the *microscopic* polarization density $\mathbf{p}(\mathbf{r}, t)$ is summed over mean atomic positions $\bar{\mathbf{R}}_i$:

$$\mathbf{p}(\mathbf{r}, t) = \sum_i \delta(\mathbf{r} - \bar{\mathbf{R}}_i)\mathbf{p}_i . \tag{1.103}$$

Here $\mathbf{p}_i = -e\hat{\mathbf{e}} \cdot \langle(\mathbf{r}_i - \mathbf{R}_i)\rangle$ is the local dipole moment of the ith atom, which we have calculated in the previous section. The electric field in this calculation is the microscopic local field $\mathbf{E}_{loc}(\bar{\mathbf{R}}_i)$, which includes contributions from the dipoles at all locations *except* $\mathbf{r}_i$. The details are given in Chapter 4.

In the traditional approach in nonlinear optics, the local field is calculated from the macroscopic average electric field $\mathbf{E}$. To do this, we assume that there is a correction field $\mathbf{E}_{corr}$ that includes all the inhomogeneous behavior:

$$\mathbf{E}_{loc} = \mathbf{E} + \mathbf{E}_{corr} . \tag{1.104}$$

The solution for $\mathbf{E}_{corr}$ is obtained most simply by the approach of dividing the dipole response into two parts: a sum over dipoles outside an imaginary sphere of radius r_0, treated as a continuous approximation, and a sum over discrete terms from atoms nearby. Here we assume $\lambda \gg r_0 \gg x_0$, where λ is the optical wavelength, and x_0 is the interatomic separation. We can assume that the true polarization $\mathbf{p}(\mathbf{r}, t)$ is approximately constant and equal to its macroscopic average value of $\mathbf{P}$ outside the sphere. As retardation effects are negligible over short distances of size r_0, the problem reduces to a static dielectric response. An accurate calculation shows that effects of nearby dipoles in a uniform lattice cancel to a good approximation, and can be neglected. Hence, we are left with the problem of calculating the electric field at the center of a spherical hole in a uniformly polarized medium.

A simple way to compute this is to apply the Maxwell boundary conditions, that the tangential electric field is continuous, so that $E_{t1} = E_{t2}$, and the normal displacement field is continuous, so that $D_{n1} = D_{n2}$. Intuitively, we expect that the local field interpolates between the two. Since two of the three electric field components are continuous, a not unreasonable guess is that, in the center,

$$\begin{aligned} \mathbf{E}_{loc} &= \frac{2}{3}\mathbf{E} + \frac{1}{3\epsilon_0}\mathbf{D} \\ &= \mathbf{E} + \frac{1}{3\epsilon_0}\mathbf{P}. \end{aligned} \tag{1.105}$$

This intuitive argument can be verified exactly by a full calculation using potentials from classical electrostatic theory. It is important to note that the result does not depend on the spherical radius r_0, owing to the long-range nature of dipole forces. Similar results also hold for magnetic fields, where the normal component of B, and the transverse components of H, are continuous. This leads to a result analogous to Eq. (1.105), namely

$$\mathbf{H}_{loc} = \frac{2}{3}\mathbf{H} + \frac{1}{3\mu_0}\mathbf{B} = \mathbf{H} + \frac{1}{3}\mathbf{M}. \tag{1.106}$$

1.7.2 Electric field expansion

In the case of a dielectric response, the linear polarization for a single atom is given by $\mathbf{p} = \epsilon_0\boldsymbol{\alpha}^{(1)} \cdot \mathbf{E}_{loc} + \cdots$, from Eq. (1.97) in the previous section. For an atomic density ρ_n, the macroscopic polarization can therefore be expanded in terms of the local fields to give

$$\begin{aligned} \mathbf{P} &= \rho_n\epsilon_0 \sum_{n>0} \boldsymbol{\alpha}^{(n)} : \mathbf{E}_{loc}^{\otimes n} \\ &= \mathbf{P}_{loc}^{L} + \mathbf{P}_{loc}^{NL}. \end{aligned} \tag{1.107}$$

This is still not in the form of the Bloembergen expansion, as it involves the local fields, not the macroscopic averages. Combining this with Eq. (1.105), assuming the response is scalar, and dropping the frequency arguments for ease of notation, we obtain

$$\mathbf{P} = n_\rho\epsilon_0 \sum_{n>0} \boldsymbol{\alpha}^{(n)} : \left(\mathbf{E} + \frac{1}{3\epsilon_0}\mathbf{P}\right)^{\otimes n}. \tag{1.108}$$

This means that the expansion for the *local* linear polarization in terms of the *macroscopic* electric field includes a nonlinear contribution. This is due to an interaction between the nonlinear dipole moment at one site, and the linear dipole moment at another site. The series now has to be re-summed, to give an expansion in terms of the macroscopic electric field. Rearranging the linear response terms, and assuming this term is a scalar, one immediately obtains

$$\mathbf{P} = \frac{1}{1 - \rho_n \alpha^{(1)}/3} [\rho_n \epsilon_0 \alpha^{(1)} \mathbf{E} + \mathbf{P}_{loc}^{NL}]. \tag{1.109}$$

Hence, we get for the linear susceptibility:

$$\chi^{(1)} = \frac{\rho_n \alpha^{(1)}}{1 - \rho_n \alpha^{(1)}/3}. \tag{1.110}$$

This shows that the linear response or linear permittivity is given by the famous Clausius–Mosotti relationship:

$$\epsilon = \epsilon_0 (1 + \chi^{(1)}) = \epsilon_0 \frac{1 + 2\rho_n \alpha^{(1)}/3}{1 - \rho_n \alpha^{(1)}/3}. \tag{1.111}$$

For the case of the two-level atom studied in Eq. (1.98), the local-field corrections shift the resonance frequency in a density-dependent way. This can be viewed as a change in the oscillator energy due to dipole–dipole interactions, and gives the result

$$\chi^{(1)}(\omega) = \frac{d^2 \rho_n / \epsilon_0 \hbar}{\nu_0 - \Delta\omega - \omega}. \tag{1.112}$$

The overall effect is to shift the resonance frequency ν_0 of the two-level transition by $\Delta\omega = d^2 \rho_n / (3\epsilon_0 \hbar)$ to a lower value of $\nu_0 - \Delta\omega$ that depends linearly on the atomic density. This is called the Lorentz red-shift, and is experimentally observable, as demonstrated in Figure 1.3. Note that in these experiments additional corrections $\Delta\omega_{coll}$ were needed to account for collisional effects due to van der Waals interatomic forces in a vapor, which we have neglected. We treat atom–field interactions in greater detail in Chapter 4.

Next, we assume the linear response is scalar and frequency-independent, and to a first approximation neglect the interactions between the higher-order terms in Eq. (1.109). That is, we only keep linear corrections to the macroscopic field (see Eq. (1.105))

$$\mathbf{E}_{loc} = \mathbf{E} + \frac{1}{3\epsilon_0} \epsilon_0 \chi^{(1)} \mathbf{E} = \left(\frac{2\epsilon_0 + \epsilon}{3\epsilon_0} \right) \mathbf{E}. \tag{1.113}$$

We can now calculate the local-field corrections to the nonlinear Bloembergen electric field susceptibilities, giving

$$\chi^{(n)} \approx \rho_n \alpha^{(n)} \left[\frac{\epsilon + 2\epsilon_0}{3\epsilon_0} \right]^n. \tag{1.114}$$

More generally, the corrections involve the tensorial character of the polarizabilities, and the crystal symmetry group. These details are available in texts specializing in nonlinear optics, given in the recommended reading list.

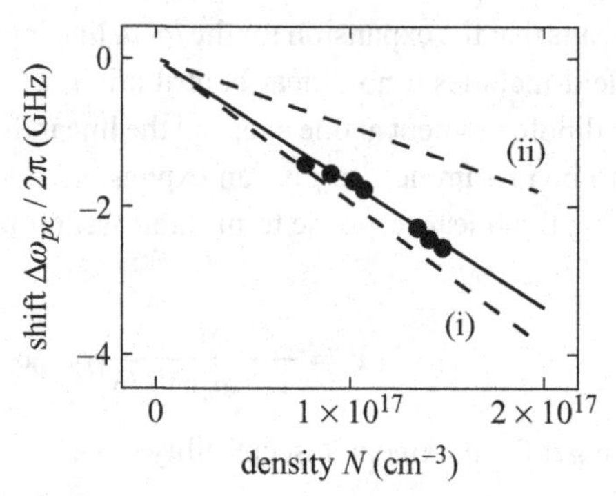

Fig. 1.3 Measured density-dependent frequency shifts, $\Delta\omega$, versus density in potassium vapor, for the $4^2S_{1/2} \longleftrightarrow 4^2P_{1/2}$ transition. The solid line gives the Lorentz local-field prediction, together with an additional small collisional red-shift of $\Delta\omega_{coll} = \beta\rho_n$, where $\beta = 5.0 \times 10^{-8}\ \mathrm{s}^{-1}\ \mathrm{cm}^3$. Full circles give experimental measurements using a nonlinear optical technique employing phase conjugation. Dashed lines (i) and (ii) show alternative theories, which do not reproduce the experimental data. Copied (with permission) from: J. J. Maki *et al.*, *Phys. Rev. Lett.* **67**, 972 (1991).

1.7.3 Displacement field expansion

We will see in Chapter 4 that the microscopic field $\mathbf{E}_{loc}$ that occurs in the dipole-coupled interaction Hamiltonian is actually the displacement field, i.e. $\mathbf{E}_{loc} \equiv \mathbf{D}(\mathbf{r}_j)/\epsilon_0$. This motivates an alternative approach to calculating dielectric response, using the inverse permittivity expansion in terms of the displacement field. These coefficients can also be worked out from the local-field corrections. From Eq. (1.105), we see that

$$\mathbf{P} = \rho_n \sum_{n>0} \epsilon_0^{1-n} \boldsymbol{\alpha}^{(n)} : (\mathbf{D} - \tfrac{2}{3}\mathbf{P})^{\otimes n}. \tag{1.115}$$

Just as for the electric field expansion, the series can be re-summed to give an expansion in terms of the macroscopic displacement field. Next, we assume the linear response is scalar and frequency-independent. Rearranging the linear response terms, we obtain

$$\mathbf{P} = \frac{\rho_n \alpha^{(1)} \mathbf{D} + \mathbf{P}_{loc}^{NL}}{1 + 2\rho_n \alpha^{(1)}/3}. \tag{1.116}$$

For the linear term, the inverse permittivity is

$$\eta^{(1)} = \frac{1}{\epsilon_0} \frac{1 - \rho_n \alpha^{(1)}/3}{1 + 2\rho_n \alpha^{(1)}/3} = \epsilon^{-1}. \tag{1.117}$$

This shows that the inverse permittivity is exactly the inverse of the usual permittivity in Eq. (1.111). We can now calculate the local-field corrections to the nonlinear displacement field susceptibilities, in the approximation where we make only the linear corrections to the local field. This gives, for $n > 1$,

$$\eta^{(n)} \approx -\chi^{(n)} \epsilon^{-n}. \tag{1.118}$$

In general, either expansion can be used to describe nonlinear response, and a more systematic way to convert between the two types of series is given in Eq. (1.29). We note that, for $\epsilon > \epsilon_0$, the local-field corrections *increase* the magnitude of $\chi^{(n)}$, while they *decrease* the magnitude of $\eta^{(n)}$. This simply compensates for the fact that $\mathbf{D} = \epsilon\mathbf{E}$ to lowest order, so the displacement field is relatively larger than the electric field inside a dielectric.

1.8 Propagation in a nonlinear medium

Let us now have a more detailed look at the propagation of waves in a nonlinear dielectric medium, starting from the commonly employed Eq. (1.3), so that the electric field expansion is the most useful one. In addition, we will split the polarization into a linear and a nonlinear part, $\mathbf{P} = \mathbf{P}^L + \mathbf{P}^{NL}$. This gives us

$$-\nabla^2\mathbf{E} + \mu_0\frac{\partial^2\mathbf{D}^L}{\partial t^2} = -\mu_0\frac{\partial^2\mathbf{P}^{NL}}{\partial t^2}, \tag{1.119}$$

where $\mathbf{D}^L = \epsilon_0\mathbf{E} + \mathbf{P}^L$. We now expand the field and polarization in frequency components,

$$\begin{aligned}
\mathbf{E}(\mathbf{r}, t) &= \sum_n (\mathcal{E}_n(\mathbf{r})e^{-i\omega_n t} + c.c.), \\
\mathbf{D}^L(\mathbf{r}, t) &= \sum_n (\mathcal{D}_n^L(\mathbf{r})e^{-i\omega_n t} + c.c.), \\
\mathbf{P}^{NL}(\mathbf{r}, t) &= \sum_n (\mathcal{P}_n^{NL}(\mathbf{r})e^{-i\omega_n t} + c.c.),
\end{aligned} \tag{1.120}$$

where $c.c.$ denotes complex conjugate. If we now make use of the frequency-dependent linear dielectric function, $\epsilon(\omega)$, and set $\mathcal{D}_n^L = \epsilon(\omega_n)\mathcal{E}_n$, we have, after substituting these expressions into the wave equation and equating the coefficients of terms with the same frequency, that

$$\nabla^2\mathcal{E}_n + \frac{\omega_n^2\epsilon(\omega_n)}{c^2\epsilon_0}\mathcal{E}_n = -\frac{\omega_n^2}{c^2\epsilon_0}\mathcal{P}_n^{NL}. \tag{1.121}$$

1.8.1 Three-wave mixing

Let us use this equation to consider a three-wave mixing process. We shall assume that each of the waves is propagating in the x direction and is polarized in a fixed direction perpendicular to x, either y or z. Because all of the fields and the polarizations of a given index will point in the same direction, we can drop the vector notation and treat everything as a scalar. We will also assume that the fields are close to plane waves. In particular, we can express $\mathcal{E}_n(x)$ as $\mathcal{A}_n(x)e^{ik_n x}$, where $k_n = [\epsilon(\omega_n)/\epsilon_0]^{1/2}(\omega_n/c)$ and $\mathcal{A}_n(x)$ is slowly varying on the scale of an optical wavelength. The equation for $\mathcal{E}_n$ then becomes

$$\left\{\frac{d^2\mathcal{A}_n}{dx^2} + 2ik_n\frac{d\mathcal{A}_n}{dx}\right\}e^{ik_n x} = -\frac{\omega_n^2}{c^2\epsilon_0}\mathcal{P}_n^{NL}(x). \tag{1.122}$$

Because $\mathcal{A}_n$ is slowly varying on the scale of a wavelength, we have that

$$\left|\frac{d^2\mathcal{A}_n}{dx^2}\right| \ll \left|k_n\frac{d\mathcal{A}_n}{dx}\right|, \tag{1.123}$$

so the second-derivative term in the above equation can be neglected. Now let us assume that $\omega_3 = \omega_1 + \omega_2$. The frequency components of the nonlinear polarization are then given by

$$\begin{aligned} \mathcal{P}_1^{nl} &= \epsilon_0\chi^{(2)}\mathcal{A}_3\mathcal{A}_2^* e^{i(k_3-k_2)x}, \\ \mathcal{P}_2^{nl} &= \epsilon_0\chi^{(2)}\mathcal{A}_3\mathcal{A}_1^* e^{i(k_3-k_1)x}, \\ \mathcal{P}_3^{nl} &= \epsilon_0\chi^{(2)}\mathcal{A}_1\mathcal{A}_2 e^{i(k_1+k_2)x}, \end{aligned} \tag{1.124}$$

where we have used the fact that, since the spatial dependence of the frequency component at ω_n is given by $\mathcal{A}_n e^{ik_n x}$, the spatial dependence of the component at $-\omega_n$ is given by $\mathcal{A}_n^* e^{-ik_n x}$. We note that it is possible to have $\mathcal{P}_1^{nl}$ and $\mathcal{P}_2^{nl}$ with orthogonal polarizations, provided the nonlinear susceptibility, $\chi^{(2)}$, has the appropriate symmetry properties. This is called type II phase matching. Finally, for the three different waves, we have the equations

$$\begin{aligned} \frac{d\mathcal{A}_1}{dx} &= \frac{i\omega_1^2\chi^{(2)}}{2k_1c^2}\mathcal{A}_3\mathcal{A}_2^* e^{i\Delta kx}, \\ \frac{d\mathcal{A}_2}{dx} &= \frac{i\omega_2^2\chi^{(2)}}{2k_2c^2}\mathcal{A}_3\mathcal{A}_1^* e^{i\Delta kx}, \\ \frac{d\mathcal{A}_3}{dx} &= \frac{i\omega_3^2\chi^{(2)}}{2k_3c^2}\mathcal{A}_1\mathcal{A}_2 e^{-i\Delta kx}, \end{aligned} \tag{1.125}$$

where $\Delta k = k_3 - k_1 - k_2$.

1.8.2 Parametric amplifier

The above equations can be solved in terms of Jacobi elliptic functions, but let us look at a simpler case, which will be of considerable interest to us, the parametric amplifier. In that case, the field at ω_3 is much stronger than the other two fields. This allows us to ignore depletion of this field, and assume that $\mathcal{A}_3$ is constant. We can, therefore, drop the third equation above, and express the first two as

$$\begin{aligned} \frac{d\mathcal{A}_1}{dx} &= i\kappa_1 e^{i\Delta kx}\mathcal{A}_2^*, \\ \frac{d\mathcal{A}_2}{dx} &= i\kappa_2 e^{i\Delta kx}\mathcal{A}_1^*, \end{aligned} \tag{1.126}$$

where $\kappa_j = (\omega_j^2\chi^{(2)}\mathcal{A}_3)/(2k_jc^2)$, for $j = 1, 2$, and, for simplicity, we shall assume that $\mathcal{A}_3$ is real. In order to solve these equations, we first set $\tilde{\mathcal{A}}_j = \mathcal{A}_j \exp(-i\Delta kx/2)$, for $j = 1, 2$. We then find that

$$\begin{aligned} \frac{d\tilde{\mathcal{A}}_1}{dx} &= -i\left(\frac{\Delta k}{2}\right)\tilde{\mathcal{A}}_1 + i\kappa_1\tilde{\mathcal{A}}_2^*, \\ \frac{dB\tilde{\mathcal{A}}_2}{dx} &= -i\left(\frac{\Delta k}{2}\right)\tilde{\mathcal{A}}_2 + i\kappa_2\tilde{\mathcal{A}}_1^*. \end{aligned} \tag{1.127}$$

Taking the complex conjugate of the second equation, we find that the resulting two differential equations can be expressed as

$$\frac{d}{dx}\begin{pmatrix}\tilde{A}_1\\ \tilde{A}_2^*\end{pmatrix} = \begin{pmatrix}-i\Delta k/2 & i\kappa_1\\ -i\kappa_2 & i\Delta k/2\end{pmatrix}\begin{pmatrix}\tilde{A}_1\\ \tilde{A}_2^*\end{pmatrix}. \tag{1.128}$$

The eigenvalues of the above matrix are $\pm\bar{\kappa}$, where $\bar{\kappa} = [\kappa_1\kappa_2 - (\Delta k/2)^2]^{1/2}$, so we will have one solution proportional to $\exp(\bar{\kappa}x)$ and one proportional to $\exp(-\bar{\kappa}x)$. The simplest case, where $\Delta k = 0$ and $\kappa_1 = \kappa_2$, gives simply that $\bar{\kappa} = \kappa_1 = \kappa_2$. The resulting solutions for $\tilde{A}_1$ and $\tilde{A}_2$ are

$$\begin{aligned}\tilde{A}_1(x) &= e^{i\Delta kx/2}\left\{\tilde{A}_1(0)\left[\cosh(\bar{\kappa}x) - \frac{i\Delta k}{2g}\sinh(\bar{\kappa}x)\right] + \frac{i\kappa_1}{g}\tilde{A}_2^*(0)\sinh(\bar{\kappa}x)\right\}\\ \tilde{A}_2(x) &= e^{i\Delta kx/2}\left\{\tilde{A}_2(0)\left[\cosh(\bar{\kappa}x) - \frac{i\Delta k}{2g}\sinh(\bar{\kappa}x)\right] + \frac{i\kappa_2}{g}\tilde{A}_1^*(0)\sinh(\bar{\kappa}x)\right\}.\end{aligned} \tag{1.129}$$

Examining these two equations we see that the waves at ω_1 and ω_2 are amplified provided $\kappa_1\kappa_2 > (\Delta k/2)^2$. For large x, they are both proportional to $\exp(\bar{\kappa}x)$. What is happening is that the medium is transferring energy from the field at ω_3, known as the pump, to the fields at ω_1 and ω_2. In reality, this exponential growth would stop, because the pump would be depleted, but this is an effect we did not take into account in our analysis.

It should also be noted that the largest gain occurs when $\bar{\kappa}$ is greatest, which happens when $\Delta k = 0$. This condition is known as phase matching. It is generally the case that nonlinear optical processes are most efficient under conditions of phase matching, and so an effort is usually made to guarantee that this condition is satisfied.

1.8.3 Parametric oscillator

We can turn the amplifier into an oscillator if we put the nonlinear medium between two mirrors. If the gain is high enough, in particular high enough to overcome losses due to the mirrors and absorption by the medium, a steady-state field will be sustained between the mirrors. This will occur if the fields reproduce themselves after one round trip. We can apply this requirement to find the required gain, and the result is known as the threshold condition.

Let us suppose, for simplicity, that the reflectivity of the mirrors, R, is the same at both ω_1 and ω_2, and also assume that the linear absorption coefficient (absorption per unit length) of the medium, γ, is also the same for both frequencies. We will also assume that the phase-matching condition is satisfied, which implies that

$$\begin{aligned}\tilde{A}_1(x) &= \tilde{A}_1(0)\cosh(\bar{\kappa}x) + \frac{i\kappa_1}{g}\tilde{A}_2^*(0)\sinh(\bar{\kappa}x),\\ \tilde{A}_2^*(x) &= \tilde{A}_2^*(0)\cosh(\bar{\kappa}x) - \frac{i\kappa_2}{g}\tilde{A}_1(0)\sinh(\bar{\kappa}x).\end{aligned} \tag{1.130}$$

In one round trip, the fields pass through the medium twice, and are reflected from each mirror. The condition that both fields are the same after one round trip is, therefore,

$$\tilde{A}_1(0) = \left[\tilde{A}_1(0)\cosh(\bar{\kappa}x) + \frac{i\kappa_1}{g}\tilde{A}_2^*(0)\sinh(\bar{\kappa}x)\right]R^2 e^{-2\gamma L},$$
$$\tilde{A}_2^*(0) = \left[\tilde{A}_2^*(0)\cosh(\bar{\kappa}x) - \frac{i\kappa_2}{g}\tilde{A}_1(0)\sinh(\bar{\kappa}x)\right]R^2 e^{-2\gamma L}. \quad (1.131)$$

These equations will have nonzero solutions for $A_1(0)$ and $A_2(0)^*$ if

$$1 = R^2 e^{2(\bar{\kappa}-\alpha)L} = e^{2[\ln R + (\bar{\kappa}-\gamma)L]}. \quad (1.132)$$

Therefore, we require that

$$\ln R + L(\bar{\kappa} - \gamma) = 0, \quad (1.133)$$

If R is close to 1, then $\ln R \cong -(1 - R)$, and the threshold condition becomes

$$(1 - R) = L(\bar{\kappa} - \gamma). \quad (1.134)$$

This ends our discussion of the classical parametric amplifier and oscillator, but we will have much more to say about both of these systems after we quantize the field. In the quantum regime, every time a photon at ω_1 is created, so is one at ω_2, and this leads to strong correlation between these two modes. The correlation is sufficiently strong that they cannot be described classically. As a result, parametric oscillators and amplifiers have been employed extensively as sources of nonclassical fields and in quantum communication schemes.

1.9 Raman processes

Finally, we would like to briefly discuss Raman processes, which play a significant role in optical fibers. In Raman scattering, the optical field scatters off an excitation in the medium. The simplest situation is when the atoms or molecules of the medium are in a state $|a\rangle$ with energy $\hbar\omega_a$, and the optical field has frequency ω. As a result of the interaction with the field, the atom or molecule is left in a state $|b\rangle$, which has energy $\hbar\omega_b$. Setting $\Delta\omega = \omega_a - \omega_b$, we see, by energy conservation, that the frequency of the outgoing optical field is now $\omega + \Delta\omega$, and that energy has either been deposited in (if $\Delta\omega < 0$) or extracted from (if $\Delta\omega > 0$) the medium.

The first case is called Stokes scattering and the second anti-Stokes scattering (Figure 1.4). The excitations can be, for example, electronic or vibrational levels in molecules or phonons in solids. If the phonons are acoustic, the process is referred to as Brillouin scattering. This process can be either stimulated or spontaneous. For example, in the case of Stokes scattering, there may be no Stokes wave present initially, and the Stokes wave then originates from spontaneous emission, or there may be an initial Stokes wave present, in which case the output Stokes wave is a result of stimulated emission. This leads to coherent gain and loss.

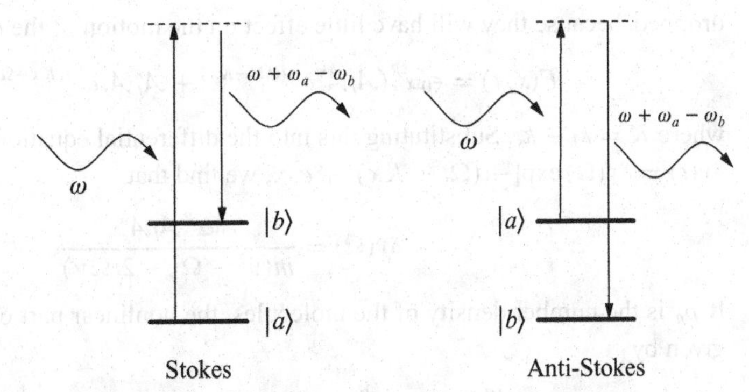

Fig. 1.4 There are two forms of Raman scattering. In a Stokes process, energy is deposited in the medium and the output field is at a lower frequency than that of the input field. In an anti-Stokes process, energy is extracted from the medium and the frequency of the output field is higher than that of the input field.

1.9.1 Diatomic model

In order to get an idea of the kind of nonlinear polarization to which this leads, let us consider a simple model for stimulated Raman scattering due to Garmire *et al.*, in which the material excitation is a vibrational state of a diatomic molecule. Suppose the equilibrium distance between the two nuclei of our molecule is y_0 and the deviation from this distance is $\delta y = y - y_0$. We will assume that the molecule is a driven, damped oscillator described by the equation

$$\frac{d^2(\delta y)}{dt^2} + 2\gamma \frac{d(\delta y)}{dt} + \nu^2 \delta y = \frac{F(t)}{m}, \tag{1.135}$$

where γ is the damping constant, ν is the vibrational frequency and $F(t)$ is the driving force, which is a result of the electromagnetic waves. We now assume that the linear polarizability of the molecule depends on the internuclear separation y. We expand the linear polarizability α in a Taylor series in the displacement $\delta y = y - y_0$ from equilibrium y_0, keeping only the first two terms: $\alpha^{(1)} = \alpha_0^{(1)} + \alpha^R \delta y$, where $\alpha^R = (d\alpha^{(1)}/dy)_{y=y_0}$. To find the force, we note that the energy of an induced dipole in an external field is $\frac{1}{2}\epsilon_0 \alpha^{(1)} E_{loc}^2$, where E_{loc} is the electric field at the molecular location. Its derivative with respect to the displacement y is the electromagnetic force. Neglecting local-field corrections, and expanding in terms of the average field, we obtain

$$F = \tfrac{1}{2}\epsilon_0 \alpha^R E^2. \tag{1.136}$$

Now assume that the electric field contains a wave at the frequency of the incident laser, ω_k, and one at the Stokes frequency, ω_s, where $\Omega = \omega_k - \omega_s$ is the detuning, which is comparable to the vibrational frequency, ν, and

$$E(x,t) = \mathcal{A}_l e^{-i(\omega_k t - k_l x)} + \mathcal{A}_s e^{-i(\omega_s t - k_s x)} + c.c. \tag{1.137}$$

We can now substitute this into the equation for the force, but we shall assume that the vibrational frequency is much smaller than the laser and Stokes frequencies, and, consequently, that terms oscillating rapidly compared to the vibrational frequency can be dropped, because they will have little effect on the motion of the nuclei. This gives us that

$$F(x,t) = \epsilon_0 \alpha^R (\mathcal{A}_l \mathcal{A}_s^* e^{-i(\Omega t - Kx)} + \mathcal{A}_l^* \mathcal{A}_s e^{-i(Kx - \Omega t)}) + c.c., \tag{1.138}$$

where $K = k_l - k_s$. Substituting this into the differential equation for $\delta y(t)$, and assuming $\delta y(t) = \delta y(\Omega) \exp[-i(\Omega t - Kx)] + c.c.$, we find that

$$\delta y(\Omega) = \frac{\epsilon_0 \alpha^R \mathcal{A}_l \mathcal{A}_s^*}{m(\nu^2 - \Omega^2 - 2i\Omega\gamma)}. \tag{1.139}$$

If ρ_n is the number density of the molecules, the nonlinear part of the polarization is then given by

$$\begin{aligned} P^{NL} &= \epsilon_0 \rho_n \alpha^R [\delta y(\Omega) e^{-i(\Omega t - Kx)} + \delta y^*(\Omega) e^{-i(Kx - \Omega t)}] \\ &\quad \times (\mathcal{A}_l e^{-i(\omega_l t - k_l x)} + \mathcal{A}_s e^{-i(\omega_s t - k_s x)} + c.c.). \end{aligned} \tag{1.140}$$

The component of the nonlinear polarization at the Stokes frequency, which serves as a source for the Stokes field, is

$$\begin{aligned} P_S^{NL} &= \epsilon_0 \alpha^R [\delta y^*(\Omega) \mathcal{A}_l e^{i(\Omega t - Kx) + i(k_l x - \omega_l t)} + c.c.] \\ &= \left[\frac{\rho_n (\epsilon_0 \alpha^R)^2 |\mathcal{A}_l|^2 \mathcal{A}_s}{m(\nu^2 - \Omega^2 + 2i\Omega\gamma)} e^{-i(\omega_s t - k_s x)} + c.c. \right]. \end{aligned} \tag{1.141}$$

This polarization leads to gain in the Stokes field. Note that the polarization, and hence the gain, is independent of the phase of the laser field. That means that the phase-matching condition is always satisfied, leading to efficient amplification of the Stokes field.

Additional reading

Books

N. Bloembergen, *Nonlinear Optics* (W. A. Benjamin, Reading, MA, 1965).
R. W. Boyd, *Nonlinear Optics* (Academic Press, San Diego, 2003).
P. N. Butcher and D. Cotter, *The Elements of Nonlinear Optics* (Cambridge University Press, Cambridge, 1990).
Y. R. Shen, *The Principles of Nonlinear Optics* (Wiley, New York, 1984).

Articles

J. A. Armstrong, N. Bloembergen, J. Ducuing, and P. S. Pershan, *Phys. Rev.* **127**, 1918 (1962).
C. M. Caves and P. D. Drummond, *Rev. Mod. Phys.* **66**, 481 (1994).
P. A. Franken, A. E. Hill, C. W. Peters, and G. Weinreich, *Phys. Rev. Lett.* **7**, 118 (1961).
E. Garmire, F. Pandarese, and C. H. Townes, *Phys. Rev. Lett.* **11**, 160 (1963).
W. Louisell, A. Yariv, and A. Siegman, *Phys. Rev.* **124**, 1646 (1961).
T. Popmintchev *et al.*, *Science* **336**, 1287 (2012).

Problems

1.1 Consider the equations describing the three-wave mixing process.

(a) Show that the intensity at each frequency is

$$I_j \propto (k_j/\omega_j)|\mathcal{A}_j|^2.$$

(b) Show that

$$\frac{d}{dx}(I_1 + I_2 + I_3) = 0.$$

(c) Show that

$$\frac{d}{dx}\left(\frac{I_1}{\omega_1}\right) = \frac{d}{dx}\left(\frac{I_2}{\omega_2}\right) = -\frac{d}{dx}\left(\frac{I_3}{\omega_3}\right).$$

1.2 Again, consider the three-wave mixing process, but now consider the case when the field at ω_1 is strong while those at ω_2 and ω_3 are much weaker. This implies that we can ignore depletion in the field at ω_1 so that $\mathcal{A}_1$ can be treated as a constant. Solve the remaining equations for $\mathcal{A}_2$ and $\mathcal{A}_3$ and find $\mathcal{A}_2(x)$ and $\mathcal{A}_3(x)$ in terms of $\mathcal{A}_2(0)$ and $\mathcal{A}_3(0)$. Do this first for the case $\Delta k = 0$ and then for the general case.

1.3 Assume that the field in a medium with $\chi^{(2)} \neq 0$ has two frequency components, one at ω and one at 2ω. Find the differential equations for $\mathcal{A}_1(x)$ and $\mathcal{A}_2(x)$, the slowly varying parts of the electric field at ω and 2ω, respectively. In the case in which the fundamental at ω is much stronger than the second harmonic at 2ω, find an expression for $\mathcal{A}_2(x)$ in terms of $\mathcal{A}_2(0)$.

1.4 A simple model for a linear medium is one composed of harmonic oscillators. Each oscillator has a resonant frequency ν_0, mass m, damping γ and charge $-e$, so that its equation of motion in the presence of an electric field $E(t)$ in the x direction is

$$\frac{d^2x}{dt^2} + 2\gamma\frac{dx}{dt} + \nu_0^2 x = \frac{-eE(t)}{m},$$

where $x(t)$ is the coordinate of the oscillator. There are ρ_n oscillators per unit volume.

Setting $E(t) = \mathcal{E}e^{-i\omega t} + c.c.$ and $x(t) = x(\omega)e^{-i\omega t} + c.c.$ (where $c.c.$ denotes complex conjugate terms), show that the low-density linear susceptibility of this model is

$$\chi^{(1)}(\omega) = \left(\frac{\rho_n e^2}{m\epsilon_0}\right)\frac{1}{\nu_0^2 - \omega^2 - 2i\gamma\omega}.$$

1.5 We can make the medium nonlinear by making the oscillator in the previous question anharmonic. Suppose now that, for small displacements, it obeys the equation

$$\frac{d^2x}{dt^2} + 2\gamma\frac{dx}{dt} + \nu_0^2 x + ax^2 = \frac{-eE(t)}{m},$$

where a is small. We can solve this equation perturbatively. We set $x(t) = x_0(t) + x_1(t)$, where $x_0(t)$ is the solution for $a = 0$ and $x_1(t)$ is linear in a. Show that this results in an equation for $x_1(t)$ of form

$$\frac{d^2x_1}{dt^2} + 2\gamma\frac{dx_1}{dt} + \nu_0^2 x_1 = -a x_0^2.$$

Again, setting $E(t) = \mathcal{E}e^{-i\omega t} + c.c.$ and using your solution for the first part of this problem for x_0 and setting $x_1(t) = x_1(2\omega)e^{-2i\omega t}$, find $x_1(2\omega)$ and then use it to find the nonlinear susceptibility $\chi^{(2)}(2\omega)$.

2 Field quantization

In Chapter 1, we treated the electromagnetic field as classical. Henceforth, we will want to treat it as a quantum field. In order to do so, we will first present some of the formalism of quantum field theory. This formalism is very useful in describing many-particle systems and processes in which the number of particles changes. Why this is important to a quantum description of nonlinear optics can be seen by considering a parametric amplifier of the type discussed in the last chapter. A pump field, consisting of many photons, amplifies idler and signal fields by means of a process in which a pump photon splits into two lower-energy photons, one at the idler frequency and one at the signal frequency. Therefore, what we would like to do in this chapter is to provide a discussion of some of the basics of quantum field theory that will be useful in the treatment of the quantization of the electromagnetic field.

In particular, we will begin with a summary of quantum theory notation, and a discussion of many-particle Hilbert spaces. These provide the arena in which all of the action takes place. We will then move on to a treatment of the canonical quantization procedure for fields. This will allow us to develop a scattering theory for fields, which is ideally what we need. This relates the properties of a field entering a medium to those of the field leaving it, and this corresponds to what is done in an experiment.

Quantum field theory is a huge subject, and the discussion here will be limited to what is needed. For those readers who want to delve into the subject in greater depth, we provide a list of further reading at the end of the chapter.

2.1 Quantum theory

We start by recalling the elementary properties of quantum theory, to establish our notation. The heart of quantum mechanics is that an experimental system is represented by a mathematical state vector $|\Psi(t)\rangle$, which is an element of a linear vector space – technically, a Hilbert space $\mathcal{H}$. Measurable properties of the physical system are associated with Hermitian operators acting on the state vectors, called observables. Time evolution in standard quantum theory is of two types:

- deterministic or *unitary* evolution occurs in the absence of an observer;
- random evolution of the quantum state occurs when measured.

We can intuitively understand these two situations as being caused by a clear physical difference. An observer must be coupled to the physical system in order to carry out measurements, which gives rise to a larger joint system. Not unexpectedly, the observer perturbs the observed quantum system, which is changed unpredictably as a consequence. We generally (but not always) assume that the quantum state is projected into the operator eigenstate corresponding to the observed eigenvalue, after measurement.

While this heuristic procedure certainly *works* – it agrees with experiments performed to date – these assumptions lead to many questions about the foundations of quantum mechanics and the nature of reality. Such questions were explored initially by Bohr, Einstein and Schrödinger, the founders of quantum mechanics, leading to the Einstein–Podolsky–Rosen (EPR) and Schrödinger cat paradoxes. Later investigators, including de Broglie, Bohm and Bell, fine-tuned these early investigations. This resulted in Bell's theorem, which famously encapsulates the quantum paradoxes related to local realism in an experimentally testable form.

Since many of these experimental tests and applications of the foundations of quantum mechanics are carried out with techniques of nonlinear quantum optics, we will return to this subject in greater detail in Chapter 12.

2.1.1 Unitary time evolution

The unitary evolution can be calculated in different time-evolution 'pictures' for convenience, but this does not change the observable physics. We use subscripts on the operators and states to distinguish the different pictures.

Schrödinger picture

In this picture, the operators are time-invariant, unless they describe explicit time dependence, for example, a coupling to an external time-varying force. The state $|\Psi(t)\rangle_S$ that describes the system during a time in which there is no observation will evolve according to a unitary operator $U(t, t_0)$. Thus, given an initial state $|\Psi(t_0)\rangle$, one has

$$|\Psi(t)\rangle_S = U(t, t_0)|\Psi(t_0)\rangle. \tag{2.1}$$

In quantum theory, this time-translation operator is generated from the quantum Hamiltonian operator, which has a close algebraic similarity to the Hamiltonian function in classical mechanics. Time evolution can also be written directly in the form of a differential equation called the Schrödinger equation:

$$i\hbar\frac{d}{dt}|\Psi(t)\rangle_S = H_S(t)|\Psi(t)\rangle_S. \tag{2.2}$$

In the simplest case of a system with time-translation symmetry, $H_S(t) = H$, and the unitary operator is simply

$$U(t, t_0) = e^{-i(t-t_0)H/\hbar}. \tag{2.3}$$

With a time-dependent Hamiltonian $H(t)$, it is straightforward to check that the solution for U is given by

$$U(t, t_0) = \lim_{\epsilon\to 0}\{e^{-i\epsilon H(t-\epsilon)/\hbar} \dots e^{-i\epsilon H(t_0+\epsilon)/\hbar} e^{-i\epsilon H(t_0)/\hbar}\}$$
$$\equiv T\left\{\exp\left[-i\int_{t_0}^{t} dt'\, H(t')/\hbar\right]\right\}, \tag{2.4}$$

where the ordering symbol T indicates a time-ordered product with later times placed to the left.

More generally, in systems described using a density matrix

$$\rho_S(t) = \sum_i P_i|\Psi_i(t)\rangle_S\langle\Psi_i(t)|_S = U(t, t_0)\rho(t_0)U^\dagger(t, t_0). \tag{2.5}$$

Here P_i is the probability of observing the state i, and the density matrix evolves according to

$$\frac{d\rho_S(t)}{dt} = \frac{i}{\hbar}[\rho_S(t), H_S(t)]. \tag{2.6}$$

Heisenberg picture

In the Heisenberg picture, the density operator and quantum states are all constant in time, while operators or observables O_H that describe measurements on the system will evolve. This can also be described by a differential equation,

$$\frac{dO_H(t)}{dt} = \frac{i}{\hbar}[H_H(t), O_H(t)] + \frac{\partial O_H(t)}{\partial t}. \tag{2.7}$$

Here $\partial O_H(t)/\partial t$ describes any *explicit* time variation in the operator. If $\partial O_H(t)/\partial t = 0$, one obtains the inverse time evolution to Eq. (2.5):

$$O_H(t) = U^\dagger(t, t_0)O(t_0)U(t, t_0). \tag{2.8}$$

Interaction picture

In this hybrid approach, the Hamiltonian is divided into two parts, $H = H_0 + H'$. The operators evolve in time according to the free-field evolution H_0, and the density matrix evolves according to the interaction Hamiltonian H'. We suppose that H_0 has no explicit time evolution. In this picture, we define a free-field unitary evolution operator as

$$U_0(t, t_0) = e^{-iH_0(t-t_0)/\hbar}. \tag{2.9}$$

The interaction-picture operators now evolve according to the free-field part of the Hamiltonian,

$$O_I(t) = U_0^\dagger(t, t_0)O(t_0)\, U_0(t, t_0), \tag{2.10}$$

and quantum states (and density matrix) evolve according to the interaction term H', which must also be expressed in terms of interaction-picture operators. This is necessary so that all observable averages are still calculated correctly, hence we must obtain

$$|\Psi(t)\rangle_I = U_0^\dagger(t,t_0)|\Psi(t)\rangle_S. \tag{2.11}$$

Differentiating the state equation (2.11), and noting that H_0 commutes with itself, so $H_0 = H_{0I}$, we arrive at the quantum-state time-evolution equation in the interaction picture:

$$i\hbar\frac{d}{dt}|\Psi(t)\rangle_I = U_0^\dagger[-H_0 + H]|\Psi(t)\rangle_S = H_I'(t)|\Psi(t)\rangle_I. \tag{2.12}$$

This is solved by the interaction-picture time-evolution operator,

$$|\Psi(t)\rangle_I = U_I(t,t_0)|\Psi(t_0)\rangle,$$

where the time-evolution operator must satisfy the initial condition, $U_I(t_0,t_0) = I$, together with the evolution equation

$$i\hbar\frac{d}{dt}U_I(t,t_0) = H_I'(t)U_I(t,t_0). \tag{2.13}$$

This is readily solved using the interaction Hamiltonian written using interaction-picture operators:

$$U_I(t,t_0) = \mathcal{T}\left\{\exp\left[-i\int_{t_0}^{t} dt'\, H_I'(t')/\hbar\right]\right\} \tag{2.14}$$

This approach has the advantage that one only has to calculate the effects of interactions, with free-field evolution removed. Where necessary, of course, free-field evolution can be included via a unitary transformation.

Rotating-frame picture

This picture is widely used in systems driven with an external laser, which provides a coherent frequency reference at an externally fixed frequency, ω_L. The Hamiltonian is divided into two parts, as in the interaction picture,

$$H = H_{0L} + H', \tag{2.15}$$

but the choice of how to split the Hamiltonian is different. In this case, H_{0L} is chosen so that it describes free evolution of a field mode at the frequency ω_L, and, by making this choice, ω_L serves as a reference for the frequencies of all other field modes. In this picture, the operators evolve in time according to H_{0L}, and the density matrix evolves according to the interaction Hamiltonian H'.

In the rotating frame, it is common to work with time-invariant operators, $a_j = a_j(t)e^{i\omega_{Lj}t}$, with the understanding that the true time evolution of the operators in the interaction picture will be added explicitly where necessary, depending on the type of observation. Thus, for example, in observations with local oscillators, at the reference frequency, the time dependence at the optical frequency cancels. In addition, any time dependence typically cancels in the Hamiltonian, due to conjugate terms being multiplied together in operator products. To indicate that an expectation value is evaluated in the rotating frame with time-invariant operators, we use the notation $\langle O\rangle_R$.

2.1.2 Measurement and uncertainty

In any of these pictures, it is conventional to assume that the act of measurement projects the system randomly into an eigenstate of the measurement operator. This assumption is not always correct, and depends on the measurement apparatus. The average or *expectation* value of a Hermitian operator or *observable* O is given by

$$\langle O \rangle = \mathrm{Tr}[\rho(t)O(t)], \tag{2.16}$$

and we see by comparing Eqs (2.5) and (2.8) that this is independent of the picture.

Fluctuations in quantum measurements have a lower bound given by the *Heisenberg uncertainty principle*, which will enter much of the material in later chapters. This states that, for two operators A and B having a commutator $C = [A, B] = AB - BA$, then

$$\Delta A \Delta B \geq \frac{|\langle C \rangle|}{2}, \tag{2.17}$$

where we define $\Delta A = \sqrt{\Delta^2 A} = \sqrt{\langle A^2 \rangle - \langle A \rangle^2}$.

This uncertainty relation was derived by Robertson, and a stronger version by Schrödinger; but, following common usage, we will call it the Heisenberg uncertainty principle.

2.2 Fock space for bosons

The Fock space or particle-number formalism, which we shall now explore, provides a useful description for both fermions and bosons. We will begin by describing the boson Fock space, and then move on to fermions in the next section. This formalism allows us to understand the relationship between the first-quantized theory of single particles, and the multi-particle space described by using field operators in later parts of this book. An important requirement is that the method we employ should properly take into account the symmetry properties of the quantum states. If the particles we are describing are bosons, then the quantum states must be completely symmetric under the interchange of particles, and, if they are fermions, the quantum states must be completely antisymmetric under particle interchange. We will see later that the field quantization method guarantees this property; in first-quantized methods, it must be imposed by hand.

2.2.1 Many-particle Hilbert space

To illustrate the Fock space formalism, let us recall nonrelativistic quantum mechanics. The Hilbert space of a single particle is $\mathcal{H}_1$, with an orthonormal basis $|f_j\rangle$, where $j = 1, 2, \ldots$. These are often chosen to be the eigenstates of the single-particle Hamiltonian, i.e. $h_{SP}|f_j\rangle = E_j|f_j\rangle$, where in nonrelativistic physics

$$h_{SP} = \frac{-\hbar^2}{2m}\nabla_{\mathbf{r}}^2 + V(\mathbf{r}). \tag{2.18}$$

If we construct the many-body states from single-particle eigenstates, we would use these eigenstates together with symmetrization requirements. We will show that this approach gives the same results as field quantization. Where there are both field and single-particle versions of operators, to distinguish the two types, we use capitals for quantum field operators and lower-case letters for single-particle operators.

Because we are now considering bosons, the N-particle Hilbert space will be the symmetric subspace of the N-fold tensor product of $\mathcal{H}_1$ with itself, i.e. $\mathcal{H}_N = (\mathcal{H}_1^{\otimes N})_{sym}$. The boson Fock space, $\mathcal{H}$, is just the direct sum of spaces corresponding to different particle numbers,

$$\mathcal{H} = \mathcal{H}_0 \oplus \mathcal{H}_1 \oplus \mathcal{H}_2 \oplus \cdots . \tag{2.19}$$

Here, $\mathcal{H}_0$ is the one-dimensional space corresponding to no particles, and we call the normalized vector in it the vacuum state. The Hilbert space $\mathcal{H}$ allows us to describe states that do not have a definite number of particles. A state may have a certain amplitude to have no particles, another amplitude to have one particle, and so on. This type of state arises often in quantum optics, where the particles are photons. For example, if we start with an atom in an excited state and wait for a time equal to the lifetime of the excited state, there will be an amplitude for no photon to be present, corresponding to the atom not having decayed, and an amplitude for one photon to be present, corresponding to the atom having decayed.

We can define a basis for $\mathcal{H}$ by using the basis for $\mathcal{H}_1$. Let $\mathbf{n}$ be a vector of nonnegative numbers, n_j, where $\mathbf{n} = (n_1, n_2, \ldots)$, and

$$N(\mathbf{n}) = \sum_{j=1}^{\infty} n_j < \infty. \tag{2.20}$$

This condition implies that only a finite number of the n_j can be nonzero. We define the state $|\mathbf{n}\rangle \in \mathcal{H}$ to be the symmetric, normalized state with n_1 particles in the state $|f_1\rangle$, n_2 in the state $|f_2\rangle$, etc. In particular, the vacuum state corresponds to the vector $\mathbf{n} = \mathbf{0}$, all of whose entries are zero. The states $|\mathbf{n}\rangle$ form an orthonormal basis for $\mathcal{H}$, and, in particular, $\langle \mathbf{n}|\mathbf{n}'\rangle = 0$ if $\mathbf{n} \neq \mathbf{n}'$. The number n_j is called the occupation number of the state $|f_j\rangle$ in the state $|\mathbf{n}\rangle$.

In both fermionic and bosonic cases, it is generally necessary to include internal degrees of freedom, corresponding to spin or other internal symmetries. Each distinct internal eigenstate is regarded as a different particle, whose Hilbert space must be included also. The overall Hilbert space is then a direct product of the Hilbert space of all particle types.

2.2.2 Bosonic creation and annihilation operators

Using these basis states, we now want to define creation, annihilation and number operators on $\mathcal{H}$. For a vector $\mathbf{n}$, define $\mathbf{n}(k)^+$ to be the vector whose components are the same as those in $\mathbf{n}$ except for the kth one, in which n_k in $\mathbf{n}$ has been replaced by $n_k + 1$ in $\mathbf{n}(k)^+$. Therefore, the quantum state $|\mathbf{n}(k)^+\rangle$ is the same as $|\mathbf{n}\rangle$, except that it has one more particle in the state $|f_k\rangle$. Similarly, we can define $\mathbf{n}(k)^-$ to be the vector whose components are the same as $\mathbf{n}$ except for the kth place, but in this case n_k in $\mathbf{n}$ is replaced by $n_k - 1$ in $\mathbf{n}(k)^-$. Note that, if $n_k = 0$, then $\mathbf{n}(k)^-$ will have a negative entry, and there will be no quantum state corresponding to it.

We can now define our operators. Define the creation operator, $a_k^\dagger$, to have the action on the states $|\mathbf{n}\rangle$,

$$a_k^\dagger|\mathbf{n}\rangle = \sqrt{n_k + 1}\,|\mathbf{n}(k)^+\rangle, \tag{2.21}$$

so that $a_k^\dagger$ adds a particle in the state $|f_k\rangle$. If $n_k \neq 0$, the annihilation operator, a_k, has the following action on $|\mathbf{n}\rangle$,

$$a_k|\mathbf{n}\rangle = \sqrt{n_k}\,|\mathbf{n}(k)^-\rangle, \tag{2.22}$$

and if $n_k = 0$, then $a_k|\mathbf{n}\rangle = 0$. Finally, the number operator N acts as

$$N|\mathbf{n}\rangle = N(\mathbf{n})|\mathbf{n}\rangle. \tag{2.23}$$

The action of an operator on the basis vectors serves to define its action on any vector. In particular, the action of the operators defined above on any vector in $\mathcal{H}$ can be found by expanding the vector in terms of the basis vectors, and then making use of the above definitions and the fact that the operators are linear.

The definitions of the creation, annihilation and number operators imply a number of useful relations, all of which can be verified by applying both sides of the resulting equations to basis vectors. We find for the commutation relations of these operators:

$$\begin{aligned} [a_j, a_k] = [a_j^\dagger, a_k^\dagger] = 0, \qquad [a_j, a_k^\dagger] = \delta_{j,k}, \\ [a_j, N] = a_j, \qquad [a_j^\dagger, N] = -a_j^\dagger. \end{aligned} \tag{2.24}$$

The total number operator can be expressed in terms of the creation and annihilation operators,

$$N = \sum_{j=1}^{\infty} a_j^\dagger a_j. \tag{2.25}$$

In addition, we find that the creation operator $a_j^\dagger$ is indeed the adjoint of the corresponding annihilation operator a_j, that is, for $\psi, \psi' \in \mathcal{H}$, we have $\langle\psi|a_j\psi'\rangle = \langle a_j^\dagger\psi|\psi'\rangle$, and the basis vectors can be expressed as

$$|\mathbf{n}\rangle = \left(\prod_{k=1}^{\infty} \frac{(a_k^\dagger)^{n_k}}{\sqrt{n_k!}}\right)|0\rangle, \tag{2.26}$$

where we have denoted the normalized vacuum vector by $|0\rangle$.

It is also possible to define creation and annihilation operators for states other than the basis states, $\{|f_j\rangle \mid j = 1, 2, \ldots\}$. Suppose $|g\rangle \in \mathcal{H}_1$ and

$$|g\rangle = \sum_{j=1}^{\infty} \langle f_j|g\rangle|f_j\rangle. \tag{2.27}$$

We can define creation and annihilation operators corresponding to the state $|g\rangle$ by

$$a^\dagger(g) = \sum_{j=1}^{\infty} \langle f_j | g \rangle a_j^\dagger,$$

$$a(g) = \sum_{j=1}^{\infty} \langle f_j | g \rangle^* a_j. \tag{2.28}$$

The commutation relation between operators corresponding to different states is $[a(g_1), a^\dagger(g_2)] = \langle g_1 | g_2 \rangle$.

2.3 Many-body operators

The operators $a(g)$ and $a^\dagger(g)$ can be expressed in a different way if we make use of a coordinate representation of the basis functions. We can use this to make contact with quantum field theory, which we will develop more directly using canonical quantization in later sections. Let us assume the basis functions are wavefunctions, i.e. square integrable functions on a box, which we shall call the quantization volume, of side length L and of volume $V = L^3$. We shall assume the functions satisfy periodic boundary conditions. Define the operators

$$\psi(\mathbf{r}) = \sum_{j=0}^{\infty} f_j(\mathbf{r}) a(f_j), \qquad \psi^\dagger(\mathbf{r}) = \sum_{j=0}^{\infty} f_j^*(\mathbf{r}) a^\dagger(f_j). \tag{2.29}$$

These operators obey the commutation relations

$$[\psi(\mathbf{r}), \psi^\dagger(\mathbf{r}')] = \delta^{(3)}(\mathbf{r} - \mathbf{r}'), \tag{2.30}$$

and we have that

$$a(g) = \int_V d^3\mathbf{r}\, g^*(\mathbf{r}) \psi(\mathbf{r}), \qquad a^\dagger(g) = \int_V d^3\mathbf{r}\, g(\mathbf{r}) \psi^\dagger(\mathbf{r}). \tag{2.31}$$

A common basis to use when applying this formalism is one consisting of plane-wave states,

$$f_{\mathbf{k}}(\mathbf{r}) = \frac{1}{\sqrt{V}} e^{i\mathbf{k}\cdot\mathbf{r}}, \tag{2.32}$$

where $\mathbf{k} = (2\pi/L)(m_1, m_2, m_3)$ and m_1, m_2 and m_3 are integers. Note that m_1, m_2 and m_3 are quantum numbers labeling the basis functions, and are not occupation numbers. In this case we set $a_{\mathbf{k}} = a(f_{\mathbf{k}})$ and $a_{\mathbf{k}}^\dagger = a^\dagger(f_{\mathbf{k}})$. Clearly, from the definition, $a_{\mathbf{k}}^\dagger$ creates a particle with a momentum of $\hbar\mathbf{k}$.

For simplicity, we have ignored internal degrees of freedom. If this is included, the fields are labeled ψ_s to include an internal spin s, and the commutation relations become

$$[\psi_s(\mathbf{r}), \psi_{s'}^\dagger(\mathbf{r}')] = \delta_{ss'} \delta^{(3)}(\mathbf{r} - \mathbf{r}'). \tag{2.33}$$

2.3.1 Quantum fields

In terms of these operators we can define quantum field operators, although we will give a more general approach to this in Section 2.5, using canonical quantization methods:

$$\psi(\mathbf{r}) = \frac{1}{\sqrt{V}}\sum_{\mathbf{k}} e^{i\mathbf{k}\cdot\mathbf{r}} a_{\mathbf{k}}, \qquad \psi^{\dagger}(\mathbf{r}) = \frac{1}{\sqrt{V}}\sum_{\mathbf{k}} e^{-i\mathbf{k}\cdot\mathbf{r}} a^{\dagger}_{\mathbf{k}}. \tag{2.34}$$

Now let us see how to represent multi-particle states. Let $g(\mathbf{r}_1, \ldots, \mathbf{r}_n)$ be a symmetric, normalized n-particle wavefunction. The state in our bosonic Fock space that corresponds to it is

$$|\Phi_g\rangle = \frac{1}{\sqrt{n!}}\int_V d^3\mathbf{r}_1 \cdots \int_V d^3\mathbf{r}_n\, g(\mathbf{r}_1, \ldots, \mathbf{r}_n)\psi^{\dagger}(\mathbf{r}_1)\cdots\psi^{\dagger}(\mathbf{r}_n)|0\rangle. \tag{2.35}$$

If we want to recover $g(\mathbf{r}_1, \ldots, \mathbf{r}_n)$ from an arbitrary n-particle state $|\Phi_g\rangle$, we find that

$$g(\mathbf{r}_1, \ldots, \mathbf{r}_n) = \langle 0|\psi(\mathbf{r}_n)\cdots\psi(\mathbf{r}_1)|\Phi_g\rangle. \tag{2.36}$$

It should be noted that, even if $g(\mathbf{r}_1, \ldots, \mathbf{r}_n)$ is not symmetric, the state $|\Phi_g\rangle$ is. This follows from the fact that the state $\psi^{\dagger}(\mathbf{r}_1)\cdots\psi^{\dagger}(\mathbf{r}_n)|0\rangle$ is symmetric.

Our next step is to define the action of certain kinds of operators on $\mathcal{H}$. Let us first consider groups of unitary operations. These include operations such as translations and rotations. A translation acting on an n-particle state should just translate each of the particles. This should, in fact, hold for any symmetry operation that is a generalization of a single-particle symmetry operation. Application of the symmetry operation to an n-particle state should just be given by applying the single-particle operation to each of the particles in the state. We can define an operator that accomplishes this in the following way. Suppose U_1 is the single-particle translation operation. It acts on the single-particle Hilbert space $\mathcal{H}_1$ so that $|g\rangle \in \mathcal{H}_1 \to U_1|g\rangle$. If $U_1(\mathbf{r}_0)$ is a translation by $\mathbf{r}_0$, then we have $U_1(\mathbf{r}_0)g(\mathbf{r}) = g(\mathbf{r} - \mathbf{r}_0)$. We define the corresponding operator acting on $\mathcal{H}$ by

$$Ua^{\dagger}(g)U^{-1} = a^{\dagger}(U_1 g), \qquad Ua(g)U^{-1} = a(U_1 g), \qquad U|0\rangle = |0\rangle. \tag{2.37}$$

This definition implies that the action of U on an n-particle state is just

$$Ua^{\dagger}(g_1)\cdots a^{\dagger}(g_n)|0\rangle = a^{\dagger}(U_1 g_1)\cdots a^{\dagger}(U_1 g_n)|0\rangle. \tag{2.38}$$

The above state is not the most general n-particle state, but any n-particle state can be expressed as the sum of such states, and so the previous equation determines the action of U on any n-particle state. With this definition, the operator U is unitary – it does not change the inner products of vectors. This type of promotion of an operator from the single-particle to the full multi-particle space we shall refer to as *multiplicative* promotion.

2.3.2 Additive promotion

Not all single-particle operators are promoted to the full multi-particle Hilbert space in the same way as elements of unitary groups. In order to see this, let us again look at translations. On the single-particle space, $\mathcal{H}_1$, translations are generated by the momentum operator,

$\mathbf{p} = -i\hbar\nabla$, so that

$$U_1(\mathbf{r}_0) = e^{-i\mathbf{r}_0\cdot\mathbf{p}/\hbar}. \tag{2.39}$$

The corresponding operator, $U(\mathbf{r}_0)$, on $\mathcal{H}$ acts on an n-particle state as

$$U(\mathbf{r}_0)a^\dagger(g_1)\cdots a^\dagger(g_n)|0\rangle = a^\dagger(U_1(\mathbf{r}_0)g_1)\cdots a^\dagger(U_1(\mathbf{r}_0)g_n)|0\rangle. \tag{2.40}$$

The operator $U(\mathbf{r}_0)$ can be expressed as the exponential of an operator $\mathbf{P}$, which is the momentum operator on $\mathcal{H}$,

$$U(\mathbf{r}_0) = e^{-i\mathbf{r}_0\cdot\mathbf{P}/\hbar}. \tag{2.41}$$

We can now differentiate both sides of Eq. (2.40), with respect to $\mathbf{r}_0$, that is, apply $\nabla_{\mathbf{r}_0}$ to both sides, and evaluate the result at $\mathbf{r}_0 = 0$. What we find is that

$$\mathbf{P}a^\dagger(g_1)\cdots a^\dagger(g_n)|0\rangle = \sum_{j=1}^{n} a^\dagger(g_1)\cdots a^\dagger(g_{j-1})a^\dagger(\mathbf{p}g_j)a^\dagger(g_{j+1})\cdots a^\dagger(g_n)|0\rangle. \tag{2.42}$$

Performing the same operation, applying $\nabla_{\mathbf{r}_0}$ to both sides, and evaluating the result at $\mathbf{r}_0 = 0$ to the equation

$$e^{-i\mathbf{r}_0\cdot\mathbf{P}/\hbar}a^\dagger(g)e^{i\mathbf{r}_0\cdot\mathbf{P}/\hbar} = a^\dagger(e^{-i\mathbf{r}_0\cdot\mathbf{p}/\hbar}g) \tag{2.43}$$

gives us

$$[\mathbf{P}, a^\dagger(g)] = a^\dagger(\mathbf{p}g). \tag{2.44}$$

Finally, we note the fact that $U(\mathbf{r}_0)|0\rangle = |0\rangle$ implies that $\mathbf{P}|0\rangle = 0$. This fact and the above commutation relation actually imply Eq. (2.42). We shall call this type of promotion from the single-particle to the multi-particle space additive promotion. Other operators that generate unitary groups are also promoted to the full multi-particle space via additive promotion. Angular momentum is an example, as is the Hamiltonian, if the particles are noninteracting. For noninteracting particles, each particle evolves independently according to a single-particle Hamiltonian, h_{SP}, which is an operator on the space of single-particle wavefunctions. The single-particle Hamiltonian is the generator of time translations in the single-particle space, and the full Hamiltonian, H, obtained from h_{SP} by additive promotion, is the generator of time translations for the full multi-particle space.

There are two other expressions for an additively promoted single-particle operator that are useful. Let w be an operator on the single-particle space $\mathcal{H}_1$, and W be the additively promoted version of w on $\mathcal{H}$. Then we have that

$$W = \sum_{j,k=0}^{M} \langle f_j|w|f_k\rangle a_j^\dagger a_k. \tag{2.45}$$

We need to show that this operator has an action similar to the one in Eq. (2.42), which it will if it obeys a commutation relation similar to that in Eq. (2.44) and annihilates the vacuum. It clearly annihilates the vacuum, and we find that

$$[W, a^\dagger(g)] = \sum_{j,k=0}^{M} \langle f_j|w|f_k\rangle\langle f_k|g\rangle a_j^\dagger = a^\dagger(wg), \tag{2.46}$$

which proves the above formula. If we choose the plane-wave basis, the formula that defines the promoted operator in quantum field theory becomes

$$W = \int_V d^3\mathbf{r}\, \psi^\dagger(\mathbf{r}) w \psi(\mathbf{r}). \tag{2.47}$$

2.3.3 Many-body Hamiltonian

A common example of the above formula is the energy of a particle of mass m, which we shall denote by H_0:

$$\begin{aligned} H_0 &= \int_V d^3\mathbf{r}\, \psi^\dagger(\mathbf{r}) h_{SP} \psi(\mathbf{r}) \\ &= \int_V d^3\mathbf{r}\, \psi^\dagger(\mathbf{r}) \left[\frac{-\hbar^2}{2m} \nabla_\mathbf{r}^2 + V(\mathbf{r}) \right] \psi(\mathbf{r}). \end{aligned} \tag{2.48}$$

Now that we know how to promote single-particle operators, we need to see how to promote two-particle operators. This is essential if we want to describe interactions. Suppose our interaction is described by a two-body potential, $V(\mathbf{r} - \mathbf{r}')$. The promoted interaction operator on $\mathcal{H}$ is given by

$$H' = \frac{1}{2} \int_V d^3\mathbf{r} \int_V d^3\mathbf{r}'\, \psi^\dagger(\mathbf{r}) \psi^\dagger(\mathbf{r}') V(\mathbf{r} - \mathbf{r}') \psi(\mathbf{r}') \psi(\mathbf{r}). \tag{2.49}$$

As we can see, this is a natural generalization of the procedure for promoting single-particle operators, but we need to verify that it does what we want it to do. Let us allow the interaction Hamiltonian to act on a two-particle state,

$$|\Psi\rangle = \frac{1}{\sqrt{2}} \int_V d^3\mathbf{r}_1 \int_V d^3\mathbf{r}_2\, g(\mathbf{r}_1, \mathbf{r}_2) \psi^\dagger(\mathbf{r}_1) \psi^\dagger(\mathbf{r}_2) |0\rangle. \tag{2.50}$$

We find

$$H'|\Psi\rangle = \frac{1}{\sqrt{2}} \int_V d^3\mathbf{r} \int_V d^3\mathbf{r}'\, V(\mathbf{r} - \mathbf{r}') g(\mathbf{r}, \mathbf{r}') \psi^\dagger(\mathbf{r}) \psi^\dagger(\mathbf{r}') |0\rangle, \tag{2.51}$$

where we have assumed that $g(\mathbf{r}_1, \mathbf{r}_2)$ is a symmetric function. It would also be the two-particle wavefunction of the state in standard quantum mechanics. What corresponds to the right-hand side of Eq. (2.51) in standard quantum mechanics (see Eq. (2.35)) is just $V(\mathbf{r} - \mathbf{r}')g(\mathbf{r}, \mathbf{r}')$, which is what it should be for a two-particle potential operator acting on a two-particle wavefunction. It is easily verified that the action of H' on an N-particle state gives the result that it should. The consequence is that the total Hamiltonian of an interacting many-body quantum system has the form

$$H = H_0 + H'. \tag{2.52}$$

For photonic systems, the interaction terms are changed, as we see later, but the general division into free and interacting terms can be retained and used to define interaction or rotating-frame pictures.

2.4 Fock space for fermions

Having developed the formalism for the boson Fock space, let us now move on to fermions. We again have the decomposition of our Fock space $\mathcal{H}$ into subspaces of different particle number,

$$\mathcal{H} = \mathcal{H}_0 \oplus \mathcal{H}_1 \oplus \mathcal{H}_2 \oplus \cdots, \tag{2.53}$$

but we now have that $\mathcal{H}_N = (\mathcal{H}^{\otimes N})_{anti}$: therefore, the N-particle subspace is the antisymmetric subspace of the N-fold tensor product of the one-particle space. As before, we define a basis for $\mathcal{H}$ by making use of a vector of occupation numbers $\mathbf{n} = (n_1, n_2, \ldots)$, where

$$N(\mathbf{n}) = \sum_{j=1}^{\infty} n_j < \infty, \tag{2.54}$$

but this time $n_j = 0, 1$. This choice reflects the fact that two or more fermions cannot occupy the same quantum state. We now define $|\mathbf{n}\rangle$ to be the antisymmetric state with n_1 particles in state $|f_1\rangle$, n_2 in state $|f_2\rangle$, and so on. As with bosons, the orthonormality relation is $\langle \mathbf{n}|\mathbf{n}'\rangle = \delta_{\mathbf{n},\mathbf{n}'}$.

2.4.1 Fermionic creation and annihilation operators

We can now define creation and annihilation operators by using the basis states $|\mathbf{n}\rangle$. The sequences $\mathbf{n}(k)^+$ and $\mathbf{n}(k)^-$ are defined as before, and we define the operators c_k and $c_k^\dagger$ by

$$c_k|\mathbf{n}\rangle = n_k \left(\prod_{j=1}^{k-1} (-1)^{n_j} \right) |\mathbf{n}(k)^-\rangle \tag{2.55}$$

and

$$c_k^\dagger|\mathbf{n}\rangle = (1 - n_k) \left(\prod_{j=1}^{k-1} (-1)^{n_j} \right) |\mathbf{n}(k)^+\rangle. \tag{2.56}$$

The number operator is defined as for bosons:

$$N|\mathbf{n}\rangle = N(\mathbf{n})|\mathbf{n}\rangle. \tag{2.57}$$

These creation and annihilation operators obey the canonical anticommutation relations. Defining $[A, B]_+ = AB + BA$, one has

$$[c_j, c_k]_+ = [c_j^\dagger, c_k^\dagger]_+ = 0, \qquad [c_j, c_k^\dagger]_+ = \delta_{jk}. \tag{2.58}$$

Note that these relations imply that $(c_k^\dagger)^2 = 0$, which means that it is impossible to create two fermions in the same quantum state. The number operator can be expressed as

$$N = \sum_{j=1}^{M} c_j^\dagger c_j, \tag{2.59}$$

and the states $|\mathbf{n}\rangle$ are given by

$$|\mathbf{n}\rangle = (c_1^\dagger)^{n_1} \cdots (c_r^\dagger)^{n_r} |0\rangle, \tag{2.60}$$

where n_r is the last nonzero element of the vector $\mathbf{n}$. The vacuum state, $|0\rangle$, corresponds to the sequence all of whose elements are zero. We note that, unlike the boson case, it is important to preserve the operator order in this definition. Since the Fermi operators anticommute, swapping the order of two neighboring operators will change the sign of the state.

We can define creation and annihilation operators corresponding to a general one-particle state, $|g\rangle \in \mathcal{H}_1$, by

$$c^\dagger(g) = \sum_{j=1}^{M} \langle f_j | g\rangle c_j^\dagger,$$

$$c(g) = \sum_{j=1}^{M} \langle f_j | g\rangle^* c_j, \tag{2.61}$$

and from these we find that $[c(g), c^\dagger(g')]_+ = \langle g|g'\rangle$, and we have that $c(g)|0\rangle = 0$. We can also, as before, define field operators

$$\psi(\mathbf{r}) = \sum_{j=0}^{\infty} f_j(\mathbf{r}) c(f_j), \qquad \psi^\dagger(\mathbf{r}) = \sum_{j=0}^{\infty} f_j^*(\mathbf{r}) c^\dagger(f_j). \tag{2.62}$$

Instead of commutation relations, these operators obey the anticommutation relations

$$[\psi(\mathbf{r}), \psi^\dagger(\mathbf{r}')]_+ = \delta^{(3)}(\mathbf{r} - \mathbf{r}'), \tag{2.63}$$

and we have that

$$c(g) = \int_V d^3\mathbf{r}\, g^*(\mathbf{r})\psi(\mathbf{r}), \qquad c^\dagger(g) = \int_V d^3\mathbf{r}\, g(\mathbf{r})\psi^\dagger(\mathbf{r}). \tag{2.64}$$

In the fermion case, it is also common to use the plane-wave basis, in which case we have that

$$\psi(\mathbf{r}) = \frac{1}{\sqrt{V}} \sum_{\mathbf{k}} e^{i\mathbf{k}\cdot\mathbf{r}} c_{\mathbf{k}}, \qquad \psi^\dagger(\mathbf{r}) = \frac{1}{\sqrt{V}} \sum_{\mathbf{k}} e^{-i\mathbf{k}\cdot\mathbf{r}} c_{\mathbf{k}}^\dagger, \tag{2.65}$$

where $c_{\mathbf{k}} = c(f_{\mathbf{k}})$ and $c_{\mathbf{k}}^\dagger = c^\dagger(f_{\mathbf{k}})$.

As with the case of bosons, fermions can have internal degrees of freedom – in fact, this is essential, as fermions have half-integer spin. With this included, the fields are labeled ψ_s to include an internal spin s, and the anticommutation relations become

$$[\psi_s(\mathbf{r}), \psi_{s'}^\dagger(\mathbf{r}')]_+ = \delta_{ss'}\delta^{(3)}(\mathbf{r} - \mathbf{r}'). \tag{2.66}$$

2.4.2 Antisymmetric multi-particle states

Multi-particle states can be represented in a way similar to that in the boson case. Let $g(\mathbf{r}_1, \ldots, \mathbf{r}_n)$ be an antisymmetric, normalized n-particle wavefunction. The state in our fermionic Fock space that corresponds to it is

$$|\Phi_g\rangle = \frac{1}{\sqrt{n!}} \int_V d^3\mathbf{r}_1 \cdots \int_V d^3\mathbf{r}_n \, g(\mathbf{r}_1, \ldots, \mathbf{r}_n) \psi^\dagger(\mathbf{r}_1) \cdots \psi^\dagger(\mathbf{r}_n)|0\rangle. \tag{2.67}$$

We can recover $g(\mathbf{r}_1, \ldots, \mathbf{r}_n)$ from $|\Phi_g\rangle$ by taking

$$g(\mathbf{r}_1, \ldots, \mathbf{r}_n) = \langle 0|\psi(\mathbf{r}_n) \cdots \psi(\mathbf{r}_1)|\Phi_g\rangle. \tag{2.68}$$

The promotion of operators from the single-particle space to the overall Fock space is identical to what is done in the case of bosons. Unitary symmetry operations are promoted via multiplicative promotion, and their generators by additive promotion. In particular, if w is a single-particle operator and W its additive promotion, we will have

$$[W, c^\dagger(g)] = c^\dagger(wg) \tag{2.69}$$

and

$$W = \sum_{j,k=0}^{M} \langle f_j|w|f_k\rangle c_j^\dagger c_k. \tag{2.70}$$

Again, if we choose the plane-wave basis, the field operator can be expressed as

$$W = \int_V d^3\mathbf{r} \, \psi^\dagger(\mathbf{r}) w \psi(\mathbf{r}). \tag{2.71}$$

The expressions for the kinetic and potential energy operators are also the same as in the boson case, except that in the fermion case the operators $\psi(\mathbf{r})$ and $\psi^\dagger(\mathbf{r})$ obey anti-commutation relations instead of commutation relations.

2.5 Canonical quantization

In order to discuss the quantum theory of nonlinear optics, we will have to quantize the electromagnetic field. Before doing so, it is useful to present the general canonical formalism for field quantization. This is a top-down alternative to the procedure of the previous section. It is useful when the identification of the particles and Hamiltonian operator is not already known.

We will follow the original route of Dirac, in which a canonical Lagrangian is introduced, whose variables are the operators of the quantum theory. The *canonical quantization* procedure then gives one the Hamiltonian function H and the commutation relations obeyed by the fields. The steps required can be summarized as follows:

1. Find a classical Lagrangian that generates the proper equations of motion.
2. This gives one the canonical momentum and a canonical Hamiltonian, which must equal the system energy.
3. Canonical commutation relations are imposed to give the quantum behavior.

2.5.1 Classical Lagrangian field theory

As noted above, when formally quantizing any system, even a single particle, the detailed procedure is the following. We begin with the classical system. We first choose coordinates, and then write down a Lagrangian, which is a function of the coordinates and their time derivatives, that describes the system. From the Lagrangian, we obtain two things: the canonical momentum, and the Hamiltonian. We then impose the canonical commutation relations between the coordinates and the momenta, and the system is thereby quantized.

Fields are more complicated than single particles, because they have many more – in fact, an infinite number of – coordinates. Let us now present the canonical formalism for a scalar field, $\phi(\mathbf{r}, t)$ and treat the more complicated case of the electromagnetic field later. We start with a Lagrangian, L, which can be expressed in terms of a Lagrangian density, $\mathcal{L}$,

$$L = \int d^3\mathbf{r}\, \mathcal{L}, \tag{2.72}$$

and we shall assume for now that $\mathcal{L}$ is a function of ϕ, $\partial_t\phi$ and $\partial_j\phi$, where ∂_t is a more compact way of writing $\partial/\partial t$ and ∂_j, $j = 1, 2, 3$, corresponds to partial spatial derivatives in the x, y and z directions, respectively. This assumption will be true for most of what we do, but there will be one case, when we discuss fields in dispersive media, for which the Lagrangian density will also depend on mixed space and time derivatives. We define the action to be

$$S = \int_{t_1}^{t_2} dt\, L. \tag{2.73}$$

When we change the field, $\phi(\mathbf{r}, t) \to \phi(\mathbf{r}, t) + \delta\phi(\mathbf{r}, t)$, where we consider only variations in the field, $\delta\phi(\mathbf{r}, t)$, which vanish at $t = t_1$, at $t = t_2$ and as $|\mathbf{r}| \to \infty$, then S goes to $S + \delta S$. The classical equations of motion are determined by the condition that the action is stationary, that is, $\delta S = 0$ for any choice of $\delta\phi$ obeying the boundary conditions.

The change in the Lagrangian can be expressed in terms of functional derivatives. These are defined by the equation

$$\delta L = \int d^3\mathbf{r} \left[\frac{\delta L}{\delta\phi}\delta\phi + \frac{\delta L}{\delta(\partial_t\phi)}\partial_t(\delta\phi) \right], \tag{2.74}$$

where only first-order terms in $\delta\phi$ and its derivatives have been kept. These functional derivatives can be expressed in terms of partial derivatives of the Lagrangian density. To do so we begin by noting that

$$\delta L = \int d^3\mathbf{r} \left[\frac{\partial\mathcal{L}}{\partial\phi}\delta\phi + \sum_{j=1}^{3} \frac{\partial\mathcal{L}}{\partial(\partial_j\phi)}\partial_j(\delta\phi) + \frac{\partial\mathcal{L}}{\partial(\partial_t\phi)}\partial_t(\delta\phi) \right]. \tag{2.75}$$

We can perform a partial integration on the term with the spatial derivatives of $\delta\phi$, yielding

$$\delta L = \int d^3\mathbf{r} \left[\left(\frac{\partial\mathcal{L}}{\partial\phi} - \sum_{j=1}^{3} \partial_j \frac{\partial\mathcal{L}}{\partial(\partial_j\phi)} \right)\delta\phi + \frac{\partial\mathcal{L}}{\partial(\partial_t\phi)}\partial_t(\delta\phi) \right]. \tag{2.76}$$

From this we see that

$$\frac{\delta L}{\delta \phi} = \frac{\partial \mathcal{L}}{\partial \phi} - \sum_{j=1}^{3} \partial_j \frac{\partial \mathcal{L}}{\partial(\partial_j \phi)},$$

$$\frac{\delta L}{\delta(\partial_t \phi)} = \frac{\partial \mathcal{L}}{\partial(\partial_t \phi)}. \tag{2.77}$$

We can now go on to examine the variation of the action. We have that

$$\delta S = \int_{t_1}^{t_2} dt \int d^3\mathbf{r} \left[\frac{\delta L}{\delta \phi} \delta \phi + \frac{\delta L}{\delta(\partial_t \phi)} \partial_t(\delta \phi) \right]. \tag{2.78}$$

We can now perform a partial integration on the time integral to give

$$\delta S = \int_{t_1}^{t_2} dt \int d^3\mathbf{r} \left[\frac{\delta L}{\delta \phi} - \partial_t \frac{\delta L}{\delta(\partial_t \phi)} \right] \delta \phi. \tag{2.79}$$

If $\delta S = 0$ for any choice of $\delta \phi$, the expression in brackets must vanish, yielding the equation of motion

$$\frac{\delta L}{\delta \phi} - \partial_t \frac{\delta L}{\delta(\partial_t \phi)} = 0, \tag{2.80}$$

or, in terms of the Lagrangian density,

$$\frac{\partial \mathcal{L}}{\partial \phi} - \sum_{j=1}^{3} \partial_j \frac{\partial \mathcal{L}}{\partial(\partial_j \phi)} - \partial_t \frac{\partial \mathcal{L}}{\partial(\partial_t \phi)} = 0. \tag{2.81}$$

Most fields in physics are vector fields, with an index k that describes internal spin degrees of freedom. These satisfy the most obvious generalization of the above equation:

$$\frac{\partial \mathcal{L}}{\partial \phi_k} - \sum_{j=1}^{3} \partial_j \frac{\partial \mathcal{L}}{\partial(\partial_j \phi_k)} - \partial_t \frac{\partial \mathcal{L}}{\partial(\partial_t \phi_k)} = 0. \tag{2.82}$$

2.5.2 Hamiltonian, quantization and commutators

We can also express the equations of motion in terms of the Hamiltonian. The canonical momentum is just the functional derivative of the Lagrangian with respect to $\partial_t \phi_i$,

$$\pi_k = \frac{\delta L}{\delta(\partial_t \phi_k)}. \tag{2.83}$$

The Hamiltonian is then

$$H = \int d^3\mathbf{r} \sum_k \pi_k(\mathbf{r}, t) \partial_t \phi_k(\mathbf{r}, t) - L, \tag{2.84}$$

and the equations of motion – called Hamilton's equations – are given by

$$\frac{\delta H}{\delta \phi_k} = -\partial_t \pi_k, \qquad \frac{\delta H}{\delta \pi_k} = \partial_t \phi_k. \tag{2.85}$$

The final step in the quantization of the theory is to make all of the above quantities into operators. To achieve this, we define a quantity called the classical Poisson bracket:

$$\{A, B\} \equiv \sum_{k=1}^{N} \left[\frac{\delta A}{\delta \phi_k(\mathbf{r})} \frac{\delta B}{\delta \pi_k(\mathbf{r})} - \frac{\delta A}{\delta \pi_k(\mathbf{r})} \frac{\delta B}{\delta \phi_k(\mathbf{r})} \right]. \tag{2.86}$$

The Dirac quantization procedure consists of equating the quantum commutator with $i\hbar$ times the classical Poisson bracket, i.e.

$$[A, B] = i\hbar\{A, B\}. \tag{2.87}$$

This has the effect of replacing classical variables with quantum operators. Here, the notation

$$[A, B] \equiv AB - BA \tag{2.88}$$

indicates a commutator, used to quantize a boson field of integer-spin particles. An anti-commutator, $[A, B]_+ \equiv AB + BA$, is used to quantize a fermion field of half-integer-spin particles. We emphasize that there is a change in interpretation on quantization, and the notation ϕ now indicates a quantum field. These are operators in Hilbert space; in fact, they are operators on the boson Fock space that we constructed at the beginning of this chapter, and consequently no longer always commute with the other quantities in the theory. In general, the commutator of any two operators will either be a constant or another operator.

Using this technique, we find that the canonical equal-time commutation relations between the coordinate, which is the field, and its corresponding momentum are

$$[\phi_j(\mathbf{r}, t), \pi_k(\mathbf{r}', t)]_{(+)} = i\hbar\delta_{jk}\delta^{(3)}(\mathbf{r} - \mathbf{r}'). \tag{2.89}$$

The equal-time commutation relations between the field and itself and between the canonical momentum and itself are

$$[\phi_j(\mathbf{r}, t), \phi_k(\mathbf{r}', t)]_{(+)} = [\pi_j(\mathbf{r}, t), \pi_k(\mathbf{r}', t)]_{(+)} = 0. \tag{2.90}$$

Here the subscript $(+)$ indicates a commutator for bosons and an anticommutator for fermions. These commutation relations (or anticommutation relations for fermions), when used in conjunction with the Hamiltonian, can be used to find the Heisenberg equation of motion for the field.

The real fields we are dealing with here have rather simple canonical behavior; to make contact with the complex fields in the previous sections, we have to introduce more sophisticated quantization rules.

2.6 One-dimensional string

As an example, let us consider a classical string along the x axis between $x = 0$ and $x = l$, which can vibrate in the y direction. In order to specify the configuration of the string at time t, we have to specify the y coordinate at each value of x for $0 \leq x \leq l$, that is, we have

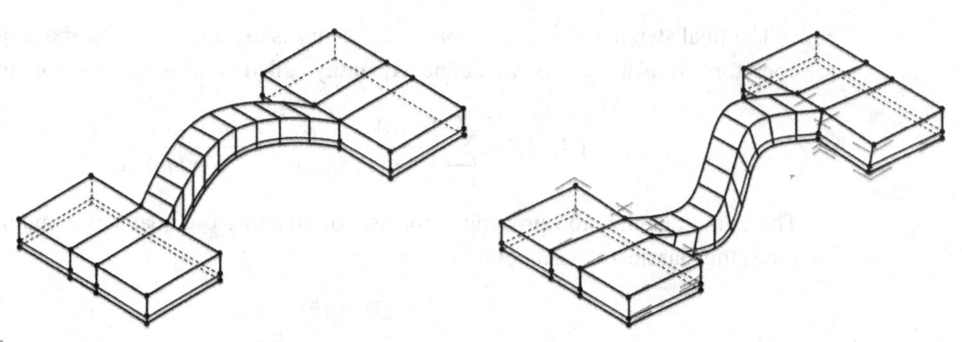

Fig. 2.1 Quantum string: two different excitation modes of a quasi-one-dimensional silicon nano-beam. The vertical wave displacements are enhanced for visibility on the right. Adapted (with permission) from: R. H. Blick *et al.*, *J. Phys.: Condens. Matter* **14** R905 (2002).

to specify a field. We shall denote the y displacement of the string at time t and position x by $\phi(x, t)$. This field is then the basic object in the theory that describes the motion of the string. A quantized theory of a vibrating string will require Lagrangians and Hamiltonians involving $\phi(x, t)$, and hence it requires a canonical formalism for fields.

While this may seem an overly simplistic example, we emphasize that it is far from being artificial. Recent advances in nanomechanics at low temperatures are now able to demonstrate clearly that such macroscopic mechanical excitations are, indeed, quantized. A diagram of excitations of a typical silicon nano-beam – functionally equivalent to a string – is shown in Figure 2.1. We will also show later that this model is mathematically identical to the dual-potential Lagrangian theory of a one-dimensional electromagnetic waveguide or optical fiber. It is also found in other fields of quantum physics, including 'circuit quantum electrodynamics', which studies the properties of superconducting microwave circuits.

The equation of motion for $\phi(x, t)$ is given by Newton's laws, which for small displacements give us the wave equation

$$v^2 \frac{\partial^2 \phi}{\partial x^2} = \frac{\partial^2 \phi}{\partial t^2}, \tag{2.91}$$

where $v = \sqrt{T/\mu}$ is the wave velocity for the string, T is its tension and μ is its mass per unit length. The Lagrangian density for this system is given by

$$\mathcal{L} = K.E. - P.E. = \frac{\mu}{2}\left(\frac{\partial \phi}{\partial t}\right)^2 - \frac{T}{2}\left(\frac{\partial \phi}{\partial x}\right)^2. \tag{2.92}$$

Application of Eq. (2.81) to this Lagrangian density yields the equation of motion above. The canonical momentum is found to be

$$\pi(x, t) = \frac{\partial \mathcal{L}}{\partial(\partial_t \phi)} = \mu \frac{\partial \phi}{\partial t}, \tag{2.93}$$

and this gives us the Hamiltonian

$$H = \int_0^l dx \left[\frac{\mu}{2}\left(\frac{\partial \phi}{\partial t}\right)^2 + \frac{T}{2}\left(\frac{\partial \phi}{\partial x}\right)^2\right]. \tag{2.94}$$

Now that we have the full canonical classical theory, we can quantize it by applying the canonical commutation relation. Since there is only one space dimension, the equal-time commutator between the field and its canonical momentum is equivalent to the following commutator of the field and its time derivative:

$$[\phi(x,t), \partial_t \phi(x',t)] = \frac{i\hbar}{\mu}\delta(x-x'). \tag{2.95}$$

We can use this to define creation and annihilation operators, which have the interpretation of creating and destroying particle-like excitations. As these excitations are spinless, these are bosonic excitations, which correspond to commutators, not anticommutators. They can have arbitrary excitations or particle numbers, provided the amplitude remains in the linear regime. The normal modes of the string are characterized by a wavenumber $k = 2\pi n/l$, where n is an integer, and a frequency $\omega_k = |k|v$. For simplicity, we assume that we have periodic boundary conditions, although other conditions may apply in practice, as in the doubly clamped nano-beam shown in Figure 2.1.

2.6.1 Creation and annihilation operators

We can define an annihilation operator corresponding to the scalar field mode with wavenumber $k = 2\pi n/l$ by

$$a_k = \sqrt{\frac{\mu}{2l\hbar}} \int_0^l dx\, e^{-ikx} \left[\sqrt{\omega_k}\phi(x,t) + \frac{i}{\sqrt{\omega_k}} \partial_t \phi(x,t) \right]. \tag{2.96}$$

With this definition, it is straightforward to verify that

$$[a_k, a_{k'}^\dagger] = \delta_{k,k'}. \tag{2.97}$$

It is possible to invert the equations for a_k and $a_k^\dagger$, in order to find them in terms of ϕ and $\partial_t \phi$, by summing over k to give

$$\sum_k \sqrt{\frac{1}{\omega_k}} (e^{ikx} a_k + e^{-ikx} a_k^\dagger) = \sqrt{\frac{\mu}{2l\hbar}} \left[\int_0^l dx' \sum_k e^{ik(x-x')} \phi(x',t) + h.c. \right] \tag{2.98}$$

(where $h.c.$ denotes Hermitian conjugate terms). Note that the terms in $\partial_t \phi$ cancel in the above equation. Next, making use of the fact that

$$\sum_k e^{ik(x-x')} = l\delta(x-x'), \tag{2.99}$$

and applying a similar procedure for the derivative terms, we obtain

$$\phi(x,t) = \sum_k \sqrt{\frac{\hbar}{2\mu l \omega_k}} (e^{ikx} a_k + e^{-ikx} a_k^\dagger),$$

$$\partial_t \phi(x,t) = i \sum_k \sqrt{\frac{\hbar \omega_k}{2\mu l}} (e^{-ikx} a_k^\dagger - e^{ikx} a_k). \tag{2.100}$$

These equations can now be inserted into the expression for the classical Hamiltonian, with the result that it, too, is now an operator,

$$H = \frac{1}{2}\sum_k \hbar\omega_k(a_k^\dagger a_k + a_k a_k^\dagger). \tag{2.101}$$

Finally, using the commutation relations and dropping an (infinite) constant, the Hamiltonian becomes

$$H = \sum_k \hbar\omega_k a_k^\dagger a_k. \tag{2.102}$$

Just as with the harmonic oscillator, the ground state of the string is the vacuum state, $|0\rangle$, which is the state that is annihilated by all of the annihilation operators, so that $a_k|0\rangle = 0$. Other states of the string, those containing excitations, are given by applying creation operators to the vacuum state. The dynamics of this theory is very simple. In the Heisenberg picture, in which the operators carry the time dependence, we see from applying Eq. (2.7) that

$$i\hbar\frac{da_k}{dt} = [a_k, H] = \hbar\omega_k a_k, \tag{2.103}$$

so that $a_k(t) = e^{-i\omega_k t}a_k(0)$. Therefore, the excitation content of states in this theory does not change with time. For example, the state with one excitation of wavenumber k at time t_0, $a_k^\dagger(t_0)|0\rangle$, is still a state with one excitation of wavenumber k at any other time, since $a_k^\dagger(t) = e^{i\omega_k(t-t_0)}a_k^\dagger(t_0)$.

Not much happens in this theory, because the excitations do not interact – they simply propagate along the string. In order to create an interaction between the excitations, we would have to add terms to the Hamiltonian. One possibility is to add terms consisting of products of three or more creation and annihilation operators. These are, in fact, the types of terms that occur when describing a quantum theory of electrodynamics in nonlinear media.

2.6.2 Infinite string quantization

Before proceeding, let us mention that it is possible to consider an infinite string. In that case, the modes are delta-function normalized. For mode functions, we can choose

$$u_k(x) = \frac{1}{\sqrt{2\pi}}e^{ikx}, \tag{2.104}$$

where $-\infty < k < \infty$ is now continuous. These functions obey the relation

$$\int_{-\infty}^{\infty} dk\, u_k^*(x)u_k(x') = \delta(x - x'). \tag{2.105}$$

The creation and annihilation operators, in this case, obey $[a_k, a^\dagger_{k'}] = \delta(k - k')$, and the field and its derivative become

$$\phi(x,t) = \int_{-\infty}^{\infty} dk \sqrt{\frac{\hbar}{4\pi\mu\omega_k}} (e^{ikx} a_k + e^{-ikx} a^\dagger_k),$$

$$\partial_t \phi(x,t) = i \int_{-\infty}^{\infty} dk \sqrt{\frac{\hbar\omega_k}{4\pi\mu}} (e^{-ikx} a^\dagger_k - e^{ikx} a_k). \tag{2.106}$$

It is straightforward to verify that these have the proper commutation relations.

2.7 Scattering matrix

We first wish to define scattering. This is a process that starts and ends with noninteracting fields. At intermediate times, interactions with other particles or external forces can occur. Many experiments in quantum nonlinear optics follow this pattern, and it is useful therefore to start with the standard definitions of scattering and S-matrices. It is obviously best to work with a definition of scattering that does not change once interactions are over, and therefore we define the scattering matrix S as the interaction-picture time-evolution operator in the long-time limit:

$$S = \lim_{t_0 \to \infty} U_I(t_0, -t_0). \tag{2.107}$$

There are many possible ways to calculate this quantity. The most traditional is by using perturbation theory, in which the time-ordered product, Eq. (2.14), is expanded in powers of the interaction Hamiltonian. Suppose the interaction is only nonvanishing between times $-t_0$ and t_0, where $t_0 > 0$. This leads to the famous result that is the basis of Feynman diagram theory,

$$S = \hat{1} - \frac{i}{\hbar} \int_{-t_0}^{t_0} dt\, H'_I(t) - \frac{1}{\hbar^2} \int_{-t_0}^{t_0} dt\, H'_I(t) \int_{-t_0}^{t} dt'\, H'_I(t') + \cdots . \tag{2.108}$$

While this expansion is usually too slowly convergent to be applied in nonlinear quantum optics, it serves as a useful point of reference. We note here that interactions are generally time-invariant, so that turning the interactions off for very early and long times is typically achieved through use of wavepackets of finite time duration, rather than an actual change in the Hamiltonian.

Let us illustrate scattering in terms of fields by looking at our 'toy model', a scalar field in one dimension. We will only treat a linear field theory at this stage, but it will allow us to illustrate the usefulness of the scattering matrix.

To obtain a scattering interaction, we add a classical external force to our quantized string. In particular, suppose that we apply a force per unit length of $j(x,t)$ to the point x at time t, where $j(x,t)$ is a c-number. We are going to assume that $j(x,t)$ is nonzero only in the range $-t_0 \le t \le t_0$ for some positive t_0. The equation of motion of the field becomes

$$\mu \frac{\partial^2 \phi}{\partial t^2} = T \frac{\partial^2 \phi}{\partial x^2} + j(x, t), \tag{2.109}$$

and the Hamiltonian is now

$$\begin{aligned} H &= \int_0^l dx \left[\frac{\mu}{2} \left(\frac{\partial \phi}{\partial t} \right)^2 + \frac{T}{2} \left(\frac{\partial \phi}{\partial x} \right)^2 + j(x, t)\phi(x, t) \right] \\ &= \sum_k [\hbar \omega_k a_k^\dagger a_k + \hbar (j_k^*(t) a_k + j_k(t) a_k^\dagger)], \end{aligned} \tag{2.110}$$

where

$$j_k(t) = \sqrt{\frac{1}{2\hbar \mu l \omega_k}} \int_0^l dx\, e^{-ikx} j(x, t). \tag{2.111}$$

2.7.1 Scattering matrix in the interaction picture

We now wish to use the interaction-picture Hamiltonian to calculate the scattering matrix. From Eq. (2.7), the equation of motion for the annihilation operator a_k in the interaction picture is

$$i\hbar \frac{da_k}{dt} = [a_k, H_0] = \hbar \omega_k a_k. \tag{2.112}$$

This equation has the solution

$$a_k(t) = e^{-i\omega_k t} a_k(0), \tag{2.113}$$

and from now on we will write $a_k \equiv a_k(0)$ for brevity. Hence, the interaction Hamiltonian H_I' in the interaction picture is

$$H_I'(t) = \hbar \sum_k [j_k^*(t) e^{-i\omega_k t} a_k + j_k(t) e^{i\omega_k t} a_k^\dagger]. \tag{2.114}$$

To obtain the scattering matrix, we will show that the interaction-picture time-evolution operator is given by the unitary operator

$$U(t, -t_0) = \exp\left(i\theta(t) - \frac{i}{\hbar} \int_{-t_0}^{t} H_I'(t')dt' \right) \tag{2.115}$$

$$= \exp\left(i\theta(t) + \sum_k [\alpha_k(t) a_k^\dagger - \alpha_k^*(t) a_k] \right), \tag{2.116}$$

where $\theta(t)$ is a phase factor and $\alpha_k(t)$ is the temporal Fourier transform of the external force,

$$\alpha_k(t) = -i \int_{-t_0}^{t} dt'\, e^{i\omega_k t'} j_k(t'). \tag{2.117}$$

We can now show that we have the correct expression for the time-evolution operator by making use of the Baker–Hausdorff theorem.

Theorem 2.1 (Baker–Hausdorff theorem) *If X and Y are operators, then, in the sense of a formal power series, we have that*

$$e^{sX}Ye^{-sX} = \sum_{n=0}^{\infty} \frac{s^n}{n!} \nu_n(X;Y),$$

$$e^{sX}e^{rY}e^{-sX} = \exp\left(r\sum_{n=0}^{\infty} \frac{s^n}{n!} \nu_n(X;Y)\right), \tag{2.118}$$

where $\nu_0(X;Y) = Y$ and $\nu_{n+1}(X;Y) = [X, \nu_n(X;Y)]$.

Corollary 2.1 *Suppose that the operators X and Y, where $[X, Y] = C$, each commute with their commutator, so that $[X, C] = [Y, C] = 0$. Then, in the sense of formal power series, we have*

$$e^{tX+sY} = e^{tX}e^{sY}e^{-tsC/2}. \tag{2.119}$$

From the corollary, we see that

$$U(t, -t_0) = e^{i\theta(t)} e^{\sum_k \alpha_k(t)a_k^\dagger} e^{-\sum_k \alpha_k^*(t)a_k} e^{-\frac{1}{2}\sum_k |\alpha_k(t)|^2}.$$

Now we can take the derivative, which gives

$$\begin{aligned}\frac{dU(t,-t_0)}{dt} &= \left\{ i\dot\theta(t) + \sum_k [\dot\alpha_k(t)a_k^\dagger - \tfrac{1}{2}(\dot\alpha_k(t)\alpha_k^* + \dot\alpha_k^*(t)\alpha_k)] \right\} U(t,-t_0) \\ &\quad + e^{i\theta(t)} e^{\sum_k \alpha_k(t)a_k^\dagger} \left(-\sum_k \dot\alpha_k^*(t) a_k \right) e^{-\sum_k \alpha_k^*(t)a_k} e^{-\frac{1}{2}\sum_k |\alpha_k(t)|^2}.\end{aligned} \tag{2.120}$$

Now use the Baker–Hausdorff theorem to commute the a_k term through the first exponential. This finally gives

$$\begin{aligned}\frac{dU(t,-t_0)}{dt} = \Bigg\{ & i\dot\theta(t) + \sum_k [\dot\alpha_k(t)a_k^\dagger - \dot\alpha_k^*(t)a_k \\ & + \tfrac{1}{2}(\dot\alpha_k^*(t)\alpha_k(t) - \dot\alpha_k(t)\alpha_k^*(t))] \Bigg\} U(t,-t_0).\end{aligned} \tag{2.121}$$

Choosing the phase term, which comes from the commutators, to satisfy

$$i\dot\theta(t) = -\tfrac{1}{2}(\dot\alpha_k^*(t)\alpha_k(t) - \dot\alpha_k(t)\alpha_k^*(t)) \tag{2.122}$$

shows that this gives a solution of

$$i\hbar\frac{d}{dt}U(t,-t_0) = H_I'(t)U(t,-t_0). \tag{2.123}$$

The S-matrix is then simply given by

$$S = \exp\left(i\theta + \sum_k \left[\alpha_k a_k^\dagger - \alpha_k^* a_k\right]\right), \tag{2.124}$$

where $\theta = \theta(t_0)$, and $\alpha_k = \alpha_k(t_0)$ is known as the coherent displacement,

$$\alpha_k = -i \int_{-t_0}^{t_0} dt'\, e^{i\omega_k t'} j_k(t'). \tag{2.125}$$

2.7.2 Heisenberg equations

We now wish to treat the system from a different point of view. We will examine the time evolution of the operators in the Heisenberg picture, and introduce the idea of asymptotic input and output operators. From Eq. (2.7), the equation of motion for the annihilation operator a_k in this picture is

$$i\hbar \frac{da_k}{dt} = [a_k, H] = \hbar\omega_k a_k + \hbar j_k(t). \tag{2.126}$$

As previously, this equation is easily solved to give, for $t < -t_0$,

$$a_k(t) = e^{-i\omega_k(t+t_0)} a_k(-t_0). \tag{2.127}$$

For $-t_0 \leq t \leq t_0$ one can verify from differentiation that

$$a_k(t) = e^{-i\omega_k(t+t_0)} a_k(-t_0) - i \int_{-t_0}^{t} dt'\, e^{-i\omega_k(t-t')} j_k(t'), \tag{2.128}$$

and, for $t > t_0$,

$$a_k(t) = e^{-i\omega_k(t+t_0)} a_k(-t_0) - i \int_{-t_0}^{t_0} dt'\, e^{-i\omega_k(t-t')} j_k(t'). \tag{2.129}$$

Now let us define two new annihilation operators, a_k^{in} and a_k^{out}, as

$$\begin{aligned} a_k^{in}(t) &= e^{-i\omega_k(t+t_0)} a_k(-t_0), \\ a_k^{out}(t) &= e^{-i\omega_k(t+t_0)} a_k(-t_0) - i \int_{-t_0}^{t_0} dt'\, e^{-i\omega_k(t-t')} j_k(t'), \end{aligned} \tag{2.130}$$

where these equations hold for all times. Both of these operators are free-field operators, that is, they have a time dependence given by $e^{-i\omega_k t}$, and they obey $[a_k^{in}, a_{k'}^{in\dagger}] = \delta_{k,k'}$ and $[a_k^{out}, a_{k'}^{out\dagger}] = \delta_{k,k'}$.

We also have that $a_k(t) = a_{k,in}(t)$ for $t < -t_0$, and that $a_k(t) = a_{k,out}$ for $t > t_0$. We can also define in and out fields by

$$\begin{aligned} \phi^{in}(x,t) &= \sum_k \sqrt{\frac{\hbar}{2\mu l \omega_k}} \left[e^{ikx} a_k^{in}(t) + e^{-ikx} a_k^{in\dagger}(t)\right], \\ \phi^{out}(x,t) &= \sum_k \sqrt{\frac{\hbar}{2\mu l \omega_k}} \left[e^{ikx} a_k^{out}(t) + e^{-ikx} a_k^{out\dagger}(t)\right], \end{aligned} \tag{2.131}$$

and we have that $\phi(x,t) = \phi^{in}(x,t)$ for $x < -t_0$ and $\phi(x,t) = \phi^{out}(x,t)$ for $t > t_0$.

2.7.3 Scattering matrix in the Heisenberg picture

Now that we have solved our model and defined in and out fields, we need to interpret what we have. First, note that we are in the Heisenberg picture, so that the states stay the same and the operators evolve with time.

Let us first consider the in vacuum state, $|0\rangle^{in}$, which is annihilated by the in annihilation operators, $a_k^{in}|0\rangle^{in} = 0$. Because, from Eq. (2.130),

$$a_k^{out}(t) = a_k^{in}(t) - i \int_{-t_0}^{t_0} dt' \, e^{-i\omega_k(t-t')} j_k(t'), \tag{2.132}$$

this state is not annihilated by the out annihilation operators. We can, however, define an out vacuum state, $|0\rangle^{out}$, which is annihilated by a_k^{out}. The in vacuum state corresponds to the state with no excitations for times $t < -t_0$, and the out vacuum state corresponds to the state with no excitations for $t > t_0$. In order to create states with one or more excitations, we apply the in or out creation operators to their respective vacuum states. For example, the state $a_k^{in\dagger}(0)|0\rangle^{in}$ is a state with one excitation with wavenumber k for times $t < -t_0$, while the state $a_k^{out\dagger}(0)|0\rangle^{out}$ is a state with a single excitation with wavenumber k for times $t > t_0$. Note that we have chosen to define these states in terms of operators at time $t = 0$, but this choice is arbitrary. If we had chosen another time, the resulting states would differ from the ones defined in terms of operators at $t = 0$ by an overall phase factor.

Now suppose we start with the state $|0\rangle^{in}$ at a time $t_1 < -t_0$, and that we would like to find the amplitude for having no excitations at a time $t_2 > t_0$. This will be given by ${}^{out}\langle 0|0\rangle^{in}$. We begin by expressing $a_k^{out}(t)$ as

$$a_k^{out}(t) = a_k^{in}(t) + e^{-i\omega_k t}\alpha_k, \tag{2.133}$$

where α_k is defined, as in Eq. (2.125), to be the temporal Fourier transform of the external force,

$$\alpha_k = -i \int_{-t_0}^{t_0} dt' \, e^{i\omega_k t'} j_k(t'). \tag{2.134}$$

We will now use the simplified notation $a_k^{in,out} = a_k^{in,out}(0)$. We then note that the unitary operator $S_k = \exp(\alpha_k a_k^{in\dagger} - \alpha_k^* a_k^{in})$ has, according to the Baker–Hausdorff theorem, the following action:

$$\begin{aligned} S_k^{-1} a_k^{in} S_k &= a_k^{in} + [\alpha_k^* a_k^{in} - \alpha_k a_k^{in\dagger}, a_k^{in}] \\ &= a_k^{in} + \alpha_k. \end{aligned} \tag{2.135}$$

Multiplying the above equation by $e^{-i\omega(t+t_0)}$ and setting $S = e^{i\theta}\prod_k S_k$, we have that

$$S^{-1} a_k^{in}(t) S = a_k^{out}(t). \tag{2.136}$$

Since, from Eq. (2.124), the unitary operator S is the S-matrix of the system, we have shown that this operator relates in states to out states. In the case of the vacuum states, we have that $a_k^{in}(t)|0\rangle^{in} = 0$ implies that $S^{-1}a_k^{in}(t)SS^{-1}|0\rangle^{in} = 0$, which can be expressed as $a_k^{out}(t)S^{-1}|0\rangle^{in} = 0$. This, in turn, implies that $|0\rangle^{out} = S^{-1}|0\rangle^{in}$. The operator S will, in fact, turn any in state into the corresponding out state. If we define

$$\begin{aligned} |k_1, k_2, \ldots, k_n\rangle^{in} &= a_{k_1}^{in\dagger}(0) a_{k_2}^{in\dagger} \cdots a_{k_n}^{in\dagger} |0\rangle^{in}, \\ |k_1, k_2, \ldots, k_n\rangle^{out} &= a_{k_1}^{out\dagger}(0) a_{k_2}^{out\dagger} \cdots a_{k_n}^{out\dagger} |0\rangle^{out}, \end{aligned} \tag{2.137}$$

then we have that $|k_1, k_2, \ldots, k_n\rangle^{out} = S^{-1}|k_1, k_2, \ldots, k_n\rangle^{in}$. The amplitude that a state that has no excitations for $t < -t_0$ still has no excitations for $t > t_0$ can be expressed in terms of the S-matrix

$$^{out}\langle 0|0\rangle^{in} = {}^{in}\langle 0|S|0\rangle^{in}, \tag{2.138}$$

and the amplitude $S_{\mathbf{k}\to\mathbf{k}'}$ to start with the excitations $k_1, \ldots, k_n$ and to finish with the excitations $k'_1, \ldots, k'_m$ can be similarly expressed as ${}^{in}\langle k'_1, \ldots, k'_m|S|k_1, \ldots, k_n\rangle^{in}$.

In order to complete the calculation of the vacuum-to-vacuum matrix element, we need the corollary to the Baker–Hausdorff theorem. We can make use of this result to express S_k as

$$S_k = e^{\alpha_k a_k^{in\dagger}} e^{-\alpha_k^* a_k^{in}} e^{-|\alpha_k|^2/2}, \tag{2.139}$$

which finally gives us

$$^{out}\langle 0|0\rangle^{in} = \exp\left(-\frac{1}{2}\sum_k |\alpha_k|^2\right). \tag{2.140}$$

The matrix elements corresponding to different processes, going from a certain set of excitations for $t < -t_0$ to another set of excitations for $t > t_0$, can be calculated in a similar fashion. This result is slightly counter-intuitive, since it is the *inverse* of the S-matrix that transforms the in states to out states. The reason for this is simple, once one remembers that the out states are not simply the time-evolved in states. Instead, they are just the opposite; they are the states that retain the relevant properties (like having no excitation) *despite* the interaction with an external force.

2.7.4 Physical interpretation

We now want to draw some conclusions from our study of this simple model and generalize them. In an experiment described by a field theory, we initially have incoming noninteracting particles. In the case of a typical nonlinear optical process, these would be photons produced by a laser. The fields then enter a region in which they interact, and then, after leaving it, propagate freely. In an optics experiment, the interaction region would be the region occupied by a linear or nonlinear medium. Therefore, for very early ($t \to -\infty$) and very late ($t \to \infty$) times, the fields propagate freely, but for intermediate times their time dependence is more complicated. We then expect that for $t \to -\infty$ we will have $\phi(x, t) \to \phi^{in}(x, t)$, and for $t \to \infty$ we will have $\phi(x, t) \to \phi^{out}(x, t)$, where $\phi^{in}(x, t)$ and $\phi^{out}(x, t)$ are free fields.

In the Heisenberg picture, what is important is the time evolution of the operators, so the essential feature of the S-matrix from this point of view is that it tells us how the operators are transformed from very early to very late times,

$$\phi^{out}(x,t) = S^{-1}\phi^{in}(x,t)S. \tag{2.141}$$

This follows the 'input–output' prescription, which will turn out to be very useful in treating quantum optical systems.

For linear optical systems, the input–output relations that transform input-mode operators into output-mode operators are particularly simple. In general, one obtains the following linear input–output relations:

$$a_k^{out} \equiv \alpha_k + \sum_{k'} [\mathcal{M}_{kk'} a_{k'}^{in} + \mathcal{L}_{kk'} a_{k'}^{in\dagger}]. \tag{2.142}$$

In most cases, unlike in our example, the interaction does not turn on for only a fixed amount of time. The finite interaction time is a result of the fact that the incoming particles are prepared in wavepackets of finite extent. The wavepackets only interact in a finite region, e.g. inside a medium. Thus, during times before the wavepackets enter the interaction region and after they leave it, they propagate freely. Formally, the description of scattering is often phrased, as we did in our example, in terms of states with well-defined momentum, which have infinite spatial extent, rather than in terms of wavepackets. We were able to do this, because the interaction was only on for a finite time. In the general case, when making use of momentum states, the interaction is assumed to be turned on adiabatically between times $-\infty$ and 0, and turned off adiabatically between times 0 and $+\infty$. Both procedures, wavepackets and adiabatically turning the interaction on and off, lead to the same results.

So far, we have discussed the operators. Now let us discuss the states. We are working in the Heisenberg picture, so our states are independent of time. Their physical content, however, does change with time, because the observables that describe the properties of a state, being operators, do change with time. In our example, for instance, the state $|0\rangle^{in}$ contains no excitations for $t < -t_0$, but does contain excitations for $t > t_0$. The state $|0\rangle^{out}$ has exactly the opposite property. In general, we define in states to be states that describe a collection of noninteracting particles as $t \to -\infty$. These are just the states that will be created by products of in creation operators acting on the in vacuum. Out states are states that describe a collection of noninteracting particles at $t \to \infty$, and these are just the states that are created by products of out creation operators acting on the out vacuum.

Our objective, when studying a scattering process, is to find the amplitude for a particular in state, $|\Psi\rangle^{in}$, which describes the initial preparation of the experiment, with particles heading toward an interaction region, to turn into a particular out state, $|\Phi\rangle^{out}$, which describes the final condition of the experiment, with particles heading away from the interaction region. This amplitude is just ${}^{out}\langle\Phi|\Psi\rangle^{in}$, but, in order to find it, we need to find the relation between the in and out operators, or, equivalently, find the S-matrix. For example, the amplitude for n particles initially with momenta $\hbar k_1, \hbar k_2, \ldots, \hbar k_n$ to emerge from the scattering process with momenta $\hbar k_1', \hbar k_2', \ldots, \hbar k_n'$ is given, from Eq. (2.137), by

$${}^{out}\langle k_1', k_2', \ldots, k_n' | k_1, k_2, \ldots, k_n\rangle^{in} = {}^{in}\langle k_1', k_2', \ldots, k_n' | S | k_1, k_2, \ldots, k_n\rangle^{in}. \tag{2.143}$$

Finding the S-matrix is usually, however, a far from simple problem.

2.7.5 Green's functions

Another way of phrasing the relation between the in and out fields and the field appearing in the Lagrangian, which is often called the interpolating field, is in terms of Green's functions. Let us split the classical Lagrangian density into two parts, a free part, $\mathcal{L}_0$, given by

$$\mathcal{L}_0 = \frac{\mu}{2}\left(\frac{\partial \phi}{\partial t}\right)^2 - \frac{T}{2}\left(\frac{\partial \phi}{\partial x}\right)^2, \tag{2.144}$$

which just describes the free propagation of excitations, and an interaction part, $\mathcal{L}_{int}$, which describes the interaction between excitations. The equation of motion for our field can then be expressed as

$$\frac{\partial^2 \phi}{\partial t^2} - v^2 \frac{\partial^2 \phi}{\partial x^2} = J(x,t), \tag{2.145}$$

where J represents the interaction terms,

$$J(x,t) = \frac{1}{\mu}\left(\frac{\partial \mathcal{L}_{int}}{\partial \phi} - \partial_x \frac{\partial \mathcal{L}_{int}}{\partial(\partial_x \phi)} - \partial_t \frac{\partial \mathcal{L}_{int}}{\partial(\partial_t \phi)}\right). \tag{2.146}$$

We note that the equation given above is the same for either quantum or classical fields. A Green's function, $G(x,t)$, for the wave equation is a solution of the inhomogeneous equation

$$\frac{\partial^2 G}{\partial t^2} - v^2 \frac{\partial^2 G}{\partial x^2} = \delta(x)\delta(t), \tag{2.147}$$

but this does not uniquely specify G. In addition, we will have to specify boundary conditions. We will, in particular, be interested in advanced and retarded Green's functions. The retarded Green's function vanishes for $t < 0$, and the advanced Green's function vanishes for $t > 0$.

2.7.6 Solutions for the Green's functions

We can find expressions for these Green's functions as follows. Let

$$G(x,t) = \frac{1}{2l\pi}\sum_k \int_{-\infty}^{\infty} dk_0\, e^{i(kx-k_0 t)} g(k_0,k), \tag{2.148}$$

and note that

$$\delta(x)\delta(t) = \frac{1}{2l\pi}\sum_k \int_{-\infty}^{\infty} dk_0\, e^{i(kx-k_0 t)}. \tag{2.149}$$

Substituting these expressions into Eq. (2.147) we find

$$g(k_0,k) = \frac{1}{v^2k^2 - k_0^2}. \tag{2.150}$$

Now in finding $G(x,t)$ from $g(k_0,k)$ we are faced with the fact that, in performing the k_0 integral,

$$\int_{-\infty}^{\infty} dk_0\, e^{-ik_0 t} \frac{1}{v^2k^2 - k_0^2}, \tag{2.151}$$

there are singularities at $k_0 = \pm kv$.

2.7.7 Retarded and advanced Green's functions

If we consider k_0 as a complex variable and go around the singularities in the upper half-plane, or, alternatively, we let

$$\frac{1}{(kv)^2 - k_0^2} \to \frac{1}{(kv)^2 - (k_0 + i\epsilon)^2} \tag{2.152}$$

and take the limit as $\epsilon \to 0^+$, we get the retarded Green's function, which we shall denote by $\Delta_{ret}(x,t)$. For $t < 0$ we can close the contour in the upper half-plane, so that the poles are outside of the contour and the integral is zero. For $t > 0$ we close the contour in the lower half-plane, and pick up contributions from the two poles. The result for $t > 0$ is

$$\Delta_{ret}(x,t) = \frac{1}{l}\sum_k e^{ikx}\frac{\sin(kvt)}{kv} \cong \frac{1}{2\pi}\int_{-\infty}^{\infty} dk\, e^{ikx}\frac{\sin(kvt)}{kv}, \tag{2.153}$$

where we have replaced the sum by an integral by using $\Delta k \sum_k \to \int dk$ with $\Delta k = 2\pi/l$. The remaining integral can be performed by expressing the sine as a sum of exponentials, and the final result is

$$\Delta_{ret}(x,t) = \frac{1}{4v}[\varepsilon(x+vt) - \varepsilon(x-vt)]\theta(t), \tag{2.154}$$

where

$$\varepsilon(s) = \begin{cases} +1, & s \geq 0, \\ -1, & s < 0, \end{cases} \qquad \theta(s) = \begin{cases} +1, & s \geq 0, \\ 0, & s < 0. \end{cases} \tag{2.155}$$

If, instead of going around the singularities at $k_0 = \pm kv$ in the upper half-plane, we go around them in the lower half-plane, we obtain the advanced Green's function, $\Delta_{adv}(x,t)$, which vanishes for $t > 0$. A calculation similar to the one above gives us that $\Delta_{adv}(x,t) = \Delta_{ret}(x,-t)$.

We can now express the equation of motion of the field, Eq. (2.145), as an integral equation. If we use the retarded Green's function, we have a solution that can be written as

$$\phi(x,t) = \phi_0(x,t) + \int_{-\infty}^{\infty} dt' \int_{-\infty}^{\infty} dx'\, \Delta_{ret}(x-x', t-t')J(x',t'), \tag{2.156}$$

where $\phi_0(x,t)$ is a free-field solution to the homogeneous equation. If $J(x,t)$ does not depend on $\phi(x,t)$, then this is a solution to our problem, as it was in the example with which we started this section. In general, $J(x,t)$ does depend on $\phi(x,t)$, so what we have is an integral equation for the interpolating field. Note that, because of the properties of

Δ_{ret}, the t' integration actually only goes up to t. Now consider the limit $t \to -\infty$. Since we are assuming that the interaction is turned on adiabatically, we have that $J(x, t)$ goes to zero as $t \to -\infty$, so that $\phi(x, t) \to \phi_0(x, t)$. But we should have $\phi(x, t) \to \phi^{in}(x, t)$ as $t \to -\infty$, so $\phi_0(x, t) = \phi^{in}(x, t)$, and our integral equation becomes

$$\phi(x, t) = \phi^{in}(x, t) + \int_{-\infty}^{\infty} dt' \int_{-\infty}^{\infty} dx' \, \Delta_{ret}(x - x', t - t') J(x', t'). \tag{2.157}$$

Similar arguments, using instead $\Delta_{adv}(x, t)$ and considering the limit $t \to \infty$, give us that

$$\phi(x, t) = \phi^{out}(x, t) + \int_{-\infty}^{\infty} dt' \int_{-\infty}^{\infty} dx' \, \Delta_{adv}(x - x', t - t') J(x', t'). \tag{2.158}$$

2.8 Quantized free electromagnetic field

In Chapter 1 we treated the electromagnetic field classically, but we now need to apply what we have learned about field quantization to treat it as a quantum mechanical system. We shall begin with the free field, and we shall treat its quantization in some detail, because, when we move to situations involving dielectric media, the quantization becomes more complicated. It will be easier to quantize the field when dielectric media are present if we have a solid understanding of its quantization in free space. There are two ways to do this. One is to break up the field into modes and treat each mode as a harmonic oscillator. This is the approach we shall use when quantizing the field in the presence of a linear dielectric. The second, which we shall use here, is the more quantum field-theoretic approach, which has been discussed in this chapter. This method of quantization will prove useful when quantizing the field in the presence of a nonlinear dielectric medium.

2.8.1 Lagrangian

Our ultimate goal is to find a Hamiltonian and commutation relations that produce the free-space Maxwell equations

$$\nabla \cdot \mathbf{E} = 0, \qquad \nabla \times \mathbf{E} = -\frac{\partial \mathbf{B}}{\partial t},$$
$$\nabla \cdot \mathbf{B} = 0, \qquad \nabla \times \mathbf{B} = \epsilon_0 \mu_0 \frac{\partial \mathbf{E}}{\partial t}, \tag{2.159}$$

as the Heisenberg-picture equations of motion. We begin with classical fields, and first find a Lagrangian, then a Hamiltonian, that leads to the above equations.

A Lagrangian requires coordinates, and ours will be the components of the vector potential $A = (A_0, \mathbf{A})$, defined so that

$$\mathbf{E} = -\frac{\partial \mathbf{A}}{\partial t} - \nabla A_0, \qquad \mathbf{B} = \nabla \times \mathbf{A}. \tag{2.160}$$

Note that, with this definition, we automatically satisfy two of the Maxwell equations,

$$\nabla \times \mathbf{E} = -\frac{\partial \mathbf{B}}{\partial t}, \qquad \nabla \cdot \mathbf{B} = 0. \tag{2.161}$$

The Lagrangian, which is a function of the vector potential and its time derivative, is expressed as the integral of a Lagrangian density,

$$L\left(A_0, \frac{\partial A_0}{\partial t}, \mathbf{A}, \frac{\partial \mathbf{A}}{\partial t}\right) = \int d^3\mathbf{r}\, \mathcal{L}\left(A_0, \frac{\partial A_0}{\partial t}, \mathbf{A}, \frac{\partial \mathbf{A}}{\partial t}\right). \tag{2.162}$$

The equations of motion that come from this Lagrangian can be expressed in terms of the Lagrangian density as

$$\partial_t \left(\frac{\partial \mathcal{L}}{\partial(\partial_t A_\mu)}\right) + \sum_{j=1}^{3} \partial_j \left(\frac{\partial \mathcal{L}}{\partial(\partial_j A_\mu)}\right) - \frac{\partial \mathcal{L}}{\partial A_\mu} = 0, \tag{2.163}$$

where $\mu = 0, 1, 2, 3$. With the correct choice of Lagrangian density, these four equations will be the remaining Maxwell equations. We choose

$$\begin{aligned} \mathcal{L} &= \frac{1}{2}\epsilon_0 |\mathbf{E}|^2 - \frac{1}{2\mu_0}|\mathbf{B}|^2 \\ &= \frac{1}{2}\epsilon_0 \left|\frac{\partial \mathbf{A}}{\partial t} + \nabla A_0\right|^2 - \frac{1}{2\mu_0}|\nabla \times \mathbf{A}|^2, \end{aligned} \tag{2.164}$$

and now need to confirm that this choice does, in fact, give us the remaining Maxwell equations. If we substitute this density into Eq. (2.163) with $\mu = 0$, we obtain Gauss' law, $\nabla \cdot \mathbf{E} = 0$, while doing so with $\mu = 1, 2, 3$ gives us

$$\nabla \times \mathbf{B} = \epsilon_0 \mu_0 \frac{\partial \mathbf{E}}{\partial t}. \tag{2.165}$$

Having found the proper Lagrangian, the next thing to do is to find the corresponding Hamiltonian. In order to do so, we first need to find the canonical momentum. Its components are given by

$$\Pi_0 = \frac{\partial \mathcal{L}}{\partial(\partial_t A_0)} = 0, \qquad \Pi_j = \frac{\partial \mathcal{L}}{\partial(\partial_t A_j)} = -\epsilon_0 E_j. \tag{2.166}$$

2.8.2 Coulomb gauge and constraints

The vanishing of Π_0, the momentum canonically conjugate to A_0, means that we lose A_0 as an independent field, and the Hamiltonian will be a function of $\mathbf{A}$ and Π only. This result is partly related to the choice of a particular Lagrangian. It is important to emphasize that there are other 'gauge' choices in which the quantization is carried out differently.

The Hamiltonian is expressed in terms of a Hamiltonian density,

$$H = \int d^3\mathbf{r}\, \mathcal{H}(\mathbf{A}, \Pi), \tag{2.167}$$

which is itself given by

$$\mathcal{H} = \sum_{j=1}^{3} (\partial_t A_j)\Pi_j - \mathcal{L} = \frac{1}{2}\epsilon_0|\mathbf{E}|^2 + \frac{1}{2\mu_0}|\mathbf{B}|^2 + \epsilon_0 \mathbf{E} \cdot \nabla A_0. \tag{2.168}$$

The last term can be eliminated by noting that, when the Hamiltonian density is substituted into the equation for the Hamiltonian, we can integrate it by parts, yielding $-\epsilon_0 \int d^3\mathbf{r} A_0 \nabla \cdot \mathbf{E}$, which is zero by Gauss' law. Therefore, the last term in the Hamiltonian density makes no contribution to the Hamiltonian, and it can be dropped, so that our final result agrees with the usual classical expression of Eq. (1.16), given the vacuum permeability and susceptibility:

$$\mathcal{H} = \frac{1}{2}\epsilon_0|\mathbf{E}|^2 + \frac{1}{2\mu_0}|\mathbf{B}|^2. \tag{2.169}$$

The equation of motion for Π_j is given by

$$\partial_t \Pi_j = -\frac{\delta H}{\delta A_j}. \tag{2.170}$$

The left-hand side of this equation is the variational derivative of H with respect to A_j, and we can find it in the same way that we found the functional derivative of the Lagrangian. Let $A_j(\mathbf{r}, t) \to A_j(\mathbf{r}, t) + \delta A_j(\mathbf{r}, t)$. We then have that

$$H \to H + \int d^3\mathbf{r} \frac{\delta H}{\delta A_j} \delta A_j(\mathbf{r}, t) + \cdots, \tag{2.171}$$

where the dots indicate terms that are of higher order in $\delta A_j(\mathbf{r}, t)$. In our case, upon letting $A_j(\mathbf{r}, t) \to A_j(\mathbf{r}, t) + \delta A_j(\mathbf{r}, t)$, we have (keeping in mind that $\mathbf{E} = -\Pi$ is an independent variable in the Hamiltonian formulation)

$$H \to H + \frac{1}{\mu_0} \int d^3\mathbf{r}\, \delta\mathbf{A} \cdot \nabla \times \mathbf{B} + \cdots, \tag{2.172}$$

after an integration by parts. We therefore find that, for the free electromagnetic field,

$$\frac{\delta H}{\delta A_j} = \frac{1}{\mu_0} (\nabla \times \mathbf{B})_j, \tag{2.173}$$

and Eq. (2.170) becomes

$$\nabla \times \mathbf{B} = \epsilon_0 \mu_0 \frac{\partial \mathbf{E}}{\partial t}. \tag{2.174}$$

Notice that, in the Hamiltonian formulation, we have lost Gauss' law as an equation of motion. This is because A_0 is no longer an independent field in this formulation. We will impose it as a constraint on the theory. This can be done, because the remaining equations of motion imply that

$$\frac{\partial}{\partial t} \nabla \cdot \mathbf{E} = \frac{1}{\epsilon_0 \mu_0} \nabla \cdot (\nabla \times \mathbf{B}) = 0. \tag{2.175}$$

Therefore, if the fields obey Gauss' law at an initial time, they will continue to obey it at later times.

We will impose an additional constraint by fixing the gauge. The physical fields are invariant under the gauge transformation

$$\mathbf{A} \to \mathbf{A} - \nabla\xi,$$
$$A_0 \to A_0 + \frac{\partial\xi}{\partial t}, \tag{2.176}$$

where $\xi(\mathbf{r}, t)$ is an arbitrary function of space and time. We shall choose the radiation gauge. This incorporates the Coulomb gauge, which requires that $\mathbf{A}$ be transverse, i.e. $\nabla \cdot \mathbf{A} = 0$, and, in addition, requires that $A_0 = 0$. We first show that we can eliminate A_0 by a choice of gauge. We start with $(A_0'', \mathbf{A}'')$, and we assume that $A_0'' \neq 0$. If we choose

$$\xi(\mathbf{r}, t) = -\int_{t_0}^{t} dt' A_0''(\mathbf{r}, t'), \tag{2.177}$$

then our new vector potential, which we shall call $(A_0', \mathbf{A}')$, *does* obey the condition $A_0' = 0$.

Now we need to impose the Coulomb-gauge condition. Starting from $\mathbf{A}'$, we choose a new function, $\xi(\mathbf{r})$, which is independent of time. This guarantees that the zero component of the new vector potential, which we shall call $\mathbf{A}$, will remain zero. If we choose

$$\xi(\mathbf{r}) = -\int d^3\mathbf{r}' \frac{1}{4\pi|\mathbf{r} - \mathbf{r}'|} \nabla' \cdot \mathbf{A}'(\mathbf{r}, t), \tag{2.178}$$

then we have $\nabla^2\xi = \nabla \cdot \mathbf{A}'$, which implies that $\nabla \cdot \mathbf{A} = 0$. Note that, though it appears that the right-hand side of the above equation depends on time, it does not. Gauss' law along with the fact that $A_0' = 0$ implies that $\partial_t \nabla \cdot \mathbf{A}' = 0$.

Finally, at the end of this sequence of gauge transformations, we have a vector potential that satisfies $A_0 = 0$ and $\nabla \cdot \mathbf{A} = 0$.

2.8.3 Electromagnetic commutators

In order to quantize the theory, we would normally – in the absence of constraints – impose the equal-time commutation relations, to give operator electromagnetic fields instead of c-number fields, i.e.

$$[A_j(\mathbf{r}, t), A_l(\mathbf{r}', t)] = 0,$$
$$[\Pi_j(\mathbf{r}, t), \Pi_l(\mathbf{r}', t)] = [E_j(\mathbf{r}, t), E_l(\mathbf{r}', t)] = 0, \tag{2.179}$$
$$[A_j(\mathbf{r}, t), \Pi_l(\mathbf{r}', t)] = -[A_j(\mathbf{r}, t), \epsilon_0 E_l(\mathbf{r}', t)] = i\hbar\delta_{jk}\delta^{(3)}(\mathbf{r} - \mathbf{r}').$$

However, the last of these commutation relations violates both the Coulomb-gauge condition and Gauss' law, so we modify it to reflect the fact that both the vector potential and the electric field are transverse fields, i.e. both have a vanishing divergence. This is an example of a constrained quantization procedure, which we explain in greater detail later in this chapter. What we need on the right-hand side of the last commutator is a kind of delta-function with a vanishing divergence, which acts in the same way that a standard delta-function does for transverse fields. The transverse delta-function has exactly these properties.

In order to define the transverse delta-function, let us assume that we are quantizing the fields in a box of volume V using periodic boundary conditions. Then we can express the transverse delta-function as

$$\delta_{lm}^{(tr)}(\mathbf{r}) = \frac{1}{V} \sum_{\mathbf{k}} (\delta_{lm} - \hat{k}_l \hat{k}_m) e^{i\mathbf{k}\cdot\mathbf{r}}. \tag{2.180}$$

Here, the wavevectors are given by

$$\mathbf{k} = \left(\frac{2\pi n_x}{L_x}, \frac{2\pi n_y}{L_y}, \frac{2\pi n_z}{L_z} \right), \tag{2.181}$$

where $V = L_x L_y L_z$ and n_x, n_y and n_z are integers, and $\hat{\mathbf{k}} = \mathbf{k}/|\mathbf{k}|$ is a unit vector. From this definition it is clear that $\sum_{l=1}^{3} \partial_l \delta_{lm}^{(tr)}(\mathbf{r}) = 0$. In addition, the transverse delta-function has the property that, for any transverse vector field, i.e. one that satisfies $\nabla \cdot \mathbf{V} = 0$, we have

$$V_l(\mathbf{r}) = \sum_{m=1}^{3} \int d^3\mathbf{r}' \, \delta_{lm}^{(tr)}(\mathbf{r} - \mathbf{r}') V_m(\mathbf{r}'). \tag{2.182}$$

Finally, we can now modify the commutation relation for the vector potential and the electric field to be

$$[A_j(\mathbf{r}, t), \epsilon_0 E_l(\mathbf{r}', t)] = -i\hbar \delta_{jl}^{(tr)}(\mathbf{r} - \mathbf{r}'). \tag{2.183}$$

The other commutators in Eq. (2.179) remain the same.

2.8.4 Mode expansion

Our next step will be to expand the vector potential and electric field in plane-wave modes, and to define creation and annihilation operators for these modes. The transverse plane-wave modes have the mode functions

$$\mathbf{u}_{\mathbf{k},\alpha}(\mathbf{r}) = \frac{1}{\sqrt{V}} \hat{\mathbf{e}}_{\mathbf{k},\alpha} e^{i\mathbf{k}\cdot\mathbf{r}}, \tag{2.184}$$

where $\alpha = 1, 2$. The vectors $\hat{\mathbf{k}}$, $\hat{\mathbf{e}}_{\mathbf{k},1}$, and $\hat{\mathbf{e}}_{\mathbf{k},2}$ form an orthonormal set of vectors, where

$$\hat{\mathbf{e}}_{\mathbf{k},1} \times \hat{\mathbf{e}}_{\mathbf{k},2} = \hat{\mathbf{k}}, \qquad \hat{\mathbf{e}}_{\mathbf{k},2} \times \hat{\mathbf{k}} = \hat{\mathbf{e}}_{\mathbf{k},1}, \qquad \hat{\mathbf{k}} \times \hat{\mathbf{e}}_{\mathbf{k},1} = \hat{\mathbf{e}}_{\mathbf{k},2}. \tag{2.185}$$

We also choose $\hat{\mathbf{e}}_{-\mathbf{k},\alpha} = -(-1)^\alpha \hat{\mathbf{e}}_{\mathbf{k},\alpha}$, which is consistent with the above equations. We now define the operator

$$a_{\mathbf{k},\alpha}(t) = \int d^3\mathbf{r} \, \mathbf{u}_{\mathbf{k},\alpha}^*(\mathbf{r}) \cdot \left[\sqrt{\frac{\epsilon_0 \omega_k}{2\hbar}} \mathbf{A}(\mathbf{r}, t) - i \sqrt{\frac{\epsilon_0}{2\hbar \omega_k}} \mathbf{E}(\mathbf{r}, t) \right], \tag{2.186}$$

where $\omega_k = |\mathbf{k}|c$. Making use of the equal-time commutation relations for the field operators, we find that

$$[a_{\mathbf{k},\alpha}, a_{\mathbf{k}',\alpha'}] = 0, \qquad [a_{\mathbf{k},\alpha}, a_{\mathbf{k}',\alpha'}^\dagger] = \delta_{\mathbf{k},\mathbf{k}'} \delta_{\alpha,\alpha'}, \tag{2.187}$$

so that $a_{\mathbf{k},\alpha}^\dagger$ and $a_{\mathbf{k},\alpha}$ can clearly be interpreted as creation and annihilation operators.

It is also possible to invert the relation between the fields and the creation and annihilation operators by making use of the relation

$$\sum_{\mathbf{k},\alpha} u_{\mathbf{k},\alpha}(\mathbf{r})_l u^*_{\mathbf{k},\alpha}(\mathbf{r}')_m = \delta^{(tr)}_{lm}(\mathbf{r}-\mathbf{r}'). \tag{2.188}$$

We find that

$$\mathbf{A}(\mathbf{r},t) = \sum_{\mathbf{k},\alpha} \sqrt{\frac{\hbar}{2\epsilon_0 \omega_k V}}\, \hat{\mathbf{e}}_{\mathbf{k},\alpha}(a_{\mathbf{k},\alpha} e^{i\mathbf{k}\cdot\mathbf{r}} + a^\dagger_{\mathbf{k},\alpha} e^{-i\mathbf{k}\cdot\mathbf{r}}),$$

$$\mathbf{E}(\mathbf{r},t) = \sum_{\mathbf{k},\alpha} i \sqrt{\frac{\hbar\omega_k}{2\epsilon_0 V}}\, \hat{\mathbf{e}}_{\mathbf{k},\alpha}(a_{\mathbf{k},\alpha} e^{i\mathbf{k}\cdot\mathbf{r}} - a^\dagger_{\mathbf{k},\alpha} e^{-i\mathbf{k}\cdot\mathbf{r}}). \tag{2.189}$$

These expressions can then be substituted into the Hamiltonian to express it in terms of the creation and annihilation operators. The result is

$$H = \frac{1}{2}\sum_{\mathbf{k},\alpha} \hbar\omega_k (a^\dagger_{\mathbf{k},\alpha} a_{\mathbf{k},\alpha} + a_{\mathbf{k},\alpha} a^\dagger_{\mathbf{k},\alpha}). \tag{2.190}$$

The normally ordered form of the Hamiltonian is

$$H = \sum_{\mathbf{k},\alpha} \hbar\omega_k a^\dagger_{\mathbf{k},\alpha} a_{\mathbf{k},\alpha}, \tag{2.191}$$

which is equivalent to the one above it, because the two Hamiltonians differ from each other only by a constant. The ground state of this Hamiltonian is the vacuum state $|0\rangle$, the state that is annihilated by all of the annihilation operators, i.e. $a_{\mathbf{k},\alpha}|0\rangle = 0$ for all $\mathbf{k}$ and $\alpha = 1, 2$. From Heisenberg's equation, Eq. (2.8), the resulting equations of motion are simply the equations of motion for a set of uncoupled harmonic oscillators,

$$\frac{\partial a_{\mathbf{k},\alpha}}{\partial t} = \frac{i}{\hbar}[H, a_{\mathbf{k},\alpha}] = -i\omega_k a_{\mathbf{k},\alpha}, \tag{2.192}$$

which have the usual oscillatory solutions one would obtain classically, i.e.

$$a_{\mathbf{k},\alpha}(t) = a_{\mathbf{k},\alpha}(0) e^{-i\omega_k t}. \tag{2.193}$$

In other words, apart from commutators – which correspond physically to vacuum fluctuations – the dynamical evolution of the quantized, free electromagnetic field is identical to the classical version.

2.8.5 Continuum modes

In the case of quantum propagation outside of cavities, is often convenient to use an infinite-volume limit. This is similar to the discrete mode expansion, except with sums replaced by Fourier integrals. In later chapters, some examples will include waveguides with reduced effective dimensionality, so we will treat the general case of d continuum dimensions and $3-d$ transverse dimensions.

In this limit, we define a continuous momentum expansion in D dimensions so that

$$\mathbf{u}_\alpha(\mathbf{k},\mathbf{r}) = \frac{1}{\sqrt{A}(2\pi)^{D/2}}\,\hat{\mathbf{e}}_{\mathbf{k},\alpha} e^{i\mathbf{k}\cdot\mathbf{r}}, \tag{2.194}$$

where $\alpha = 1, 2$, and A is a $(3-D)$-dimensional transverse area. We suppose for simplicity that there is one transverse mode, with a uniform profile; otherwise there will be a countable number of transverse mode functions, as explained in Chapter 11. The vectors $\hat{\mathbf{k}}$, $\hat{\mathbf{e}}_{\mathbf{k},1}$ and $\hat{\mathbf{e}}_{\mathbf{k},2}$ are an orthonormal set of vectors, as in the discrete mode case. We now define the operator

$$a_\alpha(\mathbf{k},t) = \int d^D\mathbf{r}\,\mathbf{u}^*_\alpha(\mathbf{k},\mathbf{r})\cdot\left[\sqrt{\frac{\epsilon_0\omega_k}{2\hbar}}\mathbf{A}(\mathbf{r},t) - i\sqrt{\frac{\epsilon_0}{2\hbar\omega_k}}\mathbf{E}(\mathbf{r},t)\right], \tag{2.195}$$

where $\omega_k = |\mathbf{k}|c$. Making use of the equal-time commutation relations for the field operators, we find that

$$[a_\alpha(\mathbf{k},t), a_{\alpha'}(\mathbf{k}',t)] = 0, \qquad [a_\alpha(\mathbf{k},t), a^\dagger_{\alpha'}(\mathbf{k}',t)] = \delta^D(\mathbf{k}-\mathbf{k}')\delta_{\alpha,\alpha'}. \tag{2.196}$$

It is also possible to invert the relation between the fields and the creation and annihilation operators by making use of the relation

$$\sum_\alpha \int d^D\mathbf{k}\,\mathbf{u}^*_\alpha(\mathbf{k},\mathbf{r})\mathbf{u}^*_\alpha(\mathbf{k},\mathbf{r}') = \delta^{(\mathrm{tr})}_{lm}(\mathbf{r}-\mathbf{r}')/A. \tag{2.197}$$

We find that, in terms of the field expansions, the two fields and potentials are, in general,

$$\begin{aligned}
\mathbf{A}(\mathbf{r},t) &= \sum_\alpha \int d^D\mathbf{k}\,\frac{\hbar}{\sqrt{2\epsilon_0\omega_k A}}\,\hat{\mathbf{e}}_{\mathbf{k},\alpha}[a_\alpha(\mathbf{k},t)e^{i\mathbf{k}\cdot\mathbf{r}} + a^\dagger_\alpha(\mathbf{k},t)e^{-i\mathbf{k}\cdot\mathbf{r}}],\\
\mathbf{E}(\mathbf{r},t) &= i\sum_\alpha \int d^D\mathbf{k}\sqrt{\frac{\hbar\omega_k}{2\epsilon_0 A}}\,\hat{\mathbf{e}}_{\mathbf{k},\alpha}[a_\alpha(\mathbf{k},t)e^{i\mathbf{k}\cdot\mathbf{r}} - a^\dagger_\alpha(\mathbf{k},t)e^{-i\mathbf{k}\cdot\mathbf{r}}],\\
\mathbf{B}(\mathbf{r},t) &= i\sum_\alpha \int d^D\mathbf{k}\sqrt{\frac{\hbar\omega_k}{2\epsilon_0 A}}\,\mathbf{k}\times\hat{\mathbf{e}}_{\mathbf{k},\alpha}[a_\alpha(\mathbf{k},t)e^{i\mathbf{k}\cdot\mathbf{r}} - a^\dagger_\alpha(\mathbf{k},t)e^{-i\mathbf{k}\cdot\mathbf{r}}].
\end{aligned} \tag{2.198}$$

These expressions can then be substituted into the Hamiltonian to express it using a continuum of creation and annihilation operators, to give

$$H = \frac{1}{2}\sum_\alpha \int d^D\mathbf{k}\,\hbar\omega(\mathbf{k})[a^\dagger_\alpha(\mathbf{k},t)a_\alpha(\mathbf{k},t) + a_\alpha(\mathbf{k},t)a^\dagger_\alpha(\mathbf{k},t)]. \tag{2.199}$$

The normally ordered form is

$$H = \sum_\alpha \int d^D\mathbf{k}\,\hbar\omega(\mathbf{k})a^\dagger_\alpha(\mathbf{k},t)a_\alpha(\mathbf{k},t). \tag{2.200}$$

The ground state of this Hamiltonian is the state that is annihilated by all of the annihilation operators, i.e. $a_\alpha(\mathbf{k},t)|0\rangle = 0$ for all $\mathbf{k}$ and $\alpha = 1, 2$. From Heisenberg's equation, Eq. (2.8), the resulting equations of motion are, as previously the equations of motion for a set of uncoupled harmonic oscillators,

$$\frac{\partial a_\alpha(\mathbf{k}, t)}{\partial t} = -i\omega(\mathbf{k})a_\alpha(\mathbf{k}, t), \tag{2.201}$$

with the usual oscillatory solutions.

2.9 Constrained quantization

In this section for advanced students, we give a brief treatment of a more rigorous approach to constrained quantization. In the simplest case, all the canonical momenta are nonzero and the time derivatives of the field can be expressed in terms of the momenta and the fields. A more general situation is one in which the canonical momenta can have algebraic dependences, for example, one or more of them can be zero, and we cannot express the time derivatives of the field in terms of the momenta and the fields. Such systems are called constrained. The resulting problem of constrained quantization occurs very widely in quantum field theory, including the quantization of the electromagnetic field. In the earlier part of this chapter, we have taken a somewhat intuitive approach to constraints. In this section, we give a more systematic procedure.

Dirac was the first to recognize and deal with the problem of constraints, by introducing a quantity called the Dirac bracket to replace the Poisson bracket. Modern methods now exist that are simpler yet equivalent to Dirac's original approach. We will summarize an approach due to Jackiw and Faddeev. It is able to extend the traditional treatment of quantization summarized in this chapter to the more general case of constrained quantization. The main objective of such methods is to derive the appropriate form of the Dirac bracket, which then defines the commutation relations of the theory.

2.9.1 First-order Lagrangians

Following Faddeev and Jackiw, we note that it is always possible to write the standard Lagrangian for N variables or field components in a form that directly generates a set of first-order equations of motion, rather than the usual second-order ones. In the case of the Lagrangian for relativistic fermions, this is the only known form of the Lagrangian. The new Lagrangian that produces the first-order equations is a function of $2N$ fields or canonical coordinates and momenta. That is, what had been the momenta now become coordinates in the new theory.

To demonstrate how this works, we will work with discrete coordinates rather than fields. In the usual case where the original Lagrangian, $L(\mathbf{x}, \dot{\mathbf{x}})$ – where $\mathbf{x}$ is a vector of canonical coordinates – gives rise to second-order equations of motion, we recall that the result for the canonical momenta and the Hamiltonian is

$$p_i = \frac{\partial L(\mathbf{x}, \dot{\mathbf{x}})}{\partial \dot{x}_i},$$
$$H(\mathbf{p}, \mathbf{x}) = \sum_i p_i \dot{x}_i - L(\mathbf{x}, \dot{\mathbf{x}}). \tag{2.202}$$

These equations are not necessarily for the Cartesian coordinates of a mechanical particle; they can be any coordinates, including amplitudes of field modes. We note that $H(\mathbf{p}, \mathbf{x})$ is an explicit function of $\mathbf{p}$ and $\mathbf{x}$ only, and is not an explicit function of any time derivatives. The Lagrangian can be rewritten as a new 'first-order' Lagrangian, L_1, which is now a function of a larger number of phase-space variables, $\mathbf{x}$ and $\mathbf{p}$:

$$L_1 = \sum_i p_i \dot{x}_i - H(\mathbf{p}, \mathbf{x}). \tag{2.203}$$

It is straightforward to see from rearranging Eq. (2.202) that this new Lagrangian is identical in value to the original one. However, it is an explicit function of $\mathbf{p}$ as well as $\mathbf{x}$, and these are now equivalent canonical coordinates. If we differentiate with respect to $\dot{x}_i$, we see that the p_i play a dual role: besides being coordinates, they are also canonical momenta conjugate to x_i. Yet there are no new canonical 'momenta' conjugate to p_i, even though these are now canonical coordinates, since the derivatives with respect to $\dot{p}_i$ are all equal to zero. On minimizing the action, this Lagrangian generates equations of motion corresponding to those of the original Lagrangian, except that they are in the form we would usually call Hamilton's equations, i.e.

$$\dot{x}_i = \frac{\partial H}{\partial p_i},$$
$$\dot{p}_i = -\frac{\partial H}{\partial x_i}. \tag{2.204}$$

2.9.2 Jackiw quantization

Quantizing first-order Lagrangians like L_1 using the older Poisson bracket method is *not* generally possible, as some of the canonical momenta vanish. The Dirac method solves this problem, as well as providing a route for treating more general constraints. In the most general case, we have a vector of discrete variables labeled $\boldsymbol{\xi}$, and the Lagrangian has the generic form

$$L_1[\boldsymbol{\xi}] = \frac{1}{2} \sum_{i,j} \xi_j C_{ji} \dot{\xi}_i - H[\boldsymbol{\xi}]. \tag{2.205}$$

In the simplest cases, C_{ji} is a constant antisymmetric matrix (it can be taken to be antisymmetric, because any symmetric part contributes a total time derivative to the Lagrangian, which can be dropped). In principle, it can be a function of the canonical variables, although we shall not treat this case here. It is covered in the extended reading material. The canonical

Hamiltonian is H, and the evolution equations are Lagrange's equations,

$$C_{ji}\dot{\xi}_j = \frac{\partial}{\partial \xi_i} H(\boldsymbol{\xi}). \tag{2.206}$$

Provided C_{ij} is invertible, the commutators obtained by quantization with the Faddeev–Jackiw method are just proportional to the matrix inverse of C_{ji}, which we label C_{ij}^{-1}, so that

$$[\xi_i, \xi_j] = i\hbar C_{ij}^{-1}. \tag{2.207}$$

In many cases C_{ij} takes a block off-diagonal form with $C_{ij} = -C_{ij}^{-1}$, and

$$C_{ij} = \begin{bmatrix} 0 & I \\ -I & 0 \end{bmatrix}_{ij}. \tag{2.208}$$

In this form, we regain the usual canonical commutation relations. We can see this by choosing $\boldsymbol{\xi} = (\mathbf{p}, \mathbf{x})$. It is then immediately obvious that the corresponding commutators are just the standard canonical commutators, corresponding to the discrete-variable version of Eq. (2.89), i.e.

$$[x_i, p_j] = i\hbar\delta_{ij}. \tag{2.209}$$

In the Faddeev–Jackiw approach, it is not actually necessary to introduce the idea of canonical momenta. We simply have classical coordinates with first-order equations of motion. The Hamiltonian is automatically defined by the first-order Lagrangian structure. The commutators arise naturally out of the relationships between the variables, so there is no need to make the usual distinction between canonical coordinates and momenta.

The particular advantage of the method is that it is a more flexible approach when quantizing unusual dynamical systems.

2.9.3 Quantizing the harmonic oscillator

An initial application of this method is to quantize a typical first-order Lagrangian, e.g. Eq. (2.205). This is nontrivial with the Poisson bracket method, because of algebraic relations between the variables and momenta, which prevents quantization following the standard route. The difficulty can be seen by trying to quantize the example given above. If we attempt to treat the mechanical momentum p_i as a coordinate in the Lagrangian of Eq. (2.203), it has a zero 'canonical' momentum, since $\partial L/\partial \dot{p}_i = 0$. Therefore, this cannot be quantized directly in the standard Dirac approach, as it leads to an inconsistent commutation relation.

However, as we saw from the result given in Eq. (2.207), quantization of this example is straightforward in the Jackiw approach. The relationships between coordinates and canonical momenta in first-order Lagrangians are in fact not true constraints, and can be easily quantized.

As an example, we will quantize the equations of motion for a harmonic oscillator, linear string or radiation field mode, which is

$$\frac{da}{dt} = -i\omega a, \tag{2.210}$$

in the complex form equivalent to Eq. (2.192), with energy equal to $\hbar\omega a^* a$. Note that here $\hbar$ is introduced in the classical energy expression to ensure that the dimensions are correct.

For a single mode, the classical first-order Lagrangian for the harmonic oscillator is

$$L_1 = \hbar\left\{\tfrac{1}{2}i\left[\dot{a}a^* - \dot{a}^* a\right] - \omega a^* a\right\}. \tag{2.211}$$

To demonstrate that this is the correct Lagrangian, we can proceed by using either complex coordinates, or real and imaginary parts as distinct canonical variables. The latter approach leads to real field quadrature variables, which are used in later chapters. However, let us see what the results look like in the case of complex variables, in which case we can regard a^* and a as formally independent canonical variables.

First, we will check that we have the correct equations of motion. The Lagrangian equations of motion (equivalent to Hamilton's equations) are

$$\frac{\partial L_1}{\partial a^*} = \partial_t \frac{\partial L_1}{\partial \dot{a}^*}, \tag{2.212}$$

which in this case reduce to

$$i\dot{a} = \omega a. \tag{2.213}$$

This is the expected equation of motion for a harmonic oscillator.

Next, since the complex canonical momenta are

$$\pi = \frac{\partial L_1}{\partial \dot{a}} = \frac{i\hbar}{2} a^*, \qquad \pi^* = -\frac{i\hbar}{2} a, \tag{2.214}$$

the Hamiltonian is simply

$$\pi\dot{a} + \pi^*\dot{a}^* - L_1 = \hbar\omega a^* a = H(a, a^*). \tag{2.215}$$

This is the classical expression for the energy, so we have identified the Lagrangian correctly, in complex form.

Finally, we evaluate the commutators. Choosing $\xi = (a, a^*)$, we can rewrite the Lagrangian in matrix form as

$$L_1 = \hbar\left\{\frac{i}{2}[a, a^*]\begin{bmatrix} 0 & -1 \\ 1 & 0 \end{bmatrix}\begin{bmatrix} \dot{a} \\ \dot{a}^* \end{bmatrix} - \omega a^* a\right\}, \tag{2.216}$$

which means that the Jackiw matrix and its inverse are

$$\begin{aligned} C_{ij} &= i\hbar \begin{bmatrix} 0 & -1 \\ 1 & 0 \end{bmatrix}_{ij}, \\ C^{-1}_{ij} &= \frac{i}{\hbar}\begin{bmatrix} 0 & -1 \\ 1 & 0 \end{bmatrix}_{ij}. \end{aligned} \tag{2.217}$$

Hence we have the following commutation relation, once a and $a^\dagger$ are 'promoted' to operators:

$$[a, a^\dagger] = [\xi_1, \xi_2] = i\hbar C_{12}^{-1} = 1. \tag{2.218}$$

This is the identical commutation relation and Hamiltonian to that obtained from the usual second-order Lagrangian approach in Eq. (2.187). This result shows that, in simple cases like this, the first-order Jackiw method is equivalent to the second-order Dirac quantization method using Poisson brackets.

2.9.4 Quantizing a complex field

The harmonic oscillator is easily extended to a multi-mode, spinor free-field case, just by summing over the independent field modes, to give

$$L = \hbar \sum_k \left\{ \frac{i}{2}[\dot{a}_{ks} a_{ks}^* - \dot{a}_{ks}^* a_{ks}] - \omega_{ks} a_{ks}^* a_{ks} \right\}. \tag{2.219}$$

This gives us a way to quantize complex fields using Lagrangian methods, which allows us to provide an alternative technique for quantizing particle fields using Fock state methods. We simply use the expansion of Eq. (2.34) to obtain the complex field Lagrangian of the form

$$L = \sum_s \int d^3\mathbf{r} \left\{ \frac{i\hbar}{2}[\dot{\psi}_s(\mathbf{r})\psi_s^*(\mathbf{r}) - \psi_s(\mathbf{r})\dot{\psi}_s^*(\mathbf{r})] - \psi_s^*(\mathbf{r}) h_{SP} \psi_s(\mathbf{r}) \right\}. \tag{2.220}$$

It is straightforward to verify that the canonical Hamiltonian that results is H, where

$$H = \sum_s \int d^3\mathbf{r}\, [\psi_s^*(\mathbf{r}) h_{SP} \psi_s(\mathbf{r})], \tag{2.221}$$

and that the commutators are the complex field commutators we found in Section 2.3, using Fock state methods:

$$[\psi_s(\mathbf{r}), \psi_{s'}^\dagger(\mathbf{r}')] = \delta_{ss'} \delta^{(3)}(\mathbf{r} - \mathbf{r}'). \tag{2.222}$$

2.9.5 Constraints

Finally, let consider the case of true constraints. This occurs if the matrix C_{ji} is not invertible, which means that there is no clear commutation relation defined using the techniques given above. Accordingly, the number of independent variables must be reduced. The matrix is now divided up into an invertible part with a reduced set of dynamical variables $\tilde{\boldsymbol{\xi}}$, and a noninvertible part, corresponding to N' 'zero' modes Z_a, for which there are no time derivatives in Hamilton's equations. In the case that C_{ij} is a constant matrix, if we have $\sum_i z_i^{(a)} C_{ij} = 0$, then multiplying both sides of Eq. (2.206) by $z_i^{(a)}$ and summing, we find that

$$\sum_i z_i^{(a)} \frac{\partial}{\partial \xi_i} H(\boldsymbol{\xi}) = 0. \tag{2.223}$$

This is a constraint on the variables $\boldsymbol{\xi}$. If we now define the new variables $\tilde{\boldsymbol{\xi}}$ and Z_a, which in this case are linear combinations of the old ones, we can rephrase the constraint. In particular, if $\partial\xi/\partial Z_a = z_i^{(a)}$, then the above equation becomes

$$\frac{\partial}{\partial Z^a} H(\tilde{\boldsymbol{\xi}}, \mathbf{Z}) = 0, \tag{2.224}$$

where $\mathbf{Z}$ is an N'-component vector with components Z_a. One can obtain an equation of this type also in the case that C_{ij} is not constant.

If the zero modes occur nonlinearly in $H(\tilde{\boldsymbol{\xi}}, \mathbf{Z})$, then each of the $a = 1, 2, \ldots, N'$ equations given above will contain the zero modes, Z_a. One can solve them directly for the zero-mode coordinates in terms of the $\tilde{\boldsymbol{\xi}}$, and thereby eliminate these modes from the problem. Otherwise, these equations imply a relation between the remaining canonical variables $\tilde{\boldsymbol{\xi}}$, in which $\mathbf{Z}$ is absent. After elimination of as many nonlinear $\mathbf{Z}$ variables as possible, one is left with a Hamiltonian depending linearly on the remaining Z_a, which can be expressed in the form

$$H(\tilde{\boldsymbol{\xi}}, \mathbf{Z}) = \sum_i Z_i \Phi_i(\tilde{\boldsymbol{\xi}})\, d\mathbf{x} + H'[\tilde{\boldsymbol{\xi}}], \tag{2.225}$$

where the Z_a act as Lagrange multipliers. Differentiating the above equation with respect to Z_a and setting the result equal to zero implies that $\Phi_i(\tilde{\boldsymbol{\xi}}) = 0$, which requires solving for some of the remaining variables $\tilde{\boldsymbol{\xi}}$ in terms of the others, thus reducing the number of variables even further. We then return to the first stage of the Lagrangian above with a reduced set of variables, and iterate the process until we have a new, invertible matrix $\tilde{C}_{ij}$, generally of reduced dimension, to define the commutators.

In electromagnetic quantization, the usual problem is that some fields will be constrained to be transverse. To deal with this, we can either expand in canonical variables that already satisfy the constraints, or use the procedure of taking transverse functional derivatives. Either procedure takes care of the constraints, and leads to a transverse delta-function commutation relation, as explained earlier. This is equivalent to the Faddeev–Jackiw constrained quantization approach in these simple cases, but the systematic Faddeev–Jackiw quantization procedure can also be used in more general cases.

2.10 Exponential complexity

In order to understand the implications of these quantization procedures, we will change our perspective and ask what would be involved in trying to numerically simulate a quantum mechanical system, in particular, one with many particles and modes, such as those we have been discussing here. For example, one might wish to simulate the dynamics of photons propagating in a nonlinear medium. In order to do so, it is necessary to drop down from the infinite dimension of the Hilbert space for the ideal theory and go to a finite-dimensional one, by restricting our attention to a finite number of modes and a finite number of particles. To get an idea of the feasibility of such a simulation, we need to know the size of our truncated Hilbert space. Here we will estimate typical Hilbert space dimension, $N_{\mathcal{H}}$. This will serve as a motivation for the techniques we will introduce later, in Chapter 7.

If there is no analytic solution, an obvious way to solve any quantum mechanical time-evolution problem is to numerically diagonalize the Hamiltonian on a basis of many-body wavefunctions, so that

$$H|E_n\rangle = E_n|E_n\rangle, \tag{2.226}$$

then expand the initial state $|\Psi_0\rangle$ at $t = 0$ in terms of the eigenstates,

$$|\Psi_0\rangle = \sum_n C_n|E_n\rangle. \tag{2.227}$$

Finally, one can then numerically add up the contribution of each eigenfunction:

$$|\Psi(t)\rangle = \sum_n e^{-iE_nt/\hbar}C_n|E_n\rangle. \tag{2.228}$$

The difficulty with this approach is, as we shall see, that the Hilbert space is typically enormous. Exact solutions are usually of little help, because they are generally not known except for the free-field examples treated in this chapter, and some related special cases. This leads us to the conclusion that one of the simplest conceptual problems in many-body physics, finding the time evolution of a system, is much less trivial than it first seems.

Despite the fundamental nature of many-body quantum problems, exponential complexity (meaning the exponential growth of the Hilbert space dimension with the size of the problem) is the reason that they are often regarded as inherently insoluble. This type of complexity grows so rapidly that – unless approximations are used – a finite computer is faced with the problem of trying to solve problems that can easily become nearly infinite in dimensionality. It was this difficulty that motivated Feynman to propose the development of the quantum computer. In this proposal, quantum problems are solved by an analog method, namely the physical system of interest is simulated by another quantum system, usually consisting of 'qubits', or two-state physical systems and quantum gates. There has been very substantial progress, both theoretical and experimental. Some developments employ the results described in Chapter 12. However, we do not yet have universal, large-scale quantum computers. We explore the reason for Feynman's disquiet about the complexity of quantum many-body theory in the remainder of this section.

2.10.1 Combinatoric estimates

To explain the theoretical issues that are the source of the complexity problem, all that is necessary is to count up the number of available states in a typical many-body Hilbert space. A photonic signal with $N = 10\,000$ photons in an optical fiber may have $M = 10\,000$ relevant longitudinal fiber modes. Similarly, an ultra-cold bosonic quantum Bose gas (Bose–Einstein condensate, BEC) with $N = 10\,000$ atoms, in a micrometer-sized magnetic trap, might have $M = 10\,000$ available trap modes. Each mode can have all, some, or none of the particles. Every possible state obtained in this way is described by a separate probability amplitude, according to standard quantum mechanics. From De Moivre's combinatoric theorem, the total number of ways to combine states, which is the relevant Hilbert subspace dimension, is given by

$$N_{\mathcal{H}} = \frac{(M+N-1)!}{(M-1)!N!}. \tag{2.229}$$

The logarithm of the number of states can be estimated from Stirling's formula, $\ln n! \approx (n+\frac{1}{2})\ln n - n$, so that

$$\ln N_{\mathcal{H}} = \ln\left(\frac{(M+N-1)!}{(M-1)!N!}\right) \simeq (M+N)\ln(M+N) - M\ln M - N\ln N. \tag{2.230}$$

In this rather small-scale example, where $M \approx N$, the result is that $N_{\mathcal{H}} \simeq 2^{2M} > 10^{10\,000}$. This number far exceeds the number of atoms in the visible Universe, which is perhaps 10^{100} or so. One obtains similar results for fermionic Hilbert spaces, where the total available Hilbert space dimension is $N_{\mathcal{H}} \simeq 2^M$, if we allow all possible particle numbers.

In some cases, analytic solutions exist for interacting many-body systems, as in the eigenstates of the interacting Bose gas in one dimension, treated in Chapter 11. More generally, one has to solve these problems numerically. Even for this relatively small example, there is no prospect of storing $N_{\mathcal{H}}$ complex amplitudes in the memory of any digital computer, let alone of carrying out direct matrix computations involving an $N_{\mathcal{H}} \times N_{\mathcal{H}}$ matrix.

This problem of many-body quantum complexity is perhaps the most significant outstanding mathematical problem in theoretical physics. After all, quantum mechanics is vital in many areas of modern science and technology, ranging from photonics and atom lasers, to nanotechnology and biophysics. It is also a fundamental problem. If we wish to test quantum mechanics for mesoscopic or macroscopic systems, we need to have a means to calculate what quantum mechanics predicts. One cannot falsify a scientific theory without knowing its predictions. We emphasize here that, in the examples treated in this book, it is relatively uncommon that approximation methods like perturbation theory can be used successfully, and they are generally not employed except in special cases of weak interactions and low powers.

Fortunately, exponential complexity is more tractable than it seems. One does not have to store every state vector in a computer. The reason for this is that, unlike a mathematical theorem, the solution to a scientific problem simply requires that computational errors can be reduced below a required level. Hence, a sampled or Monte Carlo numerical approximation using random numbers is sufficient for any scientific purpose, provided we can estimate and control the sampling errors. Thus, the Mount Everest of first-principles quantum many-body theory may not be unclimbable. Some routes up the foothills are described in Chapter 7.

Additional reading

Books

J. D. Bjorken and S. D. Drell, *Relativistic Quantum Fields* (McGraw-Hill, New York, 1965). Now available from Dover Press.

E. M. Henley and W. Thirring, *Elementary Quantum Field Theory* (McGraw-Hill, New York, 1962).

S. Weinberg, *The Quantum Theory of Fields*, vol. 1 (Cambridge University Press, Cambridge, 2005).

Articles

L. Faddeev and R. Jackiw, Hamiltonian reduction of unconstrained and constrained systems, *Phys. Rev. Lett.* **60**, 1692 (1988).

R. P. Feynman, Simulating physics with computers, *Int. J. Theor. Phys.* **21**, 467 (1982).

Problems

2.1 Consider the two-boson state $|\tilde{\Phi}\rangle = a^\dagger(f_1)a^\dagger(f_2)|0\rangle$, where $\int d^3\mathbf{r}\,|f_j(\mathbf{r})|^2 = 1$, for $j = 1, 2$.

(a) This state is not normalized. Find the proper normalization constant, and call the normalized state $|\Phi\rangle$.

(b) Calculate the quantity $\langle\Phi|\psi^\dagger(\mathbf{r})\psi(\mathbf{r})|\Phi\rangle$. If the particles are photons, this quantity is related to the probability to detect one photon at the position $\mathbf{r}$.

(c) Now suppose the particles are fermions so that $|\tilde{\Phi}\rangle = c^\dagger(f_1)c^\dagger(f_2)|0\rangle$. Properly normalize the state and find $\langle\Phi|\psi^\dagger(\mathbf{r})\psi(\mathbf{r})|\Phi\rangle$.

2.2 (a) Using the method of promotion of single-particle operators, show that the momentum operator for free bosons can be expressed as $\mathbf{P} = \sum_{\mathbf{k}} \hbar\mathbf{k}a^\dagger_{\mathbf{k}}a_{\mathbf{k}}$.

(b) Using the expression from part (a) and the commutation relations for creation and annihilation operators, find $\mathbf{P}\prod_{j=1}^{n} a^\dagger_{\mathbf{k}_j}|0\rangle$.

(c) Find an expression similar to the one in part (a) for the total angular momentum operator, $\mathbf{L}$.

2.3 For our field describing a free string, find both $[\phi(x,t), \phi(x',t')]$ and $[\phi(x,t), \partial_t\phi(x',t')]$.

2.4 (a) In our discussion of scattering, suppose that our force per unit length is static, that is, the only time dependence is an adiabatic turning on and off so that $j(x,t) = \rho(x)e^{-\epsilon|t|}$, and that we will be taking the limit $\epsilon \to 0^+$. Furthermore, assume that j_k is zero for $k = 0$. Show that, in the limit $\epsilon \to 0^+$, there is no scattering, that is, the in and out fields are the same.

(b) Now suppose that $j(x,t) = g\sin(k_0 x)\cos^2(\pi t/2t_0)$ for $-t_0 \le t \le t_0$ and zero otherwise, and that $k_0 = 2\pi n_0/l$ for some integer n_0. Find the relation between the in and out fields for this particular force per unit length and find ${}^{out}\langle 0|0\rangle^{in}$.

2.5 Calculate the commutation relations between the components of the electric field evaluated at different space-time points for the free electromagnetic field. Show that these commutators vanish when the points are space-like separated.

2.6 Show that the Jackiw constrained quantization procedure leads to the transverse delta-function commutator in the case of the free electromagnetic field.

3 Quantized fields in dielectric media

We have seen how to quantize the electromagnetic field in free space, so let us now look at how to quantize it in the presence of a medium. We will first consider a linear medium without dispersion, and then move on to a nonlinear medium without dispersion. Incorporating dispersion is not straightforward, because it is nonlocal in time; the value of the field at a given time depends on its values at previous times. This is due to the finite response time of the medium. We will, nonetheless, present a theory that does include the effects of linear dispersion.

It is worth noting that the quantization of a form of nonlinear electrodynamics was first explored in a model of elementary particles by Born and Infeld. Their theory was meant to be a fundamental one, not like the ones we are exploring, which are effective theories for electromagnetic fields in media. However, many of the issues explored in the Born–Infeld theory reappear in nonlinear quantum optics.

The approach adopted in this chapter is to quantize the macroscopic theory, that is, the theory that has the macroscopic Maxwell equations as its equations of motion. A different approach, which will be explored in a subsequent chapter, begins with the electromagnetic field coupled to matter, i.e. the matter degrees of freedom are explicitly included in the theory. One is then able to derive an effective theory whose basic objects are mixed matter–field modes called polaritons. The macroscopic theory is implicitly based on this idea as well. As we shall see, the canonical momentum is the displacement field, not the electric field. This implies that the creation operators that appear in the Hamiltonian create mixed matter–field modes, because they are linear combinations of a vector potential and the displacement field, and the displacement field includes the polarization of the medium.

3.1 Dispersionless linear quantization

We begin with the quantization of the electromagnetic field in a linear medium for which there is no dispersion. This was originally treated for the case of a homogeneous medium in a series of papers by Jauch and Watson, and was extended to the case of a single dielectric interface by Carniglia and Mandel. We shall follow the treatment due to Glauber and Lewenstein, who considered a general inhomogeneous medium (see the additional reading for citations to all three papers). The medium is described by a spatially varying dielectric function, $\epsilon(\mathbf{r}) = \epsilon_0(1 + \chi^{(1)}(\mathbf{r}))$, where $\chi^{(1)}(\mathbf{r})$ is the linear polarizability of the medium, which we shall assume to be a scalar, i.e. we are assuming that the medium is isotropic.

The equations of motion for the fields are now

$$\nabla \cdot \mathbf{D} = 0, \qquad \nabla \times \mathbf{E} = -\frac{\partial \mathbf{B}}{\partial t},$$
$$\nabla \cdot \mathbf{B} = 0, \qquad \nabla \times \mathbf{B} = \mu_0 \frac{\partial \mathbf{D}}{\partial t}, \tag{3.1}$$

where $\mathbf{D} = \epsilon(\mathbf{r})\mathbf{E}$. As in the case of the free-space theory, the basic field in the theory will be the vector potential, and the electric and magnetic fields will be expressed in terms of it in the same way. To show this is possible, we define $\mathbf{B} = \nabla \times \mathbf{A}$, which means that

$$\nabla \times \mathbf{E} = -\frac{\partial \mathbf{B}}{\partial t} = -\nabla \times \frac{\partial \mathbf{A}}{\partial t}. \tag{3.2}$$

These equations are the same as for a free field, which means that we can choose the vector potential so that

$$\mathbf{E} = -\frac{\partial \mathbf{A}}{\partial t} - \nabla A_0. \tag{3.3}$$

Our gauge choice will, however, be different. We can still choose $A_0 = 0$, but the Coulomb-gauge condition is modified. We now choose

$$\nabla \cdot [\epsilon(\mathbf{r})\mathbf{A}] = 0. \tag{3.4}$$

These gauge conditions and the definitions of the electric and magnetic fields in terms of the vector potential guarantee that the first three of the above equations are satisfied. The remaining equation implies that, provided the dielectric constant is time-invariant,

$$\nabla \times (\nabla \times \mathbf{A}) = \mu_0 \frac{\partial \mathbf{D}}{\partial t}$$
$$= \mu_0 \epsilon(\mathbf{r}) \frac{\partial \mathbf{E}}{\partial t}, \tag{3.5}$$

and using the gauge condition of $A_0 = 0$, we see that the vector potential satisfies

$$\mu_0 \epsilon(\mathbf{r}) \frac{\partial^2 \mathbf{A}}{\partial t^2} + \nabla \times (\nabla \times \mathbf{A}) = 0. \tag{3.6}$$

3.1.1 Lagrangian and Hamiltonian

Our next task is to find the Lagrangian and Hamiltonian for this theory. The Lagrangian density is given by replacing ϵ_0 by $\epsilon(\mathbf{r})$ in the free-field Lagrangian density,

$$\mathcal{L} = \frac{1}{2}\epsilon(\mathbf{r})|\mathbf{E}|^2 - \frac{1}{2\mu_0}|\mathbf{B}|^2. \tag{3.7}$$

From this we find the canonical momentum

$$\Pi_j = \frac{\partial \mathcal{L}}{\partial(\partial_t A_j)} = -D_j. \tag{3.8}$$

Note that the canonical momentum is not the same as in the free theory. This will have consequences when we get to the quantum theory, because it is the canonical momentum that appears in the canonical commutation relations. Consequently, the free-space theory and the dielectric theory will not have the same commutation relations. The Hamiltonian density is now

$$\begin{aligned}\mathcal{H} &= \sum_{j=1}^{3}(\partial_t A_j)\Pi_j - \mathcal{L} = \frac{1}{2\epsilon(\mathbf{r})}|\mathbf{\Pi}|^2 + \frac{1}{2\mu_0}|\nabla\times\mathbf{A}|^2 \\ &= \frac{1}{2}\epsilon(\mathbf{r})|\mathbf{E}|^2 + \frac{1}{2\mu_0}|\mathbf{B}|^2. \end{aligned} \tag{3.9}$$

We shall quantize this theory by decomposing the fields in terms of modes and treating each mode as a harmonic oscillator. The kth mode is given by

$$\mathbf{A}(\mathbf{r},t) = e^{-i\omega_k t}\mathbf{f}_k(\mathbf{r}), \tag{3.10}$$

where $\nabla\cdot[\epsilon(\mathbf{r})\mathbf{f}_k(\mathbf{r})] = 0$. The label k here does not denote the wavenumber of a mode as it did in the free-field case; here it simply serves as an index to label the mode. It may be discrete or it may be continuous, depending on the boundary conditions imposed on the mode functions. Here we shall assume it is discrete. The functions $\mathbf{f}_k(\mathbf{r})$ satisfy the equation

$$\mu_0\epsilon(\mathbf{r})\omega_k^2\mathbf{f}_k - \nabla\times(\nabla\times\mathbf{f}_k) = 0, \tag{3.11}$$

which follows from Eq. (3.6). These modes obey an orthogonality relation. To see this, we first define

$$\mathbf{g}_k(\mathbf{r}) = \sqrt{\epsilon(\mathbf{r})}\,\mathbf{f}_k(\mathbf{r}), \tag{3.12}$$

and note that $\mathbf{g}_k$ satisfies the equation

$$\mu_0\omega_k^2\mathbf{g}_k - \frac{1}{\sqrt{\epsilon(\mathbf{r})}}\nabla\times\left(\nabla\times\frac{\mathbf{g}_k}{\sqrt{\epsilon(\mathbf{r})}}\right) = 0. \tag{3.13}$$

This implies that $\mathbf{g}_k$ is the eigenfunction of a Hermitian operator, so that these functions, when suitably normalized, satisfy

$$\delta_{k,k'} = \int d^3\mathbf{r}\,\mathbf{g}_k^*(\mathbf{r})\cdot\mathbf{g}_{k'}(\mathbf{r}) = \int d^3\mathbf{r}\,\epsilon(\mathbf{r})\mathbf{f}_k^*(\mathbf{r})\cdot\mathbf{f}_{k'}(\mathbf{r}). \tag{3.14}$$

The functions $\mathbf{g}_k$ are complete in the space of functions satisfying $\nabla\cdot[\sqrt{\epsilon(\mathbf{r})}\,\mathbf{g}(\mathbf{r})] = 0$. We can use the functions $\mathbf{f}_k(\mathbf{r})$ to define a distribution

$$\delta_{mn}^{(\epsilon)}(\mathbf{r},\mathbf{r}') = \sum_k f_{km}(\mathbf{r})f_{kn}^*(\mathbf{r}'), \tag{3.15}$$

which, in the absence of a dielectric medium, reduces to the transverse delta-function.

3.1.2 Quantization and commutators

We can now start the quantization procedure by expanding the vector potential in terms of the mode functions $\mathbf{f}_k$,

$$\mathbf{A}(\mathbf{r},t) = \sum_k Q_k(t)\mathbf{f}_k(\mathbf{r}), \tag{3.16}$$

where the Q_k are operators. The fact that $\mathbf{A}$ is a Hermitian operator implies that

$$\sum_k Q_k(t)\mathbf{f}_k(\mathbf{r}) = \sum_k Q_k^\dagger(t)\mathbf{f}_k^*(\mathbf{r}), \tag{3.17}$$

which, along with the orthogonality condition for the $\mathbf{f}_k$, implies that

$$Q_k = \sum_{k'} Q_{k'}^\dagger U_{k'k}^*, \tag{3.18}$$

where the matrix $U_{k'k}^*$ is defined by

$$U_{k'k} = \int d^3\mathbf{r}\, \epsilon(\mathbf{r})\mathbf{f}_{k'}(\mathbf{r}) \cdot \mathbf{f}_k(\mathbf{r}). \tag{3.19}$$

This matrix is clearly symmetric, $U_{k'k} = U_{kk'}$, and is also unitary, i.e. it satisfies

$$\sum_{k'} U_{kk'} U_{k''k'}^* = \delta_{k,k''}. \tag{3.20}$$

To prove this, we assume that we can expand $\mathbf{f}_k^*(\mathbf{r})$ in terms of the $\mathbf{f}_{k'}(\mathbf{r})$. Using Eq. (3.14), which is the basic equation for orthogonality given above, we find that the expansion coefficients are just the $U_{kk'}^*$, i.e.

$$\mathbf{f}_k^*(\mathbf{r}) = \sum_{k'} U_{kk'}^* \mathbf{f}_{k'}(\mathbf{r}). \tag{3.21}$$

Next, we multiply both sides of this equation by $\epsilon(\mathbf{r})\mathbf{f}_{k''}(\mathbf{r})$ and integrate over $\mathbf{r}$. The result is Eq. (3.20). Finally, $U_{k'k} = 0$ unless k and k' correspond to modes of the same energy, because modes corresponding to solutions of Eq. (3.11) with different values of ω are orthogonal. We can also expand the canonical momentum in terms of the modes,

$$\mathbf{\Pi}(\mathbf{r}, t) = \sum_k P_k(t)\epsilon(\mathbf{r})\mathbf{f}_k^*(\mathbf{r}), \tag{3.22}$$

where, as before, the P_k are operators. The factor of $\epsilon(\mathbf{r})$ is necessary so that $\mathbf{\Pi}$ satisfies $\nabla \cdot \mathbf{\Pi} = -\nabla \cdot \mathbf{D} = 0$. The fact that $\mathbf{\Pi}$ is Hermitian and the orthogonality relation for the $\mathbf{f}_k$ imply that

$$P_k^\dagger = \sum_{k'} P_{k'} U_{k'k}^*. \tag{3.23}$$

We can now insert these expressions into the Hamiltonian. We find, making use of the properties of $U_{k'k}$, that

$$\int d^3\mathbf{r}\, \frac{1}{2\epsilon(\mathbf{r})} |\mathbf{\Pi}(\mathbf{r})|^2 = \frac{1}{2}\sum_k P_k^\dagger P_k. \tag{3.24}$$

The second term in the Hamiltonian requires more work. We have

$$\begin{aligned}\frac{1}{2\mu_0}\int d^3\mathbf{r}\,|\nabla\times\mathbf{A}|^2 &= \frac{1}{2\mu_0}\int d^3\mathbf{r}\sum_k\sum_{k'} Q_k Q_{k'}(\nabla\times\mathbf{f}_k)\cdot(\nabla\times\mathbf{f}_{k'})\\ &= \frac{1}{2\mu_0}\sum_k\sum_{k'} Q_k Q_{k'}\int d^3\mathbf{r}\,\mathbf{f}_k\cdot\nabla\times(\nabla\times\mathbf{f}_{k'})\\ &= \frac{1}{2\mu_0}\sum_k\sum_{k'}\mu_0\omega_{k'}^2 U_{kk'}Q_k Q_{k'},\end{aligned} \tag{3.25}$$

where use has been made of Eq. (3.11). Finally, we can make use of the properties of $U_{kk'}$ to give

$$\frac{1}{2\mu_0}\int d^3\mathbf{r}\,|\nabla\times\mathbf{A}|^2 = \frac{1}{2}\sum_k \omega_k^2 Q_k^\dagger Q_k, \tag{3.26}$$

so that the entire Hamiltonian is

$$H = \frac{1}{2}\sum_k (P_k^\dagger P_k + \omega_k^2 Q_k^\dagger Q_k). \tag{3.27}$$

If we now impose the equal-time commutation relations,

$$\begin{aligned}[][Q_k, Q_{k'}] = [Q_k^\dagger, Q_{k'}^\dagger] = [Q_k, Q_{k'}^\dagger] = 0,\\ [P_k, P_{k'}] = [P_k^\dagger, P_{k'}^\dagger] = [P_k, P_{k'}^\dagger] = 0,\\ [Q_k, P_{k'}] = i\hbar\delta_{k,k'},\end{aligned} \tag{3.28}$$

then the resulting Heisenberg equations of motion are identical to the Maxwell equations. By expressing $P_k^\dagger$ in terms of P_k, we obtain one final commutation relation,

$$[Q_k, P_{k'}^\dagger] = i\hbar U_{kk'}^*. \tag{3.29}$$

These commutation relations then give us the equal-time commutation relation between the vector potential and its canonical momentum,

$$[A_m(\mathbf{r},t), \Pi_n(\mathbf{r}',t)] = -[A_m(\mathbf{r},t), D_n(\mathbf{r}',t)] = i\hbar\delta_{mn}^{(\mathrm{tr})}(\mathbf{r},\mathbf{r}'). \tag{3.30}$$

Note that, if $\epsilon \neq \epsilon_0$, these are *not* the same as the free-space commutation relations given in Eq. (2.183), since they involve the displacement field as the canonical momentum, not the electric field as in the free-field case. When $\epsilon = \epsilon_0$, this gives exactly the same free-space commutators as previously. The procedure of quantization through mode expansion – when the modes are defined to satisfy constraints *a priori* – is identical to field quantization with constraints.

3.1.3 Annihilation and creation operators

Next, we would like to define annihilation and creation operators for the modes, $\mathbf{f}_k(\mathbf{r})$, and then express the fields in terms of them. These operators should satisfy the commutation relations

$$[a_k, a_{k'}^\dagger] = \delta_{k,k'}, \qquad [a_k, a_{k'}] = 0. \tag{3.31}$$

We begin by assuming that

$$a_k = \alpha_k Q_k + \beta_k P_k^\dagger, \tag{3.32}$$

where α_k and β_k are constants to be determined. We will assume, however, that they only depend on k through the frequency, ω_k, that is, if $\omega_k = \omega_{k'}$, then $\alpha_k = \alpha_{k'}$ and $\beta_k = \beta_{k'}$. By making use of the commutation relations for Q_k, P_k and their adjoints, we then find

$$[a_k, a_{k'}^\dagger] = i\hbar(\alpha_k \beta_k^* - \alpha_k^* \beta_k)\delta_{k,k'}, \tag{3.33}$$

so that we must have

$$i\hbar(\alpha_k \beta_k^* - \alpha_k^* \beta_k) = 1. \tag{3.34}$$

We also find that

$$[a_k, a_{k'}] = i\hbar(\alpha_k \beta_{k'} U_{kk'}^* - \alpha_{k'} \beta_k U_{kk'}^*). \tag{3.35}$$

Because $U_{kk'}$ vanishes unless $\omega_k = \omega_{k'}$, and α_k and β_k only depend on k through ω_k, the right-hand side of the above equation vanishes. Our next step is to express Q_k and P_k in terms of the creation and annihilation operators. If we take the adjoint of Eq. (3.32), multiply by $U_{k'k}^*$ and sum over k', we find

$$\begin{aligned}\sum_{k'} U_{k'k}^* a_{k'}^\dagger &= \sum_{k'} U_{k'k}^* (\alpha_{k'}^* Q_{k'}^\dagger + \beta_{k'}^* P_{k'}) \\ &= \alpha_k^* Q_k + \beta_k^* P_k^\dagger,\end{aligned} \tag{3.36}$$

where we have again made use of the fact that $U_{kk'}$ vanishes unless $\omega_k = \omega_{k'}$, and α_k and β_k only depend on k through ω_k. We can solve the above equation and Eq. (3.32) for Q_k and $P_k^\dagger$, and then take the adjoint to find P_k. This gives us

$$\begin{aligned}Q_k &= i\hbar\left(\beta_k^* a_k - \beta_k \sum_{k'} U_{k'k}^* a_{k'}^\dagger\right), \\ P_k &= i\hbar\left(\alpha_k a_k^\dagger - \alpha_k^* \sum_{k'} U_{k'k} a_{k'}\right).\end{aligned} \tag{3.37}$$

These expressions can now be inserted into the equations for $\mathbf{A}$ and $\mathbf{D}$. Choosing

$$\alpha_k = \left(\frac{\omega_k}{2\hbar}\right)^{1/2}, \qquad \beta_k = i\left(\frac{1}{2\hbar\omega_k}\right)^{1/2}, \tag{3.38}$$

we find that

$$\begin{aligned}\mathbf{A}(\mathbf{r}, t) &= \sum_k \left(\frac{\hbar}{2\omega_k}\right)^{1/2} [a_k \mathbf{f}_k(\mathbf{r}) + a_k^\dagger \mathbf{f}_k^*(\mathbf{r})], \\ \mathbf{D}(\mathbf{r}, t) &= i\epsilon(\mathbf{r}) \sum_k \left(\frac{\hbar\omega_k}{2}\right)^{1/2} [a_k \mathbf{f}_k(\mathbf{r}) - a_k^\dagger \mathbf{f}_k^*(\mathbf{r})].\end{aligned} \tag{3.39}$$

Finally, the expressions for Q_k and P_k in terms of the creation and annihilation operators can be inserted into the Hamiltonian, yielding

$$H = \frac{1}{2}\sum_k \hbar\omega_k(a_k^\dagger a_k + a_k a_k^\dagger)$$
$$= \sum_k \hbar\omega_k a_k^\dagger a_k + C(\epsilon), \tag{3.40}$$

where $C(\epsilon)$ is a formally infinite constant, which can be dropped.

The creation operator $a_k^\dagger$ creates a photon in the state $\mathbf{f}_k(\mathbf{r})$. Note that the fact that $\langle 0|a_k a_{k'}^\dagger|0\rangle = \langle \mathbf{f}_k|\mathbf{f}_{k'}\rangle = \delta_{kk'}$ means that the single-particle space from which the Fock space is constructed has an inner product given by

$$\langle \mathbf{f}_k|\mathbf{f}_{k'}\rangle = \int d^3\mathbf{r}\,\epsilon(\mathbf{r})\mathbf{f}_k^*(\mathbf{r})\cdot\mathbf{f}_{k'}(\mathbf{r}). \tag{3.41}$$

Looking back over the derivation, we see that, when a medium is present, the quantization procedure is different than it is in free space. As we have noted, the canonical momentum for the theory with a dielectric is different from that without one, and this has consequences for the commutation relations of the theory. In this case, the commutation relations depend not just upon the presence of a dielectric medium but also upon the spatial dependence of the polarizability as well.

3.2 Scattering in linear media

Now that we have quantized the electromagnetic field in the presence of a linear medium, let us consider the scattering problem that results. We consider a bounded medium, and waves propagating through free space, striking the medium and emerging again into free space. In particular, we want to find the out fields in terms of the in fields.

We begin by considering two sets of solutions to the wave equation Eq. (3.11). These sets differ in the boundary conditions they obey for large values of $|\mathbf{r}|$. One set contains solutions that correspond to an incoming plane wave with well-defined momentum and polarization, and an outgoing spherical wave. These are in modes. They can be thought of as an incoming plane wave, which is produced by a source far from the medium, and an outgoing spherical wave that is a result of the scattering process. If we were to let the polarizability go adiabatically to zero as $|t| \to \infty$, these modes would go to an incoming plane wave with well-defined momentum and polarization as $t \to -\infty$. We shall denote these solutions by $\mathbf{f}_k^{in}(\mathbf{r})$. The second set contains solutions that correspond to an incoming spherical wave and an outgoing plane wave with well-defined momentum and polarization. These are the out modes, $\mathbf{f}_k^{out}$. They can be thought of as an incoming spherical wave produced by sources far from the medium that result in an outgoing plane wave after scattering off the medium. In the adiabatic limit, these modes would go to outgoing plane-wave fields with well-defined momenta and polarization as $t \to \infty$. The in states for the field theory are obtained by applying creation operators corresponding to the modes $\mathbf{f}_k^{in}$ to the vacuum, and the out states are obtained by applying creation operators corresponding to the modes $\mathbf{f}_k^{out}$ to the vacuum.

3.2.1 One-dimensional example

In order to make this explicit, let us look at a simple one-dimensional example. We will consider a field on the entire x axis. The medium, which has a dielectric constant ϵ, lies between $-x_0$ and x_0. We shall assume that the vector potential points in the y direction. Since our vector potential points in the y direction and only depends on x, our gauge condition is satisfied. With these assumptions, the wave equation for the modes becomes

$$\frac{n_r^2(x)}{c^2}\omega^2 f_k^2 + \frac{d^2 f_k}{dx^2} = 0, \tag{3.42}$$

where the refractive index is given by $n_r(x) = 1$ for $|x| > x_0$ and $n_r^2(x) = \epsilon/\epsilon_0$ for $|x| \leq x_0$. We now need to find solutions to this equation for which f_k and its first derivative are continuous. Let $k = \omega/c$ and $k_1 = (\omega/c)\sqrt{\epsilon_0/\epsilon}$. Waves with a positive value of k (or k_1) are moving to the right, and waves with a negative value of k (or k_1) are moving to the left.

Now let us consider a solution of Eq. (3.42) of the following form for $k > 0$:

$$g_k(x) = \begin{cases} e^{ikx} + re^{-ikx}, & x < x_0, \\ Ae^{ik_1x} + Be^{-ik_1x}, & |x| \leq x_0, \\ \tau e^{ikx}, & x > x_0. \end{cases} \tag{3.43}$$

This solution describes an incoming wave and two outgoing scattered waves, one reflected with coefficient r and one transmitted with coefficient τ. This is, in fact, an in mode. Suppose we were to let $n_r(x)$ become a function of both x and t, and have $n_r(x,t)$ go adiabatically to one as $t \to -\infty$. Then, because the time dependence of $n_r(x,t)$ is adiabatic, the solutions will simply follow $n_r(x,t)$, that is, they will be solutions of

$$\frac{n_r(x,t)}{c^2}\omega^2 g_k + \frac{d^2 g_k}{dx^2} = 0. \tag{3.44}$$

This implies that as $t \to -\infty$ the solution given above, $g(x)$, will go to just the incoming wave, e^{ikx}, for all space. The remaining task is to find r, τ, A and B from the continuity conditions the solution must satisfy. Defining

$$D(k) = (k+k_1)^2 e^{-2ik_1x_0} - (k-k_1)^2 e^{2ik_1x_0}, \tag{3.45}$$

we find that

$$\begin{aligned} A(k) &= \frac{2k(k+k_1)}{D}e^{-i(k+k_1)x_0}, \\ B(k) &= \frac{2k(k_1-k)}{D}e^{-i(k_1-k)x_0}, \\ \tau(k) &= \frac{4kk_1}{D}e^{-2ikx_0}, \\ r(k) &= \frac{-2i(k^2-k_1^2)}{D}e^{-2ikx_0}\sin(2k_1x_0). \end{aligned} \tag{3.46}$$

The in state for $k < 0$, i.e. for $k = -\omega/c$ and $k_1 = -(\omega/c)\sqrt{\epsilon_0/\epsilon}$, is given by

$$g_k(x) = \begin{cases} \tau(|k|)e^{ikx}, & x < -x_0, \\ A(|k|)e^{ik_1x} + B(|k|)e^{-k_1x}, & |x| \le x_0, \\ e^{ikx} + r(|k|)e^{-ikx}, & x > x_0. \end{cases} \tag{3.47}$$

This mode corresponds to an incoming wave from the right, and two scattered waves. It is obtained from our previous solution by simply applying the parity operator, that is, the operator that sends x to $-x$. Henceforth we shall refer to these solutions as $g_k^{(in)}(x)$. It is simple to check that

$$|\tau|^2 + |r|^2 = 1, \tag{3.48}$$

which, of course, is the usual condition for energy conservation in a beam-splitter, as expected.

The out states are just the time-reversed in states. The solution, for $k > 0$, is given by

$$g_k^{(out)}(x) = \begin{cases} \tau^*(k)e^{ikx}, & x < -x_0, \\ A^*(k)e^{ik_1x} + B^*(k)e^{-ik_1x}, & |x| \le x_0, \\ e^{ikx} + r^*(k)e^{-ikx} & x > x_0. \end{cases} \tag{3.49}$$

This solution describes two incoming waves and a single outgoing wave. If we let $n_r(x,t) \to 1$ adiabatically as $t \to \infty$, this solution will go to e^{ikx} for all space. The out state for $k < 0$ is obtained by applying the parity operator to the above solution, with the result

$$g_k^{(out)}(x) = \begin{cases} e^{ikx} + r^*(|k|)e^{-ikx}, & x < -x_0, \\ A^*(|k|)e^{ik_1x} + B^*(|k|)e^{-ik_1x}, & |x| \le x_0, \\ \tau^*(|k|)e^{ikx}, & x > x_0. \end{cases} \tag{3.50}$$

3.2.2 Normalization

So far, we have ignored the fact that our solutions are not only not normalized, but also not normalizable. In order to get around this problem, we can put in convergence factors. For example, define

$$f_k(x;\delta) = \begin{cases} N(\delta)(e^{ikx} + re^{-ikx})e^{\delta(x+x_0)}, & x < x_0, \\ N(\delta)(Ae^{ik_1x} + Be^{-ik_1x}), & |x| \le x_0, \\ N(\delta)\tau e^{ikx}e^{\delta(x-x_0)}, & x > x_0. \end{cases} \tag{3.51}$$

The parameter $\delta > 0$ guarantees that f_k is normalizable, and $N(\delta)$ is chosen so that $\int_{-\infty}^{\infty} dx\, \eta(x)|f_k(x;\delta)|^2 = 1$, which implies that

$$\frac{1}{N^2} = \frac{1}{\delta} + \frac{1}{2(k^2+\delta^2)}[ik(re^{2ikx_0} - r^*e^{-2ikx_0}) + \delta(re^{2ikx_0} + r^*e^{-2ikx_0})] + \frac{\epsilon}{\epsilon_0}\left[2x_0(|A|^2 + |B|^2) + \frac{1}{k_1}(A^*B + AB^*)\sin(2k_1x_0)\right]. \tag{3.52}$$

In order to calculate physical quantities, such as S-matrix elements, we will take the limit $\delta \to 0$.

We can express the out states in terms of the in states. We find that the in and out states with the same value of $|k|$ can be expressed in terms of each other. For $k > 0$, for example, we find that

$$f_k^{(out)}(x;\delta) = r^* f_{-k}^{(in)}(x;\delta) + \tau^* f_k^{(in)}(x;\delta). \tag{3.53}$$

If a_k^{in} and a_k^{out} are the annihilation operators corresponding to the modes f_k^{in} and f_k^{out}, respectively, then the above equation implies that

$$(a_k^{out})^\dagger = \tau^*(a_k^{in})^\dagger + r^*(a_{-k}^{in})^\dagger, \qquad a_k^{out} = \tau a_k^{in} + r a_{-k}^{in}. \tag{3.54}$$

In the limit $\delta \to 0$ we find that $\langle f_k | f_{-k} \rangle$ goes to zero, and so, in this limit, $a_k^{(in)}$ and $(a_{-k}^{(in)})^\dagger$ commute. Therefore, the amplitude that a photon with wavenumber $k > 0$ emerges from the scattering with the same wavenumber is given by the S-matrix element

$$\begin{aligned} S_{kk} &= \langle 0 | a_k^{out} (a_k^{in})^\dagger | 0 \rangle \\ &= \langle 0 | (\tau a_k^{in} + r a_{-k}^{in})(a_k^{in})^\dagger | 0 \rangle = \tau. \end{aligned} \tag{3.55}$$

This scattering theory approach, although simplified, is the basis for how one treats the problem of an elementary beam-splitter operation, in which two input modes are transformed into two output modes. This is described by a 2×2 unitary matrix $\mathcal{M}$, using Eq. (2.141), so that

$$\begin{aligned} \begin{bmatrix} a_k^{out} \\ a_{-k}^{out} \end{bmatrix} &= \mathcal{M} \begin{bmatrix} a_k^{in} \\ a_{-k}^{in} \end{bmatrix} \\ &= \begin{bmatrix} \tau & r \\ -r^* & \tau^* \end{bmatrix} \begin{bmatrix} a_k^{in} \\ a_{-k}^{in} \end{bmatrix}. \end{aligned} \tag{3.56}$$

This must be unitary in order that the commutation relations are unchanged. It is easy to check that the matrix defined above is unitary (i.e. $\mathcal{M}^\dagger \mathcal{M} = I$), provided $|\tau|^2 + |r|^2 = 1$, which we have already shown in the previous subsection. In the case of a linear multi-port, which is an M-mode generalization of a beam-splitter that transforms M input modes into M output modes, the linear $\mathcal{M}$-matrix is an $M \times M$ unitary matrix, and can be obtained in a similar way.

3.3 Quantizing a nonlinear dielectric

After tackling linear media, let us now discuss the quantization of electrodynamics in nonlinear media. To do so, we will follow the program that has been laid out earlier. We will first choose the fundamental fields of our theory. These are usually taken to be the vector and scalar magnetic potentials, but we shall find it convenient to make another choice. We then find a Lagrangian that gives us the equations of motion, which in this case are Maxwell's equations. From the Lagrangian, we first find the canonical momentum, and then find the Hamiltonian. Finally, we impose the canonical commutation relations on the fundamental field and the canonical momentum.

Let us now carry out the steps sketched out in the previous paragraph in detail. The equations of motion for our theory are Maxwell's equations,

$$\nabla \cdot \mathbf{D} = 0, \qquad \nabla \times \mathbf{E} = -\frac{\partial \mathbf{B}}{\partial t},$$
$$\nabla \cdot \mathbf{B} = 0, \qquad \nabla \times \mathbf{B} = \mu_0 \frac{\partial \mathbf{D}}{\partial t}, \tag{3.57}$$

in the absence of external charges and currents. Here $\mathbf{D} = \epsilon_0 \mathbf{E} + \mathbf{P}$ is the displacement field. We shall assume that the medium is lossless and nondispersive, but it may be inhomogeneous, i.e. the susceptibilities can be functions of position. We want to find a Lagrangian that has Eqs (3.57) as its equations of motion, and the known energy of a dielectric, Eq. (3.60) or (3.61), as its Hamiltonian. We need to choose a particular field, which is to be the basic dynamical variable in the problem. We will carry this out using two alternative procedures:

1. Use the conventional vector potential $A = (A_0, \mathbf{A})$, where

$$\mathbf{E} = -\frac{\partial \mathbf{A}}{\partial t} - \nabla A_0, \qquad \mathbf{B} = \nabla \times \mathbf{A}. \tag{3.58}$$

   However, this quantization procedure turns out to be rather complicated, and only useful in practice for a homogeneous medium.
2. Use the dual potential $\Lambda = (\Lambda_0, \mathbf{\Lambda})$, where

$$\mathbf{B} = \mu_0 \left[\frac{\partial \mathbf{\Lambda}}{\partial t} + \nabla \Lambda_0 \right], \qquad \mathbf{D} = \nabla \times \mathbf{\Lambda}. \tag{3.59}$$

   The dual potential is a much simpler route to quantization, and can be applied to a general inhomogeneous medium. This approach is simplest if external charges and currents are absent. When this is the case, the fact that $\nabla \cdot \mathbf{D} = 0$ implies that $\mathbf{D}$ can be expressed as the curl of some vector field, and that vector field we call the dual potential, $\mathbf{\Lambda}$. Otherwise we need to separate the longitudinal and transverse parts of the displacement field, since only the transverse part can be expanded as the curl of a vector potential.

Finally, as in all methods of quantization, it is important that the final Hamiltonian density equals the classical dielectric energy density, which was treated for the electromagnetic field in a medium in Section 1.4. For a nonlinear, nondispersive medium, with a power series expansion of $\mathbf{E}(\mathbf{D}')$, we can write the energy density explicitly as

$$\mathcal{H} = \frac{1}{2\mu}|\mathbf{B}|^2 + \sum_n \frac{1}{n+1}\mathbf{D} \cdot \eta^{(n)} : \mathbf{D}^n. \tag{3.60}$$

An alternative second equivalent form can be readily obtained by inverting the nonlinear functional and expanding the power series to obtain

$$\mathcal{H} = \frac{1}{2\mu}|\mathbf{B}|^2 + \sum_n \frac{n}{n+1}\mathbf{E} \cdot \epsilon^{(n)} : \mathbf{E}^n. \tag{3.61}$$

Consequently, we should expect the Hamiltonians we find by applying the canonical formalism to be of this form.

3.4 Homogeneous nonlinear dielectric

We will start by quantizing the field in a *homogeneous* nonlinear dielectric using the standard vector potential. This follows similar techniques to the previous section, which dealt with the linear case. As we will see in the next section, if we use the dual potential, we can drop the requirement of homogeneity.

An appropriate Lagrangian density for the theory in the vector potential case is a simple generalization of the standard free-field form,

$$\mathcal{L}(\mathbf{A}, \dot{\mathbf{A}}) = \int_0^{\mathbf{D}} \mathbf{E}(\mathbf{D}') \cdot d\mathbf{D}' - \int_0^{\mathbf{B}} \mathbf{H}(\mathbf{B}') \cdot d\mathbf{B}', \tag{3.62}$$

which can be rewritten as

$$\mathcal{L}(A, \dot{A}) = \sum_n \frac{1}{n+1} \mathbf{E} \cdot \boldsymbol{\epsilon}^{(n)} : \mathbf{E}^n - \frac{1}{2\mu_0} |\mathbf{B}|^2. \tag{3.63}$$

As in the free-field case, the Lagrange equations are

$$\partial_t \left(\frac{\partial \mathcal{L}}{\partial(\partial_t A_\nu)} \right) + \sum_{j=1}^{3} \partial_j \left(\frac{\partial \mathcal{L}}{\partial(\partial_j A_\nu)} \right) - \frac{\partial \mathcal{L}}{\partial A_\nu} = 0, \tag{3.64}$$

where $\nu = 0, \ldots, 3$, but now the Lagrangian density is different. This equation with $\nu = 0$ gives us $\nabla \cdot \mathbf{D} = 0$, and the three remaining equations give us $\nabla \times \mathbf{B} = \mu(\partial \mathbf{D}/\partial t)$. The other two Maxwell equations follow from the definition of electric and magnetic fields in terms of the vector potential.

We now want to proceed to the Hamiltonian formalism, and the first thing we need to do is to find the canonical momentum. From the above Lagrangian density, we find that the canonical momentum corresponding to A, which we denote by $\Pi = (\Pi_0, \boldsymbol{\Pi})$, is

$$\Pi_0 = \frac{\partial \mathcal{L}}{\partial(\partial_0 A_0)} = 0, \qquad \Pi_i = \frac{\partial \mathcal{L}}{\partial(\partial_0 A_i)} = -D_i. \tag{3.65}$$

Here we point out two things.

1. Just as in the case of a linear dielectric, the canonical momentum is different from that in the noninteracting theory, where $\Pi_i = -E_i$. This is a consequence of the fact that the interaction depends on $\dot{\mathbf{A}} = \partial_t \mathbf{A}$.
2. The vanishing of Π_0 implies that A_0 is not an independent field. In the case of free fields, if we choose the Coulomb gauge, it is also possible to choose $A_0 = 0$. This follows from the fact that, for the free theory, $A_0 = 0$ and $\nabla \cdot \mathbf{E} = 0$ imply that the time derivative of $\nabla \cdot \mathbf{A}$ is zero, so that, if $\nabla \cdot \mathbf{A} = 0$ initially, it will remain zero. In this case the Coulomb and temporal ($A_0 = 0$) gauges are consistent. This is no longer true when a nonlinear interaction is present, because now, instead of $\nabla \cdot \mathbf{E} = 0$, we have $\nabla \cdot \mathbf{D} = 0$. Hence, if we choose $A_0 = 0$, the time derivative of $\nabla \cdot \mathbf{A}$ is no longer zero.

If we choose the Coulomb gauge, which is what we shall do, then A_0 must be determined by solving the equation

$$\nabla^2 A_0 = -\nabla \cdot \mathbf{E}, \tag{3.66}$$

where $\mathbf{E}$ will be expressed in terms of the canonical momentum, $-\mathbf{D}$.

Another consequence of the fact that A_0 is not an independent field in the Hamiltonian formulation is that we lose Gauss' law as an equation of motion. However, the equation $\nabla \times \mathbf{B} = \mu_0 \dot{\mathbf{D}}$, which is a result of the theory, implies that $\nabla \cdot \mathbf{D}$ is time-independent, and this allows us to impose Gauss' law as an initial condition.

3.4.1 Hamiltonian

For the Hamiltonian density we have

$$H(\mathbf{A}, \mathbf{\Pi}) = \int d^3\mathbf{r} \left[\frac{1}{2\mu} |\mathbf{B}|^2 + \sum_n \frac{n}{n+1} \mathbf{E} \cdot \boldsymbol{\epsilon}^{(n)} : \mathbf{E}^n \right] + \int d^3\mathbf{r}\, \mathbf{D} \cdot \nabla A_0. \tag{3.67}$$

Performing an integration by parts in the last term and using the initial condition $\nabla \cdot \mathbf{D} = 0$ allows us to eliminate the second integral. We see that it agrees, as it must if the Lagrangian is to be correct, with the known classical energy from Eq. (3.61). It is useful to express the Hamiltonian directly in terms of the canonical momenta, D_i. Making use of the equivalence between the two expressions for the energy derived in Eqs (3.60) and (3.61), we find for the Hamiltonian

$$H(\mathbf{A}, \mathbf{\Pi}) = \int d^3\mathbf{r} \left[\frac{1}{2\mu} |\mathbf{B}|^2 + \sum_n \frac{1}{n+1} \mathbf{D} \cdot \boldsymbol{\eta}^{(n)} : \mathbf{D}^n \right]. \tag{3.68}$$

The theory is quantized by imposing the equal-time commutation relations

$$[A_j(\mathbf{r}, t), \Pi_k(\mathbf{r}', t)] = i\hbar \delta^{(\mathrm{tr})}_{jk}(\mathbf{r} - \mathbf{r}'). \tag{3.69}$$

Here, as in standard quantum electrodynamics (QED), we use the transverse delta-function in order to be consistent with both the Coulomb-gauge condition, $\nabla \cdot \mathbf{A} = 0$, and Gauss' law, $\nabla \cdot \mathbf{D} = 0$. As in the case of free QED, it is possible to perform a mode expansion for the field and to define creation and annihilation operators. In particular, for the mode with momentum $\mathbf{k}$ and polarization $\hat{\mathbf{e}}_\alpha(\mathbf{k})$, we have the annihilation operator

$$a_{\mathbf{k},\alpha}(t) = \frac{1}{\sqrt{\hbar V}} \int d^3\mathbf{r}\, e^{-i\mathbf{k}\cdot\mathbf{r}} \hat{\mathbf{e}}_\alpha(\mathbf{k}) \cdot \left[\sqrt{\frac{\epsilon\omega_k}{2}} \mathbf{A}(\mathbf{r}, t) - \frac{i}{\sqrt{2\epsilon\omega_k}} \mathbf{D}(\mathbf{r}, t) \right]. \tag{3.70}$$

Because $a_{\mathbf{k},\alpha}$ depends on $\mathbf{D}$, and consequently contains matter degrees of freedom, it is not a pure photon operator. It represents a collective matter–field mode.

3.4.2 How not to quantize electromagnetic fields

We conclude this section with a word of warning. It is common in the literature to see the interacting Hamiltonian in Eq. (3.67) quantized by naively substituting into the Hamiltonian the expressions in Eq. (2.189), that is, the free-field expressions for the electric and magnetic fields are used in terms of creation and annihilation operators. This is incorrect, since only the canonical fields have invariant commutation relations when interactions are included.

As we have seen, the canonical fields depend on the Lagrangian. Unless the quantization is carried out following Dirac's prescription, the resulting commutation relations will not lead to the correct time-evolution equations. Consequently, the naive quantization procedure of substituting the free-field expressions for the electric and magnetic fields in the final Hamiltonian for the dielectric does *not* lead to Maxwell's equations as the equations of motion for the fields.

The errors from this wrong approach are substantial. Maxwell's equations are very well verified experimentally at the mean-field level of large occupation number, and quantum theory must lead to these equations at both microscopic and macroscopic levels. The naive approach therefore predicts absurd results, in complete contradiction to experiment. For example, it is not hard to verify from substituting these commutators into the Heisenberg equations of motion that the predicted speed of light even in a *linear* dielectric is completely different from the known value of $v_p = 1/\sqrt{\mu\epsilon}$.

The reason why this approach is sometimes used is that the commutators do *not* change in a microscopic approach to quantization, as we will see in Chapter 4. However, a macroscopic field involves averaging in both time and space. Technically, there is a momentum cutoff, and macroscopic quantization only treats Fourier components whose wavelength is much greater than the lattice spacing. Therefore, we should not expect the same commutation relations to apply. The averaging process means that the macroscopic fields we are treating here are not identical to the microscopic quantized fields, and they have different commutators.

This subtlety is often overlooked in a naive approach to macroscopic quantization.

3.5 Inhomogeneous nonlinear dielectric

Dielectric structures in nonlinear optics are seldom homogeneous. They have boundaries, with transmission and reflection, even in the simplest experiments. It is common to insert nonlinear crystals into interferometers or propagating beams. In much recent technology, the dielectric is a waveguide, fiber or integrated optical device, as shown schematically later in Figure 3.1. In most cases, the dielectric structure has no sharp boundaries, but rather has an index of refraction whose spatial variation is an integral functional component of the design. Sophisticated metamaterials whose macroscopic properties are caused by microscopic inhomogeneity are becoming more common.

In all these cases, it is essential to quantize an inhomogeneous dielectric. This is much simpler if we use the dual potential of the displacement field as the basic field variable. Therefore, in this section, we deal with the theory of macroscopic electromagnetic quantization in an *inhomogeneous* nonlinear dielectric. This allows us to quantize typical experimental configurations with input and output modes, as well as waveguides, fibers and other common devices. We will also see later that, from the point of view of microscopic

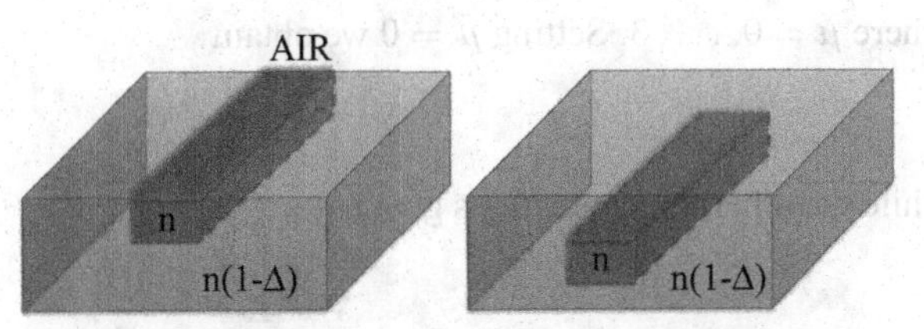

Fig. 3.1 Optical waveguide: surface and embedded rectangular waveguides. Waveguiding is achieved through having a slightly higher refractive index in the waveguide than in the surrounding optical medium.

Lagrangian field theory, it is important to use a single canonical field that interacts with the dipoles that cause a dielectric response. When using the electric dipole interaction form of the Hamiltonian, this single field is actually the Maxwell displacement field $\mathbf{D}(\mathbf{r})$. This is also microscopically identical to the local field (apart from a factor of ϵ_0, due to the SI unit system), which gives an additional motivation for this approach.

Recapitulating, the dual potential, $\Lambda = (\Lambda_0, \mathbf{\Lambda})$ is defined so that

$$\mathbf{B} = \mu\left[\frac{\partial \mathbf{\Lambda}}{\partial t} + \nabla\Lambda_0\right], \qquad \mathbf{D} = \nabla\times\mathbf{\Lambda}. \tag{3.71}$$

Note that this definition of $\mathbf{D}$ and $\mathbf{B}$ in terms of $\mathbf{\Lambda}$ and Λ_0 already guarantees that two of Maxwell's equations are satisfied, i.e.

$$\nabla\cdot\mathbf{D} = 0, \qquad \nabla\times\mathbf{B} = \mu\frac{\partial\mathbf{D}}{\partial t}. \tag{3.72}$$

3.5.1 Nonlinear Lagrangian and Hamiltonian

We now need a Lagrangian, or actually a Lagrangian density, that gives us the remaining two Maxwell equations. The expression we will use now has the opposite sign to the previous case, as we have used dual symmetry to exchange the roles of the electric and magnetic fields:

$$\mathcal{L}(\mathbf{\Lambda},\dot{\mathbf{\Lambda}}) = -\mathcal{L}(A,\dot{A}) = \int_0^{\mathbf{B}}\mathbf{H}(\mathbf{B}')\cdot d\mathbf{B}' - \int_0^{\mathbf{D}}\mathbf{E}(\mathbf{D}')\cdot d\mathbf{D}'. \tag{3.73}$$

It is convenient to express the nonlinear relation between the $\mathbf{D}$ and $\mathbf{E}$ fields as an expansion in terms of the displacement field, since the relation between $\mathbf{D}$ and $\mathbf{\Lambda}$ is relatively simple. Employing the same techniques as used to calculate the classical energy in Eq. (3.60), we obtain

$$\mathcal{L} = \frac{1}{2\mu}|\mathbf{B}|^2 - \sum_n \frac{1}{n+1}\mathbf{D}\cdot\eta^{(n)}(\mathbf{r}) : \mathbf{D}\cdots\mathbf{D}.$$

The equations of motion that come from this Lagrangian density are given by

$$\partial_t\left(\frac{\partial\mathcal{L}}{\partial(\partial_t\Lambda_\mu)}\right) + \sum_{j=1}^{3}\partial_j\left(\frac{\partial\mathcal{L}}{\partial(\partial_j\Lambda_\mu)}\right) - \frac{\partial\mathcal{L}}{\partial\Lambda_\mu} = 0, \tag{3.74}$$

where $\mu = 0, \ldots, 3$. Setting $\mu = 0$ we obtain

$$\nabla \cdot \mathbf{B} = 0, \tag{3.75}$$

while the other three equations give us

$$\nabla \times \mathbf{E} = -\frac{\partial \mathbf{B}}{\partial t}. \tag{3.76}$$

Deriving the first of these equations is straightforward, but it is useful to fill in a few steps in the derivation of the second one. We first note that

$$\begin{aligned} \frac{\partial \mathcal{L}}{\partial(\partial_t \Lambda_k)} &= B_k, \\ \frac{\partial D_l}{\partial(\partial_j \Lambda_k)} &= \varepsilon_{ljk}, \end{aligned} \tag{3.77}$$

where $k \in \{1, 2, 3\}$ and ε_{ljk} is the completely antisymmetric Levi-Civita tensor of rank three. Now, for simplicity, let us look at the case $\eta^{(j)} = 0$ for $j \geq 2$. Making use of the above relations, we find that

$$\frac{\partial \mathbf{B}}{\partial t} + \nabla \times (\boldsymbol{\eta}(\mathbf{r}) : \mathbf{D}) = 0. \tag{3.78}$$

For the case we are considering, this is just the last Maxwell equation in the previous paragraph, and it is simple to verify that Lagrange's equations correspond to the nonlinear Maxwell equations to all orders. Therefore, we now have a Lagrangian formulation of the theory.

The next step is to find the Hamiltonian formulation. The canonical momentum is given by

$$\Pi_0 = \frac{\partial \mathcal{L}}{\partial(\partial_t \Lambda_0)} = 0, \qquad \Pi_j = \frac{\partial \mathcal{L}}{\partial(\partial_t \Lambda_j)} = B_j. \tag{3.79}$$

The Hamiltonian density is then

$$\begin{aligned} \mathcal{H} &= \sum_{j=1}^{3} \Pi_j \partial_t \Lambda_j - \mathcal{L} \\ &= \int_0^{\mathbf{B}} \mathbf{H}(\mathbf{B}') \cdot d\mathbf{B}' + \int_0^{\mathbf{D}} \mathbf{E}(\mathbf{D}') \cdot d\mathbf{D}' - B \cdot \nabla \Lambda_0. \end{aligned} \tag{3.80}$$

At this point, we notice that the Hamiltonian equations of motion,

$$\partial_t \Pi_j = -\frac{\delta H}{\delta \Lambda_j}, \tag{3.81}$$

give us only three, instead of four, equations. Because of the vanishing of Π_0, we have lost $\nabla \cdot \mathbf{B} = 0$ as an equation of motion, and, in fact, the vanishing of Π_0 means that we are dealing with a constrained Hamiltonian system. The detailed theory of the quantization of constrained Hamiltonians is given in Section 2.9 and we will give a shortened but equivalent method here. We first note that taking the divergence of both sides of Eq. (3.76) gives us that

$$\frac{\partial}{\partial t}\nabla \cdot \mathbf{B} = 0, \tag{3.82}$$

so that if $\nabla \cdot \mathbf{B} = 0$ is true initially, it will remain true. Therefore, we can recover the equation $\nabla \cdot \mathbf{B} = 0$ if we impose it as an initial condition.

Before quantizing the theory, we will fix the gauge. The physical fields are unchanged under the transformation

$$\mathbf{\Lambda} \to \mathbf{\Lambda} + \nabla\Theta, \qquad \Lambda_0 \to \Lambda_0 - \frac{\partial\Theta}{\partial t}, \tag{3.83}$$

where $\Theta(\mathbf{r}, t)$ is an arbitrary function of space and time. We can eliminate Λ_0 by choosing Θ to be a solution of

$$\frac{\partial\Theta}{\partial t} = \Lambda_0, \tag{3.84}$$

which determines Θ up to an arbitrary function of position, which we shall call $\theta(\mathbf{r})$. Since $\nabla \cdot \mathbf{\Lambda}$ is time-independent,

$$\frac{\partial}{\partial t}(\nabla \cdot \mathbf{\Lambda}) = \nabla \cdot \mathbf{B} = 0, \tag{3.85}$$

we can choose θ so that $\nabla \cdot \mathbf{\Lambda} = 0$. The result is that we have a radiation gauge for $\mathbf{\Lambda}$ in which $\Lambda_0 = 0$ and $\nabla \cdot \mathbf{\Lambda} = 0$. Note that this gauge choice eliminates the last term of the Hamiltonian density, giving the standard classical result, Eq. (3.60), for the energy of nonlinear dielectric media.

3.5.2 Nonlinear quantization

In order to quantize the theory, we now impose the canonical equal-time commutation relations, thus turning our classical fields into operator quantum fields:

$$[\Lambda_j(\mathbf{r}, t), B_k(\mathbf{r}', t)] = i\hbar\delta^{(\mathrm{tr})}_{jk}(\mathbf{r} - \mathbf{r}'), \tag{3.86}$$

where $\delta^{(\mathrm{tr})}$ is the transverse delta-function. We have again used the transverse delta-function, because this choice makes the above equation consistent with the fact that both $\nabla \cdot \mathbf{\Lambda} = 0$ and $\nabla \cdot \mathbf{B} = 0$.

As usual, we would like to express the field in terms of creation and annihilation operators for plane-wave modes. These have the mode functions

$$\mathbf{u}_\alpha(\mathbf{k}) = \frac{1}{\sqrt{V}}\hat{\mathbf{e}}_{\mathbf{k},\alpha}e^{i\mathbf{k}\cdot\mathbf{r}}, \tag{3.87}$$

where $\alpha = 1, 2$. The annihilation operators are linear combinations of the fields $\mathbf{\Lambda}$ and $\mathbf{B}$. The exact linear combination will be chosen with two requirements in mind. First, we want to obtain the usual commutation relations between the creation and annihilation operators, so the linear combination should be chosen so that these commutation relations follow from the canonical commutation relations between the fields. If we define

$$a_{\mathbf{k},\alpha} = \int d^3\mathbf{r}\, \mathbf{u}^*_\alpha(\mathbf{k}) \cdot \left(c_{\mathbf{k}}\mathbf{\Lambda}(\mathbf{r}, t) + \frac{i}{2\hbar c_{\mathbf{k}}}\mathbf{B}(\mathbf{r}, t) \right), \tag{3.88}$$

where the real numbers $c_{\mathbf{k}}$ are, for the moment, arbitrary, then we indeed find that Eq. (3.86) implies that

$$[a_{\mathbf{k},\alpha}, a^{\dagger}_{\mathbf{k}',\alpha'}] = i\hbar\delta_{\mathbf{k},\mathbf{k}'}\delta_{\alpha,\alpha'}. \tag{3.89}$$

We can now take Eq. (3.88) and its adjoint, and solve for the fields in terms of the creation and annihilation operators, following the general procedure that led to Eq. (2.98). We find that

$$\begin{aligned}\mathbf{\Lambda}(\mathbf{r},t) &= \sum_{\mathbf{k},\alpha} \frac{1}{2c_{\mathbf{k}}\sqrt{V}}\hat{\mathbf{e}}_{\mathbf{k},\alpha}(e^{i\mathbf{k}\cdot\mathbf{r}}a_{\mathbf{k},\alpha} + e^{-i\mathbf{k}\cdot\mathbf{r}}a^{\dagger}_{\mathbf{k},\alpha}),\\ \mathbf{B}(\mathbf{r},t) &= \sum_{\mathbf{k},\alpha} \frac{\hbar c_{\mathbf{k}}}{i\sqrt{V}}\hat{\mathbf{e}}_{\mathbf{k},\alpha}(e^{i\mathbf{k}\cdot\mathbf{r}}a_{\mathbf{k},\alpha} - e^{-i\mathbf{k}\cdot\mathbf{r}}a^{\dagger}_{\mathbf{k},\alpha}).\end{aligned} \tag{3.90}$$

We can now substitute these expressions into the Hamiltonian.

For a homogeneous dielectric, we can choose the numbers $c_{\mathbf{k}}$ so that the free Hamiltonian

$$H_0 = \frac{1}{2}\int d^3\mathbf{r}\left(\frac{1}{\mu}|\mathbf{B}|^2 + \frac{1}{\epsilon}|\mathbf{D}|^2\right) \tag{3.91}$$

has the form

$$H_0 = \frac{1}{2}\sum_{\mathbf{k},\alpha}\hbar\omega_{\mathbf{k}}(a^{\dagger}_{\mathbf{k},\alpha}a_{\mathbf{k},\alpha} + a_{\mathbf{k},\alpha}a^{\dagger}_{\mathbf{k},\alpha}), \tag{3.92}$$

where $\omega_k = kv_p = k/\sqrt{\epsilon\mu_0}$. For a general choice of $c_{\mathbf{k}}$, there will also be terms of the form $a^{\dagger}_{\mathbf{k},\alpha}a^{\dagger}_{-\mathbf{k},\alpha}$ and $a_{\mathbf{k},\alpha}a_{-\mathbf{k},\alpha}$ present, but if we choose

$$c_{\mathbf{k}} = \left(\frac{\mu\omega_{\mathbf{k}}}{2\hbar}\right)^{1/2}, \tag{3.93}$$

then all of these terms vanish. Summarizing, then, the macroscopic quantum fields $\mathbf{\Lambda}$ and $\mathbf{B}$ can be expressed in terms of creation and annihilation operators as

$$\begin{aligned}\mathbf{\Lambda}(\mathbf{r},t) &= \sum_{\mathbf{k},\alpha}\left(\frac{\hbar}{2\mu\omega_{\mathbf{k}}V}\right)^{1/2}\hat{\mathbf{e}}_{\mathbf{k}\alpha}(e^{i\mathbf{k}\cdot\mathbf{r}}a_{\mathbf{k},\alpha} + e^{-i\mathbf{k}\cdot\mathbf{r}}a^{\dagger}_{\mathbf{k},\alpha}),\\ \mathbf{B}(\mathbf{r},t) &= \sum_{\mathbf{k},\alpha} i\left(\frac{\mu\hbar\omega_{\mathbf{k}}}{2V}\right)^{1/2}\hat{\mathbf{e}}_{\mathbf{k}\alpha}(e^{-i\mathbf{k}\cdot\mathbf{r}}a^{\dagger}_{\mathbf{k},\alpha} - e^{i\mathbf{k}\cdot\mathbf{r}}a_{\mathbf{k},\alpha}).\end{aligned} \tag{3.94}$$

This expansion of the fields is useful in the case of a homogeneous, isotropic dielectric. In the case that the dielectric is not homogeneous, it is usually better to expand the fields in terms of the eigenmodes of the linear part of the Hamiltonian. This is discussed in the next section.

3.5.3 Quantization in cavities and waveguides

In describing the behavior of the electromagnetic field in an inhomogeneous dielectric, like a cavity or a waveguide, it is useful to expand the field operators in terms of the mode functions $\mathbf{u}_n(\mathbf{r})$ of the relevant system. These generalize the previous plane-wave modes to cases where the modes, like the dielectric environment, are not homogeneous. The mode functions are determined by the linear part of the polarization. For example, a waveguide can be constructed from a spatially varying dielectric and a cavity by surrounding a *rectangular* region of free space by dielectric sheets or slabs (as shown in Figure 3.1). Optical fibers, by contrast, use a small *cylindrical* core of increased refractive index, surrounded by a region of lower refractive index.

Let us suppose that the linear behavior of our system is described by an inverse permittivity tensor $\eta(\mathbf{r}) \equiv \epsilon(\mathbf{r})^{-1}$. Then the mode functions must obey the classical wave equation (1.8) derived in Chapter 1,

$$\nabla \times [\eta(\mathbf{r}) : \nabla \times \mathbf{A}(\mathbf{r}, t)] = -\mu \ddot{\mathbf{A}}(\mathbf{r}, t), \tag{3.95}$$

together with the proper boundary conditions, which we usually take to be either periodic or vanishing. The modes have well-defined frequencies, ω_n, so that the mode functions are of the form $\mathbf{A}(\mathbf{r}, t) \propto \mathbf{u}_n(\mathbf{r})e^{-i\omega_n t}$, with the functions $\mathbf{u}_n(\mathbf{r})$ satisfying the equation

$$\nabla \times [\eta(\mathbf{r})\nabla \times \mathbf{u}_n(\mathbf{r})] = \mu\omega_n^2 \mathbf{u}_n(\mathbf{r}). \tag{3.96}$$

Note that, because the displacement fields are transverse, we must have $\nabla \cdot \mathbf{u}_n = 0$. This is an advantage of using displacement fields as canonical variables, since this constraint is not generally valid for the electric field in a dielectric.

We can show that modes corresponding to different frequencies are orthogonal by taking the scalar product of both sides of this equation with $\mathbf{u}_{n'}^*$ and integrating over the quantization volume. This gives us

$$\mu\omega_n^2 \int d^3\mathbf{r}\, \mathbf{u}_{n'}^*(\mathbf{r}) \cdot \mathbf{u}_n(\mathbf{r}) = -\mu\omega_n^2 \int d^3\mathbf{r}\, (\nabla \times \mathbf{u}_{n'}^*(\mathbf{r})) \cdot [\eta(\mathbf{r}) : \nabla \times \mathbf{u}_n(\mathbf{r})]. \tag{3.97}$$

Here we have applied the vector identity $\nabla \cdot [\mathbf{A} \times \mathbf{A}'] = \mathbf{A}' \cdot [\nabla \times \mathbf{A}] - \mathbf{A} \cdot [\nabla \times \mathbf{A}']$ and assumed vanishing boundary terms. Repeating this for the conjugate mode function, and noting that $\eta(\mathbf{r})$ is a real symmetric matrix, gives us

$$(\omega_n^2 - \omega_{n'}^2) \int d^3\mathbf{r}\, \mathbf{u}_{n'}^*(\mathbf{r}) \cdot \mathbf{u}_n(\mathbf{r}) = 0. \tag{3.98}$$

This immediately implies that, if $\omega_n \neq \omega_{n'}$, then the integral must vanish. It should also be possible to choose different modes with the same frequency to be orthogonal, and we shall assume that this has been done. In particular, note that $\mathbf{u}_n$ and $\mathbf{u}_n^*$ are both solutions of Eq. (3.96). We will assume that these can be chosen so that they are orthogonal to each other. Therefore, for the mode functions $\mathbf{u}_n$ we have

$$\int d^3\mathbf{r}\, \mathbf{u}_{n'}^*(\mathbf{r}) \cdot \mathbf{u}_n(\mathbf{r}) = \delta_{n,n'}. \tag{3.99}$$

We will also assume that the modes form a complete set in the space of transverse functions, so that

$$\sum_n u_{i,n}^*(\mathbf{r}) u_{j,n}(\mathbf{r}') = \delta_{ij}^{(\mathrm{tr})}(\mathbf{r} - \mathbf{r}'). \tag{3.100}$$

We next define annihilation operators, a_n, by integrating the quantum fields multiplied by the mode functions

$$a_n(t) = \sqrt{\frac{\mu}{2\hbar}} \int d^3\mathbf{r}\, \mathbf{u}_n^*(\mathbf{r}) \cdot \left[c_n \mathbf{\Lambda}(\mathbf{r}, t) + \frac{i}{c_n} \dot{\mathbf{\Lambda}}(\mathbf{r}, t) \right], \tag{3.101}$$

where the c_n are arbitrary, and can be chosen, for example, to diagonalize the linear part of the Hamiltonian. We will, however, assume that c_n is a function only of ω_n, so that modes with the same frequency correspond to the same value of c_n. These operators obey the commutation relations

$$[a_n(t), a_{n'}^\dagger(t)] = \delta_{n,n'}. \tag{3.102}$$

We can also invert Eq. (3.101) to express the fields in terms of the creation and annihilation operators. Summing over all the eigenfrequencies, and making use of the fact that $\mathbf{\Lambda}$ is transverse, we can write that

$$\sum_n \frac{\mathbf{u}_n(\mathbf{r})}{c_n} a_n(t) = \sqrt{\frac{\mu}{2\hbar}} \sum_n \mathbf{u}_n(\mathbf{r}) \int d^3\mathbf{r}'\, \mathbf{u}_n^*(\mathbf{r}') \cdot \left[\mathbf{\Lambda}(\mathbf{r}', t) + \frac{i}{c_n^2} \dot{\mathbf{\Lambda}}(\mathbf{r}', t) \right]. \tag{3.103}$$

Combining this with its adjoint, and using the completeness condition, Eq. (3.100), we have

$$\begin{aligned} &\sum_n \frac{1}{c_n} [a_n(t)\mathbf{u}_n(\mathbf{r}) + a_n^\dagger(t)\mathbf{u}_n^*(\mathbf{r})] \\ &= \sqrt{\frac{\mu}{2\hbar}} \left\{ 2\mathbf{\Lambda}(\mathbf{r}, t) + \int d^3\mathbf{r}' \sum_n \frac{i}{c_n^2} [u_{i,n}^*(\mathbf{r}) u_{j,n}(\mathbf{r}') - u_{i,n}(\mathbf{r}) u_{j,n}^*(\mathbf{r}')] \dot{\Lambda}_j(\mathbf{r}', t) \right\}, \end{aligned} \tag{3.104}$$

where we have taken the c_n to be real. The second term on the right-hand side vanishes, because both $\mathbf{u}_n$ and $\mathbf{u}_n^*$ are mode functions. In particular, for each n, there is an n' such that $\mathbf{u}_{n'} = \mathbf{u}_n^*$. This implies that

$$u_{i,n}^*(\mathbf{r}) u_{j,n}(\mathbf{r}') = u_{i,n'}(\mathbf{r}) u_{j,n'}^*(\mathbf{r}'), \tag{3.105}$$

so that the first term in the brackets for n is canceled by the second term in the brackets for n'. This finally gives us our mode expansion for the field as

$$\mathbf{\Lambda}(\mathbf{r}, t) = \sqrt{\frac{\hbar}{2\mu}} \sum_n \frac{1}{c_n} [a_n(t)\mathbf{u}_n(\mathbf{r}) + a_n^\dagger(t)\mathbf{u}_n^*(\mathbf{r})]. \tag{3.106}$$

The mode expansion of the $\mathbf{D}$ field can be obtained from this equation by taking the curl of both sides. Similar reasoning gives

$$\dot{\mathbf{\Lambda}}(\mathbf{r}, t) = -i\sqrt{\frac{\hbar}{2\mu}} \sum_n c_n [a_n(t)\mathbf{u}_n(\mathbf{r}) - a_n^\dagger(t)\mathbf{u}_n^*(\mathbf{r})], \tag{3.107}$$

which immediately yields the mode expansion for the $\mathbf{B}$ field. These expressions can be substituted into $H = \int d^3\mathbf{r}\, \mathcal{H}$, where $\mathcal{H}$ is given by Eq. (3.80), which will give us the Hamiltonian in terms of the mode creation and annihilation operators.

So far, we have not said anything about the ordering of the operators in our nonlinear Hamiltonians. In the linear case, differences in the order of the operators lead to Hamiltonians that differ by an overall constant, which has no effect on the dynamics of the system. In the nonlinear case, this is no longer true, so we are faced with the question of what the proper ordering should be. Within the macroscopic theory, there does not seem to be a good answer to this question, and it has to be resolved by experiment.

What we can say is that, because the nonlinearities are small, the effects of different operator orderings in $\chi^{(2)}$ and $\chi^{(3)}$ terms will lead to small changes in the linear part of the Hamiltonian, whose effects will be, presumably, small. Consequently, we shall assume that the operators in these Hamiltonians are normally ordered, that is, with the annihilation operators to the right and the creation operators to the left. This leads to energy contributions that vanish in the vacuum state.

As we shall see, if we derive an effective macroscopic theory from a microscopic model, the operator ordering emerges in a natural way.

3.6 Dispersion

A realistic description of the propagation of fields in a nonlinear medium must include the effects of linear dispersion. The effects of nonlinear dispersion are small and can be neglected to lowest order. Dispersion, however, is difficult to incorporate into the standard canonical formulation, because it is an effect that is nonlocal in time. It arises from the fact that the polarization of the medium at time t, $\mathbf{P}(t)$, depends not only on the electric field at time t, but also on its values at previous times:

$$\mathbf{P}(t) = \epsilon_0 \int_0^{\infty} d\tau\, \tilde{\chi}^{(1)}(\tau) : \mathbf{E}(t-\tau). \tag{3.108}$$

There are two known approaches to constructing a quantized theory for nonlinear media that incorporates dispersion. The first, or macroscopic theory, has as its basic objects narrow-band fields for which it is possible to derive an approximate theory that is local. In the second, the degrees of freedom of the medium are included in the theory, and the entire theory, fields plus medium, is local. Each approach has its advantages. The first is more phenomenological, but needs only a set of measurable parameters, the polarizabilities, to describe the medium. Both are useful, and we shall consider each of them.

A useful feature of the first method is that we are not obliged to solve for every distinct atomic crystal structure or impurity. Only the measured dielectric parameters, which are readily available, are needed. Thus, a range of nonlinear optical experiments, all with the same type of nonlinearity, can be treated in a unified way. The second is certainly more fundamental. However, it requires a model for the medium, which will be complicated for many systems of interest. As an example, a silica fiber is amorphous, and so does not even have a unit cell to allow a simplified microscopic treatment using a periodic lattice.

In the case of linear theories, a number of other methods of quantizing fields in the presence of dispersive media have been developed. One, due to Matloob and Loudon, starts with the equations of motion for the field operators and then introduces frequency dependence into the susceptibilities and noise currents. A second, pioneered by Gruner and Welsch, starts from the results of a microscopic model and generalizes them to be able to treat arbitrary frequency-dependent susceptibilities. A third, developed by Tip, defines a Lagrangian and Hamiltonian for fields in a medium with arbitrary frequency response by introducing auxiliary fields. This theory can then be quantized in the usual way. These methods have not yet been used to treat nonlinear media. References for all of these approaches appear in the additional reading.

Let us first look at the approximate macroscopic theory. In this section we will give a technique that leads to a second-order Lagrangian of a familiar type. Later, in Chapter 11, we will give a more extensive account of this method, which is not restricted to narrow-band fields. The basic field in our theory is the dual potential. It is simplest to start by considering a linear, dispersive medium, in which case the electric field is related to the dual potential by

$$E_i(\mathbf{r},t) = \int_0^\infty d\tau\, \tilde{\eta}_{ij}(\mathbf{r},\tau) D_j(\mathbf{r}, t-\tau), \tag{3.109}$$

where $\mathbf{D} = \nabla \times \mathbf{\Lambda}$ (note that repeated indices are summed over). This looks different from the conventional approach of Eq. (3.108), but we show in Chapter 4 that the polarization response in a microscopic dielectric model is actually driven by the displacement field. Since the electric field can be obtained from a combination of polarization and displacement field terms, this form is actually the correct microscopic description. The approximation involved is that the same form is assumed to hold macroscopically.

As is evident from this equation, η is in general a tensor, but we shall assume, for the sake of simplicity, that the medium is isotropic, which implies that η is a scalar. The relation between $\mathbf{E}$ and $\mathbf{D}$ and Maxwell's equations imply that $\mathbf{\Lambda}(\mathbf{r},t)$ satisfies the equation

$$\nabla \times \int_0^\infty d\tau\, \tilde{\eta}(\mathbf{r},\tau)[\nabla \times \mathbf{\Lambda}(\mathbf{r}, t-\tau)] = -\mu \ddot{\mathbf{\Lambda}}(\mathbf{r},t), \tag{3.110}$$

which is clearly nonlocal in time. Now suppose that $\mathbf{\Lambda}^\nu$ is a narrow-band field with frequency components near ω_ν (that is, $\mathbf{\Lambda}^\nu \sim e^{-i\omega_\nu t}$), and that $\mathbf{\Lambda}$ can be expressed as

$$\mathbf{\Lambda} = \mathbf{\Lambda}^\nu + \mathbf{\Lambda}^{-\nu}, \tag{3.111}$$

where $\mathbf{\Lambda}^{-\nu} = (\mathbf{\Lambda}^\nu)^*$. The field $\mathbf{\Lambda}^\nu$ will also satisfy Eq. (3.110). Let us now define a frequency-dependent inverse permittivity as

$$\eta(\mathbf{r},\omega) = \int_0^\infty d\tau\, e^{i\omega\tau} \tilde{\eta}(\mathbf{r},\tau), \tag{3.112}$$

where, on inverting Eq. (1.43) in the tensor case, we see that $\eta(\mathbf{r},\omega) = \boldsymbol{\epsilon}^{-1}(\mathbf{r},\omega)$, where $\boldsymbol{\epsilon}(\mathbf{r},\omega)$ is the frequency-dependent dielectric tensor for the dispersive medium. For a nonabsorbing medium, $\boldsymbol{\epsilon}(\mathbf{r},\omega)$ is real and $\boldsymbol{\epsilon}(\mathbf{r},-\omega) = \boldsymbol{\epsilon}(\mathbf{r},\omega)$. However, this can generally only hold over a restricted frequency range. The Kramers–Kronig relationship of Eq. (1.45), which is a result of causality, means that generally the occurrence of dispersion is associated with absorption. However, they may occur at different frequencies, which is what we will assume here.

Because we are interested in frequencies near ω_ν, we expand the inverse permittivity up to second order in $\omega - \omega_\nu$,

$$\eta(\mathbf{r},\omega) \cong \eta_\nu(\mathbf{r}) + \omega\eta'_\nu(\mathbf{r}) + \tfrac{1}{2}\omega^2\eta''_\nu(\mathbf{r}), \tag{3.113}$$

where

$$\begin{aligned} \eta_\nu(\mathbf{r}) &\equiv \eta(\mathbf{r},\omega_\nu) - \omega_\nu\frac{d\eta}{d\omega}(\mathbf{r},\omega_\nu) + \frac{1}{2}\omega_\nu^2\frac{d^2\eta}{d\omega^2}(\mathbf{r},\omega_\nu), \\ \eta'_\nu(\mathbf{r}) &\equiv \frac{d\eta}{d\omega}(\mathbf{r},\omega_\nu) - \omega_\nu\frac{d^2\eta}{d\omega^2}(\mathbf{r},\omega_\nu), \\ \eta''_\nu(\mathbf{r}) &\equiv \frac{d^2\eta}{d\omega^2}(\mathbf{r},\omega_\nu). \end{aligned} \tag{3.114}$$

We now consider the wave equation that $\mathbf{\Lambda}^\nu$ satisfies, Eq. (3.110), and make the following approximation. The quantity $e^{-i\omega_\nu\tau}\mathbf{\Lambda}^\nu(t-\tau)$ is a slowly varying function of τ, so we expand it in a Taylor series in τ up to second order. Taylor expansions are often used in this way in classical dispersion theory to simplify wave equations, and this technique was first introduced into macroscopic field quantization by Kennedy and Wright (see the additional reading). On doing so, we find that the integral becomes (we shall not explicitly indicate the $\mathbf{r}$ dependence of $\mathbf{\Lambda}^\nu$ and its time derivatives)

$$\begin{aligned} \mathcal{I} &= \int_0^\infty d\tau\,\eta(\tau) : e^{i\omega_\nu\tau}[e^{-i\omega_\nu\tau}\nabla\times\mathbf{\Lambda}^\nu(t-\tau)] \\ &\cong \int_0^\infty d\tau\,\eta(\tau) : e^{i\omega_\nu\tau}\nabla\times\{\mathbf{\Lambda}^\nu(t) - \tau[\dot{\mathbf{\Lambda}}^\nu(t) + i\omega_\nu\mathbf{\Lambda}^\nu(t)] \\ &\quad + \tfrac{1}{2}\tau^2[\ddot{\mathbf{\Lambda}}^\nu(t) + 2i\omega_\nu\dot{\mathbf{\Lambda}}^\nu(t) - \omega_\nu^2\mathbf{\Lambda}^\nu(t)]\} \\ &\cong \eta_\nu : \nabla\times\mathbf{\Lambda}^\nu(t) + i\eta'_\nu : \nabla\times\dot{\mathbf{\Lambda}}^\nu(t) - \tfrac{1}{2}\eta''_\nu : \nabla\times\ddot{\mathbf{\Lambda}}^\nu(t). \end{aligned} \tag{3.115}$$

The last line is obtained using Eq. (3.114). Substituting this expansion into the wave equation, Eq. (3.110), we find

$$-\ddot{\mathbf{\Lambda}}^\nu = \nabla\times[\eta_\nu : \nabla\times\mathbf{\Lambda}^\nu + i\eta'_\nu : \nabla\times\dot{\mathbf{\Lambda}}^\nu - \tfrac{1}{2}\eta''_\nu : \nabla\times\ddot{\mathbf{\Lambda}}^\nu], \tag{3.116}$$

which is a local equation for $\mathbf{\Lambda}^\nu$.

3.6.1 Dispersive Lagrangian

Next, we wish to show that these local equations can, in turn, be derived from a local Lagrangian. The Lagrangian density from which it follows is

$$\begin{aligned} \mathcal{L} = \tfrac{1}{2}[&2\mu(\dot{\mathbf{\Lambda}}^{-\nu})\cdot\dot{\mathbf{\Lambda}}^\nu - 2(\nabla\times\mathbf{\Lambda}^{-\nu})\cdot\eta_\nu : (\nabla\times\mathbf{\Lambda}^\nu) - i(\nabla\times\mathbf{\Lambda}^{-\nu})\cdot\eta'_\nu : (\nabla\times\dot{\mathbf{\Lambda}}^\nu) \\ &+ i(\nabla\times\mathbf{\Lambda}^\nu)\cdot\eta'_\nu : (\nabla\times\dot{\mathbf{\Lambda}}^{-\nu}) - (\nabla\times\dot{\mathbf{\Lambda}}^{-\nu})\cdot\eta''_\nu : (\nabla\times\dot{\mathbf{\Lambda}}^\nu)]. \end{aligned} \tag{3.117}$$

The coordinates in this Lagrangian density are $\mathbf{\Lambda}^{\nu}$ and $\mathbf{\Lambda}^{-\nu}$. The equation of motion for $\mathbf{\Lambda}^{\nu}$, which follows from the condition $\delta \int dt \int d^3\mathbf{r}\, \mathcal{L} = 0$, is

$$0 = \frac{\partial \mathcal{L}}{\partial \Lambda_j^{\nu}} - \sum_{k=1}^{3} \partial_k \left(\frac{\partial \mathcal{L}}{\partial(\partial_k \Lambda_j^{\nu})} \right) - \partial_t \left(\frac{\partial \mathcal{L}}{\partial(\partial_t \Lambda_j^{\nu})} \right) + \sum_{k=1}^{3} \partial_t \partial_k \left(\frac{\partial \mathcal{L}}{\partial(\partial_t \partial_k \Lambda_j^{\nu})} \right). \tag{3.118}$$

Insertion of the above Lagrangian density into this equation gives Eq. (3.116). The canonical momentum, $\mathbf{\Pi}^{\nu}$, is given by

$$\Pi_j^{\nu} = \frac{\delta L}{\delta \dot{\Lambda}_j^{\nu}} = \frac{\partial \mathcal{L}}{\partial \dot{\Lambda}_j^{\nu}} - \sum_{k=1}^{3} \partial_k \left(\frac{\partial \mathcal{L}}{\partial(\partial_k \dot{\Lambda}_j^{\nu})} \right), \tag{3.119}$$

which gives us that

$$\mathbf{\Pi}^{\nu} = \mu \dot{\mathbf{\Lambda}}^{-\nu} - \tfrac{1}{2} \nabla \times [\eta_{\nu}''(\nabla \times \dot{\mathbf{\Lambda}}^{-\nu}) + i\eta_{\nu}'(\nabla \times \mathbf{\Lambda}^{-\nu})]. \tag{3.120}$$

Finally, from the Lagrangian and the canonical momentum, we can find the Hamiltonian density,

$$\mathcal{H} = \mathbf{\Pi}^{\nu} \cdot \dot{\mathbf{\Lambda}}^{\nu} + \mathbf{\Pi}^{-\nu} \cdot \dot{\mathbf{\Lambda}}^{-\nu} - \mathcal{L}. \tag{3.121}$$

The Hamiltonian is then found by integrating the Hamiltonian density over the quantization volume. It is possible to simplify the Hamiltonian density by integrating some of the terms by parts and assuming that the boundary terms vanish. In particular, it is useful to make use of the identity

$$\int d^3\mathbf{r}\, \mathbf{V}_1 \cdot \nabla \times \mathbf{V}_2 = \int d^3\mathbf{r}\, \mathbf{V}_2 \cdot \nabla \times \mathbf{V}_1. \tag{3.122}$$

This allows us to combine terms in the Hamiltonian density, and the resulting Hamiltonian is

$$H_0 = \int d^3\mathbf{r} [\mu \dot{\mathbf{\Lambda}}^{-\nu} \cdot \dot{\mathbf{\Lambda}}^{\nu} + (\nabla \times \mathbf{\Lambda}^{-\nu}) \cdot \boldsymbol{\eta}_{\nu} : (\nabla \times \mathbf{\Lambda}^{\nu}) - \tfrac{1}{2}(\nabla \times \dot{\mathbf{\Lambda}}^{-\nu}) \cdot \boldsymbol{\eta}_{\nu}'' : (\nabla \times \dot{\mathbf{\Lambda}}^{\nu})], \tag{3.123}$$

where $\dot{\mathbf{\Lambda}}^{\nu}$ is to be considered a function of $\mathbf{\Pi}^{\nu}$. This Hamiltonian is the energy for a classical field in a linear dispersive dielectric. This is an important criterion, since otherwise the Lagrangian for the theory is not unique. It can, for example, be scaled by an arbitrary factor and the equations of motion will be unaffected. Hence, the scaling of the Lagrangian should be chosen so that it does give a Hamiltonian that is the classical energy.

If the dispersion in the nonlinear interaction is ignored, so that the interaction is considered local, then nonlinear polarizability can be included in the theory by adding the terms

$$H^{NL} = \int d^3\mathbf{r} [\tfrac{1}{3}\mathbf{D} \cdot \eta^{(2)} : \mathbf{D}^2 + \tfrac{1}{4}\mathbf{D} \cdot \eta^{(3)} : \mathbf{D}^3], \tag{3.124}$$

to the Hamiltonian in Eq. (3.123). In addition, if fields with several discrete frequencies are present, they can be accommodated by adding additional fields, $\mathbf{\Lambda}^{\nu}$, centered about these frequencies, to the theory. This has the effect of adding summations over ν to the Lagrangian density and Hamiltonian in Eqs (3.117) and (3.123).

3.6.2 Dispersive quantization

In order to quantize the theory, we would like to simply impose the commutation relations

$$[\Lambda_j^\nu(\mathbf{r},t),\Pi_{j'}^\nu(\mathbf{r}',t)] = i\delta_{jj'}^{(\mathrm{tr})}(\mathbf{r}-\mathbf{r}'). \tag{3.125}$$

However, in order for this to be true, the fields must have Fourier components of arbitrarily high frequency. However, $\mathbf{\Lambda}^\nu$ is limited in bandwidth, as explained in Section 3.4.2. An alternative, and in this case better, approach is to expand the field in terms of spatial modes and to use the expansion coefficients as coordinates. One then finds the corresponding canonical momentum for each coordinate and imposes the usual commutation relations between coordinates and momenta.

To begin, let us expand $\mathbf{\Lambda}^\nu(\mathbf{r},t)$ in plane-wave modes,

$$\mathbf{\Lambda}^\nu(\mathbf{r},t) = \frac{1}{\sqrt{V}}\sum_{\mathbf{k},\alpha}\lambda_{\mathbf{k},\alpha}^\nu \hat{\mathbf{e}}_{\mathbf{k},\alpha} e^{i\mathbf{k}\cdot\mathbf{r}}, \tag{3.126}$$

where the $\lambda_{\mathbf{k},\alpha}^\nu$ will become our quantized coordinates. The facts that $\mathbf{\Lambda}^{-\nu} = (\mathbf{\Lambda}^\nu)^*$ and that $\hat{\mathbf{e}}_{-\mathbf{k},\alpha} = -(-1)^\alpha \hat{\mathbf{e}}_{\mathbf{k},\alpha}$ imply that

$$\lambda_{\mathbf{k},\alpha}^\nu = -(-1)^\alpha(\lambda_{-\mathbf{k},\alpha}^{-\nu})^*. \tag{3.127}$$

The plane-wave expansion can be inserted into the Lagrangian, yielding

$$\begin{aligned} L = \sum_{\mathbf{k},\alpha}\sum_{\mathbf{k}',\alpha'}\big[&\dot{\lambda}_{\mathbf{k}',\alpha'}^{-\nu}\dot{\lambda}_{\mathbf{k},\alpha}^{\nu} M_{(\mathbf{k}',\alpha'),(\mathbf{k},\alpha)}^{(1)} + \lambda_{\mathbf{k}',\alpha'}^{-\nu}\lambda_{\mathbf{k},\alpha}^{\nu} M_{(\mathbf{k}',\alpha'),(\mathbf{k},\alpha)}^{(2)} \\ &+ \lambda_{\mathbf{k}',\alpha'}^{-\nu}\dot{\lambda}_{\mathbf{k},\alpha}^{\nu} M_{(\mathbf{k}',\alpha'),(\mathbf{k},\alpha)}^{(3)} - \dot{\lambda}_{\mathbf{k}',\alpha'}^{\nu}\lambda_{\mathbf{k},\alpha}^{\nu} M_{(\mathbf{k}',\alpha'),(\mathbf{k},\alpha)}^{(3)}\big]. \end{aligned} \tag{3.128}$$

The matrices $M_{(\mathbf{k}',\alpha'),(\mathbf{k},\alpha)}^{(j)}$, $j = 1, 2, 3$ are given by

$$\begin{aligned} M_{(\mathbf{k}',\alpha'),(\mathbf{k},\alpha)}^{(1)} &= -\mu_0(-1)^\alpha \delta_{\mathbf{k}',-\mathbf{k}}\delta_{\alpha',\alpha} \\ &\quad + \frac{1}{2V}\int d^3\mathbf{r}\, e^{i(\mathbf{k}+\mathbf{k}')\cdot\mathbf{r}}(\mathbf{k}'\times\hat{\mathbf{e}}_{\mathbf{k}',\alpha'})\cdot\eta_\nu'' : (\mathbf{k}\times\hat{\mathbf{e}}_{\mathbf{k},\alpha}), \\ M_{(\mathbf{k}',\alpha'),(\mathbf{k},\alpha)}^{(2)} &= +\frac{1}{V}\int d^3\mathbf{r}\, e^{i(\mathbf{k}+\mathbf{k}')\cdot\mathbf{r}}(\mathbf{k}'\times\hat{\mathbf{e}}_{\mathbf{k}',\alpha'})\cdot\eta_\nu : (\mathbf{k}\times\hat{\mathbf{e}}_{\mathbf{k},\alpha}), \\ M_{(\mathbf{k}',\alpha'),(\mathbf{k},\alpha)}^{(3)} &= +\frac{i}{2V}\int d^3\mathbf{r}\, e^{i(\mathbf{k}+\mathbf{k}')\cdot\mathbf{r}}(\mathbf{k}'\times\hat{\mathbf{e}}_{\mathbf{k}',\alpha'})\cdot\eta_\nu' : (\mathbf{k}\times\hat{\mathbf{e}}_{\mathbf{k},\alpha}). \end{aligned} \tag{3.129}$$

Our next step is to find the canonical momenta. These are given by

$$\pi_{\mathbf{k},\alpha}^\nu = \frac{\partial L}{\partial\dot{\lambda}_{\mathbf{k},\alpha}^\nu} = \sum_{\mathbf{k}',\alpha'}\big[\dot{\lambda}_{\mathbf{k}',\alpha'}^{-\nu} M_{(\mathbf{k}',\alpha'),(\mathbf{k},\alpha)}^{(1)} + \lambda_{\mathbf{k}',\alpha'}^{-\nu} M_{(\mathbf{k}',\alpha'),(\mathbf{k},\alpha)}^{(3)}\big] \tag{3.130}$$

and

$$\pi_{\mathbf{k}',\alpha'}^{-\nu} = \frac{\partial L}{\partial\dot{\lambda}_{\mathbf{k}',\alpha'}^{-\nu}} = \sum_{\mathbf{k},\alpha}\big[\dot{\lambda}_{\mathbf{k},\alpha}^{\nu} M_{(\mathbf{k}',\alpha'),(\mathbf{k},\alpha)}^{(1)} - \lambda_{\mathbf{k},\alpha}^{\nu} M_{(\mathbf{k}',\alpha'),(\mathbf{k},\alpha)}^{(3)}\big]. \tag{3.131}$$

We can now find the Hamiltonian

$$H = \sum_{\mathbf{k},\alpha} \left(\pi^{\nu}_{\mathbf{k},\alpha}\dot{\lambda}^{\nu}_{\mathbf{k},\alpha} + \pi^{-\nu}_{\mathbf{k},\alpha}\dot{\lambda}^{-\nu}_{\mathbf{k},\alpha}\right) - L$$
$$= \sum_{\mathbf{k},\alpha}\sum_{\mathbf{k}',\alpha'} \left[\dot{\lambda}^{-\nu}_{\mathbf{k}',\alpha'}\dot{\lambda}^{\nu}_{\mathbf{k},\alpha} M^{(1)}_{(\mathbf{k}',\alpha'),(\mathbf{k},\alpha)} - \lambda^{-\nu}_{\mathbf{k}',\alpha'}\lambda^{\nu}_{\mathbf{k},\alpha} M^{(2)}_{(\mathbf{k}',\alpha'),(\mathbf{k},\alpha)}\right]. \tag{3.132}$$

The Hamiltonian should be expressed in terms of the coordinates and the canonical momenta, and in order to do so Eqs (3.130) and (3.131) must be inverted to find expressions for $\dot{\lambda}^{\nu}_{\mathbf{k},\alpha}$ and $\dot{\lambda}^{-\nu}_{\mathbf{k},\alpha}$ in terms of the canonical momenta and the coordinates.

3.7 One-dimensional waveguide

We shall now consider a simplified version of this theory, applicable to a simple model of a one-dimensional waveguide or optical fiber. We shall suppose that the field consists of plane waves that are polarized in the y direction and propagate in the x direction. In this case the field, $\Lambda^{\nu}(x,t)$, is a scalar and a function of only one spatial coordinate. This means that $\nabla \times \mathbf{\Lambda}^{\nu}$ becomes $\partial_x \Lambda^{\nu}\hat{\mathbf{z}}$ and integrals over the quantization volume are replaced by $A\int_0^l dx$, where the quantization volume is $V = lA$, and A is the transverse area. We note that, more generally, a detailed transverse mode function is needed to define an effective transverse area, and part of the dispersion is due to the modal dispersion properties of the transverse modes. This will be treated in detail in Chapter 11.

If, in addition, we assume that the only nonlinearity present is described by $\eta^{(3)}$ and that the medium is homogeneous, we find that the Hamiltonian is

$$H = A\int dx\,\{\mu\dot{\Lambda}^{\nu}\dot{\Lambda}^{-\nu} + \eta_{\nu}(\partial_x\Lambda^{\nu})(\partial_x\Lambda^{-\nu}) - \tfrac{1}{2}\eta''_{\nu}(\partial_x\dot{\Lambda}^{\nu})(\partial_x\dot{\Lambda}^{-\nu})$$
$$+ \tfrac{1}{4}\eta^{(3)}[\partial_x(\Lambda^{\nu} + (\Lambda^{\nu})^*)]^4\}. \tag{3.133}$$

The field $\Lambda(x,t)$ is now given by

$$\Lambda(x,t) = \frac{1}{\sqrt{V}}\sum_k \left(\lambda^{\nu}_k e^{ikx} + \lambda^{-\nu}_{-k}e^{-ikx}\right). \tag{3.134}$$

Note that, because the nonlinear term does not depend on the time derivative of $\mathbf{\Lambda}(x,t)$, its addition to the theory does not change the canonical momenta, i.e. the canonical momenta for the theory without the nonlinear term are the same as the canonical momenta for the theory with the nonlinear term.

What we would like to do is to express the Hamiltonian in terms of λ^{ν}_k and $\lambda^{-\nu}_k$ and their corresponding momenta. Note that we have suppressed the polarization subscript, because it is not necessary for the simplified theory we are considering (there is only one polarization present). For plane-wave modes propagating in the x direction in a homogeneous medium, and assuming the polarization vectors satisfy $\hat{\mathbf{e}}_k = \hat{\mathbf{e}}_{-k}$, we have that

$$\begin{aligned} M^{(1)}_{k,k'} &= \left(\mu - \tfrac{1}{2}\eta''_\nu k^2\right)\delta_{k,-k'}, \\ M^{(2)}_{k,k'} &= -k^2\eta_\nu\delta_{k,-k'}, \\ M^{(3)}_{k,k'} &= -\frac{i}{2}k^2\eta'_\nu\delta_{k,-k'}. \end{aligned} \tag{3.135}$$

From this, for the linear part of the Lagrangian, L_0, we obtain

$$L_0 = \sum_k \left[\dot{\lambda}^{-\nu}_{-k}\dot{\lambda}^\nu_k\left(\mu - \tfrac{1}{2}\eta''_\nu k^2\right) - \lambda^{-\nu}_{-k}\lambda^\nu_k k^2\eta_\nu - \tfrac{1}{2}ik^2\eta'_\nu\left(\lambda^{-\nu}_{-k}\dot{\lambda}^\nu_k - \dot{\lambda}^{-\nu}_{-k}\lambda^\nu_k\right)\right], \tag{3.136}$$

and the canonical momenta

$$\begin{aligned} \pi^\nu_k &= \dot{\lambda}^{-\nu}_{-k}\left(\mu - \tfrac{1}{2}\eta''_\nu k^2\right) - \tfrac{1}{2}ik^2\eta'_\nu\lambda^{-\nu}_{-k}, \\ \pi^{-\nu}_{-k} &= \dot{\lambda}^\nu_k\left(\mu - \tfrac{1}{2}\eta''_\nu k^2\right) + \tfrac{1}{2}ik^2\eta'_\nu\lambda^\nu_k. \end{aligned} \tag{3.137}$$

The linear part of the Hamiltonian, H_0, is now given by

$$\begin{aligned} H_0 &= \sum_k \left[\dot{\lambda}^{-\nu}_{-k}\dot{\lambda}^\nu_k\left(\mu - \tfrac{1}{2}\eta''_\nu k^2\right) + \lambda^{-\nu}_{-k}\lambda^\nu_k k^2\eta_\nu\right] \\ &= \sum_k \left[\frac{1}{\mu - \eta''_\nu k^2/2}\left(\pi^{-\nu}_{-k} - \frac{i}{2}k^2\eta'_\nu\lambda^\nu_k\right)\left(\pi^\nu_k + \frac{i}{2}k^2\eta'_\nu\lambda^{-\nu}_{-k}\right) + \lambda^{-\nu}_{-k}\lambda^\nu_k k^2\eta_\nu\right]. \end{aligned} \tag{3.138}$$

Our next step is to quantize the theory. We promote λ^ν_k and π^ν_k to operators and impose the canonical commutation relations

$$[\lambda^\nu_k, \pi^\nu_{k'}] = i\hbar\delta_{k,k'}, \qquad [\lambda^{-\nu}_k, \pi^{-\nu}_{k'}] = i\hbar\delta_{k,k'}. \tag{3.139}$$

Note that, because $\Lambda = \Lambda^\nu + \Lambda^{-\nu}$ is Hermitian, we must have $\lambda^{-\nu}_{-k} = (\lambda^\nu_k)^\dagger$, and, because $\dot{\Lambda}$ is Hermitian, we have that $\pi^{-\nu}_{-k} = (\pi^\nu_k)^\dagger$. We can now define two sets of creation and annihilation operators,

$$\begin{aligned} a_k &= \frac{1}{\sqrt{2\hbar}}\left[A_k\lambda^\nu_k + \left(\frac{i}{A^*_k}\right)(\pi^\nu_k)^\dagger\right], \\ b^\dagger_k &= \frac{1}{\sqrt{2\hbar}}\left[A_k\lambda^\nu_k - \left(\frac{i}{A^*_k}\right)(\pi^\nu_k)^\dagger\right], \end{aligned} \tag{3.140}$$

where A_k is a c-number yet to be determined. These relations can be inverted to give

$$\begin{aligned} \lambda^\nu_k &= \frac{1}{A_k}\sqrt{\frac{\hbar}{2}}(a_k + b^\dagger_k), \\ \pi^\nu_k &= iA_k\sqrt{\frac{\hbar}{2}}(a^\dagger_k - b_k), \end{aligned} \tag{3.141}$$

and the results can be substituted into the Hamiltonian. If A_k is chosen to be

$$A_k = \left[\tfrac{1}{4}k^4(\eta'_\nu)^2 + \left(\mu - \tfrac{1}{2}\eta''_\nu k^2\right)k^2\eta_\nu\right]^{1/4}, \tag{3.142}$$

then the terms proportional to $a_k b_k$ and $a_k^\dagger b_k^\dagger$ vanish, with the result that the final Hamiltonian is

$$H_0 = \hbar \sum_k \left[\omega_+(k) a_k^\dagger a_k + \omega_-(k) b_k^\dagger b_k\right], \tag{3.143}$$

where an overall constant has been dropped. The frequencies $\omega_\pm(k)$ are given by

$$\omega_\pm(k) = \frac{1}{\mu_0 - \eta_\nu'' k^2/2}\left\{ \pm \tfrac{1}{2}k^2 \eta_{nu}' + \left[\tfrac{1}{4}k^4 (\eta_\nu')^2 + (\mu - \tfrac{1}{2}\eta_\nu'' k^2) k^2 \eta_\nu\right]^{1/2}\right\} \tag{3.144}$$

and are solutions of the equation

$$\omega_\pm^2 = k^2 \left(\eta_\nu \pm \omega_\pm \eta_\nu' + \tfrac{1}{2}\omega_\pm^2 \eta_\nu''\right). \tag{3.145}$$

Note that the expression in parentheses in the above equation is identical to the expansion of $\eta(\omega)$ about ω_ν if the plus sign is used. Because $\eta(\omega) = 1/\varepsilon(\omega)$, where $\varepsilon(\omega)$ is the usual frequency-dependent dielectric function, Eq. (3.145) for ω_+ is approximately the same as

$$\omega = \frac{k}{\sqrt{\varepsilon(\omega)}}, \tag{3.146}$$

which is the relation between ω and k that we would expect for a wave traveling in a linear dielectric medium. This leaves us with the question of how to interpret ω_- and b_k.

An examination of Eq. (3.140) shows us that

$$\lambda_k = \frac{1}{A_k\sqrt{2}}(a_k + b_k^\dagger), \tag{3.147}$$

while Eq. (3.143) implies that $b_k \sim e^{-i\omega_- t}$. Now ω_- is not too far from ω_ν, which implies that the $b_k^\dagger$ term in λ_k has a time dependence given approximately by $e^{i\omega_\nu t}$. This places it outside the bandwidth for the field Λ^ν. A similar phenomenon of a second band is found in a detailed microscopic approach, in Chapter 4. We will find it is related to the properties of the oscillators causing the dispersion. Clearly, this will not be treated correctly in the simple model of this chapter, as it requires a detailed microscopic theory.

In order to be consistent, we must assume that all of the b_k modes are in the vacuum state and, thereby, drop these operators from the theory. This implies that the Hamiltonian for the full theory (as opposed to just the linear part) is

$$H = \sum_k \hbar\omega_+(k) a_k^\dagger a_k + \frac{1}{4}\eta^{(3)} A \int dx\, (\partial_x(\Lambda^\nu + \Lambda^{\nu\dagger}))^4, \tag{3.148}$$

where

$$\Lambda = \sqrt{\frac{\hbar}{2V}} \sum_k \frac{1}{A_k}(a_k e^{ikx} + a_k^\dagger e^{-ikx}). \tag{3.149}$$

We can express the quantity A_k in terms of the group velocity, $v_k = d\omega_+(k)/dk$. As we saw, $\omega_+(k)$ is a solution of the equation $k^2\eta(\omega_+(k)) = \mu_0\omega_+^2(k)$, where $\eta^{(1)}$ is given by Eq. (3.113). Differentiating both sides with respect to ω_+ gives

$$k\eta(\omega_+(k)) = \left(\mu\omega_+(k) - \frac{1}{2}k^2 \frac{d\eta^{(1)}}{d\omega_+}\right)\frac{d\omega_+(k)}{dk}. \tag{3.150}$$

Now, making use of Eqs (3.144) and (3.142), we find that

$$\mu_0\omega_+ - \frac{1}{2}k^2\frac{d\eta}{d\omega_+} = \omega_+\left(\mu - \frac{1}{2}k^2\eta_v''\right) - \frac{1}{2}k^2\eta_v' = A_k^2. \tag{3.151}$$

Combining these results, then, gives us

$$A_k = \sqrt{\frac{k\eta(\omega_+(k))}{v_k}}. \tag{3.152}$$

Finally, we have for the fields

$$\begin{aligned}\mathbf{A}(x,t) &= \sum_k \left[\frac{\hbar\varepsilon(\omega_+(k))v_k}{2Vk}\right]^{1/2}(a_k e^{ikx} + a_k^\dagger e^{-ikx})\hat{\mathbf{y}},\\ \mathbf{D}(x,t) &= \sum_k i\left[\frac{\hbar k\varepsilon(\omega_+(k))v_k}{2V}\right]^{1/2}(a_k e^{ikx} - a_k^\dagger e^{-ikx})\hat{\mathbf{z}}.\end{aligned} \tag{3.153}$$

We note that these are 'photon–polariton' excitations, with a dispersion relation corresponding to a photon propagating in a dielectric at group velocities v_k near the speed of light. The dispersive model also predicts 'phonon–polariton' excitation propagating at different (usually much slower) velocities, although, as discussed above, these will not generally be treated correctly in the absence of detailed structural information. This type of band structure is also found in more realistic microscopic models. A microscopic theory of such polariton bands is given in a simplified form in Chapter 4. However, even this approach leaves out much of the true structure of real dielectrics. In particular, the model we treat later omits the complex electronic and chemical features of real solids.

The advantage of the simplified macroscopic quantization approach presented here is that it is a minimal quantization method. This theory only requires knowledge of the phenomenological dispersion relation. The other extreme – a detailed microscopic treatment of every atom in a dielectric – is scarcely feasible or even desirable. Such a first-principles analysis is made complicated by the huge range of known dielectrics that are involved in optical science. It would clearly become necessary to treat every different chemical composition as a special case. Even worse, the fact that glasses are intrinsically amorphous means that every individual material sample could potentially require analysis, due to microscopic differences.

3.7.1 Photon–polariton field

We now have an effective theory that is capable of describing the propagation of quantum fields in nonlinear, dispersive media. It has been used to describe fields propagating through a fiber with a $\chi^{(3)}$ nonlinearity, and we shall now give the theory in the form in which it can be used for this application.

We shall assume that the field has wavenumber components near k_0. First, we define the slowly varying quantum photon–polariton field

$$\psi(x) = \frac{1}{\sqrt{L}} \sum_{|k-k_1|<k_m} e^{i(k-k_0)x} a_k, \tag{3.154}$$

with frequencies close to $\omega = \omega(k_0)$. This field has equal-time commutation relations given by

$$[\psi(x,t), \psi^\dagger(x',t)] = \tilde{\delta}_{k_m}(x - x'), \tag{3.155}$$

where $\tilde{\delta}$, because it has a band-limited Fourier transform, is a smoothed version of the Dirac delta-function:

$$\tilde{\delta}_{k_m}(x - x') \equiv \frac{1}{L} \sum_{|k|<k_m} e^{ik(x-x')}. \tag{3.156}$$

We note that, as explained in Section 3.4.2, there is an implicit momentum cutoff in *any* macroscopic quantization procedure.

We shall assume, however, that this smearing effect is not pronounced, so that we can treat $\tilde{\delta}(x - x')$ as a delta-function for all practical purposes. Such an assumption may not be valid for very short pulse durations or wavelengths less than the lattice spacing.

Inverting the relation between $\psi(x)$ and a_k, we find that

$$a_k = \frac{1}{\sqrt{L}} \int dx\, e^{-i(k-k_0)x} \psi(x). \tag{3.157}$$

We can use this equation to express the Hamiltonian, Eq. (3.148), in terms of $\psi(x)$. First, note that

$$\sum_k \omega(k) e^{i(k-k_0)(x-x')} = \frac{1}{L} \int dx \int dx' \left(\sum_k \omega(k) e^{i(k-k_0)(x-x')} \right) \psi(x,t)^\dagger \psi(x',t). \tag{3.158}$$

Expanding $\omega(k)$ around k_0, we have that

$$\omega(k) = \omega + (k - k_0)v + \tfrac{1}{2}(k - k_1)^2 \omega'' + \cdots. \tag{3.159}$$

Here, $v = \omega'$ is the group velocity at k_0 and ω'' is the second derivative of $\omega(k)$ evaluated at k_0. This then gives us

$$\begin{aligned}\sum_k \omega(k) e^{i(k-k_0)(x-x')} &\cong \sum_k \left[\omega + \frac{i}{2} v(\partial_{x'} - \partial_x) + \frac{1}{2}\omega'' \partial_x \partial_{x'} \right] e^{i(k-k_1)(x-x')} \\ &\cong \left[\omega + \frac{i}{2} v(\partial_{x'} - \partial_x) + \frac{1}{2}\omega'' \partial_x \partial_{x'} \right] \delta(x - x'). \end{aligned} \tag{3.160}$$

Omitting arguments of x in the integrals for brevity, this gives for the linear part of the Hamiltonian, H_0,

$$H_0 = \hbar \int dx \left[\omega \psi^\dagger \psi + \frac{i}{2} v \left(\frac{\partial \psi^\dagger}{\partial x} \psi - \psi^\dagger \frac{\partial \psi}{\partial x} \right) + \frac{1}{2}\omega'' \frac{\partial \psi^\dagger}{\partial x} \frac{\partial \psi}{\partial x} \right]. \tag{3.161}$$

The nonlinear part of the Hamiltonian, H_{NL}, is given by

$$\frac{\eta^{(3)}}{4} A \int dx\, D^4 \cong \frac{\eta^{(3)}}{4} A \left[\frac{\hbar k_0 v \epsilon}{2A} \right]^2 \int dx\, [e^{ik_0 x}\psi - e^{-ik_0 x}\psi^\dagger]^4, \tag{3.162}$$

where $\eta^{(3)} = -\chi^{(3)}/\epsilon_1^3$. Keeping only the slowly varying terms, we obtain

$$H_{NL} \cong \frac{3\eta^{(3)}(\hbar k v \epsilon)^2}{8A} \int dx\, \psi^{\dagger 2} \psi^2. \tag{3.163}$$

The total Hamiltonian is, of course, just the sum of the two terms, $H = H_0 + H_{NL}$.

Additional reading

Books

W. Vogel and D.-G. Welsch, *Quantum Optics* (Wiley, Berlin, 2006).

Articles

- For quantization in linear media, see:

C. K. Carniglia and L. Mandel, *Phys. Rev.* D **3**, 280 (1971).
R. J. Glauber and M. Lewenstein, *Phys. Rev.* A **43**, 467 (1991).
J. M. Jauch and K. M. Watson, *Phys. Rev.* **74**, 950 (1948); **74**, 1485 (1948); **75**, 1249 (1948).

- For quantization in nonlinear and dispersive media, see:

M. Born and L. Infeld, *Proc. R. Soc.* A **143**, 410 (1934).
P. D. Drummond, *Phys. Rev.* A **42**, 6845 (1990).
M. Hillery and L. Mlodinow, *Phys. Rev.* A **30**, 1860 (1984).

- For a different approach to quantizing the field in linear media, see:

T. Gruner and D.-G. Welsch, *Phys. Rev.* A **51**, 3246 (1995).
R. Matloob and R. Loudon, *Phys. Rev.* A **52**, 4823 (1995).
A. Tip, *Phys. Rev.* A **57**, 4818 (1998).

- For a discussion of field quantization in nonlinear media in the paraxial approximation, see:

T. A. B. Kennedy and E. M. Wright, *Phys. Rev.* A **38**, 212 (1988).

Problems

3.1 Let us look at a toy model for field quantization in a linear homogeneous dielectric. Let the coordinate be q (this could correspond to the amplitude of a single mode) and the Lagrangian be $L = \frac{1}{2}(\epsilon \dot{q}^2 - \omega^2 q^2)$.

(a) Find the Hamiltonian for this system in terms of q and its conjugate momentum p.
(b) Find a frequency ν and an annihilation operator, a, so that $[a, a^{\dagger}] = 1$ and, up to an additive constant, $H = \hbar\nu a^{\dagger}a$.
(c) Now consider the case without the dielectric, that is $\epsilon = 1$. This Hamiltonian, H_0, can be expressed in terms of creation and annihilation operators, b and $b^{\dagger}$, so that $H_0 = \hbar\omega b^{\dagger}b$ (again, up to an additive constant). Find expressions for b and $b^{\dagger}$ in terms of a and $a^{\dagger}$.
(d) Find $\langle 0|b^{\dagger}b|0\rangle$, where $|0\rangle$ is the ground state of H.

3.2 In order to illustrate quantization in the presence of an inhomogeneous linear dielectric, let us look at a simple case. Our universe consists of the one-dimensional interval $0 \le x \le L$, and there is a dielectric with dielectric constant $\epsilon(x) = \epsilon_1$ in the interval $0 \le x \le l$, where $l < L$. In the rest of the space, there is no dielectric, so $\epsilon(x) = \epsilon_0$. The vector potential is now a scalar, $A(x,t)$, and satisfies the equation

$$\epsilon(x)\mu_0 \frac{\partial^2 A}{\partial t^2} - \frac{\partial^2 A}{\partial x^2} = 0,$$

and the boundary conditions $A(0,t) = A(L,t) = 0$. In addition, $A(x,t)$ and $\partial_x A(x,t)$ are continuous at $x = l$.
(a) Find a Lagrangian that gives the equation of motion for $A(x,t)$ above. Use it to find the canonical momentum and then the Hamiltonian for this system.
(b) We want to find normal modes for this system. Let them be given by $A_k(x,t) = u_k(x)e^{-i\omega t}$, where $\int_0^L dx\, \epsilon(x)u_k^2(x) = 1$, and k is the wavenumber of the mode in the interval $l \le x \le L$. Show from the equation obeyed by $u_k(x)$ and the boundary conditions that $\int_0^L dx\, \epsilon(x)u_k(x)u_{k'}(x) = 0$ unless $k = k'$. Find $u_k(x)$ and an equation giving the relation between ω and k.
(c) In order to quantize the system, we expand the vector potential and the canonical momentum as

$$A(x,t) = \sum_k q_k u_k(x), \qquad \Pi(x,t) = \sum_k p_k \epsilon(x) u_k(x).$$

Substitute the expressions into the Hamiltonian and thereby find the Hamiltonian as a function of the q_k and the p_k.
(d) To quantize the system, the q_k and the p_k become operators obeying $[q_k, q_{k'}] = 0$, $[p_k, p_{k'}] = 0$ and $[q_k, p_{k'}] = i\hbar\delta_{kk'}$. Find creation and annihilation operators, a_k and $a_k^{\dagger}$, so that $[a_k, a_{k'}] = \delta_{kk'}$ and the Hamiltonian is given by $H = \sum_k \hbar\omega_k a_k^{\dagger}a_k$ up to an additive constant.

3.3 Another way to set up a Lagrangian and Hamiltonian for electromagnetism is to use both the usual vector potential and the dual potential combined into a complex potential. We will just look at the case of the theory in vacuum with no charges or currents present. Define

$$\mathcal{A} = \sqrt{\frac{\mu_0}{\epsilon_0}}\mathbf{\Lambda} + i\mathbf{A}.$$

We work in a gauge so that $\nabla \cdot \mathcal{A} = 0$.

(a) Two of the Maxwell equations are automatically satisfied because of the way $\mathbf{\Lambda}$ and $\mathbf{A}$ are related to the fields. Show that the other two follow from the Lagrangian density

$$\mathcal{L} = \frac{i}{2}\sqrt{\frac{\epsilon_0}{\mu_0}}(\partial_t \boldsymbol{\mathcal{A}}) \cdot (\nabla \times \boldsymbol{\mathcal{A}}^*) - \frac{1}{2\mu_0}|\nabla \times \boldsymbol{\mathcal{A}}|^2.$$

(b) From the above Lagrangian density, find the canonical momentum and use it to find the Hamiltonian. You should get the usual Hamiltonian for the free electromagnetic field. Use constrained field quantization to obtain the commutators. Show that these are the same as in the main text.

4 Microscopic description of media

So far we have adopted a macroscopic approach to electrodynamics in media, describing the medium through its polarizabilities. In this chapter, we shall take a different point of view, and look at a microscopic description of the medium in terms of optically active electrons bound in atoms or molecules. In order to begin, we will need a Hamiltonian that describes the interaction of charged particles with the electromagnetic field. We will start with the minimal-coupling Hamiltonian and from it derive the dipole-coupling Hamiltonian, which is more convenient for our purposes.

The dipole-coupling approach was originally described in the Göttingen Ph.D. thesis of Maria Goeppert-Meyer, who used it to develop the theory of two-photon transitions – possibly the first theoretical work on nonlinear optics. In this approach, it is the displacement field that emerges as a natural canonical coordinate, rather than the electric field.

Finally, by studying a medium composed of two-level atoms, we will be able to show how an effective nonlinear Hamiltonian, similar to those resulting from the macroscopic approach, emerges from the microscopic theory.

4.1 The Coulomb gauge

Let us begin with a collection of charged particles, labeled by α, where $\alpha = 1, 2, \ldots, N$. The charge of particle α is q_α, its mass is m_α, and its location is $\mathbf{r}_\alpha$. In the minimal-coupling theory in the Coulomb gauge, the coordinates are the particle locations, $\mathbf{r}_\alpha$, and the vector potential, $\mathbf{A}(\mathbf{r})$, where we choose a gauge so that $\nabla \cdot \mathbf{A} = 0$.

4.1.1 Coulomb-gauge Lagrangian

The classical Lagrangian for charged particles interacting with an electromagnetic field is

$$\begin{aligned} L = & \sum_\alpha \frac{1}{2} m_\alpha \dot{\mathbf{r}}_\alpha^2 - \sum_{\alpha<\beta} \frac{q_\alpha q_\beta}{4\pi\epsilon_0 |\mathbf{r}_\alpha - \mathbf{r}_\beta|} \\ & + \frac{\epsilon_0}{2} \int d^3\mathbf{r} \, [(\partial_t \mathbf{A})^2 - c^2 (\nabla \times \mathbf{A})^2] + \sum_\alpha q_\alpha \dot{\mathbf{r}}_\alpha \cdot \mathbf{A}(\mathbf{r}_\alpha). \end{aligned} \tag{4.1}$$

Let us verify that this Lagrangian gives us the proper equations of motion. We will first find the equation of motion resulting from variation of the particle coordinates. We have that

$$\frac{\partial L}{\partial \dot{r}_{\alpha j}} = m_\alpha \dot{r}_{\alpha j} + q_\alpha A_j(\mathbf{r}_\alpha, t),$$

$$\frac{\partial L}{\partial r_{\alpha j}} = \sum_{\beta \neq \alpha} \frac{q_\alpha q_\beta (r_{\alpha j} - r_{\beta j})}{4\pi\epsilon_0 |\mathbf{r}_\alpha - \mathbf{r}_\beta|^3} + q_\alpha \partial_{r_{\alpha j}} (\dot{\mathbf{r}}_\alpha \cdot \mathbf{A}(\mathbf{r}_{\alpha j}, t)). \tag{4.2}$$

Using the vector identity

$$\partial_j (\mathbf{a} \cdot \mathbf{b}) = (\mathbf{a} \cdot \nabla) b_j + [\mathbf{a} \times (\nabla \times \mathbf{b})]_j, \tag{4.3}$$

where $\mathbf{a}$ is a constant vector and $\mathbf{b}(\mathbf{r})$ depends on $\mathbf{r}$, we find that

$$\partial_{r_{\alpha j}} (\dot{\mathbf{r}}_\alpha \cdot \mathbf{A}(\mathbf{r}_{\alpha j}, t)) = (\dot{\mathbf{r}}_\alpha \cdot \nabla_{r_\alpha}) A_j(\mathbf{r}_\alpha, t) + [\dot{\mathbf{r}}_\alpha \times (\nabla_{r_\alpha} \times \mathbf{A}(\mathbf{r}_\alpha, t))]_j. \tag{4.4}$$

We also obtain

$$\frac{d}{dt}\left(\frac{\partial L}{\partial \dot{r}_{\alpha j}}\right) = m_\alpha \ddot{r}_{\alpha j} + q_\alpha [\dot{A}_j(\mathbf{r}_\alpha, t) + (\dot{\mathbf{r}}_\alpha \cdot \nabla_{r_\alpha}) A_j(\mathbf{r}_\alpha, t)], \tag{4.5}$$

where $\dot{A}_j(\mathbf{r}_\alpha, t) = \partial_t A_j(\mathbf{r}_\alpha, t)$. Inserting everything into the Lagrange equation

$$\frac{d}{dt}\left(\frac{\partial L}{\partial \dot{r}_{\alpha j}}\right) = \frac{\partial L}{\partial r_{\alpha j}} \tag{4.6}$$

then gives us

$$m_\alpha \ddot{r}_{\alpha j} = q_\alpha [\dot{\mathbf{r}}_\alpha \times \mathbf{B}]_j - q_\alpha \dot{A}_j + \sum_{\beta \neq \alpha} \frac{q_\alpha q_\beta (r_{\alpha j} - r_{\beta j})}{4\pi\epsilon_0 |\mathbf{r}_\alpha - \mathbf{r}_\beta|^3}, \tag{4.7}$$

which is the Lorentz force law for a particle in an electromagnetic field. Note that the last three terms are just $q_\alpha \mathbf{E}(\mathbf{r}_\alpha, t)$. The $-\partial_t \mathbf{A}$ term is the transverse part of the electric field (since we chose $\nabla \cdot \mathbf{A} = 0$), and the Coulomb terms give the longitudinal part.

When finding the equation of motion resulting from the variation of the vector potential, we must exercise some care. We are actually varying only the transverse part of $\mathbf{A}$, because the longitudinal part is zero due to the Coulomb-gauge condition. The change in the action is now given by

$$\delta S = \int_{t_1}^{t_2} dt \int d^3\mathbf{r} \sum_{j=1}^{3} \left[\frac{\delta L}{\delta A_j} - \partial_t \frac{\delta L}{\delta(\partial_t A_j)}\right] \delta A_j^{\perp}, \tag{4.8}$$

where we have inserted a superscript, $\perp$, on the variation of A_j to remind ourselves that the variation must be transverse. Because the integral of the scalar product of a transverse and a longitudinal function vanishes, we can express the variation of the action as

$$\begin{aligned} \delta S &= \int_{t_1}^{t_2} dt \int d^3\mathbf{r} \sum_{j=1}^{3} \left[\frac{\delta L}{\delta A_j} - \partial_t \frac{\delta L}{\delta(\partial_t A_j)}\right]_\perp \delta A_j^{\perp} \\ &= \int_{t_1}^{t_2} dt \int d^3\mathbf{r} \sum_{j=1}^{3} \left[\frac{\delta L}{\delta A_j} - \partial_t \frac{\delta L}{\delta(\partial_t A_j)}\right]_\perp \delta A_j, \end{aligned} \tag{4.9}$$

where the variation of A_j is now arbitrary, and the $\perp$ subscript on the first quantity denotes its transverse part. The requirement that $\delta S = 0$ for any choice of δA_j now implies that

$$\left[\frac{\delta L}{\delta A_j} - \partial_t \frac{\delta L}{\delta(\partial_t A_j)}\right]_\perp = 0 \tag{4.10}$$

for $j = 1, 2, 3$.

When finding the variation of the Lagrangian when varying $\mathbf{A}$, it is useful to express the relevant part of the Lagrangian (the part that depends on $\mathbf{A}$ and $\dot{\mathbf{A}}$) in terms of a Lagrangian density. If we define a current density by

$$\mathbf{j}(\mathbf{r}) = \sum_\alpha q_\alpha \dot{\mathbf{r}}_\alpha \delta(\mathbf{r} - \mathbf{r}_\alpha), \tag{4.11}$$

then we find that

$$L = \int d^3\mathbf{r} \left\{ \frac{\epsilon_0}{2}[(\partial_t \mathbf{A})^2 + (\nabla \times \mathbf{A})^2] + \mathbf{j} \cdot \mathbf{A} \right\} + \cdots . \tag{4.12}$$

If we substitute the above Lagrangian density into Eq. (4.10), we obtain

$$\frac{1}{c^2}\frac{\partial^2 \mathbf{A}}{\partial t^2} - \nabla^2 \mathbf{A} = \frac{1}{\epsilon_0 c^2} \mathbf{j}_\perp, \tag{4.13}$$

where $\mathbf{j}_\perp$ is the transverse part of $\mathbf{j}$. This is the correct equation of motion for the vector potential in the Coulomb gauge.

4.1.2 Coulomb-gauge Hamiltonian

From the Lagrangian, we would like to obtain the Hamiltonian. The first step is to find the canonical momenta. The momentum conjugate to $\mathbf{r}_\alpha$ is

$$\mathbf{p}_\alpha = m_\alpha \dot{\mathbf{r}}_\alpha + q_\alpha \mathbf{A}(\mathbf{r}_\alpha), \tag{4.14}$$

and the momentum conjugate to $\mathbf{A}$ is $\mathbf{\Pi} = \epsilon_0 \partial_t \mathbf{A} = -\epsilon_0 \mathbf{E}_\perp$. The Hamiltonian is then given by

$$\begin{aligned} H &= \sum_\alpha \mathbf{p}_\alpha \cdot \dot{\mathbf{r}}_\alpha + \int d^3\mathbf{r}\, \mathbf{\Pi} \cdot (\epsilon_0 \partial_t \mathbf{A}) - L \\ &= \sum_\alpha \frac{1}{2m_\alpha}[\mathbf{p}_\alpha - q_\alpha \mathbf{A}(\mathbf{r}_\alpha)]^2 + \frac{\epsilon_0}{2}\int d^3\mathbf{r}\,[\,|\mathbf{E}_\perp|^2 + c^2 |\nabla \times \mathbf{A}|^2] \\ &\quad + \sum_{\alpha<\beta} \frac{q_\alpha q_\beta}{4\pi\epsilon_0 |\mathbf{r}_\alpha - \mathbf{r}_\beta|}. \end{aligned} \tag{4.15}$$

We note that there are three terms. The physical interpretation of these is very simple. From Eq. (4.14), we see that the first term is the mechanical kinetic energy of each particle, $\frac{1}{2}m\mathbf{v}^2$. Recalling that $c^2 = 1/(\mu_0\epsilon_0)$, the second term is the transverse part of the free-field energy of the electromagnetic field, just as in Eq. (1.16). Finally, the last term gives the Coulomb interaction energy between the charged particles.

A slightly unexpected feature is that the Coulomb energy is clearly nonlocal. This makes it seem that, if one particle is moved, then the energy (and presumably the resulting electromagnetic force) experienced elsewhere will change immediately, thus violating causality. In fact, this does not happen, since the gauge condition of transversality is also nonlocal. The *combination* of the last two terms is equal to the total electromagnetic energy, which is causal. This leads to the expected time delays in the force experienced at a distant location. In the multipolar gauge, this type of apparent (but not real) acausal behavior is eliminated, due to the use of a different gauge.

4.2 The multipolar gauge

We now want to introduce a different, but completely equivalent, description of the electromagnetic field interacting with charged particles. In this description, the interaction is expressed in terms of polarization and magnetization densities, and this is a convenient description to use when describing polarizable media. We will begin by considering a single atom or molecule. In the simplest case, this consists of mobile electrons of charge $q_\alpha = -e$, located at positions $\mathbf{r}_\alpha$ for $\alpha > 0$, and a nuclear charge, $q_0 = Q$, located at $\mathbf{r}_0$. The system is assumed to be neutral overall, so that $\sum_\alpha q_\alpha = 0$.

4.2.1 Polarization density

Let us begin by defining the polarization density,

$$\mathbf{P}(\mathbf{r}) = \sum_\alpha q_\alpha \mathbf{r}_\alpha \int_0^1 du\, \delta(\mathbf{r} - u\mathbf{r}_\alpha). \tag{4.16}$$

Now, as this definition is by no means an obvious one, it bears some commenting upon. Suppose we have a charge of $-q$ located on the x axis at x_0 and a second charge of q also located on the x axis but at $x_0 + \Delta x$. The polarization density for this charge configuration from the above equation is zero if $x < x_0$ or $x > x_0 + \Delta x$, and equal to $q(\Delta x)(1/\Delta x)\delta(y)\delta(z)\hat{\mathbf{x}}$ for x between x_0 and $x_0 + \Delta x$. This last quantity can be interpreted as the dipole moment of the dipole consisting of the two charges over the volume occupied by it (the 'volume' has zero extent in the y and z directions, hence the delta-functions in those directions), which is what we would expect a polarization density for a dipole to be.

This definition of polarization density also has another property that we would expect a polarization density to have: its divergence is minus the charge density, $\rho_q(\mathbf{r})$, which is given by

$$\rho_q(\mathbf{r}) = \sum_\alpha q_\alpha \delta(\mathbf{r} - \mathbf{r}_\alpha). \tag{4.17}$$

In order to see this, we first note that we can differentiate a single component of the delta-function $\delta(x-y)$ by either the first or second of the arguments, the only difference being the resulting sign change. Applying this, we find that

$$\frac{\partial}{\partial r_j}\delta(\mathbf{r}-u\mathbf{r}_\alpha)=\frac{(-1)}{r_{\alpha j}}\frac{d}{du}\delta(r_j-ur_{\alpha j})\prod_{k\neq j}\delta(r_k-ur_{\alpha k}), \tag{4.18}$$

which implies that

$$(\mathbf{r}_\alpha\cdot\nabla_r)\delta(\mathbf{r}-u\mathbf{r}_\alpha)=-\frac{d}{du}\delta(\mathbf{r}-u\mathbf{r}_\alpha). \tag{4.19}$$

This then gives us that

$$\begin{aligned}\nabla\cdot\mathbf{P}(\mathbf{r})&=-\sum_\alpha q_\alpha\int_0^1 du\,\frac{d}{du}\delta(\mathbf{r}-u\mathbf{r}_\alpha)\\&=-\sum_\alpha q_\alpha\delta(\mathbf{r}-\mathbf{r}_\alpha),\end{aligned} \tag{4.20}$$

and the right-hand side of this equation is just minus the charge density.

4.2.2 Current and magnetization

In a medium in which the charges are bound, there are two contributions to the current, one from the polarization and one from the magnetization. The continuity equation plus the fact that the charge density is minus the divergence of the polarization density gives

$$\nabla\cdot\mathbf{j}=-\frac{\partial\rho_q}{\partial t}=\frac{\partial}{\partial t}\nabla\cdot\mathbf{P}. \tag{4.21}$$

Therefore, we can express the current density as

$$\mathbf{j}=\frac{\partial}{\partial t}\mathbf{P}+\mathbf{j}_m, \tag{4.22}$$

where $\mathbf{j}_m$, the magnetization current density, is transverse. This implies that it can be expressed as the curl of a vector field, $\mathbf{j}_m=\nabla\times\mathbf{M}$, where $\mathbf{M}$ is identified as the magnetization density. What we will now do is define $\mathbf{M}$ and then verify that its curl is, in fact, $\mathbf{j}_m$.

We define $\mathbf{M}$ to be

$$\mathbf{M}=\sum_\alpha q_\alpha\mathbf{r}_\alpha\times\dot{\mathbf{r}}_\alpha\int_0^1 du\,u\delta(\mathbf{r}-u\mathbf{r}_\alpha). \tag{4.23}$$

Setting

$$\mathbf{m}_\alpha=\mathbf{r}_\alpha\times\dot{\mathbf{r}}_\alpha\int_0^1 du\,u\delta(\mathbf{r}-u\mathbf{r}_\alpha), \tag{4.24}$$

we find that

$$(\nabla \times \mathbf{m}_\alpha)_j = \epsilon_{jkl}(\mathbf{r}_\alpha \times \dot{\mathbf{r}}_\alpha)_l \partial_k \int_0^1 du\, u\delta(\mathbf{r} - u\mathbf{r}_\alpha)$$
$$= \epsilon_{jkl}\epsilon_{lab} r_{a\alpha}\dot{r}_{b\alpha} \int_0^1 du\, u \left(\frac{-1}{r_{k\alpha}}\right) \prod_{c\neq k} \delta(r_c - ur_{c\alpha}) \frac{d}{du}\delta(r_k - ur_{k\alpha}), \tag{4.25}$$

where ϵ_{jkl} is the completely antisymmetric tensor of rank three, and repeated indices are summed over. Now, using the fact that $\epsilon_{jkl}\epsilon_{lab} = \delta_{ja}\delta_{kb} - \delta_{ka}\delta_{jb}$, we have that

$$(\nabla \times \mathbf{m}_\alpha)_j = \left[\frac{-r_{j\alpha}\dot{r}_{k\alpha}}{r_{k\alpha}} + \dot{r}_{j\alpha}\right] \int_0^1 du\, u \prod_{c\neq k} \delta(r_c - ur_{c\alpha}) \frac{d}{du}\delta(r_k - ur_{k\alpha}), \tag{4.26}$$

where all indices except j are summed over. In further evaluating the first term, we note that

$$\frac{d}{dt}\delta(r_k - ur_{k\alpha}) = \frac{u\dot{r}_{k\alpha}}{r_{k\alpha}} \frac{d}{du}\delta(r_k - ur_{k\alpha}), \tag{4.27}$$

and in evaluating the second, we can integrate by parts to obtain

$$\int_0^1 du\, u \frac{d}{du}\delta(\mathbf{r} - u\mathbf{r}_\alpha) = \delta(\mathbf{r} - \mathbf{r}_\alpha) - \int_0^1 du\, \delta(\mathbf{r} - u\mathbf{r}_\alpha). \tag{4.28}$$

Putting all of this together, we find

$$\nabla \times \mathbf{m}_\alpha = \dot{\mathbf{r}}_\alpha \delta(\mathbf{r} - \mathbf{r}_\alpha) - \mathbf{r}_\alpha \frac{d}{dt} \int_0^1 du\, \delta(\mathbf{r} - u\mathbf{r}_\alpha) - \dot{\mathbf{r}}_\alpha \int_0^1 du\, \delta(\mathbf{r} - u\mathbf{r}_\alpha). \tag{4.29}$$

Inserting this result into the expression for the magnetization density gives us

$$\nabla \times \mathbf{M} = \sum_\alpha q_\alpha \left[\dot{\mathbf{r}}_\alpha \delta(\mathbf{r} - \mathbf{r}_\alpha) - \frac{d}{dt}\mathbf{r}_\alpha \int_0^1 du\, \delta(\mathbf{r} - u\mathbf{r}_\alpha)\right]$$
$$= \mathbf{j} - \frac{\partial}{\partial t}\mathbf{P} = \mathbf{j}_m, \tag{4.30}$$

which is the desired result.

4.2.3 Multipolar Lagrangian

We now wish to modify the Lagrangian of the previous section to obtain the multipolar Lagrangian. The transformation that accomplishes this is known as the Power–Zienau–Woolley transformation. It is based on the fact that, if we add to a Lagrangian the total time derivative of a function of the coordinates, then the equations of motion are unchanged. The function chosen by Power, Zienau and Woolley is

$$F = -\int d^3\mathbf{r}\, \mathbf{P}(\mathbf{r}) \cdot \mathbf{A}(\mathbf{r}), \tag{4.31}$$

which is a function of the particle coordinates $\mathbf{r}_\alpha$ and the field coordinates $\mathbf{A}(\mathbf{r})$. Note that, since $\mathbf{A}$ is transverse, only the transverse part of $\mathbf{P}$, $\mathbf{P}_\perp$, contributes to the integral. The new

Lagrangian becomes

$$L_m = L + \frac{dF}{dt}, \tag{4.32}$$

where L is the Lagrangian in Eq. (4.1). It is useful to group the time derivative of F with the interaction of part of L, which gives us

$$\begin{aligned}\int d^3\mathbf{r}\,\mathbf{j}\cdot\mathbf{A} + \frac{dF}{dt} &= \int d^3\mathbf{r}\left(\mathbf{j}\cdot\mathbf{A} - \frac{\partial \mathbf{P}}{\partial t}\cdot\mathbf{A} - \mathbf{P}\cdot\frac{\partial \mathbf{A}}{\partial t}\right) \\ &= \int d^3\mathbf{r}\,[(\nabla\times\mathbf{M})\cdot\mathbf{A} + \mathbf{P}\cdot\mathbf{E}_\perp] \\ &= \int d^3\mathbf{r}\,[\mathbf{M}\cdot\mathbf{B} + \mathbf{P}\cdot\mathbf{E}_\perp].\end{aligned} \tag{4.33}$$

Adding these terms to the remaining ones in L gives us the multipolar Lagrangian

$$\begin{aligned}L_m &= \sum_\alpha \frac{1}{2} m_\alpha \dot{\mathbf{r}}_\alpha^2 - \sum_{\alpha<\beta} \frac{q_\alpha q_\beta}{4\pi\epsilon_0 |\mathbf{r}_\alpha - \mathbf{r}_\beta|} \\ &+ \frac{\epsilon_0}{2}\int d^3\mathbf{r}\,[\mathbf{E}_\perp^2 - c^2\mathbf{B}^2] + \int d^3\mathbf{r}\,[\mathbf{M}\cdot\mathbf{B} + \mathbf{P}\cdot\mathbf{E}_\perp].\end{aligned} \tag{4.34}$$

4.2.4 Multipolar Hamiltonian

Since we have the Lagrangian, our next step is to find the canonical momenta and then use them to find the Hamiltonian. The independent coordinates are $\mathbf{r}_\alpha$ and $\mathbf{A}$, and we have to find the momentum conjugate to each. Now $\dot{\mathbf{r}}_\alpha$ appears both in the kinetic energy term and in the term $\int d^3\mathbf{r}\,\mathbf{M}\cdot\mathbf{B}$, from Eq. (4.23). Taking the derivative of L_m with respect to $\dot{\mathbf{r}}_\alpha$ then yields

$$\mathbf{p}_\alpha = m_\alpha \dot{\mathbf{r}}_\alpha + q_\alpha \int_0^1 du\, u\mathbf{B}(u\mathbf{r}_\alpha) \times \mathbf{r}_\alpha. \tag{4.35}$$

The time derivative of the vector potential is just $-\mathbf{E}_\perp$, and this also appears in two terms in the Lagrangian. Taking the derivative of L_m with respect to $\dot{\mathbf{A}}$ then gives us

$$\mathbf{\Pi} = \epsilon_0 \partial_t \mathbf{A} - \mathbf{P}_\perp = -\mathbf{D}_\perp, \tag{4.36}$$

where we have used the fact that only the transverse part of $\mathbf{P}$ contributes to the integral in the term $\int d^3\mathbf{r}\,\mathbf{P}\cdot\mathbf{E}_\perp$, and that $\mathbf{D} = \epsilon_0\mathbf{E} + \mathbf{P}$. Note that, since

$$\nabla\cdot\mathbf{D} = \epsilon_0\nabla\cdot\mathbf{E} + \nabla\cdot\mathbf{P} = \rho - \rho = 0, \tag{4.37}$$

$\mathbf{D}$ is, in fact, transverse, in the absence of free charges, so in this case we can simply say that $\mathbf{\Pi} = -\mathbf{D}$. Here we see the displacement field appearing as a natural canonical coordinate again, just as it did for macroscopic quantization of nonlinear media in the previous chapter.

The multipolar Hamiltonian, H_m, is now given by

$$\begin{aligned} H_m &= \sum_\alpha \mathbf{p}_\alpha \cdot \dot{\mathbf{r}}_\alpha + \int d^3\mathbf{r}\, \boldsymbol{\Pi} \cdot (\partial_t \mathbf{A}) - L_m \\ &= \sum_\alpha \mathbf{p}_\alpha \cdot \frac{1}{m_\alpha}\left[\mathbf{p}_\alpha - q_\alpha \int_0^1 du\, u\mathbf{B}(u\mathbf{r}_\alpha) \times \mathbf{r}_\alpha\right] \\ &\quad + \frac{1}{\epsilon_0}\int d^3\mathbf{r}\, \boldsymbol{\Pi} \cdot (\boldsymbol{\Pi} + \mathbf{P}_\perp) - L_m. \end{aligned} \tag{4.38}$$

After expressing the magnetization in terms of the magnetic field, the particle coordinates, $\mathbf{r}_\alpha$, and canonical momenta, $\mathbf{p}_\alpha$ (see Eqs (4.23) and (4.35)), we find that

$$\begin{aligned} H_m &= \sum_\alpha \frac{1}{2m_\alpha}\left[\mathbf{p}_\alpha - q_\alpha \int_0^1 du\, u\mathbf{B}(u\mathbf{r}_\alpha) \times \mathbf{r}_\alpha\right]^2 + \frac{1}{2\epsilon_0}\int d^3\mathbf{r}\, \mathbf{P}_\perp^2 \\ &\quad + \sum_{\alpha<\beta} \frac{q_\alpha q_\beta}{4\pi\epsilon_0 |\mathbf{r}_\alpha - \mathbf{r}_\beta|} + \frac{1}{2}\int d^3\mathbf{r}\left[\frac{\boldsymbol{\Pi}^2}{\epsilon_0} + \epsilon_0 c^2 \mathbf{B}^2\right] + \frac{1}{\epsilon_0}\int d^3\mathbf{r}\, \boldsymbol{\Pi} \cdot \mathbf{P}_\perp. \end{aligned} \tag{4.39}$$

4.3 Hamiltonian for a polarizable medium

We now want to construct a model for a polarizable medium and then use what we have learned about the multipolar Hamiltonian to find a Hamiltonian for it. We now suppose that we have many atoms or molecules or other localized charge distributions, composed of nuclei and electrons, located near distinct positions $\mathbf{R}_\beta$. Each charge distribution is electrically neutral, so that $\sum_\alpha q_{\beta\alpha} = 0$. We also define the position of the particle at $\mathbf{r}_{\beta\alpha}$ relative to $\mathbf{R}_\beta$ to be $\mathbf{s}_{\beta\alpha} = \mathbf{r}_{\beta\alpha} - \mathbf{R}_\beta$, where typically $|\mathbf{s}_{\beta\alpha}| \ll \lambda$ for interactions with radiation of wavelength λ.

4.3.1 Densities and currents

We define the polarization and magnetization densities to be

$$\mathbf{P}(\mathbf{r}) = \sum_\beta \mathbf{P}_\beta(\mathbf{r}), \qquad \mathbf{M}(\mathbf{r}) = \sum_\beta \mathbf{M}_\beta(\mathbf{r}), \tag{4.40}$$

where

$$\begin{aligned} \mathbf{P}_\beta(\mathbf{r}) &= \sum_\alpha q_{\beta\alpha}\mathbf{s}_{\beta\alpha} \int_0^1 du\, \delta(\mathbf{r} - \mathbf{R}_\beta - u\mathbf{s}_{\beta\alpha}), \\ \mathbf{M}_\beta(\mathbf{r}) &= \sum_\alpha q_{\beta\alpha}\mathbf{s}_{\beta\alpha} \times \dot{\mathbf{s}}_{\beta\alpha} \int_0^1 du\, u\delta(\mathbf{r} - \mathbf{R}_\beta - u\mathbf{s}_{\beta\alpha}). \end{aligned} \tag{4.41}$$

Defining charge and current densities for each atom, we have that

$$\rho_\beta(\mathbf{r}) = \sum_\alpha q_{\beta\alpha}\delta(\mathbf{r} - \mathbf{R}_\beta - \mathbf{s}_{\beta\alpha}),$$
$$\mathbf{j}_\beta(\mathbf{r}) = \sum_\alpha q_{\beta\alpha}\dot{\mathbf{s}}_{\beta\alpha}\delta(\mathbf{r} - \mathbf{R}_\beta - \mathbf{s}_{\beta\alpha}), \tag{4.42}$$

so that the total charge and current densities are given by

$$\rho_q(\mathbf{r}) = \sum_\beta \rho_{q\beta}(\mathbf{r}), \qquad \mathbf{j}(\mathbf{r}) = \sum_\beta \mathbf{j}_\beta(\mathbf{r}). \tag{4.43}$$

We now find that

$$\nabla \cdot \mathbf{P}_\beta = -\rho_{q\beta}, \qquad \nabla \times \mathbf{M}_\beta = \mathbf{j}_\beta - \frac{\partial}{\partial t}\mathbf{P}_\beta, \tag{4.44}$$

which immediately implies that

$$\nabla \cdot \mathbf{P} = -\rho_q, \qquad \nabla \times \mathbf{M} = \mathbf{j} - \frac{\partial}{\partial t}\mathbf{P}. \tag{4.45}$$

4.3.2 Coulomb-gauge Lagrangian

Next we would like to write down a Coulomb-gauge Lagrangian for our collection of atoms. This is facilitated by defining the electric potential due to the charges in charge distribution β as

$$U_\beta(\mathbf{r}) = \frac{1}{4\pi\epsilon_0}\sum_\alpha \frac{q_{\beta\alpha}}{|\mathbf{r} - \mathbf{r}_{\beta\alpha}|}. \tag{4.46}$$

Note that the longitudinal part of the electric field due to the charges in atom β, $\mathbf{E}_{\beta\|}$, which satisfies $\nabla \cdot \mathbf{E}_{\beta\|} = \rho_\beta$, is given by $\mathbf{E}_{\beta\|} = -\nabla U_\beta$. The Lagrangian is now given by

$$L = \frac{1}{2}\sum_{\alpha,\beta} m\dot{\mathbf{s}}_{\beta\alpha}^2 - \sum_{\beta<\beta'}\int d^3\mathbf{r}\,\rho_\beta(\mathbf{r})U_{\beta'}(\mathbf{r}) - \sum_\beta\sum_{\alpha<\alpha'}\frac{q_{\beta\alpha}q_{\beta\alpha'}}{4\pi\epsilon_0|\mathbf{r}_{\beta\alpha} - \mathbf{r}_{\beta\alpha'}|}$$
$$+ \frac{\epsilon_0}{2}\int d^3\,\mathbf{r}[(\partial_t\mathbf{A})^2 - c^2(\nabla \times \mathbf{A})^2] + \int d^3\mathbf{r}\,\mathbf{j}(\mathbf{r}) \cdot \mathbf{A}(\mathbf{r}), \tag{4.47}$$

where m is the electron mass. The coordinates of the system are now $\mathbf{s}_{\beta\alpha}$ for the particles and $\mathbf{A}$, as before, for the field. As before, to obtain the multipolar Lagrangian, L_m, we add a term to the Lagrangian in the previous equation,

$$L_m = L - \frac{d}{dt}\int d^3\mathbf{r}\,\mathbf{A} \cdot \mathbf{P}, \tag{4.48}$$

which yields

$$L_m = \frac{1}{2}\sum_{\alpha,\beta} m\dot{\mathbf{s}}_{\beta\alpha}^2 - \sum_{\beta<\beta'}\int d^3\mathbf{r}\,\rho_\beta(\mathbf{r})U_{\beta'}(\mathbf{r}) - \sum_\beta\sum_{\alpha<\alpha'}\frac{q_{\beta\alpha}q_{\beta\alpha'}}{4\pi\epsilon_0|\mathbf{r}_{\beta\alpha} - \mathbf{r}_{\beta\alpha'}|}$$
$$+ \frac{\epsilon_0}{2}\int d^3\mathbf{r}\,[\mathbf{E}_\perp^2 - c^2\mathbf{B}^2] + \int d^3\mathbf{r}\,(\mathbf{M} \cdot \mathbf{B} + \mathbf{P} \cdot \mathbf{E}_\perp). \tag{4.49}$$

4.3.3 Canonical momenta and Hamiltonian

The next step is, as usual, to find the canonical momenta and then the Hamiltonian. The canonical momenta are

$$\mathbf{p}_{\beta\alpha} = m\dot{\mathbf{s}}_{\beta\alpha} + q_{\beta\alpha}\int_0^1 du\, u\mathbf{B}(\mathbf{R}_\beta + u\mathbf{s}_{\beta\alpha}) \times \mathbf{s}_{\beta\alpha},$$
$$\mathbf{\Pi} = \epsilon_0\partial_t\mathbf{A} - \mathbf{P}_\perp = -\mathbf{D}, \tag{4.50}$$

and the Hamiltonian becomes

$$\begin{aligned} H &= \sum_{\alpha,\beta}\mathbf{p}_{\beta\alpha}\cdot\dot{\mathbf{s}}_{\beta\alpha} + \int d^3\mathbf{r}\,\mathbf{\Pi}\cdot\mathbf{A} - L_m \\ &= \frac{1}{2m}\sum_{\alpha,\beta}\left[\mathbf{p}_{\beta\alpha} - q_{\beta\alpha}\int_0^1 du\, u\mathbf{B}(\mathbf{R}_\beta + u\mathbf{s}_{\beta\alpha}) \times \mathbf{s}_{\beta\alpha}\right]^2 \\ &+ \sum_{\beta<\beta'}\int d^3\mathbf{r}\,\rho_\beta(\mathbf{r})U_{\beta'}(\mathbf{r}) + \sum_\beta\sum_{\alpha<\alpha'}\frac{q_{\beta\alpha}q_{\beta\alpha'}}{4\pi\epsilon_0|\mathbf{r}_{\beta\alpha} - \mathbf{r}_{\beta\alpha'}|} \\ &+ \frac{1}{2\epsilon_0}\int d^3\mathbf{r}\,\mathbf{P}_\perp^2 + \frac{1}{2}\int d^3\mathbf{r}\left[\frac{\mathbf{\Pi}^2}{\epsilon_0} + \epsilon_0 c^2\mathbf{B}^2\right] + \frac{1}{\epsilon_0}\int d^3\mathbf{r}\,\mathbf{\Pi}\cdot\mathbf{P}_\perp. \end{aligned} \tag{4.51}$$

We can express the Hamiltonian in a somewhat different form, if we combine some terms. We first note that

$$\begin{aligned} \int d^3\mathbf{r}\,\rho_\beta(\mathbf{r})U_{\beta'}(\mathbf{r}) &= -\epsilon_0\int d^3\mathbf{r}\,\nabla^2 U_\beta(\mathbf{r})U_{\beta'}(\mathbf{r}) \\ &= \epsilon_0\int d^3\mathbf{r}\,\mathbf{E}_{\beta\|}(\mathbf{r})\cdot\mathbf{E}_{\beta'\|}(\mathbf{r}). \end{aligned} \tag{4.52}$$

We can make use of the fact that $\epsilon_0\nabla\cdot\mathbf{E}_{\beta\|} = \rho_\beta = -\nabla\cdot\mathbf{P}_{\beta\|}$, which gives us that

$$\int d^3\mathbf{r}\,\rho_\beta(\mathbf{r})U_{\beta'}(\mathbf{r}) = \frac{1}{\epsilon_0}\int d^3\mathbf{r}\,\mathbf{P}_{\beta\|}(\mathbf{r})\cdot\mathbf{P}_{\beta'\|}(\mathbf{r}). \tag{4.53}$$

The integral of the square of the transverse polarization density can be expressed in terms of the atomic polarization densities as

$$\frac{1}{2\epsilon_0}\int d^3\mathbf{r}\,\mathbf{P}_\perp^2 = \frac{1}{2\epsilon_0}\sum_\beta\int d^3\mathbf{r}\,\mathbf{P}_{\beta\perp}^2 + \frac{1}{\epsilon_0}\sum_{\beta<\beta'}\int d^3\mathbf{r}\,\mathbf{P}_{\beta\perp}\cdot\mathbf{P}_{\beta'\perp}. \tag{4.54}$$

Next, because the integral of the inner product of a transverse field and a longitudinal field is zero, we have

$$\int d^3\mathbf{r}\,\mathbf{P}_\beta\cdot\mathbf{P}_{\beta'} = \int d^3\mathbf{r}\,\mathbf{P}_{\beta\|}\cdot\mathbf{P}_{\beta'\|} + \int d^3\mathbf{r}\,\mathbf{P}_{\beta\perp}\cdot\mathbf{P}_{\beta'\perp}. \tag{4.55}$$

This now allows us to write the Hamiltonian as

$$H = \frac{1}{2m}\sum_{\alpha,\beta}\left[\mathbf{p}_{\beta\alpha} - q_{\beta\alpha}\int_0^1 du\, u\mathbf{B}(\mathbf{R}_\beta + u\mathbf{s}_{\beta\alpha}) \times \mathbf{s}_{\beta\alpha}\right]^2$$
$$+ \frac{1}{2\epsilon_0}\sum_\beta \int d^3\mathbf{r}\,\mathbf{P}^2_{\beta\perp} + \sum_\beta\sum_{\alpha<\alpha'} \frac{q_{\beta\alpha}q_{\beta\alpha'}}{4\pi\epsilon_0|\mathbf{r}_{\beta\alpha} - \mathbf{r}_{\beta\alpha'}|} + \frac{1}{\epsilon_0}\int d^3\mathbf{r}\,\mathbf{\Pi}\cdot\mathbf{P}_\perp$$
$$+ \frac{1}{\epsilon_0}\sum_{\beta<\beta'}\int d^3\mathbf{r}\,\mathbf{P}_\beta\cdot\mathbf{P}_{\beta'} + \frac{1}{2}\int d^3\mathbf{r}\left[\frac{\mathbf{\Pi}^2}{\epsilon_0} + c^2\epsilon_0\mathbf{B}^2\right]. \tag{4.56}$$

It is now clear that, apart from local Coulomb terms at each charge center β, there are no long-range Coulomb forces remaining in the Hamiltonian. Such long-range forces are now explicitly causal, and appear only due to the exchange of transverse photons.

4.3.4 Quantization

Now that we have the canonical momenta and the Hamiltonian, quantizing the theory is straightforward. The canonical momentum $\mathbf{p}_{\beta\alpha}$ becomes the operator $-i\hbar\nabla_{\mathbf{s}_{\beta\alpha}}$, and the field operators must obey the commutation relations

$$[A_j(\mathbf{r}, t), \Pi_k(\mathbf{r}', t)] = i\hbar\delta^{(\mathrm{tr})}_{jk}(\mathbf{r} - \mathbf{r}'). \tag{4.57}$$

This will be accomplished if we express the fields in terms of creation and annihilation operators, using $\mathbf{\Pi}(\mathbf{r}, t) = -\mathbf{D}(\mathbf{r}, t)$, so that

$$\mathbf{A}(\mathbf{r}, t) = \sum_{\mathbf{k},\alpha}\sqrt{\frac{\hbar}{2\epsilon_0\omega V}}\,\hat{\mathbf{e}}_{\mathbf{k},\alpha}(a_{\mathbf{k},\alpha}e^{i\mathbf{k}\cdot\mathbf{r}} + a^\dagger_{\mathbf{k},\alpha}e^{-i\mathbf{k}\cdot\mathbf{r}}),$$
$$\mathbf{D}(\mathbf{r}, t) = \sum_{\mathbf{k},\alpha} i\sqrt{\frac{\hbar\omega\epsilon_0}{2V}}\,\hat{\mathbf{e}}_{\mathbf{k},\alpha}(a_{\mathbf{k},\alpha}e^{i\mathbf{k}\cdot\mathbf{r}} - a^\dagger_{\mathbf{k},\alpha}e^{-i\mathbf{k}\cdot\mathbf{r}}). \tag{4.58}$$

We see that the natural canonical field variables are the vector potential and the displacement field. The particle coordinates and momenta have their usual commutation properties, i.e.

$$[s_{i\beta\alpha}, p_{j\beta'\alpha'}] = i\hbar\delta_{ij}\delta_{\alpha\alpha'}\delta_{\beta\beta'}. \tag{4.59}$$

In this chapter, we are employing a first-quantized, nonrelativistic approach to charged-particle quantization, for simplicity. This can be readily extended to a nonrelativistic, or relativistic, field picture for the particles. However, it is necessary to recall that electrons are fermions, and hence in the first-quantized approach must have antisymmetric wavefunctions, while in the field-theoretic approach the operators describing the electron field would satisfy anticommutation relations rather than commutation relations.

4.4 Dipole-coupling approximation

We want to construct a simplified model for a typical insulator interacting with the electromagnetic field, in order to illustrate the multipolar gauge in greater detail. In this section, we introduce the widely used dipole-coupling approximation, which results in a much more compact, approximate description of atom–field coupling. Each of our charge distributions will consist of charged particles interacting with each other via a Coulomb interaction, and we will assume that the charge distributions do not overlap with each other. Note that the only terms in the multipolar Hamiltonian explicitly describing interactions between the charge distributions are of the form $\int d^3\mathbf{r}\,\mathbf{P}_\beta \cdot \mathbf{P}_{\beta'}$ for $\beta \neq \beta'$. These are zero if the charge distributions do not overlap, because the polarization $\mathbf{P}_\beta$ is only nonzero in a small region near $\mathbf{R}_\beta$.

This at first appears somewhat counter-intuitive. After all, atoms and molecules certainly do interact at long distances. Long-range forces due to dipole–dipole interactions exist between any two polarizable charge distributions. However, in the multipolar approach, the force between nonoverlapping charge distributions occurs via the exchange of transverse photons.

This approach has both advantages and disadvantages. The advantage is that there are no explicit Coulomb terms due to longitudinal electric field components. In the minimal-coupling Hamiltonian, these terms provide an apparently unphysical instantaneous action at a distance, which must be canceled by transverse electric field components in order to obtain causal behavior. The drawback is that this approach is limited to gases, insulators and near-ground-state semiconductors. The delocalized charges in metallic conductors, excited semiconductors or plasmas cannot be grouped into nonoverlapping charge distributions.

At this point, our Hamiltonian for the medium plus field is

$$H = \frac{1}{2m}\sum_{\beta\alpha}\left[\mathbf{p}_{\beta\alpha} - q_{\beta\alpha}\int_0^1 du\, u\mathbf{B}(\mathbf{R}_\beta + u\mathbf{s}_{\beta\alpha}) \times \mathbf{s}_{\beta\alpha}\right]^2 + \sum_\beta V_\beta(\mathbf{s}_{\beta\alpha})$$
$$+ \frac{1}{2}\int d^3\mathbf{r}\left[\frac{\mathbf{D}^2}{\epsilon_0} + c^2\epsilon_0\mathbf{B}^2\right] - \frac{1}{\epsilon_0}\int d^3\mathbf{r}\,\mathbf{D}\cdot\mathbf{P}_\perp, \tag{4.60}$$

where $\mathbf{s}_{\beta\alpha}$ is the coordinate of each charged particle. The potential $V_\beta(\mathbf{s}_{\beta\alpha})$ is given by

$$V_\beta(\mathbf{s}_{\beta\alpha}) = \sum_{\alpha<\alpha'}\frac{q_{\beta\alpha}q_{\beta\alpha'}}{4\pi\epsilon_0|\mathbf{r}_{\beta\alpha} - \mathbf{r}_{\beta\alpha'}|} + \frac{1}{2\epsilon_0}\int d^3\mathbf{r}\,\mathbf{P}^2_{\beta\perp}. \tag{4.61}$$

These two terms each have a clear physical explanation:

- *Coulomb potential.* Terms with $q_{\beta\alpha}q_{\beta\alpha'}$ correspond to the Coulomb potential between charged particles at the same local charge center. This includes attractive electron–nucleus interactions and the Coulomb repulsion between electrons.
- *Polarization self-energy.* Terms with $\mathbf{P}^2_\perp$ correspond to the self-energy due to vacuum fluctuations. These must include a momentum cutoff, and hence should be renormalized

in a full quantum field-theoretic approach, to prevent divergences. This term generates a part of the well-known Lamb shift.

Next, we want to examine the interaction terms. We have two types, one involving the magnetic field and one involving the displacement field. What we want to show is that the magnetic field term is much smaller than the displacement field term, and can be neglected. First, let us examine the displacement field term. We have that

$$\begin{aligned} \int d^3\mathbf{r}\,\mathbf{P}_\perp\cdot\mathbf{D} &= \int d^3\mathbf{r}\,\mathbf{P}\cdot\mathbf{D} \\ &= \sum_{\beta\alpha}\int d^3\mathbf{r}\int_0^1 du\, q\delta(\mathbf{r}-\mathbf{R}_\beta-u\mathbf{s}_{\beta\alpha})\mathbf{s}_{\beta\alpha}\cdot\mathbf{D}(\mathbf{r}) \\ &= \sum_{\beta\alpha}\int_0^1 du\, q\mathbf{s}_{\beta\alpha}\cdot\mathbf{D}(\mathbf{R}_\beta+u\mathbf{s}_{\beta\alpha}), \end{aligned} \tag{4.62}$$

where, in the first equality, we have used the fact that $\mathbf{D}$ is transverse. We are now going to assume that the field contains only wavelengths that are much larger than the size of an individual atom. For optical fields, this is most certainly the case, with optical wavelengths being of order 500 nm while atoms are of order 0.1 nm in size. In that case, we can neglect the variation in $\mathbf{D}$ over the size of the atom and simply replace $\mathbf{D}(\mathbf{R}_\beta+u\mathbf{s}_\beta)$ by $\mathbf{D}(\mathbf{R}_\beta)$. This finally gives us that

$$\int d^3\mathbf{r}\,\mathbf{P}_\perp\cdot\mathbf{D} \cong \sum_{\beta\alpha} q_{\beta\alpha}\mathbf{s}_{\beta\alpha}\cdot\mathbf{D}(\mathbf{R}_\beta). \tag{4.63}$$

Now let us compare one term in this sum to one of the magnetic field terms. We have that $(1/\epsilon_0)qs_\beta D \sim qa_0E$, where a_0 is the Bohr radius and E is the electric field. We also have that

$$\frac{1}{2m}\left[q_{\beta\alpha}\int_0^1 du\, u\mathbf{B}(\mathbf{R}_\beta+u\mathbf{s}_\beta)\times\mathbf{s}_\beta\right]^2 \sim \frac{1}{2m}(Ba_0q)^2. \tag{4.64}$$

The ratio of the displacement field term to the magnetic field term goes like $(qa_0E)/(2mc^2)$, where we have used the fact that in an electromagnetic wave $E = cB$. Even for strong electric fields (fields of the size that occur in atoms) qa_0E will be of the order of an electronvolt (eV) or less, while mc^2 is of the order of megaelectronvolts (MeV). Thus, this ratio is small, and we can neglect the magnetic field terms.

Consequently, our Hamiltonian is now approximated by a dipole-coupling approximation, which is valid as long as the wavelength is large compared to the atomic dimension:

$$\begin{aligned} H = &\sum_\beta\left[\sum_\alpha\frac{\mathbf{p}_{\beta\alpha}^2}{2m}+V(\mathbf{s}_{\beta\alpha})\right]+\frac{1}{2}\int d^3\mathbf{r}\left[\frac{\mathbf{D}^2}{\epsilon_0}+c^2\epsilon_0\mathbf{B}^2\right] \\ &-\frac{1}{\epsilon_0}\sum_{\beta\alpha}q_{\beta\alpha}\mathbf{s}_{\beta\alpha}\cdot\mathbf{D}(\mathbf{R}_\beta). \end{aligned} \tag{4.65}$$

This is called the dipole-coupled Hamiltonian. An important point about this Hamiltonian is that the canonical electromagnetic field used in quantization is the *displacement* field, and all interatomic interactions are causal interactions mediated by the exchange of photons.

4.5 Linear medium

In this section, we wish to use the multipolar Hamiltonian approach to construct a simple microscopic model for a linear dielectric medium interacting with the electromagnetic field. Each of our 'atoms' will consist of a harmonic oscillator, so that, before we make the dipole approximation, our Hamiltonian for the medium plus field is

$$H = \frac{1}{2m}\sum_{\beta}\left[\mathbf{p}_{\beta} - q\int_0^1 du\, u\mathbf{B}(\mathbf{R}_{\beta} + u\mathbf{s}_{\beta}) \times \mathbf{s}_{\beta}\right]^2 + \frac{1}{2}\sum_{\beta} m v_0^2 \mathbf{s}_{\beta}^2$$
$$+ \frac{1}{2}\int d^3\mathbf{r}\left[\frac{\mathbf{\Pi}^2}{\epsilon_0} + c^2\epsilon_0\mathbf{B}^2\right] + \frac{1}{\epsilon_0}\int d^3\mathbf{r}\,\mathbf{\Pi}\cdot\mathbf{P}_{\perp}, \tag{4.66}$$

where $\mathbf{s}_{\beta}$ is the coordinate of the oscillator, v_0 is its frequency, and all of the oscillators have the same charge, q, on the mobile particle. Just as in the previous section, we can show that the magnetic field term is much smaller than the displacement field term, and can be neglected.

Consequently, our Hamiltonian can be approximated by a dipole-coupling approximation, which is valid as long as the wavelength is large compared to the atomic dimension:

$$H = \sum_{\beta}\left[\frac{\mathbf{p}_{\beta}^2}{2m} + \frac{1}{2}m v_0^2 \mathbf{s}_{\beta}^2\right] + \frac{1}{2}\int d^3\mathbf{r}\left[\frac{\mathbf{D}^2}{\epsilon_0} + c^2\epsilon_0\mathbf{B}^2\right]$$
$$- \frac{q}{\epsilon_0}\sum_{\beta}\mathbf{s}_{\beta}\cdot\mathbf{D}(\mathbf{R}_{\beta}). \tag{4.67}$$

4.5.1 Continuum approximation

Our next step will be to replace the discrete oscillators by a continuous distribution of them. Let $\mathbf{s}(\mathbf{r})$ be the displacement of the oscillator located at the position $\mathbf{r}$ in the volume element ΔV, and let the mass of the oscillator be $\rho_m \Delta V$, where ρ_m is the oscillator mass density. If we now express the momentum of the oscillator at $\mathbf{r}$ as $\pi(\mathbf{r})\Delta V$, we find that the oscillator part of the Hamiltonian becomes

$$\sum_{\beta}\left[\frac{\pi(\mathbf{R}_{\beta})^2}{2\rho_m} + \frac{1}{2}\rho_m v_0^2 \mathbf{s}(\mathbf{R}_{\beta})^2\right]\Delta V$$
$$\rightarrow \int d^3\mathbf{r}\left[\frac{\pi(\mathbf{r})^2}{2\rho_m} + \frac{1}{2}\rho_m v_0^2 \mathbf{s}(\mathbf{r})^2\right] \tag{4.68}$$

in the limit that $\Delta V \to 0$. The fact that $\mathbf{p}_\beta = m\dot{\mathbf{r}}_\beta$ now implies that $\boldsymbol{\pi}(\mathbf{r}) = \rho_m \partial_t \mathbf{s}(\mathbf{r})$. Note also that, when we quantize the theory and promote the variables to operators, the commutation relation $[s_j(\mathbf{R}_\beta), p_k(\mathbf{R}_{\beta'})] = i\hbar\delta_{jk}\delta_{\beta\beta'}$ becomes

$$[s_j(\mathbf{R}_\beta), \pi_k(\mathbf{R}_{\beta'})] = \frac{1}{\Delta V} i\hbar\delta_{jk}\delta_{\beta\beta'} \quad \to \quad [s_j(\mathbf{r}), \pi_k(\mathbf{r}')] = i\hbar\delta_{jk}\delta(\mathbf{r}-\mathbf{r}'), \tag{4.69}$$

in the $\Delta V \to 0$ limit. Therefore, our Hamiltonian for the medium coupled to the electromagnetic field is now

$$\begin{aligned} H = & \int d^3\mathbf{r} \left[\frac{\boldsymbol{\pi}(\mathbf{r})^2}{2\rho_m} + \frac{1}{2}\rho_m v_0^2 \mathbf{s}(\mathbf{r})^2 \right] + \frac{1}{2}\int d^3\mathbf{r} \left[\frac{\mathbf{D}^2}{\epsilon_0} + c^2\epsilon_0 \mathbf{B}^2 \right] \\ & - \rho_q \int d^3\mathbf{r}\, g\mathbf{s}(\mathbf{r}) \cdot \mathbf{D}(\mathbf{r})/\epsilon_0, \end{aligned} \tag{4.70}$$

where ρ_q is the charge density of the oscillators (the charge on the oscillating particle in volume ΔV is $Q = \rho_q \Delta V$).

The continuum approximation means that we are ignoring the fact that, physically, oscillators are atoms or molecules. Hence they are composed of fermions obeying the Pauli exclusion principle, which means that they cannot overlap; and in fact we have used an approximate dipole-coupled Hamiltonian that *requires* that atomic charge clouds do not overlap. This means that a true continuum of polarizable oscillators is not physically realizable, due to short-distance anticorrelations. We see in the next section that this leads to consequences in terms of frequency shifts. These are related to the Lorentz frequency shift or local-field correction, already treated in Section 1.7, and a density-dependent renormalization is required to compensate for the simplicity of this approximation.

4.5.2 Linear equations of motion

We can now find the equations of motion from this linear, dipole-coupled Hamiltonian. These are given by

$$\partial_t \Pi_j = -\frac{\delta H}{\delta A_j}, \qquad \partial_t A_j = \frac{\delta H}{\delta \Pi_j}, \tag{4.71}$$

where $\mathbf{B} = \nabla \times \mathbf{A}$ as usual, and the canonical field momentum is $\boldsymbol{\Pi} = -\mathbf{D}$. This leads to

$$\begin{aligned} \partial_t \mathbf{D} &= \epsilon_0 c^2 \nabla \times \mathbf{B}, \\ \partial_t \mathbf{A} &= -\frac{1}{\epsilon_0}\mathbf{D} + g\mathbf{s}. \end{aligned} \tag{4.72}$$

Similarly, the oscillator variable equations are

$$\partial_t \pi_j = -\frac{\delta H}{\delta s_j}, \qquad \partial_t s_j = \frac{\delta H}{\delta \pi_j}, \tag{4.73}$$

which give us

$$\begin{aligned} \partial_t \boldsymbol{\pi} &= -\rho_m v_0^2 \mathbf{s} + \rho_q \mathbf{D}/\epsilon_0, \\ \partial_t \mathbf{s} &= \frac{1}{\rho_m}\boldsymbol{\pi}. \end{aligned} \tag{4.74}$$

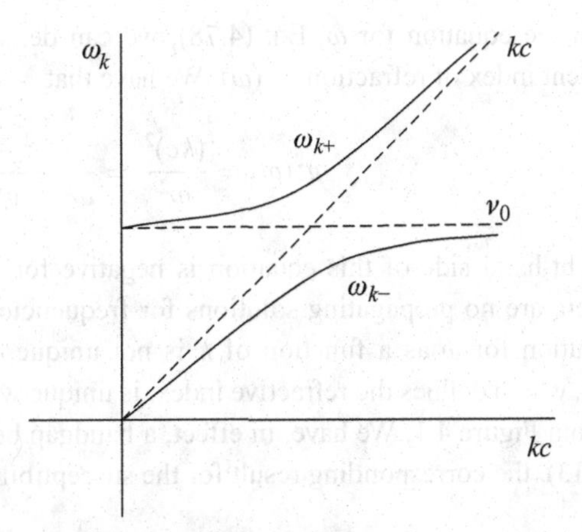

Fig. 4.1 Phonon–polariton dispersion: a schematic graph of typical solutions.

Eliminating the momenta, we find that

$$(\partial_t^2 - c^2\nabla^2)\mathbf{A} = \frac{\rho_q}{\epsilon_0}\partial_t \mathbf{s},$$
$$(\partial_t^2 + \nu_0^2)\mathbf{s} = -\frac{\rho_q}{\rho_m}(\partial_t \mathbf{A} - g\mathbf{s}). \tag{4.75}$$

We can find a propagating-wave solution to these equations by assuming that

$$\mathbf{A}(\mathbf{r}, t) = \mathbf{A}_0 e^{i(\mathbf{k}\cdot\mathbf{r}-\omega t)}, \qquad \mathbf{s}(\mathbf{r}, t) = \mathbf{s}_0 e^{i(\mathbf{k}\cdot\mathbf{r}-\omega t)}. \tag{4.76}$$

This yields

$$(\omega^2 - c^2k^2)\mathbf{A}_0 = -i\omega\frac{\rho_q}{\epsilon_0}\mathbf{s}_0,$$
$$\left(-\omega^2 + \nu_0^2 - \frac{g^2\epsilon_0}{\rho_m}\right)\mathbf{s}_0 = \frac{i\omega\rho_q}{\rho_m}\mathbf{A}_0. \tag{4.77}$$

In order for these equations to have a nonzero solution for $\mathbf{A}_0$ and $\mathbf{s}_0$, it is necessary that ω satisfy the equation

$$\omega^4 - \omega^2(k^2c^2 + \nu_0^2) + (kc)^2\left(\nu_0^2 - \frac{\rho_q^2}{\epsilon_0\rho_m}\right) = 0. \tag{4.78}$$

Here we introduce a constant with units of frequency:

$$\Delta^2 = \frac{\rho_q^2}{\epsilon_0\rho_m}.$$

This implies that we obtain a dispersion relation with two branches, as shown in Figure 4.1,

$$\omega_{k\pm}^2 = \tfrac{1}{2}\{(\nu_0^2 + k^2c^2) \pm [(\nu_0^2 - k^2c^2)^2 + 4(kc\Delta)^2]^{1/2}\}. \tag{4.79}$$

From the equation for ω, Eq. (4.78), we can derive an expression for the frequency-dependent index of refraction, $n_r(\omega)$. We have that

$$n_r^2(\omega) = \frac{(kc)^2}{\omega^2} = \frac{\omega^2 - \nu_0^2}{\omega^2 - \nu_0^2 + \Delta^2}. \tag{4.80}$$

The right-hand side of this equation is negative for $\nu_0^2 - \Delta^2 < \omega^2 < \nu_0^2$, which implies that there are no propagating solutions for frequencies in this range. It is interesting that the solution for ω as a function of k is not unique – there are two branches – but the inverse, which defines the refractive index, is unique whenever it exists. This can be clearly seen from Figure 4.1. We have, in effect, a bandgap below the resonance frequency. From Eq. (1.13), the corresponding result for the susceptibility is

$$\begin{aligned} \chi^{(1)}(\omega) &= n^2(\omega) - 1 \\ &= \frac{\Delta^2}{\nu_0^2 - \Delta^2 - \omega^2}. \end{aligned} \tag{4.81}$$

Near resonance, this gives the familiar Lorentzian lineshape,

$$\chi^{(1)}(\omega) \approx \frac{\Delta\omega_{cm}}{\nu_0 - \Delta\omega_{cm} - \omega},$$

where there is a density-dependent red-shift of

$$\Delta\omega_{cm} = \Delta^2/(2\nu_0).$$

4.5.3 Local-field red-shift

We notice that our expression is very similar to the semiclassical, local-field result given in Eq. (1.112), in that there is a density-dependent frequency shift. In order to compare the two results, we need to calculate the dipole transition matrix element $|\mathbf{d}|^2$ that appears in this equation. Although we have not yet quantized our model, we will anticipate that our quantized harmonic oscillator will have the standard behavior for a linear oscillator, for the purposes of comparison. The transition matrix element for a $0 \rightarrow 1$ transition in any of the three polarization directions of a three-dimensional harmonic oscillator is exactly the same as for a one-dimensional harmonic oscillator, namely

$$\langle 1|x|0\rangle = \sqrt{\frac{\hbar}{2m\nu_0}}. \tag{4.82}$$

This means that, if each oscillator charge is equal to the electron charge e, then

$$|\mathbf{d}\cdot\hat{\mathbf{e}}|^2 = \frac{e^2\hbar}{2m\nu_0}, \tag{4.83}$$

so that we can write, for a number density of ρ_n, that the density-dependent red-shift in the continuous oscillator model is

$$\begin{aligned}\Delta\omega_{cm} &= \frac{\rho_q^2}{2\nu_0 m \epsilon_0 \rho_n} \\ &= \frac{|\mathbf{d}\cdot\hat{\mathbf{e}}|^2 \rho_n}{\epsilon_0 \hbar}.\end{aligned} \tag{4.84}$$

This is similar to the familiar local-field Lorentz red-shift given in Eq. (1.112), which is valid for a cubic lattice of oscillators, except that the value is larger than expected by a factor of 3. We interpret this as being caused by the change of geometry, from a cubic lattice of oscillators to a continuous distribution of oscillators in the present model. In physical systems of interest, continuous distributions do not occur, due to short-distance repulsion between atoms. For this reason, the frequency shift obtained here should be regarded as an artifact of the simplified model used. For this type of model to give a physically correct red-shift, it is necessary to renormalize the resonant frequency in the Hamiltonian by setting $\nu_0 \to \nu_0 + 2\Delta\omega_{cm}/3$. This compensates for the neglect of short-range decorrelation effects in the model.

4.5.4 Sellmeir expansion

We can put our expression for the index of refraction into a more standard form. If we define a frequency ν so that $\nu^2 = \nu_0^2 - \Delta^2$ (together with the necessary renormalization described above), then the index of refraction for our model becomes

$$n_r^2(\omega) = 1 + \frac{\Delta^2}{\nu^2 - \omega^2}. \tag{4.85}$$

For a realistic polarizable medium, the actual frequency shift is geometry-dependent, and the shift calculated above in Eq. (4.84) should be replaced by a more physical model that includes a full treatment of geometry, interaction and correlation effects. The simplest model for this is the Lorenz–Lorentz red-shift, described in Eq. (1.112).

The classical Drude–Lorentz model, which also models a linear optical medium as a collection of oscillators, leads to a widely used expression for the index of refraction, n_r, of the medium known as the Sellmeir expansion. The expression we have derived for a linear polarizable medium is just the Sellmeir expansion for a medium with one resonance, though in this case the frequency of the resonance has been shifted from that of the original oscillator.

If the material has multiple resonances at the frequencies ν_j, which correspond to the frequencies of the oscillators used to model the medium, then the Sellmeir expansion gives an expression precisely of this form, except with multiple resonances,

$$n_r^2(\omega) = 1 + \sum_j \frac{\Delta_j^2}{\nu_j^2 - \omega^2}, \tag{4.86}$$

where Δ_j^2 characterizes the coupling of the jth resonance to the field. Sometimes this is written in an equivalent form as a function of (vacuum) wavelength, by defining $B_j = \Delta_j^2/\nu_j^2$:

$$n_r^2(\lambda) = 1 + \sum_j \frac{\lambda^2 B_j}{\lambda^2 - \lambda_j^2}. \tag{4.87}$$

A three-term Sellmeir expansion for optical glasses has a typical error of at most 10^{-5} in the transparency window of wavelengths from 400 nm to 2 μm.

4.6 Quantization of the linear model

Let us now quantize the model. Our ultimate object is to express the Hamiltonian in terms of creation and annihilation operators, and then to diagonalize it. Consequently, we want to begin by expressing the fundamental fields in terms of creation and annihilation operators. This has already been done for **A** and **D** in Eqs (4.58), and we now need to do something similar for **s** and $\boldsymbol{\pi}$. We will obtain the correct commutation relations between **s** and $\boldsymbol{\pi}$ (see Eq. (4.69)) if we set

$$\mathbf{s}(\mathbf{r},t) = \sum_{\mathbf{k}} \sum_{\alpha=0}^{2} \sqrt{\frac{\hbar}{2\nu_0 \rho_m V}}\, \hat{\mathbf{e}}_{\mathbf{k},\alpha}(b_{\mathbf{k},\alpha} e^{i\mathbf{k}\cdot\mathbf{r}} + b_{\mathbf{k},\alpha}^\dagger e^{-i\mathbf{k}\cdot\mathbf{r}}),$$

$$\boldsymbol{\pi}(\mathbf{r},t) = -\sum_{\mathbf{k}} \sum_{\alpha=0}^{2} i\sqrt{\frac{\hbar \nu_0 \rho_m}{2V}}\, \hat{\mathbf{e}}_{\mathbf{k},\alpha}(b_{\mathbf{k},\alpha} e^{i\mathbf{k}\cdot\mathbf{r}} - b_{\mathbf{k},\alpha}^\dagger e^{-i\mathbf{k}\cdot\mathbf{r}}), \tag{4.88}$$

where $[b_{\mathbf{k},\alpha}, b_{\mathbf{k}',\alpha'}^\dagger] = \delta_{\alpha\alpha'}\delta_{\mathbf{k}\mathbf{k}'}$. Because **s** is not a transverse field, we have added an additional polarization vector, $\hat{\mathbf{e}}_{\mathbf{k},0} = \hat{\mathbf{k}}$, to the original two. Note that with this definition we still have that $\hat{\mathbf{e}}_{-\mathbf{k},\alpha} = -(-1)^\alpha \hat{\mathbf{e}}_{\mathbf{k},\alpha}$, as in the free-field case treated in Section 2.8. Substituting these expressions into the terms in the Hamiltonian we find that

$$\int d^3\mathbf{r} \left(\frac{\boldsymbol{\pi}^2}{2\rho_m} + \frac{1}{2}\rho_m \nu_0^2 \mathbf{s}^2 \right) = \hbar\nu_0 \sum_{\mathbf{k}} \sum_{\alpha=0}^{2} \left(b_{\mathbf{k},\alpha}^\dagger b_{\mathbf{k},\alpha} + \frac{1}{2} \right). \tag{4.89}$$

For the terms in the Hamiltonian involving **A** and **D**, we find

$$\frac{1}{2}\int d^3\mathbf{r} \left(\frac{\mathbf{D}^2}{\epsilon_0} + c^2\epsilon_0 \mathbf{B}^2 \right) = \sum_{\mathbf{k}} \sum_{\alpha=1}^{2} \hbar\omega_k \left(a_{\mathbf{k},\alpha}^\dagger a_{\mathbf{k},\alpha} + \frac{1}{2} \right). \tag{4.90}$$

Here, as usual, $\omega_k \equiv c|\mathbf{k}|$ is the vacuum resonance frequency for photons with momentum **k**. Now let us move on to the interaction term. We find that

$$-g\int d^3\mathbf{r}\, \mathbf{s}\cdot\mathbf{D} = -i\sum_{\mathbf{k}} \sum_{\alpha=1}^{2} \frac{\hbar}{2}\sqrt{\frac{\epsilon_0\omega_k}{\nu_0\rho_m}}$$

$$\times [(-1)^\alpha (a_{\mathbf{k},\alpha}^\dagger b_{-\mathbf{k},\alpha}^\dagger - a_{\mathbf{k},\alpha} b_{-\mathbf{k},\alpha}) + a_{\mathbf{k},\alpha} b_{\mathbf{k},\alpha}^\dagger - a_{\mathbf{k},\alpha}^\dagger b_{\mathbf{k},\alpha}]. \tag{4.91}$$

Putting everything together and dropping constant terms, we find for the Hamiltonian for the medium plus field

$$H = \hbar \sum_{\mathbf{k}} \nu_0 b^{\dagger}_{\mathbf{k},0} b_{\mathbf{k},0} + \hbar \sum_{\mathbf{k}} \sum_{\alpha=1}^{2} \{\nu_0 b^{\dagger}_{\mathbf{k},\alpha} b_{\mathbf{k},\alpha} + \omega_k a^{\dagger}_{\mathbf{k},\alpha} a_{\mathbf{k},\alpha}$$

$$+ i g_k [(-1)^{\alpha} (a_{\mathbf{k},\alpha} b_{-\mathbf{k},\alpha} - a^{\dagger}_{\mathbf{k},\alpha} b^{\dagger}_{-\mathbf{k},\alpha}) + a^{\dagger}_{\mathbf{k},\alpha} b_{\mathbf{k},\alpha} - a_{\mathbf{k},\alpha} b^{\dagger}_{\mathbf{k},\alpha}]\}, \tag{4.92}$$

where we have set $g_k = [(\epsilon_0 \omega_k)/(4\nu_0 \rho_m)]^{1/2}$, and used the form of the dipole coupling to ensure that photons only couple to polarization excitations of the same polarization.

4.6.1 Polariton operators

We now want to diagonalize this Hamiltonian. This is done be defining new creation and annihilation operators in terms of which the Hamiltonian can be expressed as a sum of single-particle energies times number operators. These new operators, which are linear combinations of the old ones, create and annihilate excitations that are part photon and part matter excitation. These excitations are called polaritons. In order to figure out the form of the polariton operators, we note that, since our Hamiltonian is translation-invariant, momentum is conserved, and that different polarizations are not coupled. Since there are two independent types of free-field excitation, we expect there will be two different types of excitation for the coupled fields. If we call these polariton annihilation operators $c_{\mathbf{k},\alpha}$ and $d_{\mathbf{k},\alpha}$, application of these operators to a state subtracts a momentum $\hbar\mathbf{k}$ from it. The old operators that have this effect are $a_{\mathbf{k},\alpha}$, $b_{\mathbf{k},\alpha}$, $a^{\dagger}_{-\mathbf{k},\alpha}$ and $b^{\dagger}_{-\mathbf{k},\alpha}$, so $c_{\mathbf{k},\alpha}$ and $d_{\mathbf{k}}$ should be linear combinations of them. We will start by examining $c_{\mathbf{k},\alpha}$, which we assume is given by

$$c_{\mathbf{k},\alpha} = z_1 a_{\mathbf{k},\alpha} + z_2 b_{\mathbf{k},\alpha} + z_3 a^{\dagger}_{-\mathbf{k},\alpha} + z_4 b^{\dagger}_{-\mathbf{k},\alpha}. \tag{4.93}$$

The relation $[c_{\mathbf{k},\alpha}, c^{\dagger}_{\mathbf{k},\alpha}] = 1$ implies that

$$|z_1|^2 + |z_2|^2 - |z_3|^2 - |z_4|^2 = 1. \tag{4.94}$$

If we let the energy corresponding to the polariton created by $c^{\dagger}_{\mathbf{k},\alpha}$ be $\hbar\omega_{\mathbf{k}}$, then we should have

$$[c_{\mathbf{k},\alpha}, H] = \hbar\omega_{\mathbf{k}} c_{\mathbf{k},\alpha}. \tag{4.95}$$

Since this is a linear system, we also expect that the frequencies of the excitations should be the same as those of the classical plane-wave solutions, so that $\omega_{\mathbf{k}}$ should be given by Eq. (4.79). This is something we need to prove, so, for the moment, we will not assume that we know what $\omega_{\mathbf{k}}$ is. The above equation can be written in matrix form as

$$\begin{pmatrix} \omega_k & -ig_k & 0 & -i(-1)^{\alpha} g_k \\ ig_k & \nu_0 & -i(-1)^{\alpha} g_k & 0 \\ 0 & -i(-1)^{\alpha} g_k & -\omega_k & -ig_k \\ -i(-1)^{\alpha} g_k & 0 & ig_k & -\nu_0 \end{pmatrix} \begin{pmatrix} z_1 \\ z_2 \\ z_3 \\ z_4 \end{pmatrix} = \omega_{\mathbf{k}\pm} \begin{pmatrix} z_1 \\ z_2 \\ z_3 \\ z_4 \end{pmatrix}. \tag{4.96}$$

Solving this eigenvalue problem, we find that $\omega_{\mathbf{k}\pm}$ is, as we expected, given by Eq. (4.79). We will denote the solutions with the plus and minus signs as $\omega_{\mathbf{k}+}$ and $\omega_{\mathbf{k}-}$, respectively, and we will let $c_{\mathbf{k},\alpha}$ correspond to the excitation with energy $\hbar\omega_{\mathbf{k}+}$, and $d_{\mathbf{k},\alpha}$ correspond to the excitation with energy $\hbar\omega_{\mathbf{k}-}$.

4.6.2 Polariton dispersion relation

Before finding the explicit form of the operators $c_{\mathbf{k},\alpha}$ and $d_{\mathbf{k},\alpha}$ and the final form of the Hamiltonian, let us examine the single-polariton energies, $\hbar\omega_{\mathbf{k}\pm}$. For small values of k, in particular, for $kc \ll \nu_0$, we have $\omega_{k+} \cong \nu_0$ and $\omega_{k-} \cong kc$. In this regime, we see that polaritons with energy $\hbar\omega_{k+}$ are atom-like, in that they have a fixed energy equal to that of the oscillator energy-level spacing, which is independent of k. The polaritons with energy $\hbar\omega_{k-}$, however, are photon-like, because their energy has the same dependence on wavevector as does that of photons. For kc large, i.e. $kc \gg \nu_0$, the situation is reversed, so that now $\omega_{k+} \cong kc$ and $\omega_{k-} \cong \nu_0$. Therefore, ω_{k+} is atom-like for small k but becomes photon-like for large k, and ω_{k-} is photon-like for small k and becomes atom-like for large k.

Besides giving us the values of $\omega_{k\pm}$, the previous equation gives us expressions for the polariton operators in terms of the matter and field operators. Setting

$$\bar{\omega} = 2\sqrt{\omega_{k+}\omega}\left(\frac{2\omega_{k+}^2 - \omega_k^2 - \nu_0^2}{\omega_{k+}^2 - \nu_0^2}\right)^{1/2}, \tag{4.97}$$

we find that, for $c_{\mathbf{k},\alpha}$,

$$z_1 = (\omega_{k+} + \omega_k)/\bar{\omega}, \qquad z_2 = \frac{-2ig_k\omega_k}{\bar{\omega}(\nu_0 - \omega_{k+})},$$

$$z_3 = -(-1)^\alpha(\omega_k - \omega_{k+})/\bar{\omega}, \qquad z_4 = -i(-1)^\alpha\frac{(\omega_k^2 - \omega_{k+}^2)(\nu_0 - \omega_{k+})}{2\bar{\omega}\nu_0 g_k}. \tag{4.98}$$

The expressions for $d_{\mathbf{k},\alpha}$ are the same, except that ω_{k+} is replaced by ω_{k-}. The Hamiltonian is now

$$H = \sum_{\mathbf{k}}\hbar\nu_0 b_{\mathbf{k},0}^\dagger b_{\mathbf{k},0} + \sum_{\mathbf{k}}\sum_{\alpha=1}^{2}(\hbar\omega_{k+}c_{\mathbf{k},\alpha}^\dagger c_{\mathbf{k},\alpha} + \hbar\omega_{k-}d_{\mathbf{k},\alpha}^\dagger d_{\mathbf{k},\alpha}), \tag{4.99}$$

which is simply a Hamiltonian for noninteracting excitations. Therefore, a linear medium interacting with the electromagnetic field can be described in terms of noninteracting polaritons. Our next step is to show that, if the medium is nonlinear, the polaritons do interact.

4.7 Two-level atomic medium

In order to model a *nonlinear* medium, we will no longer treat the atoms as harmonic oscillators. Instead, we shall ignore all of the energy levels except for the ground state and one excited state, and treat them as two-level systems, which is a commonly made approximation in nonlinear quantum optical systems. This amounts to assuming that the effects of other levels can be ignored because they are far off-resonance or not coupled due to selection rules. We will also make the dipole approximation and assume, for simplicity, that there is one optically active electron per atom. Let the energy difference between the two states be $\hbar\nu_\beta$ and the state vectors be $|\psi_{g\beta}\rangle$ and $|\psi_{e\beta}\rangle$ for the ground and excited states, respectively. Defining the atomic operators as in Eq. (1.72),

$$\sigma^z_\beta = |\psi_{e\beta}\rangle\langle\psi_{e\beta}| - |\psi_{g\beta}\rangle\langle\psi_{g\beta}|, \qquad \sigma^{(+)}_\beta = |\psi_{e\beta}\rangle\langle\psi_{g\beta}|, \tag{4.100}$$

and $\sigma^{(-)}_\beta = (\sigma^{(+)}_\beta)^\dagger$, the dipole-coupled Hamiltonian from Eq. (4.65) becomes

$$H = \frac{1}{2}\sum_\beta \hbar\nu_\beta\sigma^z_\beta + \sum_{\mathbf{k},\alpha}\hbar\omega_k a^\dagger_{\mathbf{k},\alpha}a_{\mathbf{k},\alpha} - \frac{q}{\epsilon_0}\sum_\beta[\langle\psi_e|\mathbf{s}_\beta|\psi_g\rangle\cdot\mathbf{D}(\mathbf{R}_\beta)\sigma^{(+)}_\beta + h.c.]. \tag{4.101}$$

The interaction term was obtained by applying the operator $\sum_\beta(|\psi_{e\beta}\rangle\langle\psi_{e\beta}| + |\psi_{g\beta}\rangle\langle\psi_{g\beta}|)$ to both sides of the operator $\sum_\beta \mathbf{s}_\beta\cdot\mathbf{D}(\mathbf{R}_\beta)$, which projects this term onto the space spanned by the atomic states we are considering.

Next, we can expand the displacement field in terms of free-field annihilation and creation operators using Eq. (4.58). This leads to a model with an interaction Hamiltonian – the last term in Eq. (4.101) – given by

$$H' = -i\sum_{\mathbf{k},\alpha}\sum_\beta \hbar g_{\mathbf{k},\alpha}(e^{i\mathbf{k}\cdot\mathbf{R}_\beta}a_{\mathbf{k},\alpha} - e^{-i\mathbf{k}\cdot\mathbf{R}_\beta}a^\dagger_{\mathbf{k},\alpha})\sigma^{(+)}_\beta + h.c., \tag{4.102}$$

where we have assumed that the atoms are the same, so that their dipole matrix elements are identical:

$$g_{\mathbf{k},\alpha} = q\langle\psi_e|\mathbf{s}_\beta|\psi_g\rangle\cdot\hat{\mathbf{e}}_{\mathbf{k},\alpha}\sqrt{\frac{\omega_k}{2\hbar\epsilon_0 V}}. \tag{4.103}$$

Next, we will drop the counter-rotating terms, $a^\dagger\sigma^{(+)}_\beta$ and $a\sigma^{(-)}_\beta$. This is called the rotating-wave approximation, and it is valid near resonance. The price that is paid by using this approximation is that small frequency shifts are neglected. Doing so now results in the interaction Hamiltonian

$$H_{int} = i\sum_{\mathbf{k},\alpha}\sum_\beta \hbar(g^*_{\mathbf{k},\alpha}e^{-i\mathbf{k}\cdot\mathbf{R}_\beta}a^\dagger_{\mathbf{k},\alpha}\sigma^{(-)}_\beta - g_{\mathbf{k},\alpha}e^{i\mathbf{k}\cdot\mathbf{R}_\beta}a_{\mathbf{k},\alpha}\sigma^{(+)}_\beta). \tag{4.104}$$

What we have now is a model of a two-level atomic medium, which is widely used as a simplified model for resonant interactions of atoms and the electromagnetic field.

Since all the field modes are included, this model is often used to treat propagation in resonant media, which we treat in more detail in later sections. We will defer a more complete treatment of this Hamiltonian until we have a better understanding of the single-mode behavior, given in the next section.

4.7.1 Multi-atom Jaynes–Cummings model

In order to simplify the presentation at this stage, we will assume that only one field mode, with frequency ω and wavevector $\mathbf{k}$, is highly excited, so that all of the other modes can be ignored. The creation and annihilation operators of the excited mode will be denoted by a and $a^\dagger$ without any subscripts. We can then absorb the phases in the interaction term by redefining the atomic raising and lowering operators to be

$$\begin{aligned}\tilde{\sigma}_\beta^{(+)} &= e^{i\mathbf{k}\cdot\mathbf{R}_\beta}|\psi_{e\beta}\rangle\langle\psi_{g\beta}|,\\ \tilde{\sigma}_\beta^{(-)} &= e^{-i\mathbf{k}\cdot\mathbf{R}_\beta}|\psi_{e\beta}\rangle\langle\psi_{g\beta}|.\end{aligned} \tag{4.105}$$

This is a unitary transformation that preserves the commutation relations of the operators, σ_β^z, $\tilde{\sigma}_\beta^{(+)}$ and $\tilde{\sigma}_\beta^{(-)}$. In terms of these operators, the Hamiltonian becomes

$$H = \frac{1}{2}\sum_\beta \hbar\nu_\beta\sigma_\beta^z + \hbar\omega a^\dagger a + i\hbar\sum_\beta(g^* a^\dagger\tilde{\sigma}_\beta^{(-)} - ga\tilde{\sigma}_\beta^{(+)}). \tag{4.106}$$

Before proceeding, let us discuss one of the approximations we have made. Dropping the counter-rotating terms is justified if the detuning between the field and the atoms is not too large. This condition is also necessary for the two-level approximation to be a reasonable one. We are, however, going to want the detuning sufficiently large so that, if we start all of the atoms in their ground states, only a small number of them are in their excited states at later times. This latter condition is also necessary for the description of the medium in terms of a susceptibility that does not take into account losses to make sense. If we are too close to a resonance, absorption becomes a large effect. For detunings that are much larger than the linewidth of the upper state but much smaller than the spacing between energy levels, these conditions will hold.

Our next step is to define collective operators. Define a sum over N atoms as

$$S^z = \frac{1}{2}\sum_\beta \sigma_\beta^z, \qquad S^{(\pm)} = \sum_\beta \sigma_\beta^{(\pm)}. \tag{4.107}$$

These operators satisfy the usual angular momentum commutation relations,

$$\begin{aligned}[][S^z, S^{(\pm)}] &= \pm S^{(\pm)},\\ [S^{(+)}, S^{(-)}] &= 2S^z.\end{aligned} \tag{4.108}$$

If we also assume that all the resonant frequencies are the same, so that $\nu_\beta = \nu_0$, then in terms of these operators the Hamiltonian becomes

$$H = \hbar\nu_0 S^z + \hbar\omega a^\dagger a + i\hbar(g^* a^\dagger S^{(-)} - gaS^{(+)}). \tag{4.109}$$

Since the number of atoms is N, and because each atom has two levels, we can treat the atomic part of our system as an object with a spin of $s = N/2$. The simplest version of this model, one with only one atom, was first treated in detail by Jaynes and Cummings. It is exactly soluble, and it forms the basis for a branch of quantum optics known as cavity quantum electrodynamics (QED). The model was subsequently extended to the N-atom case by Tavis and Cummings. Since the phase of the atomic levels can be chosen arbitrarily, the phase of the coupling is not physically important, and g is often chosen as either real or imaginary for convenience.

Using the Heisenberg equation of motion, Eq. (2.7), together with the commutators given above, gives the Heisenberg-picture field and atomic equations (in the rotating-wave approximation). These are a single-mode version of the operator Maxwell–Bloch equations, as follows:

$$\frac{da}{dt} = -i\omega a + g^* S^{(-)},$$
$$\frac{dS^{(-)}}{dt} = -i\nu S^{(-)} + 2gS^z a, \tag{4.110}$$
$$\frac{dS^z}{dt} = -[g^* a^\dagger S^{(-)} + gaS^{(+)}].$$

These equations and their generalizations to include damping and driving terms form the basis of the theory of near-resonant atom–interferometer interactions, including, for example, single-mode laser theory. In the mean-field approximation, the operators are replaced by their mean values, and mean values of products are assumed to factorize so that they are replaced by the product of the mean values. This results in a set of three, nonlinear, coupled differential equations, which can be solved numerically.

4.8 Polaritonic limit

We now wish to investigate what happens in the limit of large detunings away from resonance, where we expect polariton-like behavior. To achieve this, our next step is to make use of the Holstein–Primakoff representation of spin operators. This representation, which was originally formulated to study spin waves, expresses the spin operators as functions of an auxiliary set of creation and annihilation operators. In detail, we have that

$$S^{(+)} = -ib^\dagger(2s - b^\dagger b)^{1/2}, \qquad S^z = -s + b^\dagger b, \qquad S^{(-)} = i(2s - b^\dagger b)^{1/2} b, \tag{4.111}$$

where $[b, b^\dagger] = 1$. Here we use the convention that when the argument of the square root is positive, so is the square root, and when it is negative, the square root is i times a positive number. These operators, defined in terms of creation and annihilation operators, have the same commutation relations as the spin operators. The physical spin space, on which the original spin operators act, corresponds to the space spanned by number states, $b^\dagger b|n\rangle = n|n\rangle$, for which $n \leq 2s$. A state in this subspace will remain in this subspace when acted upon by one of the operators in the above equation.

The description of a medium in terms of a nonlinear susceptibility requires that most of the atoms in the medium remain in their ground states. This implies that the fraction of atoms in their excited states is small, which further implies that, when acting on states of this type, the operator $b^\dagger b/(2s)$ will be small (produce states with a small norm). We can use this fact to expand the square roots and keep only the first two terms,

$$S^{(-)} \cong i\sqrt{2s}\left(1 - \frac{b^\dagger b}{4s}\right) b,$$
$$S^{(+)} \cong -i\sqrt{2s}\, b^\dagger \left(1 - \frac{b^\dagger b}{4s}\right), \tag{4.112}$$

with S^z remaining the same. Upon substituting these expressions into the Hamiltonian and dropping constant terms, we have

$$H = \hbar\nu_0 b^\dagger b + \hbar\omega a^\dagger a - \hbar g\sqrt{2s}(ab^\dagger + a^\dagger b) + \frac{\hbar g}{2\sqrt{2s}}(a(b^\dagger)^2 b + a^\dagger b^\dagger b^2). \quad (4.113)$$

We can now diagonalize the part of the Hamiltonian that is quadratic in the creation and annihilation operators. This is identical to the case of a linear medium, and will result in polariton modes. The difference here is that, because of the quartic terms in the Hamiltonian, the polaritons will interact. The diagonalization is very similar to the one in the previous section, and actually simpler since we now only have two modes. If we express the atomic and field operators, b and a, in terms of polariton operators, c and d, as

$$a = -\sin\theta\, d + \cos\theta\, c, \qquad b = \cos\theta\, d + \sin\theta\, c, \quad (4.114)$$

we find that the quadratic part of the Hamiltonian is diagonal when

$$\tan\theta = \frac{-(\omega - \nu_0)}{2g\sqrt{2s}}\left\{\left[1 + \left(\sqrt{2s}\frac{2g}{(\omega - \nu_0)}\right)^2\right]^{1/2} - 1\right\}. \quad (4.115)$$

Note that $\tan\theta$ is discontinuous at the point $\omega = \nu_0$. It is positive for $\omega > \nu_0$ and negative for $\omega < \nu_0$. The Hamiltonian is, in terms of the polariton operators,

$$H = \hbar\omega' c^\dagger c + \hbar\nu_0' d^\dagger d + H', \quad (4.116)$$

where $\omega' = \omega - (g\sqrt{2s})\tan\theta$ and $\nu_0' = \nu_0 + (g\sqrt{2s})\tan\theta$. The interaction term is given by

$$\begin{aligned} H_{int} = & \frac{\hbar g}{2\sqrt{2s}}[(c\cos\theta - d\sin\theta)(c^\dagger\sin\theta + d^\dagger\cos\theta)^2(c\sin\theta + d\cos\theta) \\ & + (c^\dagger\sin\theta + d^\dagger\cos\theta)(c\sin\theta + d\cos\theta)^2(c^\dagger\cos\theta - d^\dagger\sin\theta)]. \end{aligned} \quad (4.117)$$

If the c mode is highly excited and the d mode is not, then the dominant terms in the interaction will be those containing only the c-mode operators. This is a situation that results from the medium being driven by an external field, at the c-mode frequency. In that case we have

$$H_{int} \to \frac{\hbar g}{2\sqrt{2s}}\cos\theta\sin^3\theta\,(c(c^\dagger)^2 c + c^\dagger c^2 c^\dagger), \quad (4.118)$$

which is a Kerr-type interaction. Therefore, we see that a simple two-level-atom model of the medium results in a standard type of nonlinear optical interaction, but it should be noted that this is an interaction not between photons but between polaritons. Also we note that the microscopic theory gives us a specific ordering of the operators in the nonlinear terms. This is something that the quantized macroscopic theory was unable to do.

4.8.1 Comparison to macroscopic quantization

We want to compare the above expression for the interaction Hamiltonian with one derived from the macroscopic theory. These should be approximately the same in the large-detuning limit. Since the macroscopic theory does not dictate an operator ordering, we shall ignore operator ordering issues and assume that all expressions are in normally ordered form. Therefore, the above Hamiltonian will be expressed as

$$H' = \frac{\hbar g}{\sqrt{2s}} \cos\theta \sin^3\theta \, (c^\dagger)^2 c^2. \tag{4.119}$$

In the large-detuning limit, the angle θ is small, and is given approximately by

$$\theta = \frac{g\sqrt{2s}}{\nu_0 - \omega}. \tag{4.120}$$

We will express the coupling constant g as

$$g = d\sqrt{\frac{\omega}{2\hbar\epsilon_0 V}}, \tag{4.121}$$

where $d = q\langle\psi_e|\mathbf{s}_\beta|\psi_g\rangle \cdot \hat{\mathbf{e}}_{\mathbf{k},\alpha}$ is the component of the atomic dipole moment in the direction of the polarization vector of the field mode. Because the number of atoms is $2s$, the number density of atoms, ρ_n, is just $2s/V$. This gives us that

$$\frac{\hbar g}{\sqrt{2s}}\theta^3 = \frac{\omega^2}{4\hbar\epsilon_0^2 V} \frac{\rho_n d^4}{(\omega - \nu_0)^3}, \tag{4.122}$$

so that we have

$$H' = -\frac{\omega^2}{4\hbar\epsilon_0^2 V} \frac{\rho_n d^4}{(\omega - \nu_0)^3} (c^\dagger)^2 c^2. \tag{4.123}$$

Let us now see if we can extract a similar expression from the macroscopic quantization theory in Chapters 1 and 3. The interaction Hamiltonian in those chapters is now

$$H_{int} = \frac{1}{4}\eta^{(3)} \int d^3r\, D^4, \tag{4.124}$$

where the displacement field is given, from Eq. (4.58), by

$$\mathbf{D}(\mathbf{r}, t) = i\sum_{\mathbf{k},\alpha} \sqrt{\frac{\epsilon_0\hbar\omega_k}{2V}}\, \hat{\mathbf{e}}_\alpha(\mathbf{k})(e^{i\mathbf{k}\cdot\mathbf{r}} a_{\mathbf{k},\alpha} - e^{-i\mathbf{k}\cdot\mathbf{r}} a^\dagger_{\mathbf{k},\alpha}), \tag{4.125}$$

and $\eta^{(3)} = -\epsilon_0(\eta^{(1)})^4\chi^{(3)}$. Now, for a single mode of frequency ω,

$$\begin{aligned} H_{int} &\to \frac{1}{4}\eta^{(3)} \int d^3r \left(\frac{\epsilon_0\hbar\omega}{2V}\right)^2 (e^{i\mathbf{k}\cdot\mathbf{r}} a - e^{-i\mathbf{k}\cdot\mathbf{r}} a^\dagger)^4 \\ &\to -\frac{1}{4\epsilon}\chi^{(3)} \left(\frac{\epsilon_0\hbar\omega}{2\epsilon}\right)^2 \frac{6}{V} (a^\dagger)^2 a^2, \end{aligned} \tag{4.126}$$

where, in the last line, only the Kerr-type terms have been kept. Comparing this expression with Eq. (4.123), in the large-detuning limit where $\epsilon \approx \epsilon_0$, we therefore conclude that our polariton model corresponds to a $\chi^{(3)}$ coefficient of

$$\chi^{(3)} = \frac{-2d^4\rho_n}{3\epsilon_0\hbar^3(\nu_0 - \omega)^3}. \tag{4.127}$$

From Chapter 1, Eq. (1.99), we see that the semiclassical nonlinear Kerr coefficient for a medium of two-level atoms with number density ρ_n (in the low-density limit) gives the same result. This demonstrates that the macroscopic quantization procedure is able to generate essentially equivalent results to a microscopic, first-principles theory of photons and atoms.

Additional reading

Books

- For discussions of the multipolar Hamiltonian, see:

C. Cohen-Tannoudji, J. Dupont-Roc, and G. Grynberg, *Photons and Atoms: Introduction to Quantum Electrodynamics* (Wiley, New York, 1997).
W. P. Healy, *Non-Relativistic Quantum Electrodynamics* (Academic Press, New York, 1982).

Articles

- The original paper on polaritons is:

J. J. Hopfield, *Phys. Rev.* **112**, 1555 (1958).

- The original model of cavity QED was given in:

E. T. Jaynes and F. W. Cummings, *Proc. IEEE* **51**, 89 (1963).

- For more recent work on microscopic models of linear media and the issue of causality in the Coulomb gauge, see:

P. D. Drummond and M. Hillery, *Phys. Rev.* A **59**, 691 (1999).
C. W. Gardiner and P. D. Drummond, *Phys. Rev.* A **38**, 4897–4898 (1988).
B. Huttner and S. M. Barnett, *Europhys. Lett.* **18**, 487 (1992).
B. Huttner and S. M. Barnett, *Phys. Rev.* A **46**, 4306 (1992).

- For microscopic models of nonlinear media, see:

M. Hillery and L. D. Mlodinow, *Phys. Rev.* A **31**, 797 (1985).
M. Hillery and L. D. Mlodinow, *Phys. Rev.* A **55**, 678 (1997).

Problems

4.1 Hopfield started from a different Lagrangian in his paper on polaritons than the one we used. He used the Lagrangian density

$$\mathcal{L} = \frac{1}{2}\left(\epsilon_0 \mathbf{E}^2 - \frac{1}{\mu_0}\mathbf{B}^2\right) + \frac{1}{2\beta\omega^2}\left(\frac{d\mathbf{P}}{dt}\right)^2 - \frac{1}{2\beta}\mathbf{P}^2 + A_0(\nabla \cdot \mathbf{P}) + \mathbf{A} \cdot \left(\frac{d\mathbf{P}}{dt}\right).$$

Here, $\mathbf{P}(\mathbf{r}, t)$ is a polarization field, β is a matter–field coupling constant, and the coordinates of this Lagrange density are A_0, $\mathbf{A}$ and $\mathbf{P}$. Find the equations of motion generated by this Lagrangian density.

(You should get Maxwell's equations plus an equation of motion for $\mathbf{P}$. Note that this simplified model of the Lagrangian omits dipole–dipole coupling terms. As a result it is not equivalent to the form we use, and has potential problems in terms of both causality and frequency shifts.)

4.2 We want to use the equations of motion from the last problem to find the classical modes of the system. Set $\mathbf{A}(\mathbf{r}, t) = \mathbf{A}(\mathbf{k}, \omega)e^{i(\mathbf{k}\cdot\mathbf{r}-\omega t)}$, $A_0(\mathbf{r}, t) = A_0(\mathbf{k}, \omega)e^{i(\mathbf{k}\cdot\mathbf{r}-\omega t)}$ and $\mathbf{P}(\mathbf{r}, t) = \mathbf{P}(\mathbf{k}, \omega)e^{i(\mathbf{k}\cdot\mathbf{r}-\omega t)}$. Assume we are using the Coulomb gauge so that $\mathbf{k}\cdot\mathbf{A}(\mathbf{k}, \omega) = 0$. Substitute these expressions into the equations of motion in order to find the dispersion relation, that is, an equation relating ω to k.

4.3 Consider the eigenvalue problem of Eq. (4.96) What happens if we include only momenta in one direction? Show that the eigenvalue problem reduces to the form

$$\begin{pmatrix} \omega_k - \omega_{\mathbf{k}\pm} & -ig_k \\ ig_k & \nu_0 - \omega_{\mathbf{k}\pm} \end{pmatrix} \begin{pmatrix} z_1 \\ z_2 \end{pmatrix} = 0.$$

Hence, prove that the eigenvalue problem for the dispersion relation changes to

$$n_r(\omega) = \frac{(kc)}{\omega} = 1 + \frac{g_k^2/\omega}{\nu_0 - \omega}.$$

This demonstrates that the frequency shift in the linear polarization model is due to the reflected fields, and does not occur if these are omitted.

4.4 Let us have another look at the Hamiltonian

$$H = \hbar\nu_0 S^{(z)} + \hbar\omega a^\dagger a + i\hbar g(a^\dagger S^{(-)} - aS^{(+)}).$$

One way of analyzing it, in the case that the spin is large, is to replace the spin operators by c-numbers. In particular, let $S^{(x)} \to s\sin\theta\cos\phi$, $S^{(y)} \to s\sin\theta\sin\phi$ and $S^{(z)} \to s\cos\theta$, where $S^{(\pm)} = S^{(x)} \pm iS^{(y)}$. Substituting these values into the Hamiltonian, diagonalize it, find the ground-state energy, and then minimize the ground-state energy as a function of θ and ϕ.

5 Coherence and quantum dynamics in simple systems

In this chapter we will consider some simple systems arising out of the Hamiltonians in the previous chapters. Here, we will look at these systems without considering damping or reservoirs. These systems involve only a small number of field modes – typically one, two or three. We will revisit them in later chapters to consider what happens when the fields are confined to a cavity with damping, including possible input and output coupling.

Before discussing the models, however, we need to describe a number of features of fields consisting of a small number of modes. We will begin with a discussion of the quantum theory of photo-detection and optical coherence, due to Mandel and Glauber. This makes use of the microscopic model for atom–field interactions in the previous chapter, to allow us to introduce a theory of coherence and photon counting.

We will also treat the representation for a single-mode field known as a P-representation. This will be the first quasi-distribution function we will encounter. It is a c-number representation of the quantum state of a field mode. As we shall see in subsequent chapters, quasi-distribution functions are very useful in describing the dynamics of interacting quantum systems coupled to reservoirs. We will then go on to use the P-representation to define the notion of a nonclassical state of the field; such states are signified by the fact that the P-representation is no longer a well-behaved, positive distribution. Nonclassical states are natural results of nonlinear optical interactions, and we will show how that comes about for some of our models. It is also possible to define P-representations for multi-mode fields. This leads to a discussion of nonclassical correlations between modes.

In particular, we will focus on the type of nonclassical field called a squeezed field. These show important quantum properties, including noise levels reduced below the vacuum level, quantum entanglement, and the existence of an experimentally verifiable Einstein–Podolsky–Rosen (EPR) paradox. They were first observed in the pioneering experiments of Slusher, Kimble and others.

5.1 Photon counting and quantum coherence

We would like to explore the properties of the quantized electromagnetic field, and, in particular, see what properties a quantized field can have that a classical one cannot. As we shall see, this is closely connected to coherent states and the P-representation. It is also connected to nonlinear optics, because nonlinear optical processes are prime sources of light that cannot be described classically. It is, therefore, useful to start by discussing some of the novel features of quantized light, so that we know what we are looking for when we perform calculations whose goal is to find the output of nonlinear optical devices. We take advantage of the microscopic theory of atom–field interactions introduced in the previous chapter to model the action of a photo-detector interacting with a radiation field. This is essential, because quantum measurements ultimately depend on photon counting.

5.1.1 Photon counting

One of the things we can do with light is to count photons. We can open the shutter in front of a photo-detector between the times t and $t + \Delta t$ and see how many photons are detected. Photo-detection is a probabilistic process, so the number of photons detected will vary from run to run, even if the input fields are identical. Let $p(m; t, t + \Delta t)$ be the probability that m photons are detected in the time interval between t and $t + \Delta t$. The collection of these probabilities, for a fixed time interval but for arbitrary m, is called the photo-count distribution of the field in that time interval. These probabilities can be calculated by modeling the photo-detector as a collection of atoms, which then interact with the electromagnetic field. The probability $p(m; t, t + \Delta t)$ can be found by determining the probability that m of the atoms have absorbed a photon in the interval between t and $t + \Delta t$, which means finding the probability that m atoms are in an excited state at time $t + \Delta t$. The field itself can be modeled as a classical stochastic field or as the quantized electromagnetic field. The theory of photo-detection is a well-developed part of quantum optics, and we use these techniques to explain how to calculate the results of photo-detection experiments in terms of correlations of field operators.

We assume that photon counting is carried out via the measurement of an excited state of a multi-level localized quantum system – an atom, molecule or semiconductor, called the detector – coupled to the radiation field. We assume that the initial state of the detector at time $t = t_0$ is the ground state, $|g\rangle$. The initial state of the combined field–detector system is $\rho(t_0) = |g\rangle\langle g| \otimes \rho_f(t_0)$, where $\rho_f(t_0)$ is the state of the electromagnetic field that we wish to detect. We suppose that the detector has a number of upper levels, labeled by their resonance frequency ν. In other words, we use $|\nu\rangle$ to indicate an energy eigenstate with energy $E = \hbar\nu$.

The probability of observing an excited state $|\nu\rangle$ with resonance frequency ν at time t, given a ground state $|g\rangle$ at time t_0, is the expectation value of the projection operator $|\nu\rangle\langle\nu|$,

$$p(\nu, t) = \mathrm{Tr}[\rho(t)|\nu\rangle\langle\nu|]. \tag{5.1}$$

The total probability of observing a single photon, corresponding to the excitation of any of these upper levels, is therefore

$$p(1; t_0, t) = \sum_\nu \mathrm{Tr}[\rho(t)|\nu\rangle\langle\nu|]. \tag{5.2}$$

An actual extended photo-detector consists of many atoms distributed in a region of space, so that, in general, a sum (or integral) over the positions of the detector atoms will be required.

In order to calculate the detection probabilities and the count rate, we need to evolve the field–atom state from time t_0 to time $t_0 + \Delta t$. We can accomplish this by making use of the dipole-coupled Hamiltonian of Eq. (4.65), introduced in Chapter 1. In the interaction picture, the density matrix evolves according to the equation

$$\frac{d\rho(t)}{dt} = \frac{i}{\hbar}[\rho(t), H_I(t)]. \tag{5.3}$$

The interaction Hamiltonian, for a detector atom located at $\mathbf{r}$, is a multi-level extension of the usual two-level one we have seen previously,

$$H_I(t) = -\frac{1}{\epsilon_0}\mathbf{D}(\mathbf{r}, t)) \cdot \sum_\nu [\mathbf{d}_\nu \sigma^{(+)}(\nu, t) + \mathbf{d}_\nu \sigma^{(-)}(\nu, t)], \tag{5.4}$$

where $\mathbf{d}_\nu$ is a real dipole moment for the transition to the νth level, $\sigma^{(-)}(\nu, t) = [\sigma^{(+)}(\nu, t)]^\dagger$, and

$$\sigma^{(+)}(\nu, 0) = |\nu\rangle\langle g|. \tag{5.5}$$

In the interaction picture, $\sigma^{(+)}(\nu, t) = e^{i\nu t}\sigma^{(+)}(\nu, 0)$. We also make the rotating-wave approximation, of retaining only near-resonant terms, so that, if we express $\mathbf{D} = \hat{\mathbf{e}}(D^{(+)} + D^{(-)})$, where $D^{(+)}(\mathbf{r}, t) \sim e^{-i\omega t}$, then $D^{(+)}(\mathbf{r}, t)$ includes only annihilation operators like a_k. Defining $d_\nu = \mathbf{d}_\nu \cdot \hat{\mathbf{e}}$, which we assume is real for simplicity, we obtain

$$H_I(t) = -\frac{1}{\epsilon_0}\sum_\nu d_\nu [D^{(+)}(\mathbf{r}, t)\sigma^{(+)}(\nu, t) + D^{(-)}(\mathbf{r}, t)\sigma^{(-)}(\nu, t)]. \tag{5.6}$$

Formal integration of the density matrix evolution equation in the interaction picture of Section 2.1.1 yields that

$$\rho(t) = \rho(t_0) + \frac{i}{\hbar}\int_{t_0}^{t} [\rho(t_1), H_I(t_1)]\, dt_1. \tag{5.7}$$

This can be solved by iteration. We will assume here that the field has a relatively low intensity, and there is a distribution of resonant frequencies. This is essentially the situation that applies in a typical semiconductor photo-detector, where electrons in a solid are excited into a conduction band. Under these conditions, we can solve this to second order in the coupling by iterating once, and substitute $\rho(t_0)$ for $\rho(t_1)$ in the final integral of Eq. (5.7). This leads to an approximate solution:

$$\begin{aligned}\rho(t) \approx \rho(t_0) &+ \frac{i}{\hbar}\int_{t_0}^{t} [\rho(t_0), H_I(t_1)]\, dt_1 \\ &- \frac{1}{\hbar^2}\int_{t_0}^{t} dt_2 \int_{t_0}^{t_2} dt_1 [\,[\rho(t_0), H_I(t_1)], H_I(t_2)\,]\, dt_1\, dt_2.\end{aligned} \tag{5.8}$$

Including only nonvanishing terms, the probability of observing an excited state $|\nu\rangle$ is therefore

$$\begin{aligned}p(\nu, t) &= \mathrm{Tr}[\rho(t)|\nu\rangle\langle\nu|] \\ &= \frac{1}{\hbar^2}\int_{t_0}^{t} dt_2 \int_{t_0}^{t_2} dt_1\, \mathrm{Tr}_f[\langle\nu|H_I(t_2)\rho(t_0)H_I(t_1) + H_I(t_1)\rho(t_0)H_I(t_2)|\nu\rangle],\end{aligned} \tag{5.9}$$

where the subscript f on the trace indicates that it is over the field states. Noting that $\sigma^{(-)}(\nu,t)|g\rangle = 0$, this becomes

$$p(\nu,t) = \frac{d_\nu^2}{(\epsilon_0\hbar)^2}\int_{t_0}^{t}\int_{t_0}^{t} dt_1\, dt_2\, e^{i\nu(t_1-t_2)}\, \mathrm{Tr}[D^{(+)}(t_2)\rho_f(t_0)D^{(-)}(t_1)]. \tag{5.10}$$

Assuming that the field is quasi-monochromatic with frequency near ω, so that $D^{(+)}(t) \approx D^{(+)}(t_0)e^{-i\omega t}$, then this simplifies to

$$\begin{aligned} p(\nu,t) &\approx \frac{d_\nu^2}{(\epsilon_0\hbar)^2}\left|\int_{t_0}^{t} dt_1\, e^{i(\nu-\omega)t_1}\right|^2 \mathrm{Tr}_f[D^{(-)}(t_0)D^{(+)}(t_0)\rho_f(t_0)] \\ &\approx \frac{d_\nu^2}{(\epsilon_0\hbar)^2}\frac{4\sin^2[(\nu-\omega)\Delta t/2]}{(\nu-\omega)^2}\mathrm{Tr}_f[D^{(-)}(t_0)D^{(+)}(t_0)\rho_f(t_0)], \end{aligned} \tag{5.11}$$

where $\Delta t = t - t_0$. Next, we note that, at relatively long times compared to the detuning, the expression on the right-hand side becomes a sharply peaked function of frequency, and reduces to a delta-function:

$$\lim_{\Delta t\to\infty}\frac{2}{\pi}\frac{\sin^2[(\nu-\omega)\Delta t/2]}{(\nu-\omega)^2\Delta t} = \delta(\nu-\omega). \tag{5.12}$$

If we now suppose that there are $n_\nu d\nu$ excited states in a frequency interval $d\nu$, and the dipole moment has an average square value of $\langle d^2\rangle$ at the resonance frequency $\nu_0 = \omega$, then, on integrating over the excited states, the probability of exciting our detector atom is

$$\int d\nu\, n_\nu p(\nu,t) = 2\pi n_\omega \Delta t \frac{\langle d^2\rangle}{(\epsilon_0\hbar)^2}\mathrm{Tr}[D^{(-)}(t)D^{(+)}(t)\rho_f(t)]. \tag{5.13}$$

This expression can be used to find the count rate in a detector consisting of many atoms. If our detector has N atoms, each of which sees the same field, the mean number of atoms excited in a time Δt is just N times the above expression. The count rate, w, which is the number of atoms excited per unit time, will just be the mean number excited in Δt divided by Δt, or

$$w = 2\pi n_\omega N \frac{\langle d^2\rangle}{(\epsilon_0\hbar)^2}\mathrm{Tr}[D^{(-)}(t)D^{(+)}(t)\rho_f(t)]. \tag{5.14}$$

We note that there is a fixed operator ordering here, since the creation operators $a_j^\dagger$ all appear in $D^{(-)}$, and are therefore always to the left of the annihilation operators a_j, which is called normal ordering. The standard notation for this is $:\ldots:$, so that, for example, $:a^\dagger a a^\dagger: = a^\dagger a^\dagger a$. This shows that the counting rate of a photo-detector is proportional to the normally ordered product of field operators. This is related to the mean intensity I, since

$$I = \frac{2}{\epsilon_0}\mathrm{Tr}[D^{(-)}(t)D^{(+)}(t)\rho_f(t)]. \tag{5.15}$$

Our result can now be re-expressed in a simpler form as

$$w = n_w N B_E I, \tag{5.16}$$

where B_E is the Einstein B-coefficient for stimulated absorption,

$$B_E = \frac{\pi \langle d^2 \rangle}{\epsilon_0 \hbar^2}.$$

We note here that in many typical situations the atomic dipole moments are randomly oriented, due to rotational symmetry. This gives the usual relationship for the average, that

$$\langle d^2 \rangle = \langle (\mathbf{d} \cdot \hat{\mathbf{e}})^2 \rangle = |\mathbf{d}|^2 \langle \cos^2 \theta \rangle = \tfrac{1}{3} |\mathbf{d}|^2. \tag{5.17}$$

One can go on to analyze more complicated kinds of detection events, such as detecting a photon at one point at a time t_1 and subsequently detecting another at a different point at a time t_2. The probabilities for these more complicated correlation measurements are also found to be proportional to expectation values of normally ordered products of field operators.

5.1.2 Quantum coherence theory

Now that we have shown that the basic response of a photo-detector is described by a normally ordered product of field operators, let us study these objects in a bit more detail. Following Glauber, it is conventional to define the first-order quantum coherence function as a tensor quantity proportional to the count rate at an idealized single photo-detector:

$$G^{(1)}_{\mu\nu}(\mathbf{r}_1, t_1, \mathbf{r}_2, t_2) = \mathrm{Tr}\big[E^{(-)}_{\mu}(\mathbf{r}_1, t_1) E^{(+)}_{\nu}(\mathbf{r}_2, t_2) \rho_f\big]. \tag{5.18}$$

Here we use the fact that, in a vacuum, $\mathbf{E}^{(\pm)} = \mathbf{D}^{(\pm)}/\epsilon_0$. The possibility of having time-delayed arguments and different spatial locations corresponds to using more complex arrangements of frequency filters and interferometers, which we do not analyze in detail here. We note that these higher-order correlations are all normally ordered as well.

This concept can be readily generalized to the nth-order correlation function, which is proportional to the rate of counting n photons in coincidence. We introduce $x_n \equiv (\mathbf{r}_n, t_n)$, and the nth-order correlation function is then conventionally defined as

$$\begin{aligned} &G^{(n)}_{\mu_1 \ldots \mu_{2n}}(x_1, x_2, \ldots, x_{2n}) \\ &= \mathrm{Tr}\big[E^{(-)}_{\mu_1}(x_1) \cdots E^{(-)}_{\mu_n}(x_n) E^{(+)}_{\mu_{n+1}}(x_{n+1}) \cdots E^{(+)}_{\mu_{2n}}(x_{2n}) \rho_f\big]. \end{aligned} \tag{5.19}$$

The normalized first-order coherence measure is

$$g^{(1)}_{\mu\nu}(\mathbf{r}_1, t_1, \mathbf{r}_2, t_2) = \frac{G^{(1)}_{\mu\nu}(\mathbf{r}_1, t_1, \mathbf{r}_2, t_2)}{\sqrt{G^{(1)}_{\mu\mu}(\mathbf{r}_1, t_1, \mathbf{r}_1, t_1)\, G^{(1)}_{\nu\nu}(\mathbf{r}_2, t_2, \mathbf{r}_2, t_2)}}, \tag{5.20}$$

and, similarly, normalized higher-order coherence functions are then defined as

$$g^{(n)}_{\mu_1 \ldots \mu_{2n}}(x_1, x_2, \ldots, x_{2n}) = \frac{G^{(n)}_{\mu_1 \ldots \mu_{2n}}(x_1, x_2, \ldots, x_{2n})}{\sqrt{\prod_{j=1}^{2n} G^{(1)}_{\mu_j \mu_j}(x_j, x_j)}}. \tag{5.21}$$

The normalized coherence functions allow us to introduce the idea of *coherence*. The definition of nth-order coherence is that a field possesses nth-order coherence if

$$g^{(n)}_{\mu_1 \ldots \mu_n, \mu_n \ldots \mu_1}(x_1, \ldots, x_n, x_n, \ldots, x_1) = 1, \tag{5.22}$$

where we note that $g^{(n)}_{\mu_1\ldots\mu_n,\mu_n\ldots\mu_1}(x_1,\ldots,x_n,x_n,\ldots,x_1)$ is a positive, real quantity, but it is not, in general, equal to one. If this coherence condition *is* satisfied, it implies that

$$G_{\mu_1\ldots\mu_n,\mu_n\ldots\mu_1}(x_1,\ldots,x_n,x_n,\ldots,x_1)=\prod_{j=1}^{n}G^{(1)}_{\mu_j\mu_j}(x_j,x_j). \tag{5.23}$$

What this means is that the n-fold coincidence rate is just the product of the counting rates of n individual detectors, where the rate of each of the detectors is independent of all of the others. Thus, the detection events at the different space-time points are not correlated.

It is usual to suppress polarization indices in the correlation functions if only one polarization is excited. Similarly, one can even suppress space indices in a single-mode quantum system, as $g^{(2)}(x_1,\ldots,x_2,x_2,\ldots,x_1)$ is independent of position in such cases. That is what we shall do throughout the rest of this chapter, because we will be considering simple systems, and field states, with a small number of modes. In the next several sections we will see how to describe single-mode field states in a way that is motivated by the need to easily calculate the normally ordered correlation functions that appear in expressions for photon counting rates and by considerations of field coherence.

5.2 Quadratures and beam-splitters

We would next like to show how a linear device, the quantum beam-splitter, can measure quantum statistics that are not directly accessible with a photo-detector. This is an essential feature of quantum optics, and it relies on the operation of a very simple device. In the most well-known symmetric case, the quantum beam-splitter is simply a partially reflecting mirror, with a 50% transmission and 50% reflection. If we combine an unknown quantum state with an intense coherent state at this beam-splitter, we can measure phase-dependent quantum statistics. Crucially, this gives us access to quantum information that is not directly accessible using the direct detection scheme described in the previous section.

5.2.1 The quantum beam-splitter

The first thing we need to do is describe how a beam-splitter works. There are two input modes, with annihilation operators a^{in} and b^{in}, and two output modes, with annihilation operators a^{out} and b^{out}. This is an example of a linear input–output relation, introduced in Eq. (2.142). A one-dimensional beam-splitter modeled as a dielectric was treated rigorously in Eq. (3.56). In the laboratory, a beam-splitter is any device that linearly converts a set of input modes to a different set of output modes with minimal losses. The traditional instrument is a 'half-silvered' mirror set at an angle of $\pi/4$ radians to an incoming beam, but there are many different ways to achieve this goal.

In general, the constraint of unitarity and the freedom to redefine mode phases means that the operators in any general, lossless, 2×2 beam-splitter are related by

$$\begin{aligned} a^{out} &= \sqrt{T}\, a^{in} + \sqrt{R}\, b^{in}, \\ b^{out} &= -\sqrt{R}\, a^{in} + \sqrt{T}\, b^{in}, \end{aligned} \tag{5.24}$$

where T and R are the transmissivity and reflectivity of the beam-splitter, respectively. If $T = 1$, the input and output modes are the same, corresponding to complete transmission; and if $T = 0$, the input and output modes are interchanged, with appropriate phase-shifts, corresponding to complete reflection.

5.2.2 Quantum quadrature measurements

Next, we consider a typical case where the input state in the a mode is an unknown state $|\psi^{in}\rangle$, and the input state in the b mode is an intense coherent state with amplitude $\alpha = |\alpha|e^{i\phi}$ and phase ϕ (see Figure 5.1), whose density matrix we assume is factorized from the unknown state. This is actually a type of simple interferometer, and thus we do not really need to have a coherent state with a fixed absolute phase. The output only depends on the *relative* phase of the two inputs. An effective way to measure a phase-dependent field is using a balanced homodyne scheme, with equal transmission and reflection coefficients. We then take the difference of the two outputs:

$$\begin{aligned} \langle n_a^{out} - n_b^{out} \rangle &= \tfrac{1}{2}\langle (a+\alpha)^\dagger (a+\alpha) - (a-\alpha)^\dagger (a-\alpha) \rangle \\ &= |\alpha| \langle a^\dagger e^{i\phi} + a e^{-i\phi} \rangle, \end{aligned}$$

where the operators without superscripts, i.e. the ones on the right-hand side of the equation, are the *in* operators, that is, the operators at the input to the beam-splitter. We can write this in another way, by introducing the field quadrature operator, defined as

$$X(\phi) = e^{i\phi} a^\dagger + e^{-i\phi} a. \tag{5.25}$$

With this definition, we have that

$$\langle X(\phi) \rangle = \frac{1}{|\alpha|} \langle n_a^{out} - n_b^{out} \rangle.$$

In summary, as shown in Figure 5.1, the quadrature operator and its fluctuations can be measured by means of homodyne detection. In this measurement, the mode to be measured is mixed with a second mode in a strong coherent state at a beam-splitter, where ϕ is the phase of the coherent-state amplitude. The difference in the photon numbers at the two output ports of the beam-splitter is proportional to $X(\phi)$. It is essential that the coherent state is intense relative to the measured state, i.e. $|\alpha|^2 \gg n_a^{in}$, in order that the fluctuations in $\langle X(\phi) \rangle$ correspond directly to the output current fluctuations, without any contribution from the high-order coherence terms of the input.

We also have that $X(0) \equiv X$ and $X(\pi/2) \equiv Y$ are proportional to the real and imaginary parts of the annihilation operator, a, so that

$$a = \tfrac{1}{2}[X(0) + iX(\pi/2)] = \tfrac{1}{2}(X + iY). \tag{5.26}$$

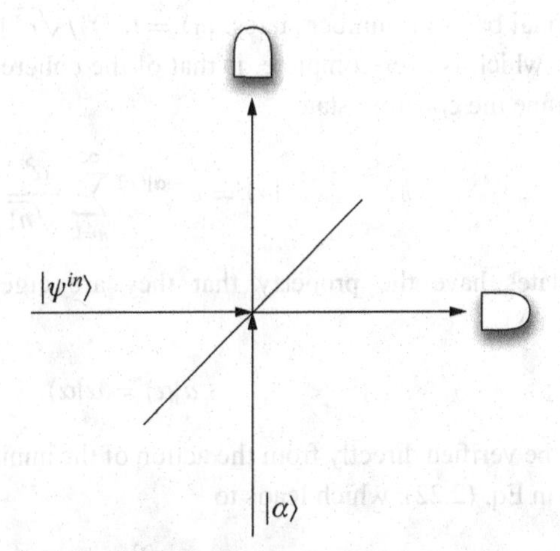

Fig. 5.1 In homodyne detection, the state whose quadrature component is to be measured, $|\psi^{in}\rangle$, is mixed at a beam-splitter with a strong coherent state, $|\alpha\rangle$. One can use either one or, as shown here, two detectors. The two-detector case is known as balanced homodyne detection. Information about the quadrature component is obtained from the difference between the photon numbers at the two detectors.

The quadrature operators X and Y have the commutator $[X, Y] = 2i$. The reason why this normalization is often used is that the Heisenberg uncertainty principle has an especially simple form. From Eq. (2.17), it follows that the quadrature operator uncertainty relation is

$$\Delta X \Delta Y \geq 1. \tag{5.27}$$

For the simplest case of a vacuum input, one has a unit vacuum noise level, as can be easily checked using the commutation relations and the result that $a|0\rangle = 0$, so that

$$\Delta X^2 = \Delta Y^2 = 1.$$

The reader should be careful about comparing results on quadratures between authors, as there are a variety of normalizations used in the literature. Since these differ by a constant factor, the resulting uncertainty principle and vacuum noise level – which is the operationally measured quantity – are correspondingly different. For simplicity, we choose the normalization that gives a vacuum noise level of unity, as in Eq. (5.27). Another common normalization is to define

$$a = x + ip. \tag{5.28}$$

With this definition, one has

$$\Delta x \Delta p \geq \tfrac{1}{4}. \tag{5.29}$$

5.3 Coherent states and *P*-representations

Let us begin an analysis of field states considering a single field mode with annihilation and creation operators a and $a^\dagger$. The Hilbert space of states of this mode is spanned by the orthonormal basis of number states, $|n\rangle = (a^\dagger)^n/\sqrt{n!}\,|0\rangle$, for $n = 0, 1, 2, \ldots$. Another set of states, which is over-complete, is that of the coherent states. For any complex number, α, we define the coherent state

$$|\alpha\rangle = e^{-|\alpha|^2/2} \sum_{n=0}^{\infty} \frac{\alpha^n}{\sqrt{n!}} |n\rangle. \tag{5.30}$$

These states have the property that they are eigenstates of the mode annihilation operator, i.e.

$$a|\alpha\rangle = \alpha|\alpha\rangle. \tag{5.31}$$

This can be verified directly from the action of the annihilation operator on a number state, as given in Eq. (2.22), which leads to

$$a|\alpha\rangle = e^{-|\alpha|^2/2} \sum_{n=1}^{\infty} \frac{\alpha\,\alpha^{n-1}}{\sqrt{(n-1)!}} |n-1\rangle. \tag{5.32}$$

The result for the eigenvalue, Eq. (5.31), then follows immediately on defining $n' = n - 1$ in the summation over number states, and removing the common factor of α.

These states can also be expressed in terms of a mode displacement operator,

$$D(\alpha) = e^{(\alpha a^\dagger - \alpha^* a)} = e^{-|\alpha|^2/2} e^{\alpha a^\dagger} e^{-\alpha^* a}, \tag{5.33}$$

as $|\alpha\rangle = D(\alpha)|0\rangle$. The mode displacement operator acquires its name because of its action on the annihilation operator, $D(\alpha)^{-1} a D(\alpha) = a + \alpha$. This result is an immediate consequence of the Baker–Hausdorff theorem. Because $a|0\rangle = 0$, we can use the above result to express a coherent state explicitly in terms of creation operators,

$$|\alpha\rangle = e^{\alpha a^\dagger - |\alpha|^2/2}|0\rangle. \tag{5.34}$$

Different coherent states are not orthogonal. In fact, it is straightforward to see that

$$\langle\beta|\alpha\rangle = \exp[-\tfrac{1}{2}(|\alpha|^2 + |\beta|^2) + \beta^*\alpha]. \tag{5.35}$$

A consequence of this is that the square of the magnitude of the inner product of two coherent states is just a Gaussian,

$$|\langle\beta|\alpha\rangle|^2 = \exp[-|\alpha - \beta|^2], \tag{5.36}$$

which is small if $|\alpha - \beta| \gg 1$. Coherent states, nonetheless, form a resolution of the identity

$$\frac{1}{\pi} \int d^2\alpha\, |\alpha\rangle\langle\alpha| = \hat{I}, \tag{5.37}$$

where $d^2\alpha = d(\Re(\alpha))d(\Im(\alpha))$. Here $\hat{I}$ is the identity operator on the bosonic Hilbert space. This relation is easily verified by expanding the coherent states in terms of number states, and changing the integration variables to polar coordinates.

We note that, because

$$\langle\alpha|(a^\dagger)^n a^n|\alpha\rangle = |\alpha|^{2n} = (\langle\alpha|a^\dagger a|\alpha\rangle)^n, \tag{5.38}$$

a coherent state is coherent to all orders. This suggests that coherent states might be useful in the description of the coherence properties of general field states. As we shall see, this supposition is correct.

5.3.1 P-representation

Suppose we have a state of our field mode given by the density matrix, ρ. Let us assume for the moment that we can express the density matrix in terms of coherent states as

$$\rho = \int d^2\alpha\, P(\alpha)|\alpha\rangle\langle\alpha|, \tag{5.39}$$

where $P(\alpha)$ is a c-number, which is called the Glauber–Sudarshan P-representation of ρ. Note that the fact that $\mathrm{Tr}(\rho) = 1$ implies that $\int d^2\alpha\, P(\alpha) = 1$. Now suppose that $O_N(a, a^\dagger)$ is an operator, which is a function of a and $a^\dagger$, expressed in normally ordered form, that is, with all annihilation operators to the right of all creation operators,

$$O_N(a, a^\dagger) = \sum_{m,n} c_{mn}(a^\dagger)^m a^n. \tag{5.40}$$

The expectation value of O_N is then given by

$$\begin{aligned} \langle O_N\rangle = \mathrm{Tr}(O_N\rho) &= \int d^2\alpha\, P(\alpha)\langle\alpha|O_N|\alpha\rangle \\ &= \int d^2\alpha\, P(\alpha)O_N(\alpha, \alpha^*), \end{aligned} \tag{5.41}$$

where $O_N(\alpha, \alpha^*) = \sum_{m,n} c_{mn}(\alpha^*)^m\alpha^n$. What this procedure has done is to allow us to express a quantum expectation value in terms of c-number quantities, one for the density matrix, $P(\alpha)$, and one for the operator, $O_N(\alpha, \alpha^*)$.

Note that the operator has to be expressed in a particular form. For a general operator $O(a, a^\dagger)$, we find $O_N(\alpha, \alpha^*)$ by first using the commutation relation, $[a, a^\dagger] = 1$, to express $O(a, a^\dagger)$ in normally ordered form, $O_N(a, a^\dagger)$, and then replacing a by α and $a^\dagger$ by α^*. This P-representation has some of the properties of a probability function, but in fact it is not a probability. As we shall see, it is not even generally a positive function.

5.3.2 Quantum characteristic function

Now we address the issue of how one would find $P(\alpha)$ from ρ. We begin by choosing $O_N(a, a^\dagger) = \exp(\xi a^\dagger)\exp(-\xi^* a)$, which implies that $O_N(\alpha, \alpha^*) = \exp(\xi\alpha^* - \xi^*\alpha)$, in our equation for P-function mean values, Eq. (5.41). This leads us to the function

$$C_1(\xi) = \int d^2\alpha\, P(\alpha)e^{\xi\alpha^* - \xi^*\alpha}. \tag{5.42}$$

The quantity $C_1(\xi)$ is known as the normally ordered characteristic function of the state ρ. In general, it is given by

$$C_1(\xi) \equiv \mathrm{Tr}(\rho e^{\xi a^\dagger} e^{-\xi^* a}). \tag{5.43}$$

It is a well-behaved function, and it is uniquely determined by the density matrix. In addition, we can use it to extract normally ordered expectation values of creation and annihilation operators. It is clear from the definition that

$$\langle a^{\dagger n} a^m \rangle = \left. \frac{\partial^{n+m} C_1(\xi)}{\partial \xi^n \partial \xi^{*m}} \right|_{\xi=0} . \tag{5.44}$$

It is possible to define different characteristic functions C_s for other operator orderings, but that issue will be discussed in Chapter 7.

Our immediate interest is in using $C_1(\xi)$ to find the P-representation in terms of the density matrix. Note that what we have in Eq. (5.42) is a type of Fourier transform. In particular, if

$$f(\alpha) = \frac{1}{\pi} \int d^2\xi \, e^{-(\xi\alpha^* - \xi^*\alpha)} \tilde{f}(\xi), \tag{5.45}$$

then, on writing out the complex variables in real and imaginary parts, one sees that, if $\alpha = \alpha' + i\alpha''$, then $\xi\alpha^* - \xi^*\alpha = 2i[\xi'\alpha'' - \xi''\alpha']$. Hence, this defines a two-dimensional Fourier transform. It follows that the inverse Fourier transform is

$$\tilde{f}(\xi) = \frac{1}{\pi} \int d^2\alpha \, e^{\xi\alpha^* - \xi^*\alpha} f(\alpha). \tag{5.46}$$

Applying this result to Eq. (5.42) implies that

$$P(\alpha) = \frac{1}{\pi^2} \int d^2\xi \, e^{\xi^*\alpha - \xi\alpha^*} C_1(\xi). \tag{5.47}$$

Now, in general, $P(\alpha)$ can be very singular. For example, if the mode is in the coherent state $|\beta\rangle$, i.e. $\rho = |\beta\rangle\langle\beta|$, then

$$P(\alpha) = \delta^2(\alpha - \beta) = \delta(\Re(\alpha) - \Re(\beta))\delta(\Im(\alpha) - \Im(\beta)).$$

For number states, derivatives of delta-functions appear, the number of derivatives increasing with the number of photons in the state. It can become so singular that, in general, it is not even a tempered distribution. There are other ways of defining quasi-distribution functions, and we shall explore them in later chapters, which produce functions that are not as singular. The P-representation is useful, however, because it is connected with a particular kind of measurement that can be performed on the electromagnetic field, counting photons.

5.4 Nonclassical states

We would like to know if there are photo-count distributions that are possible for quantized fields that are not possible for classical stochastic fields, and how these differences, if they exist, can be observed. The answer, due to Glauber, is that there are photo-count distributions that can only be the result of a quantized field. In order to show what he found, let us consider a single-mode field. A single-mode classical field is characterized by a complex field amplitude, α, which contains information about the intensity and phase of the field. If the field is a stochastic one, it is characterized by a probability distribution for this field amplitude, $P_{cl}(\alpha)$. Quantum mechanically, we have a density matrix ρ that describes the state of the field, and its P-representation, $P(\alpha)$.

What Glauber found is that the expression $p(m; t, t + \Delta t)$ for the photo-count probability, when calculated for a quantum field, is almost the same as the expression calculated for a classical stochastic field, the only difference being the replacement of $P_{cl}(\alpha)$ in the classical expression by $P(\alpha)$ in the quantum. This means that, if the quantum state ρ has a P-representation that has all of the properties of a probability distribution, that is, it is positive and has singularities no worse than a delta-function (no derivatives of delta-functions, for example), then there is a classical stochastic field that has the same photo-count distribution. However, if ρ has a P-representation that *does not* have all of the properties of a probability distribution, then there is no classical stochastic field with the same photo-count distribution. Such states are often called nonclassical. We note, however, that there are different criteria for nonclassicality in the literature, depending on what types of measurement are used. This is discussed in greater detail in Chapter 12.

5.4.1 Sub-Poissonian states

Let us look at some examples of nonclassical states. The number operator for a field mode, which is the observable corresponding to the number of photons in that mode, is given by $n_a = a^\dagger a$. The average number of photons in the mode is $\langle n_a \rangle$, and the fluctuations in the photon number are characterized by $(\Delta n_a)^2 = \langle (n_a)^2 \rangle - \langle n_a \rangle^2$. A single-mode state is said to have sub-Poissonian photon statistics if $(\Delta n_a)^2 < \langle n_a \rangle$. Such a state is nonclassical.

To see this, we first express $\langle n_a \rangle$ and $\langle n_a^2 \rangle$ in terms of the P-representation of the state:

$$\langle n_a \rangle = \int d^2\alpha\, P(\alpha)|\alpha|^2,$$

$$\langle n_a^2 \rangle = \int d^2\alpha\, P(\alpha)\langle \alpha|(a^\dagger a)^2|\alpha\rangle = \int d^2\alpha\, P(\alpha)(|\alpha|^4 + |\alpha|^2). \tag{5.48}$$

We can now calculate $(\Delta n_a)^2 - \langle n_a \rangle$ in terms of the P-representation, which gives us

$$(\Delta n_a)^2 - \langle n_a \rangle = \int d^2\alpha\, P(\alpha)(|\alpha|^2 - \langle n_a \rangle)^2. \tag{5.49}$$

From this we can see that, if the P-representation of the state behaves like a probability distribution, then the right-hand side of the above equation is greater than or equal to zero, so that the left-hand side must be as well. Therefore, for classical states, $(\Delta n_a)^2 \geq \langle n_a \rangle$, and a state that violates this condition is therefore nonclassical. Photon number states are examples of states that are sub-Poissonian. The number state, $|n\rangle$, is an eigenstate of n_a, i.e. $n_a|n\rangle = n|n\rangle$. Since, for any photon number state, $(\Delta n_a)^2 = 0$ and $\langle n_a \rangle = n$, we see that for $n > 0$ they violate the condition for classical states, which implies that any number state other than the vacuum is nonclassical.

5.4.2 Squeezed states

A second nonclassical effect is known as squeezing. This involves another kind of nonclassical field state, one with sub-Poissonian photon statistics generated from the interference between the two input beams of a quadrature detector, and is called a squeezed state. It is measured using a quadrature detector for $X(\phi)$, as described above. Let us examine the fluctuations in $X(\phi)$. For a state represented in terms of its P-representation, we have

$$\Delta^2 X(\phi) = 1 + \int d^2\alpha\, P(\alpha)[e^{i\phi}(\alpha^* - \langle a^\dagger \rangle) + e^{-i\phi}(\alpha - \langle a \rangle)]^2. \tag{5.50}$$

From this equation, we see that for a classical state the second term on the right-hand side is greater than or equal to zero, so that $\Delta^2 X(\phi) \geq 1$. The vacuum state, or, for that matter, any coherent state, has $\Delta^2 X(\phi) = 1$. This is called the shot noise or vacuum noise level. It corresponds to the noise measured in a local oscillator experiment when only the reference beam is used, and the other port has a vacuum or coherent input. The term 'shot noise' means that in a typical experiment the coherent photons in the local oscillator arrive at completely random intervals, and so are uncorrelated.

A state with $\Delta X(\phi) < 1$ for some ϕ is called squeezed, and it will be nonclassical, since clearly it cannot occur if the P-representation is positive. This implies that the photons arriving at the photo-detector are correlated in a way that reduces their variance. As an example of a squeezed state, consider

$$|\psi\rangle = c_0|0\rangle - c_2 e^{2i\phi}|2\rangle, \tag{5.51}$$

where c_0 and c_2 are real and positive, and $|c_0|^2 + |c_2|^2 = 1$. We find that

$$(\Delta X(\phi))^2 = 1 + 4c_2\left(c_2 - \frac{c_0}{\sqrt{2}}\right). \tag{5.52}$$

This state will be squeezed if $c_0 > \sqrt{2}\, c_2$, unless $c_2 = 0$, in which case we simply have a vacuum state in which $\Delta X = \Delta Y = 1$.

5.4.3 Squeeze operator

A systematic way of studying squeezing is to introduce the squeeze operator

$$S(z) = e^{[z^* a^2 - z(a^\dagger)^2]/2}. \tag{5.53}$$

(We note here that sometimes the opposite ordering is used for the terms in the exponent. This corresponds to taking $z \to -z$ in our definition.) The Baker–Hausdorff theorem can be used to show that

$$S^{-1}(z) a S(z) = a \cosh r - a^\dagger e^{i\theta} \sinh r, \tag{5.54}$$

where $z = re^{i\theta}$. This relation implies that

$$\begin{aligned} &S^{-1}(z) X(\phi) S(z) \\ &= a^\dagger\left(e^{i\phi}\cosh r - e^{i(\theta-\phi)}\sinh r\right) + a\left(e^{-i\phi}\cosh r - e^{-i(\theta-\phi)}\sinh r\right), \end{aligned} \tag{5.55}$$

so that, if $\theta = 2\phi$, then

$$S^{-1}(z)X(\phi)S(z) = e^{-r}X(\phi),$$
$$S^{-1}(z)X(\phi+\pi/2)S(z) = e^{r}X(\phi+\pi/2). \tag{5.56}$$

That means, with this choice of θ, $X(\phi)$ is squeezed and so are its fluctuations. If the fluctuations of $X(\phi)$ in the state $|\psi\rangle$ are $\Delta X(\phi)$, then those in the state $S(z)|\psi\rangle$, if $\theta = 2\phi$, are $e^{-r}\Delta X(\phi)$. Thus, by choosing r large enough, we can squeeze any state sufficiently to make $\Delta X(\phi) < 1$. If the initial state is the vacuum, then for any $r > 0$ and $\theta = 2\phi$ we will have $\Delta X(\phi) < 1$. Such states are known as squeezed vacuum states. We also note that, for a squeezed vacuum state,

$$|z\rangle_S = S(z)|0\rangle. \tag{5.57}$$

We find that

$$\Delta X(\phi) = e^{-r}, \qquad \Delta X(\phi+\pi/2) = e^{r}, \tag{5.58}$$

so that $\Delta X(\phi)\Delta X(\phi+\pi/2) = 1$. Comparing this with the Heisenberg uncertainty relation for $X(\phi)$ and $X(\phi+\pi/2)$, which is

$$\Delta X(\phi)\Delta X(\phi+\pi/2) \geq 1, \tag{5.59}$$

we see that the squeezed vacuum state achieves the lower bound in the uncertainty relation. A state that achieves the lower bound, i.e. one that satisfies the uncertainty relation as an equality, is called a minimum-uncertainty state. A squeezed vacuum state is a minimum-uncertainty state.

The squeeze operator can be interpreted as a time development transformation. If we wish to express $S(z)$ in the form $\exp(-itH/\hbar)$, then this can be accomplished by setting $z = t\varepsilon$, where ε is a coupling constant. The resulting Hamiltonian is

$$H = \frac{i\hbar}{2}(\varepsilon^* a^2 - \varepsilon a^{\dagger 2}). \tag{5.60}$$

As we shall see shortly, this Hamiltonian describes a nonlinear optical device known as a degenerate parametric amplifier, which provides an effective way to generate a squeezed state.

5.5 Two-mode states

We can easily generalize the notion of P-representation and nonclassical state to more than one mode. We shall assume that the two modes have annihilation operators a and b, respectively. A two-mode state, ρ_{ab}, can be expressed in terms of a two-mode P-representation as

$$\rho_{ab} = \int d^2\alpha \int d^2\beta\, P(\alpha,\beta)|\alpha\rangle_a\langle\alpha| \otimes |\beta\rangle_b\langle\beta|, \tag{5.61}$$

and if $P(\alpha, \beta)$ has the properties of a probability distribution, which in general it does not, the state is classical. This leads to the question of whether there exist correlations between modes that can only occur for nonclassical states.

5.5.1 Schwarz inequality

Let us start examining this question by considering the quantity

$$\langle ab^\dagger \rangle = \int d^2\alpha \int d^2\beta \, P(\alpha, \beta)\alpha\beta^*. \tag{5.62}$$

Now, if ρ_{ab} is a classical state, then $P(\alpha, \beta)$ has the properties of a probability distribution, and we can apply the Schwarz inequality to the above equation to yield

$$|\langle ab^\dagger \rangle| \leq \left[\int d^2\alpha \int d^2\beta \, P(\alpha, \beta)|\alpha|^2|\beta|^2 \right]^{1/2}, \tag{5.63}$$

which, in terms of operators, can be expressed as

$$|\langle ab^\dagger \rangle| \leq \langle a^\dagger a b^\dagger b \rangle^{1/2}. \tag{5.64}$$

This, then, is an equality that must be satisfied by classical states. It is violated by the state $(|1\rangle_a|0\rangle_b + |0\rangle_a|1\rangle_b)/\sqrt{2}$, a superposition of one photon in the a mode and none in the b mode with no photons in the a mode and one in the b mode. For this state, the right-hand side of the above inequality is zero while the left-had side is $1/\sqrt{2}$, which clearly violates it.

A second inequality that must hold for classical states can be found by considering the quantity

$$\langle ab \rangle = \int d^2\alpha \int d^2\beta \, P(\alpha, \beta)\alpha\beta. \tag{5.65}$$

In this case, if the state is classical, we can apply the Schwarz inequality to give

$$|\langle ab \rangle| \leq \left[\int d^2\alpha \int d^2\beta \, P(\alpha, \beta)|\alpha|^2 \right]^{1/2} \left[\int d^2\alpha \int d^2\beta \, P(\alpha, \beta)|\beta|^2 \right]^{1/2}, \tag{5.66}$$

which, in terms of operators, becomes

$$|\langle ab \rangle| \leq \langle a^\dagger a \rangle^{1/2} \langle b^\dagger b \rangle^{1/2}. \tag{5.67}$$

If a state violates this inequality, it is nonclassical.

5.5.2 Two-mode squeezing

Finding a state that violates the Schwarz inequality gives us a chance to introduce the two-mode squeeze operator. This is given by

$$S_{ab}(z) = e^{z^* ab - z a^\dagger b^\dagger}, \tag{5.68}$$

where, again, $z = re^{i\theta}$, and it has the property that

$$S_{ab}^{-1}(z)aS_{ab}(z) = a\cosh r - b^{\dagger}e^{i\theta}\sinh r,$$
$$S_{ab}^{-1}(z)bS_{ab}(z) = b\cosh r - a^{\dagger}e^{i\theta}\sinh r. \tag{5.69}$$

Defining

$$\begin{aligned} X_a &= a + a^{\dagger}, \\ X_b &= b + b^{\dagger}, \\ Y_a &= i(a^{\dagger} - a), \\ Y_b &= i(b^{\dagger} - b), \end{aligned} \tag{5.70}$$

we note that, if we set $\theta = 0$, then

$$\begin{aligned} S_{ab}^{-1}(r)(X_a + X_b)S_{ab}(r) &= e^{-r}(X_a + X_b), \\ S_{ab}^{-1}(r)(Y_a + Y_b)S_{ab}(r) &= e^{r}(Y_a + Y_b), \\ S_{ab}^{-1}(r)(X_a - X_b)S_{ab}(r) &= e^{r}(X_a - X_b), \\ S_{ab}^{-1}(r)(Y_a - Y_b)S_{ab}(r) &= e^{-r}(Y_a - Y_b). \end{aligned} \tag{5.71}$$

This implies that, if the uncertainty for $X_a + X_b$ in a state $|\psi\rangle_{ab}$ is $\Delta(X_a + X_b)$, then the uncertainty in the state $S_{ab}(r)|\psi\rangle_{ab}$ will be $e^{-r}\Delta(X_a + X_b)$. Therefore, the two-mode squeeze operator creates correlations between the modes, and if the squeezing is sufficiently large, the fluctuations in X_a and X_b will be approximately anticorrelated. We will see in Chapter 12 that this leads to an important consequence. Such strong correlations allow the demonstration of the EPR paradox, a famous paradox in quantum mechanics that has both fundamental importance and practical applications in quantum technologies.

Returning to our condition for nonclassical states, we find that for the two-mode squeezed vacuum state, $S_{ab}(z)|0\rangle$,

$$|\langle ab\rangle| = \cosh r \sinh r, \qquad \langle a^{\dagger}a\rangle = \langle b^{\dagger}b\rangle = \sinh^2 r. \tag{5.72}$$

Therefore, since $\cosh r > \sinh r$ for $r > 0$, we see that the condition in Eq. (5.67) is violated, and the two-mode squeezed vacuum state is nonclassical. The reason it is nonclassical is due to the correlations between the modes; the states of the individual modes, as given by their reduced density matrices, are classical. This can be seen by making use of the identity, which we shall not prove,

$$S_{ab}(z) = \exp(-a^{\dagger}b^{\dagger}e^{i\theta}\tanh r)\exp[-\ln(\cosh r)(a^{\dagger}a + b^{\dagger}b + 1)]\exp(abe^{-i\theta}\tanh r). \tag{5.73}$$

Applying this to the vacuum state we find

$$S_{ab}(z)|0\rangle = \operatorname{sech} r\sum_{n=0}^{\infty}(-e^{i\theta}\tanh r)^n|n\rangle_a|n\rangle_b, \tag{5.74}$$

so that the reduced density matrix for the a mode, which is identical to the reduced density matrix for the b mode, is

$$\rho_a = \mathrm{Tr}_b[S_{ab}(z)|0\rangle\langle 0|S_{ab}^{-1}(z)] = \mathrm{sech}^2 r \sum_{n=0}^{\infty} \tanh^{2n} r\, |n\rangle_a\langle n|. \tag{5.75}$$

The P-representation of this state is Gaussian, so it is, in fact, classical. In terms of the photon statistics, looking at just a single mode out of a two-mode squeezed vacuum is indistinguishable from looking at single-mode thermal blackbody radiation.

5.6 Mode entanglement

We now want to look at a specific kind of correlation between the modes, entanglement. Two systems are entangled if they are quantum mechanically correlated, and this is a strong condition. Correlations can exist between two systems without their being entangled. In order to define entanglement between the two modes, we first have to discuss the concept of a separable state, which is a state that is not quantum mechanically correlated. If the two-mode state can be expressed as

$$\rho_{ab} = \sum_j P_j \rho_a^{(j)} \otimes \rho_b^{(j)}, \tag{5.76}$$

where P_j are probabilities whose sum is 1, and $\rho_a^{(j)}$ and $\rho_b^{(j)}$ are density matrices for modes a and b, respectively, then ρ_{ab} is said to be separable. A separable state is one in which the correlations between the subsystems, in this case the modes a and b, are classical. A separable state can be constructed by two parties, each in possession of one of the subsystems, with each acting locally on their subsystem and communicating classically with the other party. A state that is not separable is said to be entangled, and an entangled state possesses quantum correlations between the subsystems. There is clearly a connection between classical states and separable states. In particular, all classical states are separable. Therefore, we can conclude that, in order for a two-mode state to be entangled, it must be nonclassical. However, not all nonclassical states are entangled. For example, the state with one photon in each mode, $|1\rangle_a|1\rangle_b$, is not entangled, but it is nonclassical, because the individual state of each mode is nonclassical. Therefore, being entangled is a stronger condition than being nonclassical.

5.6.1 Entanglement criteria

Deciding whether a state is entangled or not is, in general, not a simple problem. There are, however, some sufficient conditions. Perhaps the most commonly used one for field modes is the one proved by Simon and by Duan *et al.* (see the additional reading). Consider a two-mode system in which the annihilation operators for the modes are a and b. Using our previous quadrature definitions, Eq. (5.71), if a state satisfies the condition

$$\Delta^2(X_a + X_b) + \Delta^2(Y_a - Y_b) < 4, \tag{5.77}$$

then it is entangled. A related Heisenberg-like criterion is that, if a state satisfies

$$\Delta^2(X_a + X_b)\Delta^2(Y_a - Y_b) < 4, \tag{5.78}$$

then it is entangled. It should be noted that all of the quantities in these inequalities can be measured by using homodyne detection, which means that these conditions can be used operationally to determine whether fields occurring in an experiment are entangled.

Let us prove the first of these conditions. We shall assume that the state is separable, and show that the quantity on the left-hand side must be greater than or equal to 4. Hence, if this quantity is less than 4, the state must be entangled. We start by writing

$$\begin{aligned} &[\Delta(X_a + X_b)]^2 + [\Delta(Y_a - Y_b)]^2 \\ &= \sum_j P_j[\langle (X_a + X_b)^2\rangle_j + \langle (Y_a - Y_b)^2\rangle_j] - \langle X_a + X_b\rangle^2 - \langle Y_a - Y_b\rangle^2, \end{aligned} \tag{5.79}$$

where expectation values with a subscript j are taken with respect to the jth partial density matrix $\rho_a^{(j)} \otimes \rho_b^{(j)}$ in Eq. (5.76). This can be expressed as

$$\begin{aligned} &[\Delta(X_a + X_b)]^2 + [\Delta(Y_a - Y_b)]^2 \\ &= \sum_j P_j[(\Delta Y_a)_j^2 + (\Delta Y_b)_j^2 + (\Delta Y_a)_j^2 + (\Delta Y_b)_j^2] \\ &\quad + \sum_j P_j[(\langle Y_a\rangle_j + \langle Y_b\rangle_j)^2 + (\langle Y_a\rangle_j - \langle Y_b\rangle_j)^2] - \langle X_a + X_b\rangle^2 - \langle Y_a - Y_b\rangle^2. \end{aligned} \tag{5.80}$$

We now note that, because $[X_a, Y_a] = 2i$ and $[X_b, Y_b] = 2i$, we have $\Delta X_a \Delta Y_a \geq 1$ and $\Delta X_b \Delta Y_b \geq 1$. These relations imply that

$$\begin{aligned} (\Delta X_a)^2 + (\Delta Y_a)^2 &\geq 2, \\ (\Delta X_b)^2 + (\Delta Y_b)^2 &\geq 2. \end{aligned} \tag{5.81}$$

The Schwarz inequality implies that

$$\left(\sum_j P_j\right)\left(\sum_j P_j \langle X_a + X_b\rangle_j^2\right) \geq \left(\sum_j P_j \langle X_a + X_b\rangle_j\right)^2, \tag{5.82}$$

with a similar result for the momenta. These results, when substituted into Eq. (5.80), yield Eq. (5.77).

Now let us use this condition to study the entanglement of the state

$$|\Psi\rangle = c_0|0\rangle - c_1 a^\dagger b^\dagger |0\rangle, \tag{5.83}$$

where $|c_0|^2 + |c_1|^2 = 1$. This is the type of state that is produced, to lowest order in the interaction, by a two-mode nondegenerate parametric down-converter. Either the field is in the vacuum, with amplitude c_0, or a pair of photons have been emitted, with amplitude c_1. Typically, $|c_1| \ll |c_0|$. This state is entangled as long as neither $c_0 = 0$ nor $c_1 = 0$.

To show this, substituting this state into the left-hand side of the inequality appearing in Eq. (5.77), we find

$$[\Delta(X_a + X_b)]^2 + [\Delta(Y_a - Y_b)]^2 = 4 + 4(2|c_1|^2 - c_1^* c_0 - c_0^* c_1). \tag{5.84}$$

The right-hand side will be less than 4 if c_1 is nonzero and $2|c_1|^2 < c_1^* c_0 + c_0^* c_1$. This can certainly happen when $|c_1| \ll |c_0|$, for example, when c_0 is real and positive, and c_1 is real and positive. The smallest value that the right-hand side of the above equation can attain is $8 - 4\sqrt{2}$, when $c_0 = \cos(\pi/8)$ and $c_1 = \sin(\pi/8)$. Therefore, we see that this condition can be used to demonstrate that the light emerging from a parametric down-converter is entangled. However, a real parametric down-converter will have states with superpositions of higher numbers of photons as well.

5.6.2 Classicality and separability

Finally, let us show that our two Schwarz inequality conditions that a state must satisfy in order to be classical, Eqs (5.64) and (5.67), are, in fact, conditions that must be satisfied by separable states. That means that, if either of these conditions is violated, then not only must the state be nonclassical, it must be entangled as well. This gives us alternative sufficient conditions to prove entanglement.

Let us begin with Eq. (5.64). We assume that we have a separable density matrix as given in Eq. (5.76), and we set $\rho_j = \rho_a^{(j)} \otimes \rho_b^{(j)}$. Note that, because ρ_j is a product, $|\mathrm{Tr}(\rho_j a b^\dagger)| = |\mathrm{Tr}(\rho_j a b)|$, so

$$\begin{aligned} |\langle ab^\dagger \rangle| &\le \sum_j P_j |\mathrm{Tr}(\rho_j a b^\dagger)| = \sum_j P_j |\mathrm{Tr}(\rho_j a b)| \\ &\le \sum_j P_j (\langle a^\dagger a b^\dagger b \rangle_j)^{1/2}, \end{aligned} \tag{5.85}$$

where $\langle a^\dagger a b^\dagger b \rangle_j = \mathrm{Tr}(\rho_j a^\dagger a b^\dagger b)$, and we have made use of the Schwarz inequality. We can apply the Schwarz inequality again to obtain

$$\begin{aligned} |\langle ab^\dagger \rangle| &\le \left(\sum_j P_j \right)^{1/2} \left(\sum_j P_j \langle a^\dagger a b^\dagger b \rangle_j \right)^{1/2} \\ &\le (\langle a^\dagger a b^\dagger b \rangle)^{1/2}, \end{aligned} \tag{5.86}$$

which shows that the inequality in Eq. (5.64) does indeed hold for all separable states. Now let us move on to Eq. (5.67). We have that

$$\begin{aligned} |\langle ab \rangle|^2 &\le \sum_{j,k} P_j p_k |\mathrm{Tr}(\rho_j a b)| |\mathrm{Tr}(\rho_k b^\dagger a^\dagger)| \\ &\le \sum_{j,k} P_j p_k (\langle a^\dagger a \rangle_j \langle b^\dagger b \rangle_j \langle a^\dagger a \rangle_k \langle b^\dagger b \rangle_k)^{1/2}. \end{aligned} \tag{5.87}$$

In terms of the quantities $\langle a^\dagger a\rangle_j = \mathrm{Tr}(a^\dagger a\rho_j) = n_j^a$ and $\langle b^\dagger b\rangle_j = \mathrm{Tr}(b^\dagger b\rho_j) = n_j^b$, this inequality can be rewritten as

$$|\langle ab^\dagger\rangle|^2 \leq \sum_j P_j^2 n_j^a n_j^b + 2\sum_{j>k} P_j p_k (n_j^a n_j^b n_k^a n_k^b)^{1/2}. \tag{5.88}$$

Next we consider $\langle a^\dagger a\rangle\langle b^\dagger b\rangle = \sum_j P_j^2 n_j^a n_j^b + \sum_{j>k} P_j P_k (n_j^a n_k^b + n_k^a n_j^b)$. As we have $n_j^a n_k^b + n_k^a n_j^b \geq 2(n_j^a n_k^b n_k^a n_j^b)^{1/2}$, we see that the inequality in Eq. (5.67) holds for all separable states, i.e. if a state violates this inequality, it must be entangled.

Therefore, the violation of the conditions in Eqs (5.64) and (5.67) can be used to show that a state is entangled.

5.7 Parametric interactions

Before studying complicated processes such as the propagation of the electromagnetic field in a nonlinear dielectric, or developing the formalism necessary to describe fields in cavities containing nonlinear media, it is useful to study some highly simplified models to get an idea of some of the phenomena that can result. Our models will be described by Hamiltonians that couple several modes either to themselves or to each other. These Hamiltonians underlie much of the work that has been done on the quantum theory of nonlinear optics. We have already seen that there is a close relationship between the squeezing operator and certain nonlinear Hamiltonians. We now investigate this in more detail.

5.7.1 Two-mode down-conversion

We will start with a two-mode interaction. One mode, whose creation and annihilation operators are $a^\dagger$ and a, has a frequency of ω; and the second, whose creation and annihilation operators are $b^\dagger$ and b, has a frequency of 2ω. From a previous chapter, we see that the $\chi^{(2)}$ term in the Hamiltonian is cubic in the field, and hence cubic in the creation and annihilation operators. We will keep only the cubic terms that are slowly varying in time. That means we look at the time dependence of the operators without the interaction, which in this case is $a(t) = e^{-i\omega t}a(0)$ and $b(t) = e^{-2i\omega t}b(0)$, and keep only those terms in the interaction that have no time dependence. Because the interaction is weak, we expect that it will still be the case that, even in the presence of the interaction, $a(t)$ will be approximately proportional to $e^{-i\omega t}$ and $b(t)$ will be approximately proportional to $e^{-2i\omega t}$. The essential idea here is that rapidly oscillating terms have a much smaller effect on the dynamics than those that vary more slowly. Dropping rapidly oscillating terms leaves us with the Hamiltonian

$$H = \hbar\omega a^\dagger a + 2\hbar\omega b^\dagger b + \frac{i\hbar\chi}{2}[(a^\dagger)^2 b - a^2 b^\dagger]. \tag{5.89}$$

The first two terms comprise the free-field Hamiltonian of the two modes, and the term proportional to χ, which is itself proportional to $\chi^{(2)}$, describes the interaction.

The interaction in the Hamiltonian can do two things. It can combine two a photons into a b photon, and it can split one b photon into two a photons. Consequently, it can describe two processes. The first is parametric down-conversion, in which a pump at frequency 2ω produces a subharmonic (also called the signal) at frequency ω. The second is second-harmonic generation, in which a strong field at frequency ω produces its second harmonic at frequency 2ω.

5.7.2 Parametric down-conversion

Let us look at parametric down-conversion first. In doing so, we are immediately faced with a problem. The Heisenberg equations of motion resulting from the above Hamiltonian cannot be solved in any simple form. In order to get around this problem, an additional approximation is introduced, the parametric approximation. We assume that the pump field is in a large-amplitude coherent state, in particular the state $|\beta\rangle$, and that the operators in the Hamiltonian corresponding to it can be replaced by time-dependent c-numbers corresponding to the unperturbed operator expectation values.

In particular, b can be replaced by $\beta e^{-2i\omega t}$ and $b^\dagger$ can be replaced by $\beta^* e^{2i\omega t}$. This gives us the single-mode Hamiltonian (after dropping the irrelevant constant term $2\hbar\omega|\beta|^2$)

$$H = \hbar\omega a^\dagger a + \frac{i\hbar\chi}{2}[\beta e^{-2i\omega t}(a^\dagger)^2 - \beta^* e^{2i\omega t} a^2]. \tag{5.90}$$

The solution of the Heisenberg equations for this Hamiltonian is straightforward. We find that

$$\frac{da}{dt} = \frac{i}{\hbar}[H, a] = -i(\omega a + i\chi e^{-2i\omega t} a^\dagger). \tag{5.91}$$

Setting $a(t) = A(t)e^{-i\omega t}$, we obtain the simpler equation

$$\frac{dA}{dt} = \chi\beta A^\dagger, \tag{5.92}$$

which has the solution

$$A(t) = A(0)\cosh(|\beta|\chi t) + e^{i\phi} A^\dagger(0)\sinh(|\beta|\chi t), \tag{5.93}$$

where we have expressed β as $\beta = |\beta|e^{i\phi}$. Comparing this to the action of the single-mode squeeze operator, we see that the parametric down-converter squeezes the signal mode.

We can make this more obvious if we consider the case $\phi = 0$ and define the field quadrature components

$$X(t) = A(t) + A^\dagger(t), \qquad Y(t) = i(A^\dagger(t) - A(t)). \tag{5.94}$$

We then find that

$$X(t) = e^{\beta\chi t} X(0), \qquad Y(t) = e^{-\beta\chi t} Y(0). \tag{5.95}$$

Therefore, the parametric down-converter amplifies the X quadrature, and it attenuates the Y quadrature. If the input state to the a mode is a coherent state, $|\alpha\rangle$, then, if α is real, we find that $\langle A(t)\rangle = \alpha\exp(\beta\chi t)$, and the input is amplified. If, however, α is imaginary, then $\langle A(t)\rangle = \alpha\exp(-\beta\chi t)$, and the input is attenuated. This is an example of phase-sensitive amplification.

We can also look at the fluctuations in the output of the down-converter. As we have seen, the uncertainties in X and Y obey the uncertainty relation $\Delta X \Delta Y \geq 1$. Suppose we start the down-converter in the vacuum state. We then find that $\Delta X(0) = \Delta Y(0) = 1$ and

$$\Delta X(t) = e^{\beta\chi t}, \qquad \Delta Y(t) = e^{-\beta\chi t}. \tag{5.96}$$

Therefore, the state that results from the vacuum state is a minimum-uncertainty state, and it is, in fact, a squeezed vacuum state. The realization that what comes out of the output of a down-converter when the input is the vacuum state is a squeezed, minimum-uncertainty state is due to David Stoler.

Squeezed states have a number of uses. They can be used for highly accurate measurements. For example, we might wish to detect a signal by observing the shift in the amplitude of a coherent state, i.e. $|\alpha\rangle \to |\alpha + \delta\alpha\rangle$. This would be the case if we were trying to detect a phase-shift in an interferometer using a coherent state as an input state. For a coherent state, as for the vacuum, we have $\Delta X(0) = \Delta Y(0) = 1$, where $X(0)/2$ corresponds to the real part of α and $Y(0)/2$ to its imaginary part. Consequently, we can only determine $\delta\alpha$ to an accuracy of $1/2$. If we use a squeezed state, however, and we are only interested in determining the change in the Y component, then we can measure it to an accuracy of $(1/2)\exp(-\beta\chi t)$. It has been shown that the use of squeezed states in an interferometer can significantly improve its ability to detect small phase changes. Squeezed states have, more recently, found a number of applications in the field of quantum information.

5.7.3 Conservation laws

Now let us return to the Hamiltonian in Eq. (5.89). We shall have more to say about the parametric approximation shortly. While we cannot solve the equations of motion resulting from this Hamiltonian, we can make some statements about the states that are produced. This is because the Hamiltonian obeys a conservation law, in particular, it commutes with the operator

$$M = n_a + 2n_b, \tag{5.97}$$

so that M is a conserved quantity. One immediate consequence of this is a simple relation between the number of photons in the a mode and the number in the b mode,

$$\langle M(0)\rangle = 2\langle n_b(t)\rangle + \langle n_a(t)\rangle, \tag{5.98}$$

where, it should be noted, the left-hand side requires only knowledge of the initial state of the system.

If we work harder, we can find a relation between the photon number fluctuations in the two modes. Using Eq. (5.97) and its square we find

$$[\Delta M(0)]^2 = 4[\Delta n_b(t)]^2 + [\Delta n_a(t)]^2 + 4[\langle n_a(t)n_b(t)\rangle - \langle n_a(t)\rangle\langle n_b(t)\rangle]. \tag{5.99}$$

The Schwarz inequality implies that

$$|\langle(n_a - \langle n_a\rangle)(n_b - \langle n_b\rangle)\rangle| \leq \Delta n_a \Delta n_b, \tag{5.100}$$

where we have dropped the time argument for simplicity. Substituting the previous equation into this one gives

$$-4\Delta n_a \Delta n_b \leq (\Delta M)^2 - 4(\Delta n_b)^2 - (\Delta n_a)^2 \leq 4\Delta n_a \Delta n_b, \tag{5.101}$$

which implies the two inequalities

$$\begin{aligned} \Delta M(0) &\leq 2\Delta n_b(t) + \Delta n_a(t), \\ \Delta M(0) &\geq |2\Delta n_b(t) - \Delta n_a(t)|. \end{aligned} \tag{5.102}$$

Finally, these can be combined to give

$$2\Delta n_b(t) + \Delta M(0) \geq \Delta n_a(t) \geq |2\Delta n_b(t) - \Delta M(0)|, \tag{5.103}$$

which gives us the desired relation between the number fluctuations.

This relation is easiest to interpret if both the a and b modes are initially in number states. This implies that $\Delta M(0) = 0$, which in turn tells us that $\Delta n_a(t) = 2\Delta n_b(t)$. If $\Delta M(0)$ is small but not zero, for example, if the signal mode is initially in the vacuum and the photon statistics of the pump are highly sub-Poissonian, then this relation will be approximately true. Under these conditions, even without solving for the detailed dynamics, we can conclude that the number fluctuations in the signal are roughly twice those in the pump.

5.7.4 Nonclassicality from pair production

We can also use the conservation law to determine whether the signal-mode state is nonclassical. We will first give a short argument indicating why the states produced by our Hamiltonian are nonclassical, and then follow it with a more detailed discussion. Consider the case in which at $t = 0$ the pump mode is in an arbitrary state and the signal mode is in the vacuum. Because every photon that disappears from the pump produces two photons in the signal, only signal-mode number states with even photon numbers are populated. Such a state cannot be classical unless it is the vacuum state. As a result, if the signal-mode state has photons in it, then it is nonclassical.

Let us look at this argument in more detail. First, we need to show that a state with only even numbers of photons is either the vacuum or nonclassical. To that end, define the projection operators

$$\begin{aligned} Q_o &= \sum_{n=0}^{\infty} |2n+1\rangle\langle 2n+1|, \\ Q'_e &= \sum_{n=1}^{\infty} |2n\rangle\langle 2n|. \end{aligned} \tag{5.104}$$

The operator Q_o projects onto the space of odd photon number states and Q'_e projects onto the space that includes all even photon number states except the vacuum. Representing the state of the field in terms of a P-representation, we find

$$\langle Q_o \rangle - \langle Q'_e \rangle = \int d^2\alpha \, P(\alpha) e^{-|\alpha|^2} (1 + \sinh|\alpha|^2 - \cosh|\alpha|^2). \tag{5.105}$$

When the state is classical, i.e. $P(\alpha) \geq 0$, then the right-hand side of Eq. (5.105) is non-negative and

$$\langle Q_o \rangle \geq \langle Q'_e \rangle. \tag{5.106}$$

If this condition is violated, the state is nonclassical. Suppose the state contains only even photon numbers, then $\langle Q_o \rangle = 0$, so that it is nonclassical if $\langle Q'_e \rangle > 0$. Therefore, if a state contains only even photon numbers, it is either the vacuum ($\langle Q'_e \rangle = 0$) or it is nonclassical.

We now need to show that a signal-mode state containing only even photon numbers is produced. In order to do this, we will show that ${}_a\langle n_a|\rho_a|n_a\rangle_a$ is zero unless n_a is even, where ρ_a is the reduced density matrix of the signal mode. We begin by noting that, if

$$|\Psi(t)\rangle = U(t)|0\rangle_a \otimes |\Psi_b\rangle_b, \tag{5.107}$$

where $U(t)$ is the time development transformation, then, because $[M, H] = 0$, we have

$$e^{i\pi M}|\Psi(t)\rangle = U(t)e^{i\pi M}|0\rangle_a \otimes |\Psi_b\rangle_b = U(t)|0\rangle_a \otimes e^{2\pi i n_b}|\Psi_b\rangle_b. \tag{5.108}$$

The operator $\exp(2\pi i n_b)$ is just the identity operator on the b mode, as can be seen immediately by considering its action on number states. The above equation can now be expressed as

$$e^{i\pi n_a}|\Psi(t)\rangle = |\Psi(t)\rangle, \tag{5.109}$$

which implies that

$$e^{i\pi n_a}\rho_a(t) = \rho_a(t). \tag{5.110}$$

Taking the expectation of both sides of this equation in the number state $|n_a\rangle_a$ gives

$$(-1)^{n_a}{}_a\langle n_a|\rho_a(t)|n_a\rangle_a = {}_a\langle n_a|\rho_a(t)|n_a\rangle_a. \tag{5.111}$$

This implies that ${}_a\langle n_a|\rho_a(t)|n_a\rangle_a = 0$ if n_a is odd, that is, only even photon numbers are present in the signal-mode state. This completes our proof, and shows that, if the number of photons in the signal is greater than zero, the signal is in a nonclassical state.

5.7.5 Second-harmonic generation

It is also possible to show that second-harmonic generation, which is described by the same Hamiltonian, will produce nonclassical states. In that case, one starts from the state $|\alpha\rangle_a|0\rangle_b$, i.e. a coherent state in the a mode and the vacuum in the b mode. The basic idea is that it takes two a-mode photons to produce one b-mode photon. Consequently, the number distribution of the b mode is more concentrated near zero than that of the a mode. It is sufficiently concentrated, in fact, as to be nonclassical. The argument that shows this in detail will not be reproduced here, but it leads to the result that the b mode is either in the vacuum state or nonclassical.

A final property that can be deduced from the conservation law is the relation between the rotational symmetries in phase space of the initial pump state and the signal-mode state. As we shall see, the signal-mode state has twice the rotational symmetry of the pump state. This is perhaps best illustrated by means of an example. Suppose that the pump is

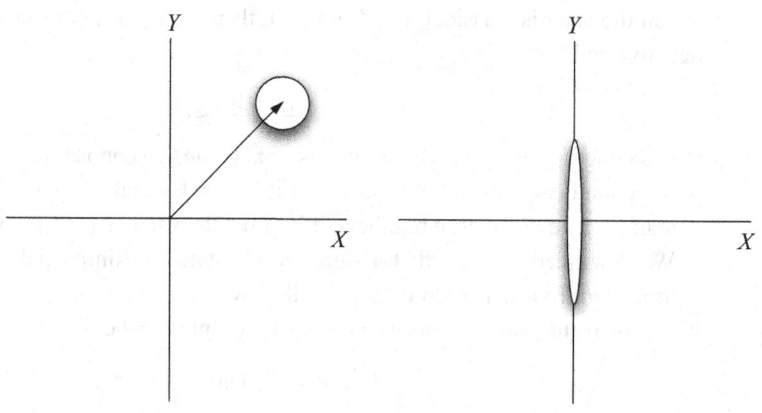

Fig. 5.2 A coherent state, pictured here on the left with an 'error box' indicating its fluctuations, is invariant under a rotation by an angle of 2π in phase space. A squeezed vacuum state, pictured here on the right and represented by an ellipse, is, however, invariant under a rotation by π. The quadrature fluctuations of a coherent state are the same in any direction, while those of a squeezed vacuum state are large in one direction but small in the orthogonal direction.

initially in a coherent state, $|\beta\rangle_b$, and the signal is in the vacuum state. The pump state can be represented as a point in the complex plane (the b-mode phase space) located at β surrounded by a circle of radius $1/2$. The circle represents the fluctuations in the real and imaginary parts of the field amplitude, which correspond to the operators $X_b = (b^\dagger + b)$ and $Y_b = i(b^\dagger - b)$, respectively. For a coherent state, the fluctuations are the same in both (and, in fact, all) directions. If this state is rotated about the origin by an angle of 2π, it is mapped back into itself (as is any state). On the other hand, the signal-mode state, because it has twice the symmetry of the pump state, is invariant under a rotation by π. This is consistent with our knowledge of the results from the parametric approximation. In that case, a pump mode in a strong coherent state produces a signal mode in a squeezed vacuum state. This state is represented in our two-dimensional phase space as an ellipse centered at the origin; a state that is invariant under a rotation by π (see Figure 5.2).

In order to prove these assertions, let us first define the operators

$$U_a(\theta) = e^{i\theta n_a}, \qquad U_b(\theta) = e^{i\theta n_b}, \qquad U_M(\theta) = e^{i\theta M/2} = U_a(\theta/2)U_b(\theta). \tag{5.112}$$

The initial state of the system is assumed to be $|\Psi\rangle = |0\rangle_a|\Psi_b\rangle_b$, where

$$U_b(2\pi/n)|\Psi_b\rangle_b = e^{i\phi}|\Psi_b\rangle_b, \tag{5.113}$$

that is, a rotation by $2\pi/n$ maps $|\Psi_b\rangle_b$ into itself multiplied by a phase factor. This implies that

$$U_M(2\pi/n)U(t)|\Psi\rangle = U(t)U_M(2\pi/n)|\Psi\rangle = e^{i\phi}U(t)|\Psi\rangle. \tag{5.114}$$

From this we find

$$\begin{aligned}
&{}_b\langle n_b|U(t)|\Psi\rangle\langle\Psi|U^{-1}(t)|n_b\rangle_b \\
&\quad = {}_b\langle n_b|U_M(2\pi/n)U(t)|\Psi\rangle\langle\Psi|U^{-1}(t)U_M^{-1}(2\pi/n)|n_b\rangle_b \\
&\quad = U_a(\pi/n)\,{}_b\langle n_b|U(t)|\Psi\rangle\langle\Psi|U^{-1}(t)|n_b\rangle_b U_a^{-1}(\pi/n).
\end{aligned} \tag{5.115}$$

Summing both sides over n_b, we obtain

$$\rho_a(t) = U_a(\pi/n)\rho_a(t)U_a^{-1}(\pi/n), \tag{5.116}$$

i.e. the a-mode state is invariant under a rotation by π/n. Summarizing, we can say that, if at $t = 0$ the signal mode is in the vacuum state and the pump-mode state is invariant under a rotation by $2\pi/n$, then at any time the signal-mode state is invariant under a rotation by π/n.

This result implies that, if the pump mode is initially in a squeezed vacuum state (invariant under a rotation by π), then the signal will be invariant under a rotation by $\pi/2$. This suggests that in phase space the signal-mode state will have a four-pronged structure. This has been verified by numerical calculations of the signal-mode state, which clearly show four prongs and a four-fold rotational symmetry.

5.7.6 Three-mode parametric interactions

There is also a three-mode version of the process we have been considering. Its Hamiltonian is given by

$$H = \hbar\omega c^\dagger c + \hbar\omega_1 a^\dagger a + \hbar\omega_2 b^\dagger b + i\hbar\chi(ca^\dagger b^\dagger - c^\dagger ab), \tag{5.117}$$

where c, a and b are the annihilation operators of the pump, signal and idler modes, respectively, and $\omega = \omega_1 + \omega_2$. This Hamiltonian can describe parametric amplification or sum-frequency generation. In parametric amplification, the pump mode is in a large-amplitude coherent state (the parametric approximation is usually employed) and strong correlations are produced between the signal and idler modes (a and b modes). These highly correlated two-mode states have found application in quantum information. For sum-frequency generation, the a and b modes are initially excited, and they give rise to photons in the c mode whose frequency is the sum of those in the a and b modes. In this case, the two operators

$$M_1 = n_a - n_b, \qquad M_2 = 2n_c + n_a + n_b, \tag{5.118}$$

or any linear combination of them, commute with H. As in the degenerate case, we can find a relation between the number fluctuations in the pump and those in either of the other two modes. We find that

$$\begin{aligned} \Delta n_c(t) + \Delta K_1(0) &\geq \Delta n_a(t) \geq |\Delta n_c(t) - \Delta K_1(0)|, \\ \Delta n_c(t) + \Delta K_2(0) &\geq \Delta n_b(t) \geq |\Delta n_c(t) - \Delta K_2(0)|, \end{aligned} \tag{5.119}$$

where

$$\begin{aligned} K_1 &= \tfrac{1}{2}(M_1 + M_2) = n_c + n_a, \\ K_2 &= \tfrac{1}{2}(M_2 - M_1) = n_c + n_b. \end{aligned} \tag{5.120}$$

These relations are most useful when $\Delta K_1(0)$ and $\Delta K_2(0)$ are small. For example, this will occur if the signal and idler modes are initially in their vacuum states and the pump is in a highly sub-Poissonian state.

In the nondegenerate case, it is also possible to find a relation between the number fluctuations of the a and b modes. We have

$$\Delta M_1(0) \geq |\Delta n_a(t) - \Delta n_b(t)|. \tag{5.121}$$

If $\Delta M_1(0)$ is small, then the number fluctuations in the signal and idler are similar. If both modes start in the vacuum state, then $\Delta M_1(0) = 0$ and $\Delta n_a(t) = \Delta n_b(t)$. This conclusion is independent of the initial pump state.

We can also use the fact that M_1 is conserved to show that, if the signal and idler modes are initially in the vacuum state, then at later times the signal–idler state is either the vacuum or nonclassical. This follows from the fact that a two-mode state is nonclassical if

$$\langle (n_a(t) - n_b(t))^2 \rangle - \langle n_a(t) - n_b(t) \rangle^2 < \langle n_a(t) \rangle + \langle n_b(t) \rangle. \tag{5.122}$$

If the a and b modes are originally in the vacuum state, the left-hand side of Eq. (5.122) will be zero at $t = 0$, and the fact that $[M_1, H] = 0$ implies that it will be zero for all time. Therefore, if either $\langle n_a(t) \rangle$ or $\langle n_b(t) \rangle$ is greater than zero, the signal–idler state will be nonclassical. The reason for this is that the number of photons in the two modes is highly correlated. For example, if we measure the photon number in one mode, we immediately know what it is in the other. Correlations that are this strong are not permitted classically. These correlations have been observed experimentally in both a continuous-wave, oscillator configuration and in a pulsed, amplifier configuration.

5.8 Anharmonic oscillator and Schrödinger's cat

Perhaps the most common phenomenon to arise out of a $\chi^{(3)}$ nonlinearity is that of the intensity-dependent refractive index. This gives an optical equivalent of the anharmonic oscillator. In a medium of this type, the refractive index consists of two terms: the first is constant, and is just the usual linear index of refraction; the second is proportional to the intensity of the field. In the case of a single mode, this leads to the quantum mechanical Hamiltonian

$$H_1 = \hbar\omega a^\dagger a + \tfrac{1}{2}\hbar\kappa (a^\dagger)^2 a^2, \tag{5.123}$$

and in the case of two modes,

$$H_2 = \hbar\omega_1 a^\dagger a + \hbar\omega_2 b^\dagger b + \tfrac{1}{2}\hbar\kappa a^\dagger b^\dagger ab, \tag{5.124}$$

where, in both cases, κ is proportional to $\chi^{(3)}$.

5.8.1 Quantum nondemolition measurements

In order to see what types of effects these Hamiltonians give rise to, let us first consider the second one and see how an initial state that is a product of a coherent state in the a mode and a number state in the b mode evolves. We have that

$$\begin{aligned} |\Psi(t)\rangle &= e^{-itH_2/\hbar}|\alpha\rangle_a|n\rangle_b \\ &= e^{-in\omega_2 t}e^{-it[\omega_1+(n\kappa/2)]a^\dagger a}|\alpha\rangle_a|n\rangle_b \\ &= e^{-in\omega_2 t}|\Psi_n(t)\rangle_a|n\rangle_b, \end{aligned} \tag{5.125}$$

where we use the property that $|n\rangle_b$ is an eigenstate of the number operator, $b^\dagger b$. Next, we must consider what happens to the mode in a coherent state. Defining a phase-shift $\phi = t[\omega_1 + (n\kappa/2)]$, we see that $|\Psi_n(t)\rangle_a$ is given by

$$\begin{aligned} |\Psi_n(t)\rangle_a &= e^{-i\phi a^\dagger a}|\alpha\rangle_a \\ &= e^{-|\alpha|^2/2}\sum_{n_a}\frac{\alpha^{n_a}}{\sqrt{n_a!}}e^{-i\phi n_a}|n_a\rangle_a. \end{aligned} \tag{5.126}$$

Therefore, we see that, for times greater than zero, we still have the product of a coherent state and a number state, but the phase of the coherent state has changed so that $\alpha(t) = \exp\{-it[\omega_1 + (n\kappa/2)]\}\alpha$. The amount of the change is proportional to the number of photons initially in the b mode. If we measure the phase-shift in the a mode, which could be done by means of an interferometer, then we can determine the number of photons in the b mode. Note also that the state of the b mode is not affected by this measurement. This represents what is known as a 'quantum nondemolition measurement' of photon number.

The reason for the use of this term is that the photon number can be measured without the state being changed. Although this seems like a sensible way to carry out a measurement, photo-detectors in the laboratory work differently. They absorb photons and create electronic excitations instead. These instruments change the state of the radiation field, so they perform 'quantum demolition measurements'. The state of the field is different after the measurement, even when the field is in an eigenstate of the photon number operator.

5.8.2 Quantum decay and revival

Let us now see what happens when we use the single-mode Hamiltonian to evolve a state that is initially a coherent state. We have that

$$|\Psi(t)\rangle = e^{-itH_1/\hbar}|\alpha\rangle = e^{-|\alpha|^2/2}\sum_{n=0}^{\infty}\frac{\alpha^n}{n!}e^{-it[n\omega+n(n-1)(\kappa/2)]}|n\rangle. \tag{5.127}$$

We can use this to calculate the expectation value of the annihilation operator, which corresponds to the complex amplitude of the field.

First, recall that the action of an annihilation operator, from Eq. (2.22), is that

$$a|n\rangle = \sqrt{n}\,|n-1\rangle. \tag{5.128}$$

Hence, on defining $n' = n - 1$, we find that

$$a|\Psi(t)\rangle = \alpha e^{-|\alpha|^2/2}\sum_{n'=0}^{\infty}e^{-it[(n'+1)\omega+n'(n'+1)(\kappa/2)]}\frac{\alpha^{n'}}{\sqrt{n'!}}|n'\rangle. \tag{5.129}$$

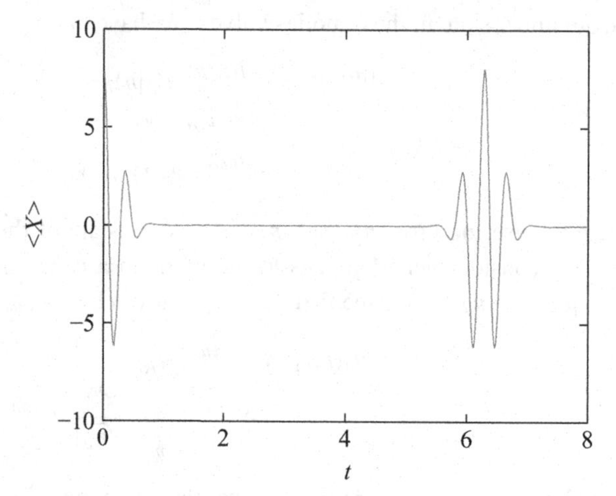

Fig. 5.3 The decay and quantum revival of the X quadrature in an anharmonic oscillator. The parameters used here are $\alpha = 4$ and $\kappa = \omega = 1$.

The expectation value is then

$$\langle a(t)\rangle = \langle\Psi(t)|a|\Psi(t)\rangle = \alpha e^{-|\alpha|^2 - i\omega t}\sum_{n=0}^{\infty} e^{-i\kappa n t}\frac{|\alpha|^{2n}}{n!}$$
$$= \alpha e^{-i\omega t}\exp[|\alpha|^2(e^{-i\kappa t} - 1)], \tag{5.130}$$

where we have relabeled the summation variable as n.

The nonlinear interaction leads to an additional phase, which is proportional to $|\alpha|^2$, and, at longer times, causes the magnitude of the complex amplitude to decay. This is a result of the fact that the different number-state components of the coherent state pick up different phases as the state evolves and this causes the overall phase uncertainty of the state to increase. That means that, when adding the contributions of the different number-state components to form the complex amplitude, there is some cancellation because of the different phases. The resulting behavior can be clearly seen on expanding $e^{-i\kappa t}$,

$$\langle a(t)\rangle = \alpha e^{-i\omega t}\exp[|\alpha|^2(-i\kappa t - (\kappa t)^2/2)], \tag{5.131}$$

where we have kept only the first three terms. We see that the oscillations of $\langle a(t)\rangle$ decay with a Gaussian envelope given by $\exp[-(\kappa|\alpha|t)^2/2]$. Because the number operator commutes with the Hamiltonian, the expectation value of the number operator, and of its moments, is not affected by the time evolution.

A further examination of Eq. (5.130) shows an interesting behavior, with three characteristic time-scales, as shown in Figure 5.3:

1. First, on very short time-scales, there is simply an oscillation with a renormalized frequency of $\omega' = \omega + \kappa\bar{n}$, where $\bar{n} = |\alpha|^2$. A similar result is obtained classically.
2. On intermediate time-scales, there is a quadratic decay in the expectation value, with a characteristic damping rate of $\kappa\sqrt{\bar{n}}$.

3. Finally, on very long time-scales, there is a succession of periodic revivals where the initial state and all its properties are regained exactly, apart from a possible phase-shift. This occurs whenever $t = 2\pi/\kappa$.

Such coherent revivals and decays are strong evidence of quantum interference, and are characteristic of small quantum systems with discrete level spacings. The anharmonic oscillator with a coherent initial condition is a very special case. Here the revivals are periodic, and continue indefinitely as long as there is no damping. More generally, revivals are aperiodic, and gradually diminish through damping. Such decays and revivals have been experimentally observed in a closely related system, the Bose–Einstein condensate of ultra-cold atoms.

5.8.3 Mesoscopic quantum superpositions

An interesting aspect of this behavior is an intermediate state that forms at times that are odd multiples of π/κ. While there is no obvious signature of something unusual in the expectation value, $\langle a(t)\rangle$, at this time, nevertheless strange behavior exists in the quantum state itself.

In the case of $t = \pi/\kappa$, if we set $\omega = \kappa/2$ to remove a trivial phase rotation, we find that

$$\left|\Psi\left(\frac{\pi}{\kappa}\right)\right\rangle = e^{-|\alpha|^2/2}\sum_{n=0}^{\infty}\frac{\alpha^n}{n!}e^{-in^2\pi/2}|n\rangle. \tag{5.132}$$

The state splits up into separate odd and even photon number parts,

$$\left|\Psi\left(\frac{\pi}{\kappa}\right)\right\rangle = e^{-|\alpha|^2/2}\sum_{n=0,2,\ldots}^{\infty}\left[\frac{\alpha^n}{n!}|n\rangle - i\frac{\alpha^{n+1}}{(n+1)!}|n+1\rangle\right], \tag{5.133}$$

which can be expressed as

$$\left|\Psi\left(\frac{\pi}{\kappa}\right)\right\rangle = \frac{1}{\sqrt{2}}[e^{-i\pi/4}|\alpha\rangle + e^{i\pi/4}|-\alpha\rangle]. \tag{5.134}$$

We see from this result that the quantum dynamics has generated a 'Schrödinger cat': a mesoscopic superposition of two quantum states that become macroscopically distinct in the limit of $|\alpha| \gg 1$. The existence of such states would result in paradoxical behavior at the macroscopic level, as measured by means of interference phenomena. Although Schrödinger proposed the possibility of such states in the early days of quantum mechanics, the debate over the properties of quantum systems in these types of states, particularly massive particles, is still continuing. We do know, however, that these quantum states are very sensitive to losses, since in this case the random loss of a single photon is enough to remove any macroscopic quantum interference. We have treated this in terms of simple optical interactions, but similar behavior can occur in Bose–Einstein condensates and ion traps. A treatment of the quantum signature of these states in multi-mode cases is given in Chapter 12.

5.8.4 Sub-Poissonian statistics using a quantum beam-splitter

Inputting the Kerr squeezed state together with a coherent state into a quantum beam-splitter generates a different type of nonclassical state. We recall that the photon statistics of a single-mode field are sub-Poissonian if $(\Delta n_a)^2 < \langle n_a \rangle$, and that such a state is nonclassical. In this case, the input state in the a mode will be the state in Eq. (5.127) and the input state in the b mode will be a coherent state with amplitude $\beta e^{-i\omega t}$. We shall drop the $e^{-i\omega t}$ factors in what follows, because they cancel for the quantities we are calculating. We shall also be interested in the case in which T is close to one and $|\beta|$ is large, with $\sqrt{R}\beta = \xi$.

Making these substitutions, together with the results of Eq. (5.2), we find that, in the a^{out} output mode,

$$\begin{aligned}(\Delta n_a^{out})^2 - \langle n_a^{out} \rangle \\ = 2[\xi \langle (a^\dagger)^2 a \rangle + \xi^* \langle a^\dagger a^2 \rangle - |\alpha|^2 (\xi \langle a^\dagger \rangle + \xi^* \langle a \rangle)] \\ + \xi^2 \langle (a^\dagger)^2 \rangle + (\xi^*)^2 \langle a^2 \rangle - \xi^2 \langle a^\dagger \rangle^2 - (\xi^*)^2 \langle a \rangle^2 + 2|\xi|^2 (|\alpha|^2 - |\langle a \rangle|^2),\end{aligned} \tag{5.135}$$

where the expectation values without subscripts, i.e. the ones on the right-hand side of the equation, are expectation values of the *in* operators, that is, the operators at the input to the beam-splitter. These are just expectation values in the state in Eq. (5.127). Setting $\phi = \kappa t$ and again neglecting the $e^{-i\omega t}$ factors, we find that

$$\begin{aligned}\langle a^2 \rangle &= \alpha^2 e^{-i\phi} e^{-2|\alpha|^2 (i\phi + \phi^2)}, \\ \langle (a^\dagger)^2 a \rangle &= (\alpha^*)^2 \alpha e^{i\phi} e^{|\alpha|^2 (i\phi - \phi^2/2)}.\end{aligned} \tag{5.136}$$

Now set $\alpha = |\alpha| e^{i\theta}$, $\xi = r e^{i\eta}$, and assume that $\phi \ll 1$, $|\alpha| \gg 1$, and $\phi|\alpha|$ is of order one, to obtain

$$\begin{aligned}(\Delta n_a^{out})^2 - \langle n_a^{out} \rangle \\ = 2r|\alpha|^3 [e^{i(\eta-\theta)} e^{|\alpha|^2 (i\phi - \phi^2/2)} (e^{i\phi} - 1) + c.c.] \\ + r^2 |\alpha|^2 [e^{2i(\eta-\theta)} e^{2i|\alpha|^2 \phi} e^{-|\alpha|^2 \phi^2} (e^{-|\alpha|^2 \phi^2} e^{i\phi} - 1) + c.c.] \\ + 2r^2 |\alpha|^2 (1 - e^{-|\alpha|^2 \phi^2}).\end{aligned} \tag{5.137}$$

We now use the assumption that $|\alpha|\phi$ is of order one. For $(e^{i\phi} - 1)$ in the first term we substitute $i\phi$, and for $e^{i\phi}$ in the second term we substitute 1. This makes both of these terms of order $|\alpha|^2$. Keeping additional terms in ϕ would lead to small corrections. Making these substitutions, we find that

$$\begin{aligned}(\Delta n_a^{out})^2 - \langle n_a^{out} \rangle \\ = -4r\phi|\alpha|^3 e^{-(|\alpha|\phi)^2/2} \sin(\eta - \theta + |\alpha|^2 \phi) \\ + 2r^2 |\alpha|^2 (1 - e^{-(|\alpha|\phi)^2}) \{1 - e^{-(|\alpha|\phi)^2} \cos[2(\eta - \theta + |\alpha|^2 \phi)]\}.\end{aligned} \tag{5.138}$$

Finally, setting $\eta - \theta + |\alpha|^2 \phi = \pi/2$ and minimizing with respect to r, we find

$$(\Delta n_a^{out})^2 - \langle n_a^{out} \rangle = -\frac{2|\alpha|^3 \phi\, e^{-(|\alpha|\phi)^2}}{1 - e^{-2(|\alpha|\phi)^2}}. \tag{5.139}$$

We see, then, that using this scheme of mixing the output from a $\chi^{(3)}$ medium with an appropriately chosen coherent state at a beam-splitter, we can create a field whose photon statistics are significantly sub-Poissonian. We note that the right-hand side of the above equation is comparable to the expectation value of the photon number in the a^{out} mode, which is just

$$\langle n_a^{out} \rangle = |\alpha|^2 + \frac{|\alpha|\phi\, e^{-(|\alpha|\phi)^2/2}}{1 - e^{-2(|\alpha|\phi)^2}}. \tag{5.140}$$

This is yet another illustration of the fact that nonlinear media provide good ways to generate nonclassical states of the electromagnetic field.

5.9 Jaynes–Cummings dynamics

Another model with interesting quantum dynamics is the Jaynes–Cummings model, which we have briefly encountered already. This is the fully quantized version of a model of atom–field interactions, which we have met in a semiclassical approximation in Section 1.6.2, and, in the limit of large numbers of atoms and large detunings, in Section 4.7.1. Now we wish to investigate the behavior of this model as we change from far off-resonance to resonant interactions, with only one atom to create the simplest nonlinearity possible.

We recall that in this model, with an appropriate choice of coupling phase, we can set

$$H_{JC} = \tfrac{1}{2}\hbar\nu_0\sigma^{(z)} + \hbar\omega a^\dagger a + \hbar|g|(a\sigma^{(+)} + a^\dagger\sigma^{(-)}). \tag{5.141}$$

It is useful to write the Hamiltonian as the sum of a free term and an interacting term, thus $H_{JC} = H_0 + H'$, with

$$\begin{aligned} H_0 &= \hbar\omega[a^\dagger a + \tfrac{1}{2}\sigma^{(z)}], \\ H' &= \tfrac{1}{2}\hbar\Delta\sigma^{(z)} + \hbar|g|(a\sigma^{(+)} + a^\dagger\sigma^{(-)}), \end{aligned} \tag{5.142}$$

where we have introduced the detuning $\Delta = \nu_0 - \omega$, as previously. Since these two parts of the Hamiltonian commute with each other, H_0 is conserved by the action of the total Hamiltonian. This can be interpreted as saying that there is a conservation law of the number of effective quanta, $N_Q = a^\dagger a + \frac{1}{2}\sigma^{(z)}$. This number N_Q has eigenvalues $n + 1/2$, for $n = 0, 1, \ldots$, and for $n > 0$ each eigenvalue is doubly degenerate. In particular, for $n > 0$, the eigenvalue $n + 1/2$ corresponds to a two-dimensional space spanned by the vectors $|n+1, g\rangle$ and $|n, e\rangle$, where n is the number of photons, and e and g indicate the ground and excited state as in Figure 1.1.

Since only the states with the same eigenvalue of N_Q are coupled to each other by the full Hamiltonian, it follows that, in order to find the eigenvalues and eigenstates of H_{JC}, we only need to diagonalize it in each of the two-dimensional subspaces. In the subspace corresponding to the eigenvalue $n + 1/2$ of N_Q, we have that

$$H_{JC} \to \hbar \begin{bmatrix} n\omega + \nu_0/2 & |g|\sqrt{n+1} \\ |g|\sqrt{n+1} & (n+1)\omega - \nu_0/2 \end{bmatrix}, \tag{5.143}$$

where the matrix is expressed in the ordered basis $\{|e, n\rangle, |g, n+1\rangle\}$. Therefore, for a given n, we have exact energy eigenvalues of $E_{n\pm} = \hbar\omega_{n\pm}$, where

$$\omega_{n\pm} = \omega(n + \tfrac{1}{2}) \pm \tfrac{1}{2}\Omega(\Delta). \tag{5.144}$$

The quantity $\Omega(\Delta) \equiv \sqrt{4|g|^2(n+1) + \Delta^2}$ is called the quantum Rabi frequency. It is the quantum version of the classical Rabi oscillation frequency introduced in Eq. (1.76). By quantizing the radiation field, we introduce a dynamics in which the photon number oscillates out of phase with the atomic excitation. This, of course, is a consequence of energy conservation.

The corresponding eigenstates are then written as

$$\begin{aligned} |n,+\rangle &= \cos(\theta_n)|n,e\rangle + \sin(\theta_n)|n+1,g\rangle, \\ |n,-\rangle &= -\sin(\theta_n)|n,e\rangle + \cos(\theta_n)|n+1,g\rangle, \end{aligned} \tag{5.145}$$

where $\theta_n = \frac{1}{2}\tan^{-1}(2|g|\sqrt{n+1}/\Delta)$.

The two types of interacting eigenstates should be familiar to the reader by now; these are just the resonant versions of the two types of polaritons we have already encountered in Chapter 4. These exact eigenvalues reduce to the approximate results we found at large detunings in Eq. (4.116), where they were shown to correspond to a photon-like and an atom-like excitation, respectively. We also see this in the present exact solutions, where large Δ implies small θ_n. In this limit,

$$\omega_{n\pm} \to \omega(n + \tfrac{1}{2}) \pm \tfrac{1}{2}(\nu_0 - \omega). \tag{5.146}$$

In other words, as one might expect intuitively, there are two modes of excitation at large detuning: we can change the field excitation, which takes $\omega_{n\pm} \to \omega_{n+1\pm}$, or we can change the atomic excitation, which takes $\omega_{n-} \to \omega_{n+}$. In the two-mode polariton Hamiltonian of Eq. (4.116), the first type of excitation corresponds to the photon–polariton operator c, the second type to the operator d.

To understand the resulting dynamical behavior in this exactly soluble model, consider an initial quantum state $|\Psi(0)\rangle = \sum_n C_n|n,e\rangle$, which corresponds to an initially excited atom injected into a state with a coherent superposition of different number eigenstates. The initial state is therefore

$$|\Psi(0)\rangle = \sum_n C_n[\cos(\theta_n)|n,+\rangle - \sin(\theta_n)|n,-\rangle]. \tag{5.147}$$

This is an expansion in terms of the eigenstates, so it follows that the dynamical solution is written as

$$|\Psi(t)\rangle = \sum_n [\cos(\theta_n)|n,+\rangle e^{-it\omega_{n+}} - \sin(\theta_n)|n,-\rangle e^{-it\omega_{n-}}]. \tag{5.148}$$

This model has similar behavior to the anharmonic oscillator model, in that it also shows collapses and revivals if the initial state of the radiation field is a coherent state. These have been experimentally observed in micromaser experiments. For large detunings, as we expect from our previous analysis, the model simply reduces to the anharmonic oscillator.

5.10 Parametric approximation

As we have seen, if one of the coupled modes we are considering is initially in a highly excited coherent state, we can assume that, at least for a certain period of time, it will act like a classical field. To that end, we can replace the operators for this field in the Hamiltonian by c-numbers. This is the parametric approximation. Once this is done, we can solve the problem exactly, if the interaction is the result of a $\chi^{(2)}$ nonlinearity, because the equations of motion for the remaining operators are linear. Because this approximation is both important and so often employed, we would now like to enquire into its history and justification.

The parametric approximation was born at the same time as the quantum mechanical study of parametric processes. In a 1961 paper Louisell, Yariv and Siegman considered a model of a parametric amplifier consisting of two modes of the electromagnetic field (the signal and the idler) coupled by a medium with an oscillating dielectric constant (see the additional reading in Chapter 1). The modulation of the dielectric constant occurs at a frequency that is the sum of the frequencies of the two modes, and it is assumed that this modulation can be described classically. This led them to the Hamiltonian

$$H = \hbar\omega_1 a^\dagger a + \hbar\omega_2 b^\dagger b + \hbar(\xi^* e^{i\omega t} ab + \xi e^{-i\omega t} a^\dagger b^\dagger), \tag{5.149}$$

where a and b are the annihilation operators of the two modes, $\omega = \omega_1 + \omega_2$, and $|\xi|$ is proportional to the amplitude of the modulation of the dielectric constant. They used this model to study the quantum noise properties of the parametric amplification process.

Mollow and Glauber were the next to analyze the parametric amplifier. They emphasized that the oscillating dielectric constant of Louisell, Yariv and Siegman is the result of an intense light wave in a nonlinear dielectric. They used the Hamiltonian in the previous paragraph and concentrated their attention on the one- and two-mode P-representations of the signal and the idler. They found that, if one of the modes starts in a finite-temperature thermal state, then the other mode, no matter what its initial state, will become classical after a sufficient period of time. The situation with the two-mode P-function is the opposite; two-mode states that are initially classical can remain so for only a finite period of time.

The intense interest in squeezing in the 1980s led to a renewal of work on the parametric amplifier. As we have seen, the degenerate parametric amplifier produces minimum-uncertainty squeezed states. In the parametric approximation, the amount of squeezing that can be obtained is arbitrarily large; one only need wait long enough. This, however, is clearly an artifact of the parametric approximation. At long enough times, the dynamical and quantum mechanical aspects of the pump, both of which are ignored in this approximation, will play a role. It became obvious that it would be necessary to go beyond the parametric approximation to determine how much squeezing is possible.

Now let us look at corrections to the parametric approximation for the degenerate parametric amplifier in more detail. A number of methods have been used to calculate these corrections, but here we will give a simple argument due to Crouch and Braunstein that gets to the heart of the matter rather quickly (see the additional reading). Following their treatment, we take for the interaction Hamiltonian

$$H = i\frac{\hbar\chi}{2}[b(a^\dagger)^2 - b^\dagger a^2]. \tag{5.150}$$

Note that by using the interaction part of the Hamiltonian as the full Hamiltonian, we are neglecting the part of the evolution due to the free-field part of the full Hamiltonian. We will be using this Hamiltonian to find equations of motion for the creation and annihilation operators, and neglecting the free-field Hamiltonian simply means that we are solving for the slowly varying behavior of these operators, that is, in the notation of a previous subsection, we are solving for $A(t)$ instead of $a(t)$. If the pump mode is in a large-amplitude coherent state with amplitude $\beta = \sqrt{N_p}\exp(i\phi_p)$, we make the parametric approximation by replacing b and $b^\dagger$ in H by β and β^*, respectively, to give

$$H_p = i\frac{\hbar\chi}{2}\sqrt{N_p}\,[e^{i\phi_p}(a^\dagger)^2 - e^{-i\phi_p}a^2]. \tag{5.151}$$

The equations of motion for a and $a^\dagger$ that follow from H_p are

$$\frac{da}{dt} = \chi\sqrt{N_p}\,e^{i\phi_p}a^\dagger, \qquad \frac{da^\dagger}{dt} = \chi\sqrt{N_p}\,e^{-i\phi_p}a, \tag{5.152}$$

and are easily solved to give

$$a(t) = a(0)\cosh u + a^\dagger(0)e^{i\phi_p}\sinh u, \tag{5.153}$$

where $u = \chi\sqrt{N_p}\,t$. This solution, and its adjoint, allow us to find the properties of the signal as a function of time. In particular, if we define the quadrature operators

$$X_a = a^\dagger + a, \qquad Y_a = i(a^\dagger - a), \tag{5.154}$$

and start at $t = 0$ with the signal mode in the vacuum state, we find

$$(\Delta X_a)^2 = e^{2u}\cos^2(\phi_p/2) + e^{-2u}\sin^2(\phi_p/2), \tag{5.155}$$

$$(\Delta Y_a)^2 = e^{-2u}\cos^2(\phi_p/2) + e^{2u}\sin^2(\phi_p/2). \tag{5.156}$$

Note that, if $\phi_p = 0$, then $\Delta X_a(u)$ grows exponentially with time and $\Delta Y_a(u)$ decreases exponentially with time, thereby becoming squeezed.

5.10.1 Limits to quantum squeezing

At this level of approximation, there is no limit to how much squeezing in the Y_a direction is possible. There are two effects, however, that have been neglected in the parametric approximation, which will enforce a limit. The first is pump depletion. In replacing the pump-mode operators by c-numbers, the parametric approximation assumes that the amplitude of the pump remains constant. This is not a bad approximation for small times, but it gets worse as time progresses. The second effect that has been ignored is pump fluctuations. Even if classical fluctuations can be eliminated, there will still be quantum fluctuations in the pump field that will affect signal-mode squeezing.

Let us now estimate the effect of the pump phase fluctuations. We shall assume that the pump is initially in a coherent state with a large, real amplitude $\sqrt{N_p}$. The mean phase of

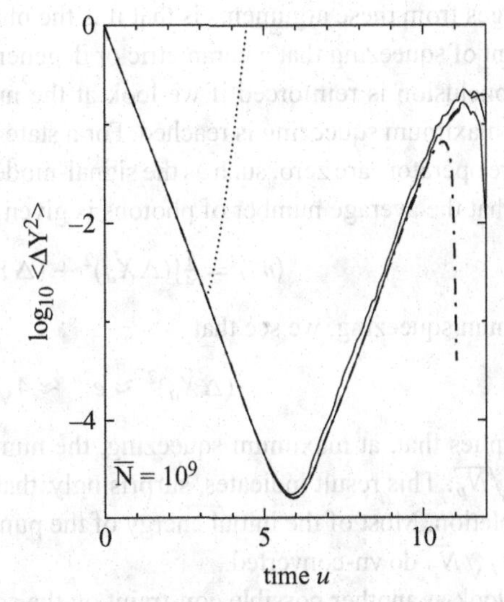

Fig. 5.4 The limits to squeezing in a driven parametric oscillator. Here $\bar{N} = 10^9$ is the mean coherent photon number and u is the dimensionless time. The dashed-dotted line is the approximate analytic theory. The solid lines represent the sampling errors in a stochastically exact positive P simulation, described in the next chapter. Taken (with permission) from P. Kinsler *et al.*, *Phys. Rev.* A **48**, 3310–3320 (1993).

the pump in this state is zero and

$$(\Delta\phi_p)^2 = \langle\phi_p^2\rangle = \frac{1}{4N_p}. \tag{5.157}$$

Here we have taken $\Delta\phi_p = \Delta X_b/(2\sqrt{N_p})$ and then set $\Delta X_b = 1$, which is its value for a coherent state. If the pump phase were exactly zero, we see from Eq. (5.155) that ΔY_a would be perfectly squeezed. The pump phase fluctuations, however, cause the squeezing direction to fluctuate, with the consequence that some of the amplified noise is mixed into ΔY_a. In order to find out how much, we use the fact that the phase fluctuations are small to replace, in Eq. (5.155), $\cos^2(\phi_p)$ by 1 and $\sin^2(\phi_p)$ by ϕ_p^2. We then use Eq. (5.157) to average over the phase noise in the pump to obtain

$$[\Delta Y_a(u)]^2 \cong e^{-2u} + \frac{1}{16N_p}e^{2u}. \tag{5.158}$$

As a function of u, ΔY_a first decreases and then increases. Its minimum value is (corresponding to maximum squeezing)

$$(\Delta Y_a)^2 \cong \frac{1}{2\sqrt{N_p}}, \tag{5.159}$$

which occurs when $u \cong (1/4)\ln(16N_p)$, as shown in Figure 5.4. This tells us that the maximum squeezing (minimum value of $(\Delta Y_a)^2$) scales as the square root of the number of photons in the pump. This argument gives the same result as one finds from more detailed calculations (the results of a numerical calculation are given in Figure 5.4). The conclusion that emerges from these arguments is that it is the phase fluctuations in the pump that limit the amount of squeezing that a parametric or degenerate parametric amplifier can produce.

This conclusion is reinforced if we look at the number of photons in the signal mode when the maximum squeezing is reached. For a state in which the expectation values of the quadrature operators are zero, such as the signal-mode state in parametric down-conversion, we have that the average number of photons is given by

$$\langle n_a \rangle = \tfrac{1}{4}[(\Delta X_a)^2 + (\Delta Y_a)^2] - \tfrac{1}{2}. \tag{5.160}$$

At maximum squeezing, we see that

$$(\Delta X_a)^2 \approx e^{2u} \approx 4\sqrt{N_p}, \tag{5.161}$$

which implies that, at maximum squeezing, the number of photons in the signal mode is of order $\sqrt{N_p}$. This result indicates, surprisingly, that, at maximum squeezing, there is not much depletion. Most of the initial energy of the pump is still present, with only a fraction of order $1/\sqrt{N_p}$ down-converted.

Let us look at another possible constraint on the squeezing of the signal mode. Suppose we want to compare the phase uncertainty of the input to the down-converter, which is the pump, to the phase uncertainty of the output, which is the signal, at maximum squeezing. We can estimate the phase uncertainty of the down-converted signal at maximum squeezing by defining

$$\Delta\phi_a \approx \frac{\Delta Y_a}{\Delta X_a} = 2e^{-2u} = \frac{1}{\sqrt{8N_p}}. \tag{5.162}$$

The input pump beam has a phase uncertainty of $O(1/\sqrt{N_p})$. The fact that these quantities are essentially the same shows the fundamental limiting process. An experiment cannot create information where none is present initially. In parametric down-conversion, the overall Hamiltonian has a symmetry under simultaneous changes in the phase of the pump and down-converted signal. Therefore, the only phase information in the output signal must come from the input signal, in this case the pump. Maximum squeezing occurs when this information limit is reached, and the down-converted signal reaches a similar level of phase uncertainty to the input signal. This 'information saturation' limit creates a stronger limit to squeezing than the more familiar 'energy saturation' limit caused by energy conservation and depletion.

5.10.2 Higher-order processes

Finally, let us briefly mention the role of the parametric approximation in higher-order down-conversion processes. The degenerate parametric amplifier is the simplest example of a down-conversion process. If the initial intensity of the pump mode, at frequency 2ω, is large and the signal mode at frequency ω is in the vacuum state, then at later times the intensity of the signal will have increased at the expense of the pump. Light at frequency 2ω is thereby converted into light at frequency ω.

Higher-order nonlinearities can be employed to generalize this process, though the strength of the interaction decreases as the order of the nonlinearity increases. In particular, we consider the process in which a pump photon at frequency $n\omega$ produces

n signal photons at frequency ω. The Hamiltonian describing this interaction is

$$H_n/\hbar = \omega a^\dagger a + n\omega b^\dagger b + \frac{i\chi_n}{n}[(a^\dagger)^n b + a^n b^\dagger]. \qquad (5.163)$$

If we assume the b mode is in a large-amplitude coherent state with amplitude β (we shall assume β is real for simplicity) and then make the parametric approximation, the Hamiltonian in the interaction picture becomes

$$H_{pn}/\hbar = \frac{i\chi_n}{n}\beta[(a^\dagger)^n + a^n]. \qquad (5.164)$$

This Hamiltonian leads to a time development transformation that has a rather unpleasant property for $n > 2$: if one starts in the vacuum state, the photon number will become infinite in a finite time. This is a consequence of the neglect of pump depletion in the parametric approximation. In the cases $n = 1$ and $n = 2$ a photon number divergence occurs, but only as $t \to \infty$. This problem does not exist for the full two-mode Hamiltonian H_n, so that it is an artifact of the parametric approximation.

Additional reading

Books

- Coherent states, P-representations, squeezed states and photon counting are standard topics in quantum optics books. So are a number of the simple processes that are discussed in this chapter. Some quantum optics books are listed below:

P. D. Drummond and Z. Ficek (eds), *Quantum Squeezing* (Springer, Berlin, 2004).

C. Gerry and P. Knight, *Introductory Quantum Optics* (Cambridge University Press, Cambridge, 2004).

L. Mandel and E. Wolf, *Optical Coherence and Quantum Optics* (Cambridge University Press, Cambridge, 1995).

M. O. Scully and M. S. Zubairy, *Quantum Optics* (Cambridge University Press, Cambridge, 1997).

D. F. Walls and G. J. Milburn, *Quantum Optics* (Springer, Heidelberg, 1994).

Articles

- Classic lectures and papers on coherence, coherent states and the theory of photon counting appear in:

R. J. Glauber, *Phys. Rev.* **130**, 2529 (1963).

R. J. Glauber, *Phys. Rev.* **131**, 2766 (1963).

R. J. Glauber, in *Quantum Optics and Electronics*, ed. C. Dewitt, A. Blandin, and C. Cohen-Tannoudji, p. 53 (Gordon and Breach, New York, 1965).

- Some papers on entanglement conditions for field modes are:

L.-M. Duan, G. Giedke, J. I. Cirac, and P. Zoller, *Phys. Rev. Lett.* **84**, 2722 (2000).

M. Hillery and M. S. Zubairy, *Phys. Rev. Lett.* **96**, 050503 (2006).
S. Mancini, V. Giovannetti, D. Vitali, and P. Tombesi, *Phys. Rev. Lett.* **88**, 120401 (2002).
E. Shchukin and W. Vogel, *Phys. Rev. Lett.* **95**, 230502 (2005).
R. Simon, *Phys. Rev. Lett.* **84**, 2726 (2000).

- The papers dealing with the properties of fields generated by $\chi^{(2)}$ processes that formed the basis of our discussion are:

M. Hillery, *Phys. Rev.* A **31**, 338 (1985).
M. Hillery, D. Yu, and J. Bergou, *Phys. Rev.* A **49**, 1288 (1994).
D. Stoler, *Phys. Rev. Lett.* **33**, 1397 (1974).

- The fact that $\chi^{(3)}$ media and a beam-splitter can be used to create states with sub-Poissonian photon statistics appears in:

M. Kitagawa and Y. Yamamoto, *Phys. Rev.* A **34**, 3974 (1986).

- The paper mentioned in our discussion of the parametric approximation by Louisell *et al.* was cited in Chapter 1. The rest of the work discussed there appears in the papers:

D. Crouch and S. Braunstein, *Phys. Rev.* A **38**, 4696 (1988).
M. Hillery, *Phys. Rev.* A **42**, 498 (1990).
B. Mollow and R. J. Glauber, *Phys. Rev.* **160**, 1076 (1967).
B. Mollow and R. J. Glauber, *Phys. Rev.* **160**, 1097 (1967).

- Two more papers bearing on when the parametric approximation is valid are:

M. Hillery and M. S. Zubairy, *Phys. Rev.* A **29**, 1275 (1984).
P. Kinsler, M. Fernee, and P. D. Drummond, *Phys. Rev.* A **48**, 3310 (1993).

Problems

5.1 We want to have some practice with coherent and squeezed states.
(a) Show that $(1/\pi)\int d^2\alpha\,|\alpha\rangle\langle\alpha| = I$.
(b) Find the P-representation of the one-photon number state and the P-representation of the thermal state

$$\rho = (1-x)\sum_{n=0}^{\infty} x^n |n\rangle\langle n|,$$

where $x = \exp(-\hbar\omega/k_B T)$, so that $0 < x < 1$. Here, ω is the mode frequency, k_B is Boltzmann's constant and T is the temperature.
(c) Prove that $S^{-1}(z)aS(z) = a\cosh r - a^{\dagger}e^{i\theta}\sinh r$, where $S(z)$ is the squeeze operator and $z = re^{i\theta}$.

5.2 Show that a two-mode state satisfying the condition

$$\langle (n_a - n_b)^2\rangle - \langle n_a - n_b\rangle^2 < \langle n_a\rangle + \langle n_b\rangle$$

is nonclassical.

5.3 Show that, for a single-mode state,

$$\langle 2n\rangle + 1 \geq \Delta X^2 + \Delta Y^2.$$

Use this result along with the uncertainty relation satisfied by X and Y to find a lower bound on ΔX in terms of the mean photon number of a state. This bound implies that, in order to obtain large squeezing, we need a large number of photons.

5.4 (a) Consider a beam-splitter with modes a and b with an input state $|\psi\rangle_a|0\rangle_b$. Apply the entanglement condition $|\langle ab^\dagger\rangle| > \langle a^\dagger ab^\dagger b\rangle$ to the output of the beam-splitter in order to find what condition the state $|\psi\rangle_a$ must satisfy in order for the output of the beam-splitter to be entangled. Note that, if $|\psi\rangle_a$ satisfies the condition you find, the output of the beam-splitter will be entangled, but, since our entanglement condition is only a sufficient condition, the output may be entangled even if the condition you find is not satisfied.

(b) Now apply the entanglement condition $|\langle ab\rangle| > \langle a^\dagger a\rangle\langle b^\dagger b\rangle$ to the output of a parametric amplifier. Assuming that the initial state is of the form $|\psi\rangle_a|0\rangle_b$, find a condition on $|\psi\rangle_a$ in order for the state at a later time to be entangled.

5.5 Let us have a closer look at homodyne detection. We start with a state $|\psi\rangle_a|\beta\rangle_b$, where we would like to measure a quadrature operator for mode a, and mode b is in a large-amplitude coherent state. We will look at two different versions of homodyne detection.

(a) First, let us look at only the a-mode output. In this case, we will assume $T \gg R$, $|\beta| \gg 1$ and $|\beta|\sqrt{R} \gg 1$. Find the leading contribution to $\langle (a_{a,out}^\dagger)^2 a_{a,out}^2\rangle - \langle n_{a,out}\rangle^2$ and show that it can be used to determine the fluctuations in a quadrature operator.

(b) Now let us look at both the a- and b-mode outputs. Set $R = T = 1/2$, and show that, by finding $\langle n_a - n_b\rangle$ and $\langle (n_a - n_b)^2\rangle$, we can determine both the expectation value and the fluctuations in a quadrature operator. This scheme is known as balanced homodyne detection.

5.6 Let us investigate the parametric approximation in a case where we can solve everything. We will start with the two-mode Hamiltonian $H = i\hbar g(a^\dagger b - ab^\dagger)$.

(a) Find the Heisenberg equations of motion for the operators a and b, and solve them for $a(t)$ and $b(t)$ in terms of $a(0)$ and $b(0)$. Now if the initial state of the system is $|0\rangle_a|\beta\rangle_b$, where, for convenience, we shall assume the amplitude of the coherent state, β, is real, use your solutions to find $\langle a(t)\rangle$.

(b) The Hamiltonian for the a mode in the parametric approximation corresponding to our two-mode Hamiltonian with the b mode in the large-amplitude coherent state $|\beta\rangle$ (we will again assume β is real) is just $H_p = i\hbar g\beta(a^\dagger - a)$. Assuming the initial state for the a mode is the vacuum state, find $\langle a(t)\rangle$. By comparing this answer to the one derived from the full two-mode Hamiltonian, find the time interval during which the parametric approximation is valid.

6 Decoherence and reservoirs

For the next several chapters we will mainly be focusing on the behavior of a medium with nonlinear susceptibility in an optical cavity. This will entail a detour through the quantum theory of open systems. All physical systems are coupled to the 'outside world' to some extent. We call such a system an open system, and the part of the 'outside world' that is coupled to it is called a reservoir. If the coupling is very weak, we can ignore it and treat the system as closed.

For the systems of interest here, we must take the coupling to the outside world into account, and, in particular, we need a way to see how this coupling affects the dynamics of the system itself. In the case of a nonlinear medium in a cavity, some of the light leaves the cavity, so the light inside the cavity is coupled to modes outside the cavity. In addition, in order to measure a system, we have to couple it to another system. There are two main ways of treating open quantum systems, namely quantum Langevin equations and master equations. Both methods will be treated here.

6.1 Reservoir Hamiltonians

Now let us begin our analysis of open systems. We have a system with degrees of freedom in which we are interested, but these degrees of freedom are coupled to excitations or modes about whose detailed dynamics we do not care. These degrees of freedom in which we are *not* interested are called reservoirs. Examples include modes of the electromagnetic field in spontaneous emission, modes outside of a cavity in transmission through a cavity mirror, dielectric density fluctuations when looking at light propagation through a dielectric, or even coupling to amplifying transitions in a laser. Often, one thinks of reservoirs as a way to describe dissipation or losses. However, they can also be used to describe gain or measurement. After all, measurement must always involve at least one other system – often referred to as the meter – which stores the measured information.

6.1.1 Finite temperatures

We will begin by considering a free reservoir as a system of bosons in thermal equilibrium at finite temperature. In finding the density matrix for this system, we make the assumption that the density matrix corresponds to the maximum entropy or maximum disorder possible at a given mean energy. This results in the canonical ensemble, or finite-temperature density matrix,

$$\rho \propto e^{-H/k_B T}, \tag{6.1}$$

where H is the Hamiltonian, k_B is the Boltzmann constant and T is the temperature. We do not use canonical ensembles for interacting systems in nonlinear optics, as these are highly nonequilibrium systems driven far from thermal equilibrium by external energy sources. However, when such interacting systems are weakly coupled to a reservoir, we will generally assume that the *reservoir* is described by a canonical ensemble at a finite temperature.

For a noninteracting system of bosons, we have that $H = \sum_k \hbar\omega_k a_k^\dagger a_k$. The fact that operators of distinct modes commute means that the density matrix can be written in a product form as

$$\rho = \prod_k \mathcal{Z}_k^{-1} \exp[-\hbar\omega_k a_k^\dagger a_k / k_B T] = \prod_k \rho_k. \tag{6.2}$$

Here $\mathcal{Z}_k$ is a normalizing factor for each mode. Since the density matrix is factorized, we can treat every mode k separately. Inserting an identity operator for the kth mode, which we can write as

$$\hat{1}_k = \sum_{n=0}^{\infty} |n\rangle_k \langle n|_k, \tag{6.3}$$

and using the fact that the number state $|n\rangle_k$ is an eigenstate of the number operator $a_k^\dagger a_k$ with eigenvalue n, we obtain

$$\begin{aligned} \rho_k &= \mathcal{Z}_k^{-1} \sum_{n=0}^{\infty} \exp[-\hbar\omega_k a_k^\dagger a_k / k_B T] |n\rangle_k \langle n|_k \\ &= \mathcal{Z}_k^{-1} \sum_{n=0}^{\infty} e^{-n\hbar\omega_k / k_B T} |n\rangle_k \langle n|_k. \end{aligned} \tag{6.4}$$

This allows us to evaluate the trace, and from the condition that $\mathrm{Tr}(\rho_k) = 1$, we see that

$$\mathcal{Z}_k = \sum_{n=0}^{\infty} e^{-n\hbar\omega_k / k_B T} = \frac{1}{1 - e^{-\hbar\omega_k / k_B T}}. \tag{6.5}$$

We wish to obtain the mean thermal occupation number of the kth mode, which is given by

$$n_{th}(\omega_k, T) = \mathrm{Tr}[a_k^\dagger a_k \rho_k]. \tag{6.6}$$

The occupation number is most readily evaluated on differentiating the expression for $\mathcal{Z}_k$, since

$$\begin{aligned} \mathrm{Tr}[a_k^\dagger a_k \rho_k] &= \mathcal{Z}_k^{-1} \sum_{n=0}^{\infty} n e^{-n\hbar\omega_k / k_B T} \\ &= -\frac{k_B T}{\hbar \mathcal{Z}_k} \frac{\partial \mathcal{Z}_k}{\partial \omega_k}. \end{aligned} \tag{6.7}$$

This gives the final expression for the well-known Bose–Einstein mean thermal occupation number for a mode of angular frequency ω at temperature T,

$$n_{th}(\omega, T) = [e^{\hbar\omega/k_B T} - 1]^{-1}. \tag{6.8}$$

6.1.2 Decoherence and damping

There are two main approaches to these types of problem. One is simply to describe the entire system – including the reservoirs – as one large quantum system. Clearly, such an approach is very general, and allows one to deal with any type of reservoir – including ones that are strongly coupled or narrow-band. This type of situation, called a non-Markovian reservoir, is of increasing importance in many applications. In general, the treatment of these problems requires consideration of colored-noise (finite-bandwidth) stochastic processes, either as quantum Langevin equations, or as the corresponding equations found using operator correspondence rules. This exact method is always possible in principle, but it is unnecessary if the details of the reservoir time evolution are not important. The second method is to describe the relevant quantum state using a density operator, traced over the set of reservoir variables This generally involves additional approximations, which are only valid for weakly coupled, broad-band reservoirs. The reservoirs cause a damping (relaxation) of energy and phase, as well as a decoherence of any superposition states. These methods lead to an equation of motion for the system density matrix called a master equation, and such equations have a number of well-developed methods for their solution.

Decoherence is a tendency for the density matrix to become diagonal with respect to the variables that are coupled to the external reservoirs. This phenomenon is a result of the entanglement between the system and the reservoir, which is caused by the system–reservoir coupling. This is, in fact, why the master equation describes the evolution of the density matrix of the system and not the evolution of a pure state. Even if the system is originally in a pure state, the coupling to the reservoir causes it to become entangled with the reservoir, and, consequently, after the reservoir degrees of freedom are traced out, the system is in a mixed state.

6.1.3 System–reservoir coupling

Now let us make this quantitative. The system of interest is coupled to a set of reservoir modes, so that the entire Hamiltonian can be written as

$$H = H^S + H^{SR} + H^R, \tag{6.9}$$

where H^S is the system Hamiltonian, H^R is the reservoir Hamiltonian and H^{SR} describes the system–reservoir coupling. A typical system–reservoir coupling can be modeled, in the rotating-wave approximation, by the Hamiltonian

$$H^{SR} = \hbar \sum_j [S_j^\dagger R_j + S_j R_j^\dagger], \tag{6.10}$$

where the S_j are system operators and the R_j are reservoir operators. In general, there are several different ways to solve for the dynamics of the system depending on the precise details of the reservoir properties and the size of the relevant Hilbert space. One method, which displays the physics clearly, is to consider the operator Heisenberg equations of motion from Eq. (2.7). Provided that $[S_{j'}, S_j] = 0$, one obtains

$$\frac{dS_j}{dt} = \frac{i}{\hbar}[H^S, S_j] + i\sum_{j'}[S^\dagger_{j'}, S_j]R_{j'}. \tag{6.11}$$

In this approach, one can simply treat the time evolution of R_j as the central problem of reservoir theory. This approach leads to a type of quantum Langevin equation, in which there are terms corresponding to a deterministic motion in the system Hilbert space, together with an additional term corresponding to random noise, which is a result of the coupling to the reservoir. While these equations are physically appealing, they are hard to solve except for linear cases. We develop other methods to solve them in later chapters.

The system itself can be any one of a number of simple quantum systems. The most typical examples are atomic systems – a two-level atom is the most common idealized model of an atom – and single-mode cavities. Another simple quantum system encountered in reservoir theory is a magnetic spin, which is mathematically equivalent to a two-level atom. The following treatments are applicable to any type of elementary quantum system.

6.2 Absorption

As our first application of reservoir theory, let us consider the absorption and emission of photons. The spontaneous emission of a photon by an excited two-level atom, in which the reservoir is the set of modes of the electromagnetic field, is perhaps the simplest example of photon emission. In this case, the coupling to the reservoir is extremely broad-band, having a bandwidth of 10^{15} Hz for optical transitions, while the atomic linewidth itself may be only 10^9 Hz.

One of the more common examples of photon absorption is the absorption of a photon traveling through a medium. Even in the most transparent of optical systems – the silica optical fiber – absorption or scattering of radiation can occur. In the case of silica, there is a relatively flat absorption profile, with a minimum absorption coefficient of 0.2 dB km^{-1} in the vicinity of the commonly used communications wavelength of $\lambda = 1.5$ µm. In this case, it is field modes – often guided modes – that constitute the system that is coupled to reservoirs. The reservoirs themselves are, in this case, typically either atomic transitions or other, nonguided field modes to which the system modes are coupled by inhomogeneities in the dielectric.

6.2.1 Absorbing reservoirs

The simplest model of absorption consists of a linear coupling to a reservoir of many bosonic modes near their ground state, with closely spaced resonant frequencies. Generally, a bosonic reservoir can have different polarizations and propagation directions. However, we wish to treat a minimal reservoir model here, appropriate to one-dimensional propagation of a single polarization direction.

An absorption reservoir is often modeled by a continuum of harmonic oscillators at different resonant frequencies. This notation allows us to treat different types of reservoirs in a similar way, without having to explicitly include dispersion:

$$H^R = \hbar \int_0^\infty d\omega\, \omega b^\dagger(\omega) b(\omega). \tag{6.12}$$

Here $[b(\omega), b^\dagger(\omega')] = \delta(\omega - \omega')$. The reservoir Hamiltonian causes rapidly varying operator evolution of each reservoir variable at its resonant frequency ω. These are coupled to the system variable with a frequency-dependent coupling constant, $A(\omega)$, with a coupling Hamiltonian

$$H^{SR} = i\hbar \int_0^\infty d\omega\, S b^\dagger(\omega) A(\omega) + h.c. \tag{6.13}$$

Here the reservoir operator has been expanded into its frequency components, so that, if we compare this to Eq. (6.10), we can see that the reservoir operator is an integral over a range of frequencies with

$$R^\dagger = i \int_0^\infty d\omega\, A(\omega) b^\dagger(\omega). \tag{6.14}$$

The equations for the absorbing reservoirs can be integrated immediately. In the Heisenberg picture, the reservoir variable $b(\omega)$ obeys

$$\begin{aligned} \frac{\partial}{\partial t} b(\omega, t) &= \frac{i}{\hbar}[H, b(\omega, t)] \\ &= -i\omega b(\omega, t) + A(\omega) S(t). \end{aligned} \tag{6.15}$$

The solution to this equation is given, for $t > t_0$, by

$$b(\omega, t) = b(\omega, t_0)\, e^{-i\omega(t-t_0)} + A(\omega) \int_{t_0}^t e^{-i\omega(t-t')} S(t')\, dt'. \tag{6.16}$$

The initial time is taken to be in the distant past ($t_0 \to -\infty$), and the initial state of the reservoir is usually assumed to be a thermal state, having

$$\begin{aligned} \langle b^\dagger(\omega, t_0) b(\omega', t_0) \rangle &= n_{th}(\omega, T)\delta(\omega - \omega'), \\ \langle b(\omega, t_0) b^\dagger(\omega', t_0) \rangle &= [n_{th}(\omega, T) + 1]\delta(\omega - \omega'). \end{aligned} \tag{6.17}$$

Here $n_{th}(\omega, T)$ is the thermal occupation number for the bosonic reservoir mode at frequency ω and temperature T. From Eq. (6.8), this is

$$n_{th}(\omega, T) = [e^{\hbar\omega/k_B T} - 1]^{-1}.$$

6.2.2 Quantum Langevin equation

We now want to investigate the effects of the reservoir on the system. This results in a Heisenberg equation for any system variable, called the quantum Langevin equation. To obtain this, the solution for the reservoir operator $b(\omega, t)$ is substituted into the Heisenberg equation for the system variable, giving

$$\frac{d}{dt}S(t) = \frac{i}{\hbar}[H^S, S] + [S^\dagger, S]\int_0^\infty A^*(\omega)b(\omega, t)\, d\omega. \tag{6.18}$$

If we assume that $[S^\dagger, S] = -1$, as would be the case for creation and annihilation operators, then the last term becomes

$$\begin{aligned} -\int_0^\infty A^*(\omega)b(\omega, t)\, d\omega &= -\int_0^\infty d\omega\, |A(\omega)|^2 \int_{t_0}^t dt'\, e^{-i\omega(t-t')}S(t') \\ &\quad - \int_0^\infty d\omega\, A^*(\omega)e^{-i\omega(t-t_0)}b(\omega, t_0) \\ &= -\int_{-\infty}^t dt'\, \gamma(t-t')S(t') + \Gamma_A(t), \end{aligned} \tag{6.19}$$

where the response function, $\gamma(t)$, is given by

$$\gamma(t) = \theta(t)\int_0^\infty d\omega\, |A(\omega)|^2\, e^{-i\omega t}, \tag{6.20}$$

and we have taken $t_0 \to -\infty$ in the first term on the right-hand side. This term, called a drift term, represents a deterministic addition to the motion of the system variable. Similarly, the quantum reservoir, or noise, operator is given by

$$\Gamma_A(t) = -\int_0^\infty d\omega\, A^*(\omega)e^{-i\omega(t-t_0)}b(\omega, t_0). \tag{6.21}$$

We emphasize that this is an *operator* stochastic term, not simply classical noise as in the conventional Langevin approach. The final quantum Langevin equation in this case has the form

$$\frac{d}{dt}S(t) = \frac{i}{\hbar}[H^S, S] - \int_{-\infty}^t dt'\, \gamma(t-t')S(t') + \Gamma_A(t). \tag{6.22}$$

6.2.3 Markovian approximation

In the case of a reservoir with a nearly flat spectral density near a system resonance at frequency ω_0, we obtain, approximately (neglecting frequency shifts), a uniform Markovian loss term with

$$\gamma(t)e^{i\omega_0 t} \approx 2\gamma\delta(t), \tag{6.23}$$

where the average amplitude loss coefficient is

$$\begin{aligned} \gamma &= \int_0^\infty \int_0^\infty d\omega\, dt\, |A(\omega)|^2\, e^{-i(\omega-\omega_0)t} \\ &\approx \pi |A(\omega_0)|^2. \end{aligned} \tag{6.24}$$

In order to make these statements plausible, suppose that the system variable goes like $e^{-i\omega_0 t}$ times a slowly varying part. The second term on the right-hand side of the Langevin equation (6.22) can then be expressed as

$$e^{-i\omega_0 t}\int_{-\infty}^{t} dt'\,\gamma(t-t')e^{i\omega_0(t-t')}[e^{i\omega_0 t'}S(t')], \tag{6.25}$$

where the term in brackets is slowly varying. We will now assume that the scale of variation of the slowly varying system variable is much slower than the correlation time of the reservoir; that is, the length of time for which $\gamma(t)$ is appreciably different from zero. In that case, we can evaluate the slowly varying term at $t' = t$ and remove it from the integral, giving

$$S(t)\int_{-\infty}^{t} dt'\,\gamma(t-t')e^{i\omega_0(t-t')}. \tag{6.26}$$

Setting $\tau = t - t'$, the remaining integral in the above equation is given by

$$\begin{aligned}\int_0^{\infty} d\tau\,\gamma(\tau)e^{i\nu_0\tau} &= \int_0^{\infty} d\omega\int_0^{\infty} d\tau\,|A(\omega)|^2 e^{-i(\omega-\omega_0)\tau}\\ &\approx \pi|A(\omega_0)|^2,\end{aligned} \tag{6.27}$$

where we have ignored small frequency shifts. To obtain this result we have used

$$\int_0^{\infty} d\tau\, e^{-i(\omega-\omega_0)\tau} = \pi\delta(\omega-\omega_0) + iP\left(\frac{1}{\omega_0-\omega}\right), \tag{6.28}$$

where P denotes principal value. For simplicity, we have dropped the principal value term, because it only produces a small frequency shift (small compared to ω_0) in the system variable. However, in precision spectroscopy, this term should be retained for greater accuracy. Finally, we note that, in the Markovian limit, the equation of motion for the system variable becomes

$$\frac{d}{dt}S(t) = \frac{i}{\hbar}[H^S, S] - \gamma S(t) + \Gamma_A(t). \tag{6.29}$$

This approximation, known as the Markov approximation, is generally rather accurate for the absorbing reservoirs, whose response does not typically vary fast with frequency. An exception to this rule would be any case involving resonant impurities, or very short pulses whose bandwidth is comparable to the frequency bandwidth of absorption processes.

The second quantity, $\Gamma_A(t)$, behaves like a stochastic term due to the random initial conditions. Its correlation functions are

$$\begin{aligned}\langle\Gamma_A(t)\Gamma_A^\dagger(t')\rangle &= \int_0^{\infty} d\omega\,|A(\omega)|^2 e^{-i\omega(t-t')}(n_{th}(\omega,T)+1)\\ &\approx [\gamma(t-t')\theta(t-t') + \gamma^*(t'-t)\theta(t'-t)][1+n_0]\end{aligned} \tag{6.30}$$

and

$$\begin{aligned}\langle\Gamma_A^\dagger(t)\Gamma_A(t')\rangle &= \int_0^{\infty} d\omega\,|A(\omega)|^2 e^{i\omega(t-t')}n_{th}(\omega,T)\\ &\approx [\gamma^*(t-t')\theta(t-t') + \gamma(t'-t)\theta(t'-t)]n_0.\end{aligned} \tag{6.31}$$

Here $n_0 \equiv n_{th}(\omega_0, T)$, and we have assumed that $A(\omega)$ has a broad peak centered at a system resonance at ω_0, and that the width of this peak is small compared to the scale of variation of $n_{th}(\omega)$. Taking the Markovian limit, this reduces to

$$\langle \Gamma_A(t)\Gamma_A^\dagger(t')\rangle = 2\gamma[1+n_0]\,\delta(t-t'),$$
$$\langle \Gamma_A^\dagger(t)\Gamma_A(t')\rangle = 2\gamma n_0\,\delta(t-t'). \tag{6.32}$$

The dimensions for the amplitude relaxation rates γ are $[\mathrm{s}]^{-1}$. It is easy to show that 2γ corresponds to the linear intensity absorption rate. To the extent that this effect is dependent on wavelength (and hence frequency), the resulting effect can be included, but this will give rise to a response function $\gamma(t)$ different from the one obtained in the Markovian case. Non-Markovian effects are often neglected completely – but this approximation is only valid for excitations of the system operators that are narrow-band compared to the reservoir bandwidths.

6.3 Gain

While lasers have many applications – and are treated in more detail later – it is sometimes useful to consider a laser amplifier as a type of reservoir. This has the merit that one can treat a rather minimal model, in which only the essential quantum features of the amplifier are included, without too many complications.

The detailed equations for gain or laser reservoirs generally involve the nonlinear response of atomic or solid-state transitions used to provide gain in the lasing medium. This also requires a pump process to maintain the lasing atoms in an inverted state. The simple reservoir theory presented here is able to provide a minimal model for quantum noise effects in amplifiers. In particular, it reproduces many results observed in experiments. The reservoir variable $\sigma_\mu^- = |1\rangle_\mu\langle 2|_\mu$ is an atomic transition operator, which induces a near-resonant atomic transition from an upper state $|2\rangle$ to a lower state $|1\rangle$, in the μth atom. The two-level transitions have an assumed number density in frequency space of $\rho(\omega) = \sum_\mu \delta(\omega-\omega_\mu)$ in resonant angular frequency ω.

We note that the results obtained here can apply to either gain or loss, since this depends on the atomic reservoir density matrix. If the atomic reservoir is near the ground state, it will give absorption, while if the atomic reservoir is near an excited state, it will lead to gain.

6.3.1 Gain reservoir coupling

We now need to specify the reservoir and system–reservoir Hamiltonians. These are given by Eq. (4.104), written as a frequency integral, so that

$$H^R = \tfrac{1}{2}\hbar\int_0^\infty d\omega\,\omega\sigma^z(\omega),$$
$$H^{SR} = i\hbar\int_0^\infty d\omega\,[S^\dagger\sigma^-(\omega)G^*(\omega) - h.c.], \tag{6.33}$$

where $G(\omega)$ is the frequency-dependent system–reservoir coupling, and the atomic raising and lowering field operators, $\sigma^{\pm}(\omega)$, are defined in terms of discrete Pauli operators by

$$\sigma^{+}(\omega) = \sum_{\mu} |2\rangle\langle 1|_{\mu}\, \delta(\omega - \omega_{\mu}),$$
$$\sigma^{-}(\omega) = \sum_{\mu} |1\rangle\langle 2|_{\mu}\, \delta(\omega - \omega_{\mu}), \tag{6.34}$$
$$\sigma^{z}(\omega) = \sum_{\mu} [|2\rangle\langle 2| - |1\rangle\langle 1|]_{\mu}\, \delta(\omega - \omega_{\mu}).$$

Noting that the elementary atomic raising and lowering Pauli operators have the commutators

$$\begin{aligned} [\sigma_{\mu}^{\dagger}, \sigma_{\nu}] &= [|2\rangle\langle 1|_{\mu}, |1\rangle\langle 2|_{\nu}] \\ &= \delta_{\mu\nu}[|2\rangle\langle 2|_{\nu} - |1\rangle\langle 1|_{\nu}] \\ &= \delta_{\mu\nu}\sigma_{\mu}^{z}, \end{aligned} \tag{6.35}$$

the frequency-dependent operators therefore must have the commutation relations

$$\begin{aligned} [\sigma^{+}(\omega), \sigma^{-}(\omega')] &= \sum_{\mu\nu} [\sigma_{\mu}^{\dagger}, \sigma_{\nu}] \delta(\omega - \omega_{\mu}) \delta(\omega' - \omega_{\nu}) \\ &= \sum_{\mu} \sigma_{\mu}^{z} \delta(\omega - \omega_{\mu}) \delta(\omega' - \omega_{\mu}). \end{aligned} \tag{6.36}$$

Making use of these commutators, we find that

$$\frac{\partial}{\partial t}\sigma^{-}(t,\omega) = -i\omega\sigma^{-}(t,\omega) + \sigma^{z}(t,\omega)G(\omega)S(t). \tag{6.37}$$

As in the case of the loss reservoir, this equation can be integrated easily, and the general, time-dependent solutions are

$$\sigma^{-}(t,\omega) = \sigma^{-}(t_0,\omega)\, e^{-i\omega(t-t_0)} + G(\omega) \int_{t_0}^{t} dt'\, e^{-i\omega(t-t')} \sigma^{z}(t',\omega) S(t'). \tag{6.38}$$

We are now going to assume that the reservoir is initially completely inverted, that is, all of the atoms are in their excited state. We are going to go further than this, however, and assume that the inversion stays complete. This approximation is similar to the parametric approximation we discussed for the parametric amplifier, and it will, like the parametric approximation, only be valid for a limited period of time. In a pumped laser amplifier, there can also be nonequilibrium dynamics to replenish the inversion, but we do not include this here.

In the case of complete inversion, since $\langle \sigma_{\mu}^{z} \rangle = \langle [|2\rangle\langle 2|_{\nu} - |1\rangle\langle 1|_{\mu}] \rangle = \langle |2\rangle\langle 2|_{\nu} \rangle = 1$, we find that $\langle \sigma^{z}(\omega) \rangle = \rho(\omega)$. We assume that, in the far past, the reservoir variables are all uncorrelated for different atoms, and that $\langle \sigma^{+}(t_0,\omega) \rangle = \langle \sigma^{-}(t_0,\omega') \rangle = 0$. The initial correlations for the reservoir variables in the far past ($t_0 \to -\infty$) are given by

$$\begin{aligned}\langle\sigma^+(t_0,\omega)\sigma^-(t_0,\omega')\rangle &= \sum_{\mu\nu}\langle\sigma_\mu^+(t_0)\sigma_\nu^-(t_0,\omega')\rangle\delta(\omega-\omega_\mu)\delta(\omega'-\omega_\nu)\\ &= \sum_{\mu}\langle\sigma_\mu^z\rangle\delta(\omega-\omega_\mu)\delta(\omega'-\omega_\mu).\end{aligned} \tag{6.39}$$

Thus, to summarize, the early-time correlations for complete inversion of the gain atoms are all given by

$$\begin{aligned}\langle\sigma^+(t_0,\omega)\sigma^-(t_0,\omega')\rangle &= \rho(\omega)\delta(\omega-\omega'),\\ \langle\sigma^-(t_0,\omega)\sigma^+(t_0,\omega')\rangle &= 0.\end{aligned} \tag{6.40}$$

Provided $[S^\dagger, S] = -1$, as before, the system variable equations are

$$\frac{d}{dt}S(t) = \frac{i}{\hbar}[H^S, S] + \int_0^\infty G^*(\omega)\sigma^-(\omega,t)\,d\omega. \tag{6.41}$$

We now substitute the solution for $\sigma^-(t,\omega)$ from Eq. (6.38) into the Heisenberg equation for the system variable, assuming no depletion of the inversion, and tracing the inversion over the atomic gain reservoirs. This amounts to replacing $\sigma^z(t,\omega)$ by $\rho(\omega)$.

The approximation that the gain inversion is time-independent is, as we have noted, a very strong one, and will not be satisfied in all cases. In the no-depletion limit, this gives rise to an extra time-dependent term in the system equation, of the form

$$\begin{aligned}\int_0^\infty G^*(\omega)\sigma^-(t,\omega)\,d\omega &= \int_0^\infty d\omega\,|G(\omega)|^2\rho(\omega)\int_{t_0}^t dt'\,e^{-i\omega(t-t')}S(t')\\ &\quad + \int_0^\infty d\omega\,G(\omega)e^{-i\omega(t-t_0)}\sigma^-(t_0,\omega)\\ &= \int_0^\infty dt''\,g(t'')S(t-t'') + \Gamma_G(t),\end{aligned} \tag{6.42}$$

where $t'' = t - t'$, the quantum noise term is

$$\Gamma_G(t) = \int_0^\infty d\omega\,G(\omega)e^{-i\omega(t-t_0)}\sigma^-(t_0,\omega) \tag{6.43}$$

and the time-dependent gain response is

$$g(t) \approx \theta(t)\int_0^\infty d\omega\,|G(\omega)|^2\,e^{-i\omega t}. \tag{6.44}$$

This leads to the final non-Markovian quantum Langevin equation for an amplifying reservoir:

$$\frac{d}{dt}S(t) = \frac{i}{\hbar}[H^S, S] + \int_{-\infty}^t dt'\,g(t-t')S(t') + \Gamma_G(t). \tag{6.45}$$

6.3.2 Markovian approximation

In the case of an amplifier with a flat spectral density near frequency ω_0, we have, as in the previous section,

$$g(t)e^{-i\omega_0 t} \approx 2g\delta(t), \tag{6.46}$$

where the average amplitude gain coefficient is

$$g = \pi |G(\omega_0)|^2. \tag{6.47}$$

As with the loss case, $\Gamma_G(t)$ behaves like a stochastic noise term due to the random initial conditions. The correlation functions in the general non-Markovian case are $\langle \Gamma_G(t)\Gamma_G^\dagger(t')\rangle = 0$ and

$$\begin{aligned}\langle \Gamma_G^\dagger(t)\Gamma_G(t')\rangle &= \int_0^\infty d\omega\, |G(\omega)|^2 \rho(\omega) e^{-i\omega(t-t')} \\ &= [g^*(t-t')\theta(t-t') + g(t'-t)\theta(t'-t)].\end{aligned} \tag{6.48}$$

This last equation is a type of fluctuation–dissipation theorem for gain, but it should be noted that it was derived for the case of complete inversion only. For this reason, it provides a lower bound on the noise in more general cases of incomplete inversion. In these more general cases (discussed later), the noise levels may rise relative to the gain.

In the Markovian limit, the equation of motion is

$$\frac{d}{dt}S(t) = \frac{i}{\hbar}[H^S, S] + gS(t) + \Gamma_G(t), \tag{6.49}$$

and the correlation functions become

$$\begin{aligned}\langle \Gamma_G(t)\Gamma_G^\dagger(t')\rangle &= 0, \\ \langle \Gamma_G^\dagger(t)\Gamma_G(t')\rangle &= 2g\delta(t-t').\end{aligned} \tag{6.50}$$

Finally, let us consider the result of incomplete inversion of an amplifier. Here, the noninverted atoms give rise to absorption, not gain, and will generate additional quantum noise and absorption response terms. These behave just as in the previous section, with non-Markovian effects if the absorption line is narrow-band. An important consequence is that the measured gain only gives the *difference* between the gain and the loss. This does not cause any problems with the deterministic response function – but it does cause difficulties in determining the amplifier quantum noise levels. These can only be uniquely determined through spontaneous fluorescence measurements. The lowest quantum noise levels, as given in the theory presented above, occur when *all* the lasing transitions are completely inverted.

6.4 Phase decoherence

An important physical effect in many quantum systems is phase decoherence. In atoms, phase decoherence is caused by elastic collisions, and is an important line-broadening effect in spectroscopy. In electromagnetic propagation, phase decoherence is due to refractive index fluctuations, which are usually attributed to Raman and/or Brillouin scattering off molecular or solid-state vibrational excitations. This is a prototype of a phase decay mechanism. Unlike the case of absorption, this type of decoherence is responsible for phase noise and frequency shifts in a laser signal, rather than actual loss of photons.

In solids, these effects start in the low-frequency kilohertz to megahertz region as a $1/f$ noise attributed to tunneling and defect structures, move into the gigahertz frequency region as acoustic Brillouin scattering of phonons, and extend up to strong, localized Raman resonances in the infrared, which occur at around 12–14 THz in the case of silica. The terahertz frequency regime is typical of molecular vibrational transitions as well. This means that it is sometimes necessary to perform low-noise quadrature phase measurements at low temperatures. Even though $kT \ll \hbar\omega$ for optical photons at room temperature, the phonon levels can be thermally populated at room temperature. However, the use of low temperatures does not entirely eliminate Raman coupling, which can occur spontaneously even at zero temperature. The fact that several different physical effects are involved, with widely varying coupling constants and resonances, means that these effects usually require a non-Markovian response function to be treated accurately.

We will look at Raman scattering as a source of phase decoherence. The Raman interaction energy in a solid or molecule, in terms of atomic displacements from their mean lattice positions, has the form

$$H^{SR} + H^R = \frac{1}{2}\sum_j \eta_j^R(\mathbf{r}^j) : \mathbf{D}(\mathbf{r}^j)\mathbf{D}(\mathbf{r}^j)\delta\mathbf{r}^j + \sum_j \frac{\mathbf{p}(\mathbf{r}_j)^2}{2m} + \frac{1}{2}\sum_{ij} \kappa_{ij} : \delta\mathbf{r}^i\delta\mathbf{r}^j . \tag{6.51}$$

Here $\mathbf{D}(\mathbf{r}^j)$ is the electric displacement at the jth mean atomic location $\mathbf{r}^j$, $\delta\mathbf{r}^j$ is the atomic displacement operator, $\mathbf{p}(\mathbf{r}_j)$ is the momentum of the jth atom, m is its mass (we assume for simplicity that all of the atoms have the same mass) and $\eta_j^R(\mathbf{r}^j)$ is a Raman coupling tensor.

In order to quantize this interaction with atomic positions, we must now take into account the existence of a corresponding set of phonon operators, $b(\omega)$ and $b^\dagger(\omega)$, which diagonalize the atomic displacement Hamiltonian H^R, and have well-defined eigenfrequencies. For large molecules or solids, we can assume there is nearly a continuum of relevant phonon frequencies, so that these can be considered a reservoir. The phonon–photon coupling induces Raman transitions and scattering from acoustic waves (the Brillouin effect), resulting in extra noise sources and an additional contribution to the nonlinearity. Like photon reservoirs, the initial state of phonons is generally thermal, with

$$\langle b(\omega)^\dagger b(\omega')\rangle = n_{th}(\omega, T)\delta(\omega - \omega') = \delta(\omega - \omega')[\exp(\hbar\omega/kT) - 1]^{-1}. \tag{6.52}$$

At high temperatures compared to the relevant phonon energies, this produces a linear temperature dependence of the phonon population on temperature, with

$$\langle b(\omega)^\dagger b(\omega')\rangle \approx \frac{kT}{\hbar\omega}\delta(\omega-\omega'). \tag{6.53}$$

6.4.1 Hamiltonian and Heisenberg equations

Expanding in terms of the phonon and photon operators, and using the rotating-wave approximation for the high-frequency single-mode photon terms, the Raman–Brillouin interaction Hamiltonian for a single field mode becomes

$$H^{SR} + H^R = \hbar \int_0^\infty d\omega \, [a^\dagger a (R^*(\omega) b(\omega) + R(\omega) b^\dagger(\omega)) + \omega b(\omega) b^\dagger(\omega)]. \tag{6.54}$$

Here, the atomic vibrations are modeled as a continuum of localized oscillators, which are coupled to the radiation modes by a Raman transition with a frequency-dependent strength of $R(\omega)$. The atomic displacement is assumed proportional to $R^*(\omega)b + R(\omega)b^\dagger$, where the phonon annihilation and creation operators, b and $b^\dagger$, have the commutation relations

$$[b(\omega), b^\dagger(\omega')] = \delta(\omega-\omega'). \tag{6.55}$$

The frequency dependence of the coupling $R(\omega)$ can be easily determined empirically through measurements of the Raman gain spectrum, or fluorescence spectrum.

Note that the type of coupling we have to the reservoir is different from what we have seen in the cases of absorption and gain. In those cases, excitations of the system could be absorbed or emitted by the reservoir. Here the reservoir coupling does not change the number of excitations, in this case photons, in the system. Its effect will be on their phase.

The coupled set of system and reservoir operator equations are

$$\begin{aligned} \frac{\partial}{\partial t} b(t,\omega) &= -i\omega b(t,\omega) - iR(\omega) a^\dagger a, \\ \frac{d}{dt} a(t) &= \frac{i}{\hbar}[H^S, a] - i\int_0^\infty d\omega \, a(t)(R^*(\omega) b(t,\omega) + R(\omega) b^\dagger(t,\omega)). \end{aligned} \tag{6.56}$$

On integrating the reservoir equations, one obtains an immediate solution, of a similar form to that previously obtained,

$$b(t,\omega) = b(t_0,\omega) e^{-i\omega(t-t_0)} - iR(\omega)\int_{t_0}^{t} e^{-i\omega(t-t')}[a^\dagger a](t')\,dt'. \tag{6.57}$$

Upon substituting this into the equation for the field operator, the result is a modified Heisenberg equation with a delayed nonlinear response to the field due to the Raman coupling,

$$\frac{d}{dt} a(t) = \frac{i}{\hbar}[H^S, a] - i\left[\int_0^\infty dt' \, \kappa_r(t')[a^\dagger a](t-t') + \Gamma_R(t)\right] a(t), \tag{6.58}$$

where

$$\kappa_r(t) = -2\int_0^{\infty} |R^2(\omega)| \sin(\omega t)\, d\omega,$$

$$\Gamma_R(t) = \int_0^{\infty} d\omega\, (R^*(\omega) b(t_0,\omega)\, e^{-i\omega(t-t_0)} + R(\omega) b^\dagger(t_0,\omega)\, e^{i\omega(t-t_0)}). \tag{6.59}$$

As before, the noise terms are stochastic operators. The correlation function, for an initially thermally populated phonon bath at temperature T, is

$$\langle \Gamma_R(t)\Gamma_R(t')\rangle = \int_0^{\infty} d\omega\, |R(\omega)|^2 \{[n_{th}(\omega,T)+1]\, e^{-i\omega(t-t')} + n_{th}(\omega,T)\, e^{i\omega(t-t')}\}. \tag{6.60}$$

To obtain the correlations in frequency space, we can define Fourier transforms, using the normal Fourier transform conventions for field operators,

$$\Gamma_R(\omega) = \frac{1}{\sqrt{2\pi}} \int_{-\infty}^{\infty} dt\, \exp(i\omega t)\Gamma_R(t), \tag{6.61}$$

whose correlations are given by

$$\langle \Gamma_R(\omega)\Gamma_R(\omega')\rangle = 2\pi\delta(\omega+\omega')|R^2(\omega)|[n_{th}(|\omega|,T) + \theta(\omega)]. \tag{6.62}$$

In some cases, one finds that the frequency dependence of the Raman–Brillouin coupling and that of the thermal population cancel each other at high temperatures to give an approximately flat phase noise spectrum over a region with a characteristic frequency ω_0. In this (special) case, one finds that

$$\langle \Gamma_R(t)\Gamma_R(t')\rangle \approx \pi |R(\omega_0)|^2 n_{th}(\omega_0,T)\delta(t-t') = 2\gamma_p \delta(t-t'). \tag{6.63}$$

More usually, the measured decoherence due to Raman effects is strongly frequency-dependent. The simplest way to parameterize this is to expand the Raman response function in terms of a multiple Lorentzian model, which can then be fitted to observed Raman fluorescence data using a nonlinear least-squares fit. Note that, if $|R(\omega)|^2$ is a single Lorentzian, defined as

$$|R(\omega)|^2 = \frac{A}{(\omega-\omega_0)^2 + \delta^2}, \tag{6.64}$$

then, for $t > 0$,

$$\kappa_r(t) = \frac{-2\pi A}{\delta} \sin(\omega_0 t) e^{-\delta t}. \tag{6.65}$$

For multiple Lorentzians, we therefore expand as

$$\kappa_r(t) = -2\pi \sum_{j=0}^{n} \frac{A_j}{\delta_j} e^{-\delta_j t} \sin(\omega_j t). \tag{6.66}$$

In the above expansion, A_j are a set of dimensionless Lorentzian strengths, and ω_j and δ_j are the resonant frequencies and widths, respectively, of the effective Raman resonances at each frequency. A result of this model is that the phonon operators do not have white-noise behavior. In fact, this colored-noise property is significant enough to invalidate the usual Markov and rotating-wave approximations in many cases of interest.

6.5 Input–output relations

Although the quantum Langevin equations – or other methods – directly result in information about the system variables, it is often the case that it is actually the reservoir variables that are measured in the end. A typical example of this is the spontaneous emission of an excited atom. We may be able to directly measure the excited-state population, but it is more common and typical that the fluorescent radiation is monitored with a photo-detector. Under these circumstances, the system variables are not measured, but rather their value is inferred from corresponding measurements on the reservoir.

Such measurements are quite informative, due to the fact that the reservoir variables often satisfy linear equations, in which the system variables appear as a type of source term. The system variables may also depend on certain initial conditions or Langevin noise sources, which are called input fields. It is usually the case that we only have access to some of the reservoir variables – called output fields. The overall relationship between input fields, system variables and output fields is generically called an input–output relation.

For example, in the case of a loss reservoir, we found that the reservoir variables were given by

$$b(\omega, t) = b(\omega, t_0)\, e^{-i\omega(t-t_0)} + A(\omega) \int_{t_0}^{t} e^{-i\omega(t-t')} S(t')\, dt'. \tag{6.67}$$

We see that the system variable is acting as a source for the reservoir. We also had

$$\frac{d}{dt} S(t) = \frac{i}{\hbar}[H^S, S] - \int_{-\infty}^{t} dt'\, \gamma(t-t') S(t') + \Gamma_A(t), \tag{6.68}$$

which shows that the system variable depends on the initial state of the reservoir through the noise operator, $\Gamma_A(t)$, which, in turn, depends on the reservoir operators at an early time, that is, on the operators $b(\omega, t)$. So, at early times the reservoir can be viewed as an input field. This input field interacts with the system and produces an output field, which depends on both the input and system operators for some sufficiently large value of t. The output field will depend on both the input field and the interaction with the system.

The simplest class of practical devices of this type are cavities with external coherent driving lasers. This is an example of input–output theory, where the input field is no longer in the vacuum state, but has a mode with a narrow-band, non-Markovian excitation present. The excited mode can be treated classically in the limit that the laser is coherent. This allows us to treat it by adding a term to the system Hamiltonian. What we want to do with systems in cavities is to relate the properties of the field inside the cavity to those of the output field. There are well-developed techniques for determining the field inside a cavity that contains a nonlinear medium and is being driven by one or more lasers. This is where the input–output formalism is useful.

Strictly speaking, input–output results are approximations, and are usually obtained from considering the boundary conditions at a single-sided cavity with one lossy, broad-band mirror and one lossless mirror. Multiple-port devices, which can have several types of loss, are more common. These can be included too, as long as it is recognized that the total loss rate corresponds to the sum of the loss rates at each port, while the boundary conditions only hold for the individual loss rates through each mirror under consideration.

The original development of input–output theory was due to a result of Bernard Yurke (see the additional reading). In the early 1980s, it was shown by Walls and Milburn (see the additional reading in Chapter 5) that the amount of intra-cavity squeezing that could be obtained by a parametric oscillator was very limited. Yurke, by using a simple input–output theory, similar to the second one we discuss below, demonstrated that the squeezing in the output field can be much larger than the squeezing inside the cavity. This led, on the experimental side, to the development of the parametric oscillator as a standard source of squeezed light, and, on the theoretical side, to the further elaboration of input–output theories.

6.5.1 Input–output theory: coupled operator approach

We will now look at three approaches to input–output theory. The first, the Collett–Gardiner theory, is derived from reservoir theory, and is the most fully developed of the three. In the following section, we present a more physically oriented derivation. Standard input–output theory relates the properties of an internal cavity mode to external modes. This theory is based on three assumptions:

1. The system–bath interaction, i.e. the internal–external mode interaction, is linear in the bath operators.
2. The rotating-wave approximation is made.
3. The system–bath coupling is independent of frequency.

Let us now consider the case of a cavity mode coupled to external modes. We denote the cavity-mode annihilation and creation operators by $a(t)$ and $a^\dagger(t)$, where $[a(t), a^\dagger(t)] = 1$, and the external mode operators by $b(\omega, t)$ and $b^\dagger(\omega, t)$, where $[b(\omega, t), b^\dagger(\omega', t)] = \delta(\omega - \omega')$. We generally follow the example of Section 6.2, except that we wish to understand the role of the reservoir modes in propagating information.

The total Hamiltonian consists of a cavity Hamiltonian, H^S, which describes the dynamics of the system in the cavity, $H^R = \int_0^\infty d\omega \hbar\omega b^\dagger(\omega)b(\omega)$, which describes the external modes, and

$$H^{SR} = i\hbar \int_0^\infty d\omega\, A(\omega)[ab^\dagger(\omega) - a^\dagger b(\omega)], \tag{6.69}$$

describing their coupling. Before proceeding, let us make an approximation. In an interaction picture, in which frequencies are measured with respect to some optical frequency, ν, the integrals in the Hamiltonian will extend from $-\nu$ to ∞. Since ν is large, we can extend the frequency integrals from $-\infty$ to $+\infty$ to a good approximation. Henceforth, we shall assume this has been done. Hence, we can do the same in the Heisenberg picture, without changing the accuracy of the approximation.

The Heisenberg equation of motion for $b(\omega)$ is

$$\frac{db(\omega)}{dt} = -i\omega b(\omega) + A(\omega)a. \tag{6.70}$$

We can formally solve this equation in two different ways. On the one hand, we can write, for $t > t_0$,

$$b(\omega, t) = e^{-i\omega(t-t_0)} b_0(\omega) + A(\omega) \int_{t_0}^{t} dt'\, e^{-i\omega(t-t')} a(t'), \tag{6.71}$$

where the operator $b_0(\omega) = b(\omega, t_0)$ is an initial value for the reservoir operators, and it is analogous to the *in* operators discussed in Chapter 2. On the other hand, for $t < t_1$, we can just as easily write down a time-reversed solution,

$$b(\omega, t) = e^{-i\omega(t-t_1)} b_1(\omega) - A(\omega) \int_{t}^{t_1} dt'\, e^{-i\omega(t-t')} a(t'). \tag{6.72}$$

Here the operator $b_1 = b(\omega, t_1)$ is a *final* value for the reservoir operators, and it is analogous to the *out* operators discussed in Chapter 2. The equation of motion for the cavity-mode operator is

$$\begin{aligned}
\frac{da}{dt} &= \frac{-i}{\hbar}[a, H^S] - \int_{-\infty}^{\infty} d\omega\, A(\omega) b(\omega) \\
&= \frac{-i}{\hbar}[a, H^S] - \int_{-\infty}^{\infty} d\omega\, A(\omega) e^{-i\omega(t-t_0)} b_0(\omega) \\
&\quad - \int_{-\infty}^{\infty} d\omega\, A(\omega)^2 \int_{t_0}^{t} dt'\, e^{-i\omega(t-t')} a(t'),
\end{aligned} \tag{6.73}$$

where we have made use of Eq. (6.71). We will now make use of the assumption that $A(\omega)$ is approximately constant and replace it by $(\gamma/\pi)^{1/2}$, following the procedure in Eq. (6.24). This then gives us

$$\frac{da}{dt} = \frac{-i}{\hbar}[a, H^S] - \gamma a(t) + \sqrt{2\gamma}\, b^{in}(t), \tag{6.74}$$

where we have defined

$$b^{in}(t) = -\frac{1}{\sqrt{2\pi}} \int_{-\infty}^{\infty} d\omega\, e^{-i\omega(t-t_0)} b_0(\omega). \tag{6.75}$$

Note that

$$[b^{in}(t), b^{in\dagger}(t')] = \delta(t - t'). \tag{6.76}$$

Similarly, one finds that

$$\frac{da}{dt} = \frac{-i}{\hbar}[a, H^S] + \gamma a(t) - \sqrt{2\gamma}\, b^{out}(t), \tag{6.77}$$

where

$$b^{out}(t) = \frac{1}{\sqrt{2\pi}} \int_{-\infty}^{\infty} d\omega\, e^{-i\omega(t-t_1)} b_1(\omega), \tag{6.78}$$

and $[b^{out}(t), b^{out\dagger}(t')] = \delta(t - t')$. Subtracting Eq. (6.74) from Eq. (6.77), we obtain

$$b^{in}(t) + b^{out}(t) = \sqrt{2\gamma}\, a(t). \tag{6.79}$$

Let us now look at the example of an empty cavity, in which case we set $H_S = \hbar\nu_0 a^\dagger a$. Defining the Fourier transform

$$a(\omega) = \frac{1}{\sqrt{2\pi}} \int_{-\infty}^{\infty} dt\, e^{i\omega t} a(t), \tag{6.80}$$

and similarly for $b^{in}(\omega)$ and $b^{out}(\omega)$, we find from Eq. (6.74) that

$$[\gamma - i(\omega - \nu_0)]a(\omega) = \sqrt{2\gamma}\, b^{in}(\omega), \tag{6.81}$$

and from Eq. (6.79) that

$$b^{in}(\omega) + b^{out}(\omega) = \sqrt{2\gamma}\, a(\omega). \tag{6.82}$$

Solving these equations for b^{out} in terms of b^{in} gives us

$$b^{out}(\omega) = \frac{\gamma + i(\omega - \nu_0)}{\gamma - i(\omega - \nu_0)} b^{in}(\omega). \tag{6.83}$$

We see that the output operator is just a phase-shift times the input operator, with the phase-shift being determined by the detuning from the cavity resonance and the cavity linewidth.

6.6 Photon flux and density

We will now consider a quantum optical theory that gives similar equations to those obtained by Collett and Gardiner, but with a different point of view based on the physics of electromagnetic waves and beam-splitters, to give a more intuitive picture. We also consider a one-dimensional space with a perfect mirror at $x = 0$ and a beam-splitter at $x = L$, and quantize the field in this space.

The results of input–output theory can be explained in terms of physical photons and waveguides. We start by considering a fundamental property, the photon flux of an electromagnetic field. In the case of a linear dispersive dielectric, we found in Eq. (3.154) that, for a one-dimensional noninteracting dispersive dielectric, the equivalent slowly varying *input* quantum field for negative wavevectors $k \approx -k_1$ was

$$\psi^{in}(x, t) = \frac{1}{\sqrt{2\pi}} \int dk\, e^{i(k+k_1)x} a^{in}(k, t), \tag{6.84}$$

where we have taken an infinite-volume limit in terms of a continuum of mode operators such that

$$[a^{in}(k, t), a^{in\dagger}(k', t)] = \delta(k - k'). \tag{6.85}$$

As input–output theory uses frequency rather than wavenumber to label the modes, we define an input field at $x = L$ so that

$$b^{in}(\omega, t) \equiv \frac{e^{i(k(\omega)+k_1)L}}{\sqrt{v(\omega)}} a^{in}(k(\omega), t), \tag{6.86}$$

where $v = d\omega(k)/dk$ is the group velocity. The commutation relations of these operators are then

$$[b^{in}(\omega, t), b^{in\dagger}(\omega', t)] = \frac{dk}{d\omega}\delta(k - k') = \delta(\omega - \omega'). \tag{6.87}$$

Using this notation, the operators $b(\omega, t)$ effectively generate photons with a given flux per unit angular frequency rather than a given density. This can be seen from calculating the total photon number N in the *in* field, given by

$$\begin{aligned} N &= \int dx\, \psi^{in\dagger}(x, t)\psi^{in}(x, t) \\ &= \int dk\, a^{in\dagger}(k, t)a^{in}(k, t) \\ &= \int d\omega\, b^{in\dagger}(\omega, t)b^{in}(\omega, t). \end{aligned} \tag{6.88}$$

Thus, we see that the fields $b^{in}(\omega, t)$ are a scaled version of the *in* quantum fields, with units of $s^{1/2}$, corresponding to the photon density in angular frequency. The fields $b^{out}(\omega, t)$ have a similar interpretation, except that they are defined in terms of output photons at positive wavevectors.

For a narrow-band input, with constant group velocity v, the input field at $x = L$ is given by

$$\psi^{in}(L, t) = \frac{1}{\sqrt{2\pi}}\int dk\, e^{i(k+k_1)L} a^{in}(k, t), \tag{6.89}$$

$$= \frac{1}{\sqrt{2\pi v}}\int d\omega\, b^{in}(\omega, t) = -\frac{b^{in}(t)}{\sqrt{v}}, \tag{6.90}$$

This implies that $b^{in\dagger}b^{in} = v\psi^{in\dagger}\psi^{in}$, so that $b^{in\dagger}b^{in}$ has the interpretation of a photon flux (units of s^{-1}), instead of photon density (units of m^{-1}), which is the interpretation of $\psi^{in\dagger}\psi^{in}$. Similarly, we can write that

$$\psi^{out}(L, t) = \frac{b^{out}(t)}{\sqrt{v}}. \tag{6.91}$$

6.6.1 Input–output theory: beam-splitter approach

Now let us take a look at input–output theory from a different point of view. Just as previously, suppose we have a cavity with a perfect mirror at $x = 0$ and a beam-splitter at $x = L$. We will consider a one-dimensional model with propagation only in the x direction. We shall denote the right face of the beam-splitter by $x = L+$ and the left face by $x = L-$. Classically, the beam-splitter will act as follows. Let $\psi^{in}(t)$ be the field propagating to the left at $x = L+$, i.e. the input field to the cavity, $\psi(t)$ the cavity field at $x = L-$ propagating to the left, and ψ^{out} the field propagating to the right at $x = L+$, i.e. the output field from the cavity.

Noting that the transmissivity and reflectivity phase is arbitrary, the beam-splitter conditions from Eq. (3.56) give us that

$$\begin{aligned}\psi^{out}(t) &= -\sqrt{T}\,\psi'(t) + \sqrt{R}\,\psi^{in}(t),\\ \psi(t) &= -\sqrt{T}\,\psi^{in}(t) - \sqrt{R}\,\psi'(t),\end{aligned} \tag{6.92}$$

where R and T are the reflectivity and transmissivity of the beam-splitter, and $R + T = 1$. Here $\psi'(t)$ is the cavity field that started at $t - \tau$, where τ is the time for a cavity round trip, and made one round trip through the cavity, and is now reaching the left face of the beam-splitter propagating to the right.

These conditions can be rephrased in terms of annihilation operators. First, note that the operators b^{in} and b^{out} have dimensions of $[\text{time}]^{-1/2}$. This follows from their commutation relations and the fact that $\delta(t - t')$ had dimensions of $[\text{time}]^{-1}$. Therefore, quantities such as $b^{in\dagger}b^{in}$ represent photon flux, that is, number of photons per second. The cavity annihilation operator, $a(t)$, is dimensionless, which follows from its commutation relations. The corresponding operator that would give the photon flux in the cavity is $a/\sqrt{\tau}$; this operator times its adjoint is just the photon number flux in the cavity.

We are now almost ready to give the operator version of Eq. (6.92), but we need to specify what operator corresponds to $\psi'(t)$. It should be

$$\psi'(t) \to -e^{itH^S\tau/\hbar}\frac{a(t-\tau)}{\sqrt{v\tau}}\,e^{-itH^S\tau/\hbar}, \tag{6.93}$$

since the field enters the cavity at $t - \tau$, propagates for a time $\tau/2$, experiences a phase change on reflection, and is at the beam-splitter to be reflected at time t after a further propagation. The operator equations for the beam-splitter relation are therefore

$$\begin{aligned}b^{out}(t) &= \sqrt{\frac{T}{\tau}}\,e^{itH^S\tau/\hbar}a(t-\tau_r)e^{-itH^S\tau/\hbar} - \sqrt{R}\,b^{in}(t),\\ \frac{a(t)}{\sqrt{\tau}} &= \sqrt{T}\,b^{in} + \sqrt{\frac{R}{\tau}}\,e^{itH^S\tau/\hbar}\frac{a(t-\tau)}{\sqrt{\tau}}\,e^{-itH^S\tau/\hbar}.\end{aligned} \tag{6.94}$$

Next, we expand the last term in the second equation to first order in τ. This gives us

$$\sqrt{R\tau}\,\frac{da}{dt} = \frac{\sqrt{R}-1}{\tau}\,a + \sqrt{T}\,b^{in} - \frac{i}{\hbar}[a, H^S]\sqrt{R\tau}. \tag{6.95}$$

Finally, we assume that $T \ll R$, and expand in T, keeping lowest-order terms. Since light in an empty cavity would lose a fraction T of its intensity every round trip, we see that $\gamma = T/(2\tau)$. This results in the equation

$$\frac{da}{dt} = -\gamma a + \sqrt{2\gamma}\,b^{in} - \frac{i}{\hbar}[a, H^S]. \tag{6.96}$$

The above equation is the same as Eq. (6.74). We also note that, when we expand in the small quantity T, keeping lowest-order terms, and assume a short round-trip time compared to the time-scale of the cavity dynamics, the first of Eqs (6.94) becomes identical to Eq. (6.79). Hence, we can recover both of the equations of the Collett–Gardiner theory. We do also see, however, that the theory is valid only for short cavities with small losses, i.e. $T \ll 1$, which was not apparent in the previous derivation.

6.6.2 Input–output theory: 'modes of the universe' approach

Finally, let us consider an approach that uses the modes of a cavity and free space to quantize the field. This is sometimes called the 'modes of the universe' approach, as the entire system is treated as one giant mode, without explicit distinctions between internal and external modes. In principle, this is the most general method.

We consider the same set-up as in the previous case, i.e. a totally reflecting mirror at $x = 0$ and a beam-splitter at $x = L$ with reflectivity R and transmissivity T. Our first task is to find the classical modes corresponding to this situation. These modes should satisfy the wave equation, vanish at $x = 0$, and satisfy beam-splitter boundary conditions at $x = L$. Setting $\omega = kc$, we have that the solutions can be expressed as $e^{i\omega t}u_k(x)$, where

$$u_k(x) = \begin{cases} 2iA_k \sin kx = A_k(e^{ikx} - e^{-ikx}),\ 0 \leq x < L, \\ B_k e^{ikx} + C_k e^{-ikx},\ x > L. \end{cases} \tag{6.97}$$

The wave proportional to B_k is a left-propagating wave incident on the cavity, which corresponds to the in field; and the wave proportional to C_k is a right-propagating wave corresponding to the out field from the cavity. The beam-splitter boundary conditions are

$$\begin{aligned} B_k e^{ikL} &= \sqrt{T} A_k e^{ikL} - \sqrt{R} C_k e^{-ikL}, \\ -A_k e^{-ikL} &= \sqrt{T} C_k e^{-ikL} + \sqrt{R} A_k e^{ikL}. \end{aligned} \tag{6.98}$$

These equations are easily solved for A_k and B_k in terms of C_k, yielding

$$\begin{aligned} A_k &= \frac{-\sqrt{T} C_k}{1 + \sqrt{R}\, e^{2ikL}}, \\ B_k &= -\left(\frac{1 + \sqrt{R}\, e^{-2ikL}}{1 + \sqrt{R}\, e^{2ikL}} \right) C_k. \end{aligned} \tag{6.99}$$

Note that cavity resonances occur when $\exp(2ikL) = -1$, or $k = (2m + 1)\pi/(2L)$, where m is an integer, because, when k satisfies this condition, the cavity field is a maximum. A long, and rather tedious, calculation shows that, if we choose $C_k = 1/\sqrt{2\pi}$, then, for $k_1, k_2 > 0$,

$$\int_0^\infty dx\, u_{k_1}^*(x) u_{k_2}(x) = \delta(k_1 - k_2). \tag{6.100}$$

When performing this calculation, convergence factors of the form $e^{-\epsilon x}$, where $\epsilon > 0$, are inserted into the integrals from L to ∞, and then the limit $\epsilon \to 0^+$ is taken. Noting that

$$u_k^*(x) = -\left(\frac{1 + \sqrt{R}\, e^{2ikL}}{1 + \sqrt{R}\, e^{-2ikL}} \right) u_k(x), \tag{6.101}$$

we see that we also have

$$\int_0^\infty dx\, u_{k_1}(x) u_{k_2}(x) = \delta(k_1 - k_2). \tag{6.102}$$

We shall assume these modes are complete.

Now for waves propagating in the x direction, we will take the dual potential to be in the y direction, $\Lambda(x,t) = \Lambda(x,t)\hat{\mathbf{y}}$. This implies that the only nonzero component of $\mathbf{D}(x,t)$ is $D_z(x,t)$, and the only nonzero component of $\mathbf{B}(x,t)$ is $B_y(x,t)$, where

$$D_z(x,t) = \frac{\partial}{\partial x}\Lambda(x,t), \qquad B_y(x,t) = \mu_0 \frac{\partial}{\partial t}\Lambda(x,t). \tag{6.103}$$

This gives us, for the Hamiltonian, that

$$H = \frac{1}{2}\int_0^\infty dx \left[\mu_0(\partial_t \Lambda)^2 + \frac{1}{\epsilon_0}(\partial_x \Lambda)^2\right]. \tag{6.104}$$

We now expand the dual potential and its time derivative in terms of the mode functions

$$\begin{aligned}\Lambda(x,t) &= \int_0^\infty dk\, \alpha_k [a_k u_k(x) + a_k^\dagger u_k^*(x)], \\ \partial_t \Lambda(x,t) &= -i\int_0^\infty dk\, \beta_k [a_k u_k(x) - a_k^\dagger u_k^*(x)],\end{aligned} \tag{6.105}$$

where α_k and β_k remain to be determined. Assuming completeness of the mode functions, the equal-time commutation relations for $\Lambda(x,t)$ and $\partial_t\Lambda(x,t)$ give us that $\alpha_k\beta_k = \hbar/(2\mu_0)$. We now substitute the above expressions for the dual potential and its time derivative into the Hamiltonian. In evaluating the spatial derivative term, we integrate by parts to give

$$\begin{aligned}\int_0^\infty dx\, \partial_x u_{k_1}^*(x)\partial_x u_{k_2}(x) &= u_{k_1}^*(L-)\partial_x u_{k_2}(L-) - \int_0^L dx\, u_{k_1}^*(x)\partial_x^2 u_{k_2}(x) \\ &\quad - u_{k_1}^*(L+)\partial_x u_{k_2}(L+) - \int_L^\infty dx\, u_{k_1}^*(x)\partial_x^2 u_{k_2}(x) \\ &= k_2^2 \delta(k_1 - k_2),\end{aligned} \tag{6.106}$$

where, as before, we have inserted convergence factors $e^{-\epsilon x}$ into the integrals from L to ∞, and then taken the limit $\epsilon \to 0^+$, and made use of the fact that

$$u_{k_1}^*(L-)\partial_x u_{k_2}(L-) = u_{k_1}^*(L+)\partial_x u_{k_2}(L+). \tag{6.107}$$

If we now demand that the terms proportional to a_k^2 vanish, we find an additional condition on α_k and β_k,

$$\mu_0 \beta_k^2 = \frac{k^2}{\epsilon}\alpha_k^2. \tag{6.108}$$

Solving our equations for α_k and β_k, we find that

$$\begin{aligned}\Lambda(x,t) &= \int_0^\infty dk \sqrt{\frac{\hbar}{2\mu_0\omega}}\, [a_k u_k(x) + a_k^\dagger u_k^*(x)], \\ \partial_t \Lambda(x,t) &= -i\int_0^\infty dk \sqrt{\frac{\hbar\omega}{2\mu_0}}\, [a_k u_k(x) - a_k^\dagger u_k^*(x)].\end{aligned} \tag{6.109}$$

This theory can be used to find the output field if a nonlinear medium is placed in the cavity, and this is something we shall do in a later chapter.

6.7 Two-time correlation functions

We will need to relate two-time correlation functions outside the cavity to those inside the cavity to compute the squeezing spectrum of the output of nonlinear devices. In order to do this, we shall make use of the Collett–Gardiner theory. We start by integrating Eq. (6.71) over ω to obtain

$$a^{in}(t) = \sqrt{\gamma}\, a(t) - \frac{1}{\sqrt{2\pi}} \int_{-\infty}^{\infty} d\omega\, b(\omega, t). \tag{6.110}$$

Because reservoir and system operators at equal times commute, this implies that, if $c(t)$ is any system operator at time t, then

$$[c(t), a^{in}(t)] = \sqrt{\gamma}\, [c(t), a(t)]. \tag{6.111}$$

In addition, because a system operator at time t can only depend on *in* field operators for times $t' < t$, we have that $[c(t), a^{in}(t')] = 0$ for $t' > t$. Finally, for $t > t'$, we have, using Eq. (6.79), that

$$\begin{aligned} [c(t), a^{in}(t')] &= [c(t), (\sqrt{2\gamma}\, a(t') - a^{out}(t'))] \\ &= [c(t), \sqrt{2\gamma}\, a(t')], \end{aligned} \tag{6.112}$$

because $a^{out}(t')$ can only be a function of a system operator evaluated at time t if $t' > t$, so this implies that, for $t > t'$, we must have $[c(t), a^{out}(t')] = 0$.

Let us now consider the case in which the input state of the system is a coherent state or the vacuum, which means that the state is an eigenstate of $a^{in}(t)$. In that case, the expectation values $\langle a^{in\dagger}(t) a^{in}(t')\rangle$, $\langle a(t) a^{in}(t')\rangle$, $\langle a^{\dagger}(t) a^{in}(t')\rangle$, $\langle a^{in\dagger}(t) a(t')\rangle$ and $\langle a^{in\dagger}(t) a^{\dagger}(t')\rangle$ factorize. This, along with Eq. (6.79), implies that

$$\langle a^{out\dagger}(t), a^{out}(t')\rangle = 2\gamma \langle a^{\dagger}(t), a(t')\rangle, \tag{6.113}$$

where we define $\langle X, Y\rangle \equiv \langle XY\rangle - \langle X\rangle\langle Y\rangle$, for any two operators X and Y. This gives us a relation between intra-cavity two-time correlation functions and *out* two-time correlation functions. We also need a relation between correlation functions containing two annihilation or two creation operators in order to evaluate squeezing spectra. We can obtain this from

$$\begin{aligned} \langle a^{out}(t), a^{out}(t')\rangle &= \langle a^{in}(t) - \sqrt{2\gamma}\, a(t),\ a^{in}(t') - \sqrt{2\gamma}\, a(t')\rangle \\ &= 2\gamma \langle a(t), a(t')\rangle - \sqrt{2\gamma} \langle [a^{in}(t), a(t')]\rangle \\ &= 2\gamma \langle a(\max(t, t')),\ a(\min(t, t'))\rangle. \end{aligned} \tag{6.114}$$

The relation between the intra-cavity and *out* two-time correlation functions involving two creation operators is obtained from the above equation simply by taking the complex conjugate of both sides.

6.8 Master equations

In this section, another technique for solving for the time development of open quantum systems is introduced: the master equation method. So far, we have dealt with solving operator Langevin equations, which is a Heisenberg-picture approach to the dynamics. The master equation method represents the corresponding Schrödinger-picture approach. The master equation itself is an equation of motion for the density matrix of the system. The system density matrix is, in fact, a reduced density matrix, which is obtained by tracing out the reservoir degrees of freedom from the quantum state for the system plus reservoir. Finding an equation for the system density matrix alone generally involves additional approximations, which are only valid for weakly coupled, broad-band reservoirs. The reservoirs cause a damping (relaxation) of energy and phase, as well as a decoherence of superposition states of the system.

Thus, as before, we begin with the system coupled to a set of reservoir modes, so that the entire Hamiltonian can be written as

$$H = H^S + H^{SR} + H^R. \tag{6.115}$$

Also, as before, we will model the reservoir–system coupling, in the rotating-wave approximation, by the following Hamiltonian:

$$H^{SR} = \hbar \sum_j [S_j^\dagger R_j + S_j R_j^\dagger]. \tag{6.116}$$

Under conditions of weakly coupled, broad-band reservoirs, the Markov approximation will be introduced, which allows us to obtain a closed-form equation for the system density matrix. This density matrix is defined by taking a trace of the total density matrix of the system and reservoir, $\rho(t)$, over the reservoir Hilbert space R, giving

$$\rho_S = \mathrm{Tr}_R[\rho(t)].$$

In order to find an equation of motion for ρ_S, we begin with the equation of motion for the total density matrix,

$$i\hbar \frac{d\rho}{dt} = [H, \rho]. \tag{6.117}$$

We now go into the interaction picture with respect to the reservoir and part of the system. It is useful to remove any fast time dependence due to the system Hamiltonian by going into an interaction picture with respect to that fast time dependence. That is, suppose $H^S = H^{S0} + H^{SV}$, where H^{S0} is the free part of the system Hamiltonian and H^{SV} is an interaction involving only system variables. For example, if the system consists of a field mode in a nonlinear medium, H^{S0} is the free evolution of the mode, which takes place at an optical frequency, and H^{SV} is the nonlinear interaction, which takes place on a much slower time-scale. Going into the interaction picture with respect to H^{S0} has the effect of removing the fast time dependence from the density matrix. We now define

$$\rho_I(t) = e^{i(H^R + H^{S0})t/\hbar} \rho(t) e^{-i(H^R + H^{S0})t/\hbar}. \tag{6.118}$$

This density matrix now obeys the equation of motion

$$i\hbar\frac{d\rho_I}{dt} = [H_I^{SV} + H_I^{SR}, \rho], \tag{6.119}$$

where

$$\begin{aligned} H_I^{SR} &= \exp(iH^R t/\hbar)H^{SR}\exp(-iH^R t/\hbar), \\ H_I^{SV} &= \exp(iH^{S0} t/\hbar)H^{SV}\exp(-iH^{S0} t/\hbar). \end{aligned} \tag{6.120}$$

Note that H_I^{SR} and H_I^{SV} are time-dependent. In the case of the system–reservoir interaction, going into the interaction picture has the effect of replacing the operator S_j by $S_{jI} = \exp(iH^{S0}t/\hbar)S_j\exp(-iH^{S0}t/\hbar)$, and the reservoir operator R_j by $R_{jI} = \exp(iH^R t/\hbar)R_j\exp(-iH^R t/\hbar)$.

We can now formally integrate our equation for the derivative of ρ_I from time t to time $t+\Delta t$ to obtain

$$\rho_I(t+\Delta t) = \rho_I(t) + \frac{1}{i\hbar}\int_t^{t+\Delta t} dt'\,[H_I^{SV}(t') + H_I^{SR}(t'), \rho_I(t')], \tag{6.121}$$

and then iterate it to obtain

$$\begin{aligned} \rho_I(t+\Delta t) = \rho_I(t) &+ \frac{1}{i\hbar}\int_t^{t+\Delta t} dt'\,[H_I^{SV}(t') + H_I^{SR}(t'), \rho_I(t)] \\ &+ \frac{1}{(i\hbar)^2}\int_t^{t+\Delta t} dt' \int_t^{t'} dt'' \\ &\times [H_I^{SV}(t') + H_I^{SR}(t'), [H_I^{SV}(t'') + H_I^{SR}(t''), \rho(t'')]]. \end{aligned} \tag{6.122}$$

So far, we have not made any approximations, but now we shall do so. We assume that system and reservoir are initially uncorrelated. In time, correlations will develop, but, if the interaction is weak, and the reservoir is large and the system is small, we expect that the effect of the time evolution on the state of the reservoir will be small. So, we set $\rho_I(t) \simeq \rho_S(t) \otimes \rho_R$, where ρ_R is a time-independent reservoir density matrix. We will insert this form of the density matrix into the above equation and then trace over the reservoir, but, before we do, we shall make one further assumption. It is usually the case that the expectation values of the reservoir variables appearing in the system–reservoir coupling vanish, so we shall assume that $\mathrm{Tr}(R_{jI}(t')\rho_R) = \mathrm{Tr}(R_{jI}^\dagger(t')\rho_R) = 0$. Doing so, we find

$$\begin{aligned} \rho_S(t+\Delta t) = \rho_S(t) &+ \frac{1}{i\hbar}\int_t^{t+\Delta t} dt'\,[H_I^{SV}(t'), \rho_S(t)] \\ &+ \frac{1}{(i\hbar)^2}\int_t^{t+\Delta t} dt' \int_t^{t'} dt''\,\mathrm{Tr}_R\{[H_I^{SR}(t'), [H_I^{SR}(t''), \rho_S(t'')\otimes\rho_R]]\}. \end{aligned} \tag{6.123}$$

We now want to use this equation to calculate the time derivative of ρ_S, which entails subtracting $\rho_S(t)$ from both sides, dividing by Δt and then taking the limit as $\Delta t \to 0$. Normally, the double integral, being of order $(\Delta t)^2$, will not make a contribution when this is done. However, as we shall see, the integrand contains reservoir correlation functions, and in some cases these can be well approximated as delta-functions. This reduces the order of the double integral term from $(\Delta t)^2$ to Δt, which means that it will contribute to the time derivative of ρ_S. The physical reason behind this is that the time-scale for the variation of reservoir correlations is much shorter than the time-scale on which the system varies. In addition, once Δt is much smaller than the scale on which the system varies with time, we can replace $\rho_S(t'')$ by $\rho_S(t)$. Finally, we have that

$$\frac{d\rho_S}{dt} = \frac{1}{i\hbar}[H_I^{SV}(t), \rho_S(t)] + \frac{1}{(i\hbar)^2}\frac{1}{\Delta t}\int_t^{t+\Delta t} dt' \int_t^{t'} dt''\, \mathrm{Tr}_R\{[H_I^{SR}(t'), [H_I^{SR}(t''), \rho_S(t)\otimes\rho_R]]\}. \tag{6.124}$$

It looks as though we should take the limit $\Delta t \to 0$, but what we are after here is a coarse-grained time derivative. We are only interested in behavior on the time-scale of the system evolution, not short-time fluctuations due to the reservoir. Therefore, we can take Δt to be small on the scale of the system evolution but large on the scale of the reservoir time-scale.

We can make the form of the last term more explicit if we make use of the form of the system–reservoir coupling we assumed in Eq. (6.116). In order to do so, let us define several reservoir correlation functions:

$$\begin{aligned} C_{jk}^{(1)}(t', t'') &= \mathrm{Tr}_R[R_{jI}(t')R_{kI}(t'')\rho_R], \\ C_{jk}^{(2)}(t', t'') &= \mathrm{Tr}_R[R_{jI}^\dagger(t')R_{kI}(t'')\rho_R], \\ C_{jk}^{(3)}(t', t'') &= \mathrm{Tr}_R[R_{jI}(t')R_{kI}^\dagger(t'')\rho_R]. \end{aligned} \tag{6.125}$$

The integrand in the last term of the master equation now becomes

$$\begin{aligned} \sum_{j,k}[&C_{kj}^{(1)*}(t'', t')(S_{jI}S_{kI}\rho_S - S_{kI}\rho_S S_{jI}) + C_{jk}^{(1)*}(t', t'')(\rho_S S_{kI}S_{jI} - S_{jI}\rho_S S_{kI}) \\ &+ C_{jk}^{(2)}(t', t'')(S_{jI}S_{kI}^\dagger\rho_S - S_{kI}^\dagger\rho_S S_{jI}) + C_{kj}^{(3)}(t'', t')(\rho_S S_{kI}^\dagger S_{jI} - S_{jI}\rho_S S_{kI}^\dagger) \\ &+ C_{jk}^{(3)}(t', t'')(S_{jI}^\dagger S_{kI}\rho_S - S_{kI}\rho_S S_{jI}^\dagger) + C_{kj}^{(2)}(t'', t')(\rho_S S_{kI}S_{jI}^\dagger - S_{jI}^\dagger\rho_S S_{kI}) \\ &+ C_{jk}^{(1)}(t', t'')(S_{jI}^\dagger S_{kI}^\dagger\rho_S - S_{kI}^\dagger\rho_S S_{jI}^\dagger) + C_{kj}^{(1)}(t'', t')(\rho_S S_{kI}^\dagger S_{jI}^\dagger - S_{jI}^\dagger\rho_S S_{kI}^\dagger)]. \end{aligned} \tag{6.126}$$

We have not indicated time dependences of the operators in the above equation, but ρ_S is evaluated at time t, S_{jI} is evaluated at time t' and $S_{j'I}$ is evaluated at time t''.

Now let us look at the case of a single cavity mode with the reservoir in the vacuum state. In that case, the Hamiltonian describing the total system (system plus reservoir) is

$$H = \hbar\nu_0 a^\dagger a + \hbar\int_0^\infty d\omega\, b^\dagger(\omega)b(\omega) + i\hbar\int_0^\infty d\omega\,[A(\omega)ab^\dagger(\omega) - A^*(\omega)a^\dagger b(\omega)], \tag{6.127}$$

where a is the annihilation operator for the cavity mode of frequency ν_0, and the $b(\omega)$ are the annihilation operators for the reservoir. In this case, only one of the correlation functions is nonzero,

$$C^{(3)}(t',t'') = \int_0^\infty d\omega\, |A(\omega)|^2 e^{-i\omega(t'-t'')}. \tag{6.128}$$

Now we expect that the cavity mode will couple primarily to external modes whose frequencies are close to that of the cavity mode, so let us assume that $|A(\omega)|^2$ is of the form

$$|A(\omega)|^2 = \frac{\eta^2 |A(\nu_0)|^2}{(\omega - \nu_0)^2 + \eta^2}, \tag{6.129}$$

where η determines the range of frequencies to which the cavity mode couples. Inserting this into the expression for $g^{(3)}(t',t'')$, we find that

$$C^{(3)}(t',t'') \simeq \int_{-\infty}^{\infty} d\omega \frac{\eta^2 |A(\nu_0)|^2}{(\omega - \nu_0)^2 + \eta^2} e^{-i\omega(t'-t'')} = \pi\eta A(\nu_0) e^{-\gamma|t'-t''|}. \tag{6.130}$$

In the broad-band limit, i.e. η large, this goes to

$$C^{(3)}(t',t'') \to 2\pi |A(\nu_0)|^2 \delta(t'-t''). \tag{6.131}$$

Substituting this into the master equation gives us

$$\frac{d\rho_S}{dt} = \pi |A(\nu_0)|^2 (2a\rho_S a^\dagger - a^\dagger a \rho_S - \rho_S a^\dagger a). \tag{6.132}$$

Setting $\gamma = \pi |A(\nu_0)|^2$, multiplying both sides of the equation by a, and taking the trace, we find that

$$\frac{d\langle a\rangle}{dt} = -\gamma \langle a\rangle, \tag{6.133}$$

which has the solution $\langle a\rangle = \langle a\rangle_0 e^{-\gamma t}$, where $\langle a\rangle_0$ is the value of $\langle a\rangle$ at $t = 0$. We see then that the field in the cavity decays exponentially, with a rate given by γ.

The above master equation is in what is known as the generalized Lindblad form,

$$\frac{d\rho}{dt} = \frac{1}{i\hbar}[H_{sys}, \rho] + \sum_j \gamma_j (2A_j \rho A_j^\dagger - A_j^\dagger A_j \rho - \rho A_j^\dagger A_j). \tag{6.134}$$

We have dropped the subscript S on the density matrix with the understanding that the density matrix appearing in any master equation is the system density matrix. The system Hamiltonian in the above equation can be either the interaction-picture system Hamiltonian, in which case ρ_S is the interaction-picture system density matrix, or the Schrödinger-picture Hamiltonian, in which case ρ_S is the Schrödinger-picture system density matrix.

Here the operators A_j represent a set of system operators that are coupled to the reservoirs, with corresponding damping (or gain) constants γ_j. Typically, these are atomic transition operators like $\sigma^\pm$, or mode operators like a or $a^\dagger$, but they may have a more general character as well, depending on the precise cause of the damping. In the case of absorption, where the reservoir can add particles at finite temperature, a more general expression can be worked out following the same techniques. This finite-temperature master equation combines both

Table 6.1 Gain or damping operators, rates and physical interpretation

Gain/damping operator, A_j	γ_j	Physical interpretation
σ, a	γ	Linear (single-photon) decay
$\sigma^+, a^\dagger$	g	Linear (single-photon) gain
$\sigma\sigma, a^2$	$\gamma^{(2)}/2$	Nonlinear (two-photon) decay
$\sigma^z, a^\dagger a$	γ_p	Phase damping

damping-like and gain-like terms, and, for thermal occupation numbers of n_j^{th} in the jth reservoir, it is given by

$$\frac{d\rho}{dt} = \frac{1}{i\hbar}[H_{sys}, \rho] + \sum_j \gamma_j(n_j^{th} + 1)(2A_j\rho A_j^\dagger - A_j^\dagger A_j\rho - \rho A_j^\dagger A_j)$$
$$+ \sum_j \gamma_j n_j^{th}(2A_j^\dagger \rho A_j - A_j A_j^\dagger \rho - \rho A_j A_j^\dagger). \tag{6.135}$$

We shall consider the effects of four main types of damping couplings, as shown in Table 6.1.

6.9 Gain and damping rates

In order to understand the effect of the master equation, consider a case where the evolution under the reversible system Hamiltonian is negligible. Then, the behavior of the expectation value of an arbitrary system operator S is as follows:

$$\left\langle \frac{dS}{dt} \right\rangle = \mathrm{Tr}\left(\sum_j \gamma_j(2A_j\rho A_j^\dagger - A_j^\dagger A_j\rho - \rho A_j^\dagger A_j)S \right)$$
$$= \sum_j \gamma_j \, \mathrm{Tr}([A_j^\dagger, S]A_j\rho - [A_j, S]\rho A_j^\dagger). \tag{6.136}$$

Setting S equal to a single-mode annihilation operator a, we obtain the results shown in Table 6.2, in the four cases for the time evolution of the mean mode amplitude. Here we can see that, as expected, the first two cases provide simple linear damping and gain, respectively. A coupling to a nonlinear (two-photon) reservoir induces nonlinear amplitude damping, while phase damping appears similar to ordinary amplitude damping as far as the mean evolution equations are concerned.

Next, suppose we consider the evolution of the total boson number $N = a^\dagger a$. We will give a detailed derivation of the differential equation in the case that the damping operator is the annihilation operator, a. Note that, in the absence of any Hamiltonian evolution, the

Table 6.2 Gain or damping operators and mean amplitude equations

Gain/damping operator, A_j	γ_j	Damping equation for a
a	γ	$\langle da/dt\rangle = -\gamma\langle a\rangle$
$a^\dagger$	g	$\langle da/dt\rangle = g\langle a\rangle$
a^2	$\gamma^{(2)}/2$	$\langle da/dt\rangle = -\gamma^{(2)}\langle a^\dagger a^2\rangle$
$a^\dagger a$	γ_p	$\langle da/dt\rangle = -\gamma_p\langle a\rangle$

Table 6.3 Damping operators and mean number equations

Damping operator, A_j	γ_j	Damping equation for $N = a^\dagger a$
a	γ	$\langle dN/dt\rangle = -2\gamma\langle N\rangle$
$a^\dagger$	g	$\langle dN/dt\rangle = 2g\langle N\rangle$
a^2	$\gamma^{(2)}/2$	$\langle dN/dt\rangle = -2\gamma^{(2)}\langle (a^\dagger)^2 a^2\rangle$
$a^\dagger a$	γ_p	$\langle dN/dt\rangle = 0$

mean photon number follows the equation

$$\begin{aligned}\frac{d}{dt}\langle a^\dagger a\rangle &= \mathrm{Tr}\left[\frac{d\rho}{dt}a^\dagger a\right]\\ &= \gamma\,\mathrm{Tr}[(2a\rho a^\dagger - a^\dagger a\rho - \rho a^\dagger a)a^\dagger a]\\ &= 2\gamma\,\mathrm{Tr}_R[\rho a^\dagger a^\dagger a^2 - \rho a^\dagger a a^\dagger a]\\ &= -2\gamma\langle N\rangle.\end{aligned} \tag{6.137}$$

In the other cases, setting $S = N$, we obtain the results summarized in Table 6.3. The first two cases provide simple linear damping and gain of the boson number, except with twice the rate of the amplitude damping and gain. Coupling to a nonlinear (two-photon) reservoir induces nonlinear damping, while phase damping has no effect at all on the mean boson number. These results are identical to those obtained from the quantum Langevin equations in the Markovian limit, which is physically the same limit as the one which allows the master equation to be used.

6.10 Driven linear cavity example

In this section, we will examine a driven empty cavity. This will serve as a first step toward studying a driven cavity containing a nonlinear medium. We will analyze the cavity using both a quantum Langevin equation and a master equation.

We begin with the quantum Langevin equation approach. The system Hamiltonian is

$$H^S = \hbar\nu_0 a^\dagger(t)a(t) + i\hbar[\mathcal{E}e^{-i\omega_L t}a^\dagger(t) - \mathcal{E}^* e^{i\omega_L t}a(t)], \tag{6.138}$$

where the terms proportional to $\mathcal{E}$ represent the driving field, which will be treated here as classical. This implies that $\mathcal{E}$ is a c-number, and, while it is related to the amplitude of the driving field, exactly what it corresponds to will be made clear shortly. We now apply the Collett–Gardiner input–output theory so that the Heisenberg equation for the cavity annihilation operator is

$$\begin{aligned}\frac{da}{dt} &= \frac{i}{\hbar}[H^S, a] - \gamma a + \sqrt{2\gamma}\, b^{in} \\ &= \mathcal{E}e^{-i\omega_L t} - i\nu_0 a - \gamma a + \sqrt{2\gamma}\, b^{in}.\end{aligned} \tag{6.139}$$

We now want to relate the driving term in the Hamiltonian to the input and output fields of the cavity. We begin by considering the growth rate of the intra-cavity photon number $\langle N\rangle = \langle a^\dagger a\rangle$, for times short compared to the cavity decay time, so that the terms proportional to γ are negligible. Calculating this from the operator equations, with $\nu_0 = \omega_L$ and a cavity initially in the vacuum state, gives the solution

$$a(t) = (a(0) + \mathcal{E}t)e^{-i\nu_0 t}. \tag{6.140}$$

Hence, the mean intra-cavity photon number is given by

$$\langle N\rangle = \langle a^\dagger(t)a(t)\rangle = |\mathcal{E}t|^2. \tag{6.141}$$

Now consider the same problem using the classical Maxwell equations, expressed in terms of the square root of the input photon flux, which we shall call Φ^{in}. This approach is certainly valid for a linear cavity with large photon numbers, as the operator and classical Maxwell equations are identical in this case. For an initially empty cavity in the vacuum state, suppose the intensity transmissivity of the input mirror is T_1. The transmitted mean intra-cavity flux amplitude must be $\Phi^{in}\sqrt{T_1}$. After n round trips, assuming constructive interference (on-resonance) and negligible decay, this intra-cavity flux will be $n\Phi^{in}\sqrt{T_1}$. The corresponding photon number after a time t is therefore $N = t^2 T_1 |\Phi^{in}|^2/\tau_r$, where τ_r is the cavity round-trip time. This means that, to obtain the correspondence of the classical and quantum theories, we must set $\mathcal{E} = \sqrt{T_1/\tau_r}\,\Phi^{in}$, thus giving us the relation between the parameter $\mathcal{E}$ and the cavity input flux.

By including the driving field in the Hamiltonian, we have removed it from the input field described by the operators b^{in} and $b^{in\dagger}$. That is, the mean value of the input field is removed from the reservoir in this treatment, so that the residual reservoir operators then have zero initial expectation value, which is a requirement of the master equation method.

Now, let us look at this system in the Schrödinger picture. The cavity field is now described by a density matrix, which obeys the master equation

$$\frac{d\rho}{dt} = \frac{1}{i\hbar}[H^S, \rho] + \gamma(2a\rho a^\dagger - a^\dagger a\rho - \rho a^\dagger a). \tag{6.142}$$

The constant γ is the cavity decay rate for the amplitude. It is determined physically by the cavity mirror reflection and transmission coefficients, together with the internal cavity losses. It is often the case that it is simpler to work in a type of interaction picture in which the density matrix only evolves according to the interactions. For convenience, the free part of the Hamiltonian is often defined as

$$H_0 = \hbar\omega_L a^\dagger a. \tag{6.143}$$

With this definition, the operators will evolve according to the laser frequency, while the states evolve according to the rest of the system Hamiltonian – including part of the cavity energy term if the laser is not on-resonance.

In this picture, the operators have a known time evolution of $a_I(t) = ae^{-i\omega_L t}$, so it is possible to cancel all of the explicit time-dependent terms, and work with operators a that are time-independent. We call this the rotating-frame picture. It gives us an interaction-picture master equation with time-independent operators

$$\frac{d\rho_I}{dt} = [-i\Delta\omega a^\dagger a + \mathcal{E}a^\dagger - \mathcal{E}^* a, \rho] + \gamma(2a\rho a^\dagger - a^\dagger a\rho - \rho a^\dagger a). \tag{6.144}$$

Here $\Delta\omega = \nu_0 - \omega_L$ represents the detuning between the cavity-mode resonance and the laser (angular or carrier) frequency. In practical terms, this must be much less than the cavity-mode spacing for the single-mode approximation to be applicable.

Let us find the expectation value $\langle a(t)\rangle$ in the case $\Delta\omega = 0$. Multiplying the above master equation by a and taking the trace, we find

$$\frac{d\langle a\rangle}{dt} = \mathcal{E} - \gamma\langle a\rangle, \tag{6.145}$$

which has the solution

$$\langle a(t)\rangle = \langle a(0)\rangle e^{-\gamma t} + \frac{\mathcal{E}}{\gamma}(1 - e^{-\gamma t}). \tag{6.146}$$

From this, we see that the expectation value of the cavity field goes from an initial value of $\langle a(0)\rangle$ to a final value of $\mathcal{E}/\gamma$ on a time-scale of γ^{-1}. We note that, to recover the full time dependence of the operator, we should multiply these expectation values by $e^{-i\omega_L t}$.

Additional reading

Books

- Many of the quantum optics books mentioned previously discuss operator Langevin equations and master equations. Let us mention here two older books that treat these methods in the context of laser theory:

H. Haken, *Laser Theory* (Springer, Berlin, 1984).

M. Sargent, M. Scully, and W. Lamb, *Laser Physics* (Addison-Wesley, Reading, MA, 1974).

The first makes extensive use of Langevin methods, while the second emphasizes the master equation approach.

- A newer book that emphasizes the master equation approach is:

H. J. Carmichael, *Statistical Methods in Quantum Optics 1: Master Equations and Fokker–Planck Equations* (Springer, New York, 2010).

- For a discussion of input–output theory, see:

C. W. Gardiner and P. Zoller, *Quantum Noise* (Springer, Berlin, 2000).
B. Yurke, Input–output theory, in *Quantum Squeezing*, ed. P. D. Drummond and Z. Ficek (Springer, Berlin, 2004).

The first of these references is also a good place to learn about reservoir theory in general.

Articles

- On input–output theory, see:

M. J. Collett and C. W. Gardiner, *Phys. Rev.* A **30**, 1386 (1984).
C. W. Gardiner and M. J. Collett, *Phys. Rev.* A **31**, 3761 (1984).
B. Yurke, *Phys. Rev.* A **29**, 408 (1984).
B. Yurke, *Phys. Rev.* A **32**, 300 (1985).
B. Yurke and J. S. Denker, *Phys. Rev.* A **29**, 1419 (1984).

- Our third version of input–output theory is a modification of the one presented in:

J. Gea-Banacloche, N. Lu, L. M. Pedrotti, S. Prasad, M. O. Scully, and K. Wodkiewicz, *Phys. Rev.* A **41**, 369 (1990).
J. Gea-Banacloche, N. Lu, L. M. Pedrotti, S. Prasad, M. O. Scully, and K. Wodkiewicz, *Phys. Rev.* A **41**, 381 (1990).

Problems

6.1 Let us use the gain Langevin equation to find the noise that an amplifier introduces into the quadrature component of a field mode. Assuming the system Hamiltonian is zero (this implies that we are in the interaction picture with respect to the free-mode Hamiltonian, $H_0 = \hbar\nu_0 a^\dagger a$), solve the gain Langevin equation in the Markovian limit for $a(t)$ in terms of $a(0)$. Use this to find $\Delta X_1(t)$ in terms of $\Delta X_1(0)$. Define the intensity gain of the amplifier to be $G = e^{2g't}$. You should find that, if G is greater than a particular value, G_0, then, even if the state of the field mode is initially squeezed, it no longer will be after being amplified. Find G_0.

6.2 Using the Collett–Gardiner formalism, let us see what happens when a pulse hits an empty cavity. Suppose that $\langle a(0)\rangle = 0$, and

$$\langle a^{in}(t)\rangle = A_0 e^{i\nu_0(t-t_0)} e^{-\eta|t-t_0|},$$

where t_0 and η are positive. Find $\langle a^{out}(t)\rangle$ for $t > 0$.

6.3 The phase-damping master equation in the case that the system Hamiltonian is equal to zero is given by

$$\frac{d\rho}{dt} = \gamma_p(2N\rho N - N^2\rho - \rho N^2),$$

where $N = a^\dagger a$. This equation is easy to solve in the number-state representation of the density matrix, $\rho = \sum_{m,n=0}^{\infty} \rho_{mn} |m\rangle\langle n|$.

(a) Find $\rho_{mn}(t)$ in terms of $\rho_{mn}(0)$.

(b) Use your solution in part (a) to find $\langle a^k \rangle = \mathrm{Tr}(a^k \rho(t))$ if the state at time $t = 0$ is the coherent state $|\alpha\rangle$.

6.4 It is possible to obtain a formal solution to the Lindblad-form master equation in the case that the system Hamiltonian is zero. That means that we want to solve

$$\frac{d\rho}{dt} = \gamma(2A\rho A^\dagger - A^\dagger A\rho - \rho A^\dagger A).$$

The operator A here is a general operator and not necessarily an annihilation operator.

(a) As a first step, define $\tilde{\rho}(t) = e^{\gamma t A^\dagger A} \rho(t) e^{\gamma t A^\dagger A}$ and find the equation satisfied by $\tilde{\rho}$. Your answer should be of the form

$$\frac{d\tilde{\rho}}{dt} = 2\gamma B(t)\tilde{\rho}(t)B^\dagger(t),$$

and you should be able to express $B(t)$ in terms of A and $A^\dagger$.

(b) Show that the expression

$$\tilde{\rho}(t) = \sum_{n=0}^{\infty} (2\gamma)^n \int_0^t dt_1 \int_0^{t_1} dt_2 \cdots \int_0^{t_{n-1}} dt_n \times B(t_1)B(t_2)\cdots B(t_n)\tilde{\rho}(0)B^\dagger(t_n)\cdots B^\dagger(t_1)$$

is a formal solution to the equation for $\tilde{\rho}$.

(c) In the case $A = a$, that is, A is a mode annihilation operator, find an explicit solution for $\rho(t)$ using the above expression. You should find

$$\rho(t) = \sum_{n=0}^{\infty} \frac{(1 - e^{-2\gamma t})^n}{n!} e^{-\gamma t a^\dagger a} a^n \rho(t) (a^\dagger)^n e^{-\gamma t a^\dagger a}.$$

6.5 (a) Using the expression given in problem 6.4(c), find the time evolution of a coherent state $|\alpha\rangle$ assuming that the system Hamiltonian is zero.

(b) Now find the time evolution of a superposition of two coherent states, $\eta_0(|\alpha\rangle + |\beta\rangle)$, where η_0 is a normalization constant. Find the time-scale for the decay of the off-diagonal elements of the density matrix.

7 Phase-space distributions

In this chapter, we will develop methods for mapping operator equations to equivalent c-number equations. This results in a continuous phase-space representation of a many-body quantum system, using phase-space distributions instead of density matrices. In order to do this, we will first find c-number representations of operators. We have already seen one such representation, the Glauber–Sudarshan P-representation for the density matrix. We now introduce several more such representations. The main focus will be on the truncated Wigner representation, valid at large photon number, and the positive P-representation, which uses a double-dimensional phase space and exists as a positive probability for all quantum density matrices.

Phase-space techniques have a great advantage over conventional matrix-type solutions to the Schrödinger equation, in that they do not have an exponential growth in complexity with mode and particle number. Instead, the equations that describe the dynamics of these c-number representations of the density matrix are Fokker–Planck equations. These have equivalent stochastic differential equations, which behave as c-number analogs of Heisenberg-picture equations of motion.

This means that problems that would be essentially impossible to solve using conventional number-state representations can be transformed into readily soluble differential equations. In many cases, no additional approximations, such as perturbation theory or factorization assumptions, are needed. The techniques described in this chapter are especially useful for open systems, which readily lend themselves to phase-space methods. The resulting equations can be treated analytically, with exact solutions in some cases. More generally, it is straightforward to solve these equations numerically, even if analytic techniques are not available. As there is widely available public-domain software even for multi-mode systems, this is generally much more feasible than a number-state expansion.

We will begin this chapter with a general discussion of statistical methods in many-body theory, and then move on to develop the theory of c-number representations of operators and equations of motion for phase-space distributions.

7.1 Diffusion processes

Stochastic equations are a way to deal with large numbers of degrees of freedom. This idea originated in the early classical work of Einstein and Langevin on random motion of molecules in solution, and was later generalized by Fokker and Planck. These equations were originally classical, but, as we shall see, they can also be extended to quantum problems.

In order to understand the nature of stochastic equations, we will first summarize the main results of this early work, before establishing more general results.

To understand the statistical and sampling techniques useful to quantum many-body theory, it is simplest to start with the diffusion equation. This was originally used, by Einstein, to treat the random thermal motion of small pollen particles immersed in fluid. It was later shown by Langevin that there is an equivalent description as a dynamical equation with random forces. In more recent terminology, a general diffusion equation is called a Fokker–Planck equation (FPE), while a general Langevin equation is called a stochastic differential equation (SDE).

7.1.1 Einstein's theory of diffusion

In general, all physical quantities can fluctuate. How do these fluctuations develop in time? Einstein's original theory of thermal motion predicted diffusive behavior of the probability distribution of observing small particles in solution near position x, provided we only look at relatively long time intervals, so that particle motion is effectively independent at different times.

In more detail, the theory developed by Einstein treats the motion of massive, suspended particles in a fluid. The simplifying feature is that only one particle at a time is treated, with the fluid being treated as a 'background' reservoir of fluctuations acting on the particle. While this is not the most general case, the resulting statistical behavior gives an insight into the kinetics of real, imperfect, interacting fluids. It also provides an understanding of how general physical systems couple to their environment.

Einstein's theory made the following crucial assumptions:

- Each particle is independent of all other particles.
- Motion is independent at different times, for long time separations.
- For simplicity, only motion in one dimension is considered.

To show what these assumptions lead to, consider N particles suspended in a fluid, subject to random forces due to the thermal motion of the fluid itself. The x coordinate of the ith individual particle changes by Δx_i in a time interval Δt. The number of particles experiencing a coordinate shift between Δx and $\Delta x + d\Delta x$ is defined as

$$dN = NP_J(\Delta x, \Delta t) \cdot d\Delta x, \tag{7.1}$$

where $P_J(\Delta x, \Delta t)$ is a jump probability of changing by Δx in time interval Δt, defined so that it is normalized to unity,

$$\int_{-\infty}^{\infty} d\Delta x \, P_J(\Delta x, \Delta t) = 1. \tag{7.2}$$

To obtain the simplest case, two further mathematical conditions are imposed on the jump probability. These are that, in equilibrium, motion is equally likely in either direction, so

$$P_J(\Delta x, \Delta t) = P_J(-\Delta x, \Delta t), \tag{7.3}$$

and that there is a vanishing probability for an infinitely large coordinate change in a finite time, which implies that

$$\lim_{|\Delta x|\to\infty} P_J(\Delta x, \Delta t) \to 0. \tag{7.4}$$

7.1.2 Einstein's solution

Let $P(x,t)$ be the number of particles per unit distance, or equivalently the probability density of observing a single particle. We must compute the distribution of particles at time $t+\Delta t$ from that at time t. From the definition of $P_J(\Delta x, \Delta t)$, the number of particles between x and $x+dx$ at time $t+\Delta t$ is given by

$$P(x, t+\Delta t)\,dx = dx \int_{-\infty}^{\infty} d\Delta x\, P_J(\Delta x, \Delta t) P(x-\Delta x, t). \tag{7.5}$$

Expanding this expression as a first-order Taylor expansion at short times, one can write

$$P(x, t+\Delta t) = P(x,t) + \Delta t \frac{\partial}{\partial t} P(x,t). \tag{7.6}$$

From a Taylor expansion to second order in space, one similarly finds that

$$P(x-\Delta x, t) = P(x,t) - \Delta x \frac{\partial P(x,t)}{\partial x} + \frac{\Delta x^2}{2}\frac{\partial^2 P(x,t)}{\partial x^2}. \tag{7.7}$$

Combining the two types of Taylor expansion in Eqs (7.6) and (7.7), we can re-express the original integral equation (7.5) as a sum of three integrals over Δx:

$$\begin{aligned} P(x,t) &+ \Delta t \frac{\partial}{\partial t} P(x,t) \\ &= P(x,t)\int_{-\infty}^{\infty} d\Delta x\, P_J(\Delta x, \Delta t) - \frac{\partial P(x,t)}{\partial x}\int_{-\infty}^{\infty} d\Delta x\, \Delta x\, P_J(\Delta x, \Delta t) \\ &+ \frac{\partial^2 P(x,t)}{\partial x^2}\int_{-\infty}^{\infty} d\Delta x\, \frac{\Delta x^2}{2} P_J(\Delta x, \Delta t). \end{aligned} \tag{7.8}$$

The first integral is unity from the normalization condition, Eq. (7.2), while the second integral vanishes due to the symmetry condition, Eq. (7.3), so that

$$\frac{\partial}{\partial t} P(x,t) \approx \frac{1}{2}\left(\int_{-\infty}^{\infty} d\Delta x\, \frac{\Delta x^2}{\Delta t} P_J(\Delta x, \Delta t)\right)\frac{\partial^2 P(x,t)}{\partial x^2}. \tag{7.9}$$

The last integral is, however, nontrivial. We treat this by introducing a new constant, called the diffusion constant, as follows:

$$D(\Delta t) = \int_{-\infty}^{\infty} d\Delta x\, \frac{\Delta x^2}{\Delta t} P_J(\Delta x, \Delta t). \tag{7.10}$$

If we assume that higher-derivative terms in x vanish, and if we define $D = \lim_{\Delta t\to 0} D(\Delta t)$, then we immediately obtain the well-known diffusion equation

$$\frac{\partial P}{\partial t} = \frac{D}{2}\frac{\partial^2 P}{\partial x^2}. \tag{7.11}$$

Einstein's brilliance lay in recognizing that this minimal solution, essentially valid only for time-scales long compared to the velocity relaxation times, nevertheless encapsulates most of the important physics of these experiments, regardless of the detailed nature of the underlying physical processes.

If the particle is initially located at $x = 0$, that is, $P(x, 0) = \delta(x)$, the probability distribution at later times is a Gaussian,

$$P(x, t) = \frac{1}{\sqrt{2\pi t D}} e^{-x^2/2Dt}. \tag{7.12}$$

Diffusive spreading in time occurs, and, from the Gaussian solution, the variance is clearly given by the usual probabilistic result

$$\langle x^2(t)\rangle = Dt. \tag{7.13}$$

The Gaussian also plays a role if we ask what jump probabilities result in $D(\Delta t)$ being time-independent for short times. If the jump probability P_J is Gaussian, i.e.

$$P_J(\Delta x, \Delta t) = \frac{1}{\sqrt{2\pi \Delta t D}} e^{-\Delta x^2/2D\Delta t}, \tag{7.14}$$

then this will be the case.

7.2 Fokker–Planck equations

If we extend the previous approach to an inhomogeneous space-dependent jump probability $P_J(x, \Delta x, \Delta t)$, for jumps starting at position x, and introduce an nth-order coefficient,

$$D^{(n)}(x) = \lim_{\Delta t \to 0} \int_{-\infty}^{\infty} d\Delta x \frac{(\Delta x)^n}{\Delta t} P_J(x, \Delta x, \Delta t), \tag{7.15}$$

then we obtain a generalized diffusion equation:

$$\frac{\partial}{\partial t} P(x, t) = \left\{ \sum_{n \geq 0} \frac{(-1)^n}{n!} D^{(n)}(x) \frac{\partial^n}{\partial x^n} \right\} P(x, t). \tag{7.16}$$

By rearranging the order of the differential operators so that the equation is explicitly probability-conserving, and truncating higher-order terms, we obtain a general class of second-order partial differential equations of the diffusion type, which are generically termed Fokker–Planck equations (FPEs). Consider a partial differential equation of the form

$$\frac{\partial}{\partial t} P(x, t) = \left\{ -\frac{\partial}{\partial x} A(x) + \frac{1}{2} \frac{\partial^2}{\partial x^2} D(x) \right\} P(x, t). \tag{7.17}$$

Suppose that D, the diffusion constant, is real and positive, while A, called the drift, is real. We notice first that, if $D(x) = 0$, then the equation is a first-order partial differential equation. This is immediately soluble by the method of characteristics. More generally, it is always possible to have a probabilistic interpretation, since the propagator can be shown to be real and positive.

7.2.1 Green's function and path integral

To demonstrate the nature of the solutions to Eq. (7.17), consider an initial condition where the particle is localized at x' and time t', so that $P_0(x, t') = \delta(x - x')$. The solution with this initial condition is called the Green's function, $G(x, t \mid x', t')$. For short times $t = t' + \Delta t$, the solution is localized in the region $x \approx x'$, so we can approximate $D(x) \approx D(x')$ and $A(x) \approx A(x')$. For these small time intervals, the Green's function is just the Gaussian solution given above to a diffusion equation with constant coefficients, except with the addition of a constant displacement proportional to $\Delta t A(x')$. This can be verified by simple differentiation.

We find

$$G(x, t \mid x', t') = \frac{1}{\sqrt{2\pi D(x')\Delta t}} \exp\left\{\left[\frac{-[\Delta x - \Delta t A(x')]^2}{2D(x')\Delta t}\right]\right\}, \tag{7.18}$$

where $\Delta x = x - x'$. Clearly G is real and positive, and the short-time solution for an arbitrary initial function $P(x', t')$ is therefore

$$P(x, t) = \int dx' \, G(x, t \mid x', t') P(x', t'). \tag{7.19}$$

The solution over a finite interval $[t^{(1)}, t]$ is then obtained by iterating this process, and taking the limit as $\Delta t \to 0$ to give

$$P(x, t) = \lim_{\Delta t \to 0} \int \cdots \int dx^{(n)} \cdots dx^{(1)} \, G(x, t \mid x^{(n)}, t^{(n)}) \cdots G(x^{(2)}, t^{(2)} \mid x^{(1)}, t^{(1)}) \times P(x^{(1)}, t^{(1)}). \tag{7.20}$$

Here, $\Delta t = (t - t^{(1)})/n$, and we fix the number of steps so that $t^{(n)} = t^{(1)} + (n-1)\Delta t$. This is an example of a path-integral form of the solution to a partial differential equation, although we have written it explicitly as a limit. A typical result of considering a finite number of these paths is shown in Figure 7.1.

7.2.2 Higher-dimensional Fokker–Planck equations

In the case of multiple phase-space dimensions, we can generalize this equation in the following way:

$$\frac{\partial}{\partial t} P(\mathbf{x}, t) = \left\{ -\sum_i \frac{\partial}{\partial x_i} A_i(\mathbf{x}) + \frac{1}{2} \sum_{ij} \frac{\partial^2}{\partial x_i \partial x_j} D_{ij}(\mathbf{x}) \right\} P(\mathbf{x}, t). \tag{7.21}$$

Here $\mathbf{x} = (x_1, \ldots, x_M)$ is an M-dimensional real vector, $\mathbf{A}(\mathbf{x})$ is a real vector function and $\mathbf{D}(\mathbf{x})$ is a symmetric, positive definite matrix function. We start by assuming that $D_{ij}(\mathbf{x})$ is diagonal, for simplicity. Similar to the case of one phase-space dimension, we can use the

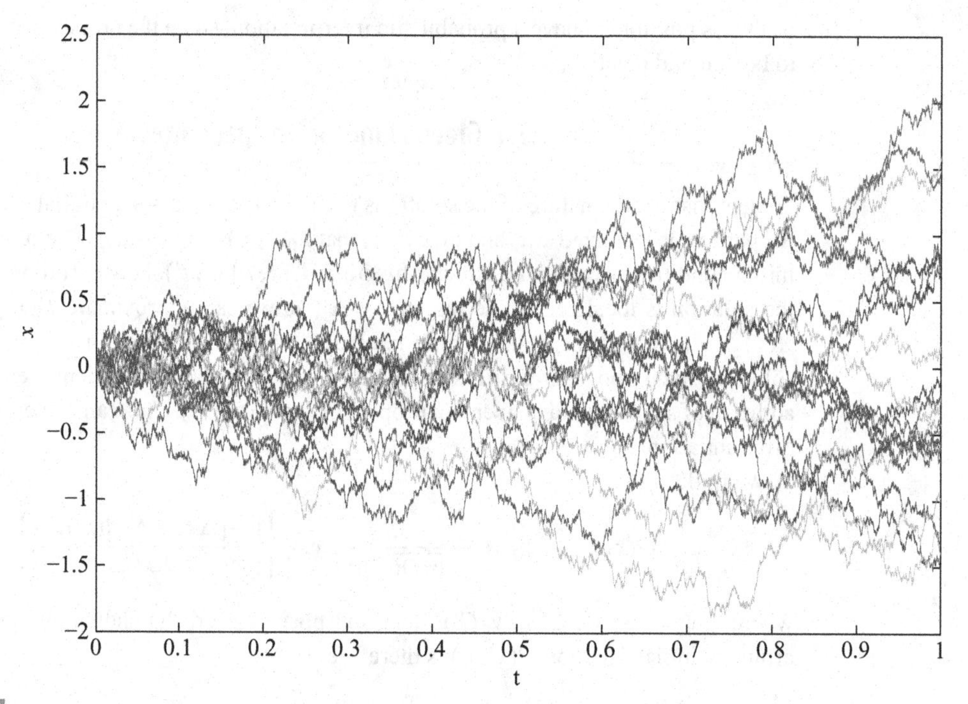

Fig. 7.1 Twenty random paths of a path integral with $D = 1$ and $A = 0$.

method of separation of variables, and write the Green's function as a product. Defining $\Delta x_i = x_i - x_i'$, we obtain, for a small interval Δt,

$$\begin{aligned} G(\mathbf{x}, t \mid \mathbf{x}', t') &= \prod_{i=1}^{M} G_i(\mathbf{x}, t \mid \mathbf{x}', t') \\ &= \prod_{i=1}^{M} \frac{1}{\sqrt{2\pi \Delta t D_i(\mathbf{x}')}} \exp\left\{ \frac{-[\Delta x_i - \Delta t A_i(\mathbf{x}')]^2}{2D_i(\mathbf{x}')\Delta t} \right\}. \end{aligned} \tag{7.22}$$

This has a simple generalization to non-diagonal cases, as can again be verified by simple differentiation:

$$\begin{aligned} G(\mathbf{x}, t \mid \mathbf{x}', t') &= \frac{1}{\sqrt{(2\pi\Delta t)^M \det[\mathbf{D}(\mathbf{x}')]}} \\ &\times \exp\left\{ \frac{-[\Delta x_i - \Delta t A_i(\mathbf{x}')] D_{ij}^{-1}(\mathbf{x}')[\Delta x_j - \Delta t A_j(\mathbf{x}')]}{2\Delta t} \right\}. \end{aligned} \tag{7.23}$$

We require that $\mathbf{D}(\mathbf{x}')$ is positive definite (only has positive eigenvalues) for this equation to be valid, and the solution for an arbitrary initial function $P(\mathbf{x}, t)$ is therefore

$$P(\mathbf{x}, t) = \int d^M\mathbf{x}\, G(\mathbf{x}, t \mid \mathbf{x}', t') P(\mathbf{x}', t'), \tag{7.24}$$

where $d^M\mathbf{x}$ is an M-dimensional real volume measure. The solution over the finite interval $[t^{(1)}, t]$ is then obtained by iterating this process, as before, to give a multi-dimensional path-integral solution:

$$P(\mathbf{x},t) = \lim_{\Delta t\to 0}\int\cdots\int d^M\mathbf{x}^{(n)}\cdots d^M\mathbf{x}^{(1)} \times G(\mathbf{x},t \mid \mathbf{x}^{(n)},t^{(n)})\cdots G(\mathbf{x}^{(2)},t^{(2)} \mid \mathbf{x}^{(1)},t^{(1)})P(\mathbf{x}^{(1)},t^{(1)}). \tag{7.25}$$

Cases where some of the eigenvalues of $\mathbf{D}(\mathbf{x})$ are zero are tractable as well, simply by taking the limit as the relevant eigenvalue approaches zero. In this case, one or more of the relevant diagonal Green's functions become a simple delta-function.

7.3 Stochastic differential equations

The Gaussian nature of the exponent in the above solutions means that the Fokker–Planck equation solution is equivalent to an average over a set of stochastic processes. This is a generalization of the method of characteristics, to include diffusion terms as well, and is the fundamental mathematical basis for the equivalence between the Einstein diffusion theory and Langevin theory. These two approaches appear different, yet we shall show that they are, in fact, two sides of the same coin. There is a deep equivalence between the Einstein diffusion equation and the Langevin equation. This is embodied in the theory of the Fokker–Planck and stochastic equations, which we turn to next.

We consider the diagonal case first, and note that the multi-dimensional path-integral solution of Eq. (7.25) can be rewritten as

$$P(\mathbf{x},t) = \lim_{\Delta t\to 0}\int\cdots\int d^M\Delta\mathbf{x}^{(n)}\cdots d^M\mathbf{x}^{(1)} \times G(\mathbf{x},t \mid \mathbf{x}^{(n)},t^{(n)})\cdots G(\mathbf{x}^{(2)},t^{(2)} \mid \mathbf{x}^{(1)},t^{(1)})P(\mathbf{x}^{(1)},t^{(1)}), \tag{7.26}$$

where $\Delta\mathbf{x}^{(k)} = \mathbf{x}^{(k+1)} - \mathbf{x}^{(k)}$, and $G(\mathbf{x}^{(k)},t^{(k)} \mid \mathbf{x}^{(k-1)},t^{(k-1)})$ can be considered as a function of $\Delta\mathbf{x}^{(k)}$ and Δt. This finite-dimensional integral can be integrated as a Monte Carlo integral by first sampling over the initial position $\mathbf{x}$ with probability density $P(\mathbf{x},t)$. Next, one must sample over each possible sample increment $\Delta\mathbf{x}^{(k)}$ by choosing it randomly according to a Gaussian distribution corresponding to each successive Green's function.

From the formula for the Green's function, in the diagonal case, we can see that the mean of the Gaussian increment in each dimension is $-A_i(\mathbf{x})\Delta t$. Similarly, the variance is $D_i(\mathbf{x})\Delta t$. The equation for each increment can therefore be written as follows:

$$\Delta x_i^{(k)} = x_i^{(k+1)} - x_i^{(k)} = A_i(\mathbf{x}^{(k)})\Delta t + B_i(\mathbf{x}^{(k)})\Delta W_i. \tag{7.27}$$

Here ΔW_i is a set of real Gaussian random numbers with zero mean and unit variance, generated independently at each time step, with the property that $\langle\Delta W_i\Delta W_j\rangle = \Delta t\delta_{ij}$ and $B_i(\mathbf{x}) = \sqrt{D_i(\mathbf{x})}$. This generates one discretized trajectory, and each set of points obtained in this way represents a possible stochastic trajectory. An estimate of $P(\mathbf{x}',t')$ is then obtained by taking the double limit of $\Delta t \to 0$ and $N_s \to \infty$, where N_s is the number of samples. The set of trajectories obtained in this way is technically known as an Ito stochastic process with Gaussian random noise. It is clear from this construction that the functions

$A_i(\mathbf{x})$ and $B_i(\mathbf{x})$ are always evaluated with their arguments determined from the value of $\mathbf{x}$ at the *start* of each interval.

There is no need to restrict the calculation to diagonal cases, since the coordinate rotations can be trivially restored, giving a general result valid for nondiagonal diffusion matrices D. This more general result is usually written in a compact form, valid for the limit of small step sizes,

$$dx_i = A_i(\mathbf{x})\,dt + \sum_{j=1}^{M'} B_{ij}(\mathbf{x})\,dW_j. \tag{7.28}$$

Here dW_i is a real Gaussian – distributed random number with $\langle dW_i dW_j\rangle = \delta_{ij}$, and a 'noise' matrix, $\mathbf{B}$, is introduced, which is a (nonunique) matrix square root, defined so that $\mathbf{D} = \mathbf{B}\mathbf{B}^{\mathrm{T}}$. Although one may assume that $\mathbf{B}$ is a square matrix, this is not necessary as long as it gives rise to a matrix square root in the specified form. For example, if $\mathbf{D}$ has some vanishing eigenvalues, then $\mathbf{B}$ will be an $M \times M'$ matrix, where $M' < M$, as should be clear from the derivation.

This is called an Ito stochastic differential equation (SDE). When the Ito form is used, the simplest equivalent numerical technique is the Euler (one-step) method, which is simply a direct application of Eq. (7.28) for a finite value of dt. While the Ito form of SDE is mathematically simple, this is slightly deceptive. The sample paths of such an equation are nondifferentiable. As a result, the ordinary rules of calculus of differentiable curves do not apply. For example, variable changes do not follow the customary chain rule, so that for a function $g(\mathbf{x})$ one obtains

$$dg(\mathbf{x}) = \sum_i \frac{\partial g(\mathbf{x})}{\partial x_i} dx_i + \frac{dt}{2} \sum_{i,j} \frac{\partial^2 g(\mathbf{x})}{\partial x_i \partial x_j} D_{ij}(\mathbf{x}), \tag{7.29}$$

and the application of the more advanced numerical techniques, such as Runge–Kutta methods, can result in systematic errors if $\mathbf{B}(\mathbf{x})$ is space-dependent.

7.3.1 Langevin equation

As the most famous example of this formalism, we turn to Langevin's theory of particle motion in a fluid, which predicts Newtonian yet random trajectories for individual particles over short time intervals, with random forces due to molecular collisions. This theory gives a phase-space trajectory for each particle with position x and momentum p. It is of the form (including external forces F, like gravity)

$$\begin{aligned} \frac{dx}{dt} &= \frac{p}{m}, \\ \frac{dp}{dt} &= F - \gamma p + b\xi(t). \end{aligned} \tag{7.30}$$

Here γ is a damping constant and $\xi(t) = dW(t)/dt$ is a random force, which we assume is Gaussian and delta-correlated, so that

$$\langle \xi(t)\xi(t')\rangle = \delta(t - t'). \tag{7.31}$$

What this equation says is that a particle immersed in a medium, such as a fluid or a gas, experiences, besides deterministic forces described by F, friction (damping) and random forces due to the particles of the medium. Writing the phase-space variable as $\mathbf{x} = (x, p)$, this has the standard form of an Ito equation. This is now written in a differential form that is to be interpreted in the same way as the more formal Eq. (7.28), so that

$$\frac{d\mathbf{x}}{dt} = \mathbf{A} + \mathbf{B} \cdot \xi(t). \tag{7.32}$$

Here we see that the drift and diffusion terms have the form

$$\mathbf{A} = \begin{pmatrix} p/m \\ F - \gamma p \end{pmatrix} \tag{7.33}$$

and

$$\mathbf{B} = \begin{pmatrix} 0 \\ b \end{pmatrix}. \tag{7.34}$$

Substituting to obtain the Fokker–Planck equation, we find that

$$\frac{\partial}{\partial t} P(\mathbf{x}, t) = \left\{ -\frac{\partial}{\partial x}\frac{p}{m} - \frac{\partial}{\partial p}(F - \gamma p) + \frac{b^2}{2}\frac{\partial^2}{\partial p^2} \right\} P(\mathbf{x}, t). \tag{7.35}$$

7.3.2 Fluctuation–dissipation theorem

Suppose that $F = -dV/dx$, i.e. the force is derived from a potential, like gravity or an external Coulomb field. Then this equation has a steady-state solution, given at finite temperature T by

$$P(\mathbf{x}, \infty) = C \exp[-\beta H], \tag{7.36}$$

where $H = p^2/2m + V(x)$ is the total energy, and $\beta = 1/k_B T$. To show that this is a solution, we simply differentiate, to give

$$\frac{\partial}{\partial t} P(\mathbf{x}, \infty) = C \left\{ -\frac{\partial}{\partial x}\frac{p}{m} + \frac{\partial}{\partial p}\left(\frac{dV}{dx} + \gamma p\right) + \frac{b^2}{2}\frac{\partial^2}{\partial p^2} \right\} e^{-\beta(p^2/2m+V)} = 0. \tag{7.37}$$

Substituting, we find that this always gives a steady-state solution, provided that $V(x)$ leads to an integrable solution (for example, it should be bounded below for stability), and one satisfies the following *fluctuation–dissipation* relation, which relates the noise coefficient to the damping and the temperature,

$$b = \sqrt{2mk_B T\gamma}. \tag{7.38}$$

We note that one can also interpret this as a way to define the temperature, in terms of the observed noise levels. This result can be generalized in many ways, both to higher-dimensional mechanical systems, and to electronic and/or photonic cases. The fluctuation–dissipation theorem can easily be related to Einstein's original displacement result of Eq. (7.13). If there is no external potential, then

$$x(t) = \int_0^t \frac{p}{m}\, dt'. \tag{7.39}$$

However, for strong damping and long times, so that $\gamma t \gg 1$, we can neglect the time derivative in the Langevin momentum equation, Eq. (7.30), so that $\gamma p = b\zeta(t)$. This is an approximation called adiabatic elimination, and it leads to the result that

$$\begin{aligned}\langle x^2(t)\rangle &= \left(\frac{b}{\gamma m}\right)^2 \left\langle \int_0^t \int_0^t \xi(t_1)\xi(t_2)\,dt_1\,dt_2 \right\rangle \\ &= \left(\frac{b}{\gamma m}\right)^2 \int_0^t dt_2 = D^E t.\end{aligned} \tag{7.40}$$

Here we have used the delta-correlation property of Eq. (7.31). This shows that the Einstein effective diffusion coefficient for position, D^E, has the well-known temperature-dependent form $D^E = 2k_B T/\gamma m$.

7.3.3 Real Stratonovich process

While the Ito method gives correct results, it is often not a very robust or stable procedure for numerical integration. To obtain higher-order convergence, additional correction terms are needed. Rather than investigate these in detail, we now turn to an alternative form of SDE that has more familiar rules and behavior. When the noise coefficient $\mathbf{B}(\mathbf{x})$ is not constant, a substantial practical improvement is encountered if the path-integral increments are evaluated at the midpoint of each spatial interval, to reduce the discretization error. This symmetric version of the Green's function requires that the Green's function arguments are all evaluated at the midpoint, $\bar{\mathbf{x}}_M^{(k)} = (\mathbf{x}^{(k+1)} + \mathbf{x}^{(k)})/2$, giving what is called a Stratonovich stochastic differential equation. This improvement results in a more efficient way of evaluating the stochastic equations, since it converges more rapidly as $\Delta t \to 0$.

The transformation to a Stratonovich form of stochastic calculus involves some subtle effects. The stochastic drift term $\bar{A}_i(\bar{\mathbf{x}})$ must now be evaluated at the midpoint, even though this is not known initially. The simplest corresponding numerical algorithm is therefore a semi-implicit algorithm, which involves solving for the derivative, rather than evaluating it directly. This leads to a correction in the drift term and gives a resulting stochastic increment for the path of the form

$$dx_i = \bar{A}_i(\bar{\mathbf{x}})\,dt + \sum_{j=1}^{M'} B_{ij}(\bar{\mathbf{x}})\,dW_j, \tag{7.41}$$

where the modified drift term is given by

$$\begin{aligned}\bar{A}_i(\mathbf{x}) &= A_i(\mathbf{x}) - \frac{1}{2}\sum_{k=1}^{M}\sum_{j=1}^{M'} \frac{\partial B_{ij}(\mathbf{x})}{\partial x_k} B_{kj}(\mathbf{x}) \\ &= A_i(\mathbf{x}) - \Delta^s A_i(\mathbf{x}).\end{aligned} \tag{7.42}$$

One can verify this result in a straightforward way, by considering a time-reversed solution to the Fokker–Planck equation Green's function, to obtain the midpoint solution, followed by an ordinary forward time step. Full details of this can be found in texts on stochastic differential equations.

The advantage of this procedure is that, unlike the Ito process, the Stratonovich form of stochastic differential equation obeys the ordinary variable-change rules of calculus. In fact, it is the wide-band limit of an ordinary differential equation, driven by finite-bandwidth (colored) noise. This means that Stratonovich equations can be treated without error using techniques from the toolbox of ordinary differential equations. As discussed in the additional reading, a particularly stable form of algorithm is found using a semi-implicit midpoint method. Standard Runge–Kutta methods can also be used, although their convergence rate is different from what is expected for ordinary differential equations.

7.3.4 Complex stochastic processes

We will generally have to deal with complex stochastic processes in quantum optics, as bosonic modes have both amplitude and phase. These are very similar to real stochastic processes, since any complex number $z = x + iy$ can simply be written as two real numbers. However, the complex notation is often more compact. We define $z_i = x_i + iy_i$, $z_{i+M} = z_i^* = x_i - iy_i$, and similarly

$$\begin{aligned}\frac{\partial}{\partial z_i} &= \frac{1}{2}\left[\frac{\partial}{\partial x_i} + \frac{\partial}{i\partial y_i}\right],\\ \frac{\partial}{\partial z_i^*} &= \frac{1}{2}\left[\frac{\partial}{\partial x_i} - \frac{\partial}{i\partial y_i}\right].\end{aligned} \tag{7.43}$$

Suppose the complex Fokker–Planck equation has the form

$$\frac{\partial}{\partial t}P(\mathbf{z},t) = \left\{-\sum_{i=1}^{2M}\frac{\partial}{\partial z_i}A_i(\mathbf{z}) + \frac{1}{2}\sum_{ij=1}^{2M}\frac{\partial^2}{\partial z_i \partial z_j}D_{ij}(\mathbf{z})\right\}P(\mathbf{z},t), \tag{7.44}$$

where $\mathbf{D} = \mathbf{B}\mathbf{B}^{\mathrm{T}}$, $A_{i+M} = A_i^*$ and $B_{i+M,j} = B_{i,j}^*$. We note that this gives a real-valued differential operator, which is necessary if we wish to have a positive probability. Next, writing this in terms of real variables, we define $x_{i+M} = y_i$, and similarly

$$\begin{aligned}A_i &= A_i^x + iA_{i+M}^x = A_i^x + iA_i^y,\\ B_{ij} &= B_{ij}^x + iB_{i+M,j}^x = B_{ij}^x + iB_{i,j}^y.\end{aligned} \tag{7.45}$$

Substituting into Eq. (7.44), we find that this complex Fokker–Planck equation can be rewritten in the form of a real Fokker–Planck equation, with $2M$ real variables:

$$\frac{\partial}{\partial t}P(\mathbf{x},t) = \left\{-\sum_{i=1}^{2M}\frac{\partial}{\partial x_i}A_i^x(\mathbf{x}) + \frac{1}{2}\sum_{ij=1}^{2M}\frac{\partial^2}{\partial x_i \partial x_j}D_{ij}^x(\mathbf{x})\right\}P(\mathbf{x},t), \tag{7.46}$$

where $\mathbf{D}^x = \mathbf{B}^x\mathbf{B}^{x\mathrm{T}}$. This leads to a set of SDEs in Ito form, using real variables, with $\mathbf{z} = \mathbf{x} + i\mathbf{y}$:

$$d\mathbf{x} = \mathbf{A}^x\, dt + \mathbf{B}^x \cdot d\mathbf{W}. \tag{7.47}$$

Transforming back to the $\mathbf{z}$ variables, the complex-valued SDE has the form one would expect from naively using the real variable formalism, i.e.

$$d\mathbf{z} = \mathbf{A}\,dt + \mathbf{B}\cdot d\mathbf{W}. \tag{7.48}$$

We note that, although the complex formalism is $2M$-dimensional, there are only M independent complex equations, as the conjugate equations do not give any new behavior.

7.3.5 Complex Stratonovich process

Just as with real variables, it is often more convenient to integrate complex stochastic processes using the Stratonovich form, which requires a correction to the complex drift coefficients. We start with the Stratonovich correction term in real variables, which is

$$\Delta^s A_i^x(\mathbf{x}) = \frac{1}{2}\sum_{k=1}^{2M}\sum_{j=1}^{M'} \frac{\partial B_{ij}^x(\mathbf{x})}{\partial x_k} B_{kj}^x(\mathbf{x}). \tag{7.49}$$

After the transformation to complex variables, this results in a complex Stratonovich correction of the form

$$\begin{aligned}\Delta^s A_i(\mathbf{x}) &= \Delta^s A_i^x(\mathbf{x}) + i\Delta^s A_i^y(\mathbf{x}) \\ &= \frac{1}{2}\sum_{k=1}^{M}\sum_{j=1}^{M'}\left[\frac{\partial B_{ij}(\mathbf{x})}{\partial x_k} B_{kj}^x(\mathbf{x}) + \frac{\partial B_{ij}(\mathbf{x})}{\partial y_k} B_{kj}^y(\mathbf{x})\right].\end{aligned} \tag{7.50}$$

This can also be re-expressed in terms of complex derivatives, using Eq. (7.43), to give

$$\begin{aligned}\Delta^s A_i(\mathbf{x}) &= \frac{1}{2}\sum_{k=1}^{M}\sum_{j=1}^{M'}\left[\left(B_{kj}(\mathbf{x})\frac{\partial}{\partial z_k} + B_{kj}^*(\mathbf{x})\frac{\partial}{\partial z_k^*}\right) B_{ij}(\mathbf{x})\right] \\ &= \frac{1}{2}\sum_{k=1}^{2M}\sum_{j=1}^{M'} B_{kj}(\mathbf{x})\frac{\partial B_{ij}(\mathbf{x})}{\partial z_k}.\end{aligned} \tag{7.51}$$

In summary, even the Stratonovich correction has exactly the same form as in the real-variable case, although it is necessary in general to sum over conjugate terms. If $B_{ij}(\mathbf{x})$ is holonomic for $i \le M$, so it is only a function of z_i for $i \le M$, then half the derivative terms will vanish. The final result is analogous to the real Stratonovich correction, giving

$$\bar{A}_i(\mathbf{z}) = A_i(\mathbf{z}) - \frac{1}{2}\sum_{k=1}^{M}\sum_{j=1}^{M'} \frac{\partial B_{ij}(\mathbf{z})}{\partial z_k} B_{kj}(\mathbf{z}). \tag{7.52}$$

7.4 Phase-space representations

We now want to explore the use of phase-space methods to calculate quantum dynamics. In these equations, the random events occurring in quantum theory are translated into stochastic differential equations. In optics, phase-space methods use coherent states, which comprise an over-complete, nonorthogonal basis. As originally pointed out by Schrödinger, the dynamics of certain types of motion in a coherent-state basis are inherently classical. Thus, using only a classical set of degrees of freedom, we can describe the complete quantum dynamics of certain linear systems. Techniques of this type were originally developed by Wigner, Moyal and Husimi, then extended further by Glauber, Sudarshan, Lax, Haken and many others. These techniques are called phase-space representations.

Such methods map a state in a quantum Hilbert space into a quasi-probability distribution over an equivalent *classical* phase space. Most commonly, the technique is applied to a quantum field mode operator, so that a single bosonic mode degree of freedom generates a single complex amplitude. The Glauber P-function is a typical example of this type of representation and has proved most useful in treating the approximate quantum dynamics of lasers, which are strongly coupled to damping reservoirs. This technique was introduced in Chapter 5, and is defined in Eq. (5.39) of Section 5.3. It was shown, however, that the P-function is very singular for nonclassical states. Therefore, other techniques are preferable for calculating quantum dynamics when the quantum states become inherently nonclassical.

7.4.1 Classical and nonclassical phase space

The phase-space representations of Glauber, Wigner and Husimi correspond to different operator orderings, but, as we will see, they have similar problems. There is no stochastic process in a classical phase space that can represent a quantum system with a nonlinear coupling. This greatly reduces the utility of these techniques for treating the real-time dynamics of interacting quantum systems. Despite this, these methods are often employed as approximate techniques, under conditions where the additional (nondiffusive) terms that arise can be neglected.

A common approach to using a phase-space representation involves analyzing the problem in terms of the Wigner function. To obtain probabilistic results, it is necessary to truncate third- and higher-order derivatives in the corresponding evolution equation for the Wigner distribution. The truncation approximation generates c-number stochastic equations for the field. These closely resemble classical equations. However, in most cases, the truncation approximation restricts this technique to very large photon numbers. Provided that truncation is permitted, this method is very simple to use, having a classical-like phase space with added noise.

A more rigorous approach uses the idea of dimension doubling. Think of this as a classical universe, but with twice as many dimensions as the usual classical universe we are familiar with, in order to give space to define quantum superposition states in addition to the classical states. This technique is called the positive P-representation. In this case, the fundamental expansion uses off-diagonal coherent-state projection operators, and a positive distribution function always exists. By mapping quantum-state evolution into a (nonclassical) phase space of double the classical dimension, it is possible to obtain a positive definite probability distribution with a positive diffusion. This can be numerically simulated using stochastic methods. A necessary technical requirement is that the distribution is strongly bounded at infinity, so that certain boundary terms vanish.

The positive P-representation is able to treat all types of nonclassical radiation as a positive distribution on a nonclassical phase space. Using this method, the operator equations are transformed to complex stochastic equations, which only involve c-number (commuting) variables. These equations reproduce the *classical* equations in the limit of small quantum effects, or large photon number. With this technique, quantum dynamics can be numerically simulated using stochastic equations. We note that, as in any numerical method, there can be computational errors. Large quantum noise effects will lead to an increased sampling error in finite ensembles of stochastic trajectories. In the worst case, the distribution is so spread out in phase space that 'boundary term corrections', due to partial integrations in the derivation of the representation, can start to appear – this issue is discussed in later sections.

In summary, it is often more useful to employ phase-space distributions or operator representations, rather than other more traditional methods, to calculate observable quantities. Ideally, the resulting distributions should have a positive definite character. If so, they can be regarded as probability distributions, which can possibly be numerically simulated using stochastic equations. However, obtaining a stochastic dynamics equivalent to quantum theory is a rather more subtle problem than just finding an equivalent positive probability distribution.

7.4.2 General phase-space representations

We now introduce a class of general phase-space representations for an M-mode bosonic quantum system. The s-ordering approach, introduced by Cahill and Glauber, associates a characteristic function with phase-space mappings including the P-representation, Wigner representation ($W(\boldsymbol{\alpha})$) and Husimi Q-representation ($Q(\boldsymbol{\alpha})$). First, the phase-space representation for an arbitrary operator ordering is defined using a distribution $P_s(\boldsymbol{\alpha})$ over a nonorthogonal operator basis:

$$\rho = \int P_s(\boldsymbol{\alpha})\hat{\lambda}_s(\boldsymbol{\alpha})\, d^M\boldsymbol{\alpha}. \tag{7.53}$$

Here $\boldsymbol{\alpha} = (\alpha_1, \ldots, \alpha_M)$ is an M-dimensional complex vector of mode amplitudes, and $\hat{\lambda}_s(\boldsymbol{\alpha})$ is a corresponding basis element in the space of operators acting on $\mathcal{H}$. To define this in detail, we use general orderings of the quantum characteristic function that was defined earlier in Section 5.3.2. The bosonic quantum characteristic function for arbitrary orderings is a c-number function given by

$$C_s(\boldsymbol{\zeta}) = \mathrm{Tr}[\rho \hat{C}_s(\boldsymbol{\zeta})]. \tag{7.54}$$

Here $\boldsymbol{\zeta}$ is another M-dimensional complex vector, and the operators $\hat{C}_s(\boldsymbol{\zeta})$ are s-ordered characteristic function generators, which are the Fourier transforms of $\hat{\lambda}_s(\boldsymbol{\alpha})$:

$$\hat{C}_s(\boldsymbol{\zeta}) = \,:\exp[\boldsymbol{\zeta} \cdot \mathbf{a}^\dagger - \boldsymbol{\zeta}^* \cdot \mathbf{a}]:_s\,. \tag{7.55}$$

The notation $: \ :_s$ indicates an s-ordered product, where $s = 1, \ 0, \ -1$ for normal, symmetric and anti-normal ordering, respectively. To explain what this means, we write explicitly that

$$\begin{aligned} \hat{C}_1(\boldsymbol{\zeta}) &= \exp[\boldsymbol{\zeta} \cdot \mathbf{a}^\dagger] \exp[-\boldsymbol{\zeta}^* \cdot \mathbf{a}], \\ \hat{C}_0(\boldsymbol{\zeta}) &= \exp[\boldsymbol{\zeta} \cdot \mathbf{a}^\dagger - \boldsymbol{\zeta}^* \cdot \mathbf{a}], \\ \hat{C}_{-1}(\boldsymbol{\zeta}) &= \exp[-\boldsymbol{\zeta}^* \cdot \mathbf{a}] \exp[\boldsymbol{\zeta} \cdot \mathbf{a}^\dagger]. \end{aligned} \tag{7.56}$$

From the Baker–Hausdorff theorem, all these characteristic functions are closely related to each other, since, defining the generic characteristic function $\hat{C}(\boldsymbol{\zeta}) \equiv \hat{C}_0(\boldsymbol{\zeta})$ as the symmetric case:

$$\hat{C}_s(\boldsymbol{\zeta}) = e^{s|\boldsymbol{\zeta}|^2/2} \hat{C}(\boldsymbol{\zeta}). \tag{7.57}$$

The characteristic functions of phase-space basis elements can be identified as

$$\mathrm{Tr}[\hat{\Lambda}_s(\boldsymbol{\alpha}) \hat{C}_s(\boldsymbol{\zeta})] = e^{\boldsymbol{\zeta} \cdot \boldsymbol{\alpha}^* - \boldsymbol{\zeta}^* \cdot \boldsymbol{\alpha}}. \tag{7.58}$$

The corresponding distributions are labeled P_s for the three different orderings, and hence from the phase-space expansion the characteristic functions can be calculated as

$$C_s(\boldsymbol{\zeta}) = \int d\boldsymbol{\alpha} \, P_s(\boldsymbol{\alpha}) e^{\boldsymbol{\zeta} \cdot \boldsymbol{\alpha}^* - \boldsymbol{\zeta}^* \cdot \boldsymbol{\alpha}}. \tag{7.59}$$

Using the complex inverse Fourier transform relation of Eq. (5.46) then gives an expression for the distribution function:

$$P_s(\boldsymbol{\alpha}) = \frac{1}{\pi^{2M}} \int d\boldsymbol{\zeta} \, e^{\boldsymbol{\zeta}^* \cdot \boldsymbol{\alpha} - \boldsymbol{\zeta} \cdot \boldsymbol{\alpha}^*} \, C_s(\boldsymbol{\zeta}). \tag{7.60}$$

In the normally ordered case of $s = 1$, this is simply the multi-dimensional version of the Glauber P-function, already introduced using characteristic functions in Section 5.3.2. More general, dimension-doubled representations can be treated as well. For example, the positive P-representation is a normally ordered distribution that is always positive for *any* density matrix. This is a function of a double-dimensional phase space, $P_+(\boldsymbol{\alpha}, \boldsymbol{\beta})$. In fact, any ordering can be used to obtain a positive distribution $P_{+s}(\boldsymbol{\alpha}, \boldsymbol{\beta})$ in a double-dimensional phase space, with the s-ordered characteristic function

$$C_s(\boldsymbol{\zeta}) = \int d^M\boldsymbol{\alpha} \, d^M\boldsymbol{\beta} \, P_{+s}(\boldsymbol{\alpha}, \boldsymbol{\beta}) e^{\boldsymbol{\zeta} \cdot \boldsymbol{\beta} - \boldsymbol{\zeta}^* \cdot \boldsymbol{\alpha}}. \tag{7.61}$$

In the following sections, we give the detailed properties of representations that use these different orderings.

7.5 Wigner and *Q*-representations

The earliest approach to quantum phase space is the Wigner representation. The Wigner function was originally introduced by E. P. Wigner in 1932 to facilitate the calculation of symmetrically ordered moments. In quantum optics, it has the effect of introducing vacuum noise equivalent to half a photon into each mode of the field. The Wigner function is thus initially more spread out on its phase space than the Glauber P-function of Eq. (5.39). In fact, the Wigner function can be written as a Gaussian convolution of the Glauber P-function. Similarly, the Husimi Q-representation is a Gaussian convolution of the Wigner function. This is even more spread out still, and in fact is always a positive function. In this section, we will use a single-mode theory for simplicity. These results can all be generalized to multi-mode cases.

7.5.1 Symmetric ordering

The Wigner function $W(\alpha)$ for a single-mode field is a function of a single complex variable, α, and is defined from Eq. (7.60) as

$$W(\alpha) = \frac{1}{\pi^2} \int d^2\zeta \, \mathrm{Tr}(\rho e^{\zeta a^\dagger - \zeta^* a}) e^{\zeta^* \alpha - \zeta \alpha^*}. \tag{7.62}$$

The Wigner distribution can be used, as we mentioned, to evaluate symmetrically ordered products of creation and annihilation operators. Let us prove that statement.

A symmetrically ordered product of creation and annihilation operators is defined in the following way. A product of m annihilation operators and n creation operators can be ordered in $(n+m)!/n!m!$ ways. The symmetrically ordered product of these operators, which we shall denote by $:(a^\dagger)^n a^m :_0$ is just the average of all of these different orderings. For example,

$$\begin{aligned} :a^\dagger a :_0 &= \tfrac{1}{2}(a^\dagger a + a a^\dagger), \\ :a^\dagger a^2 :_0 &= \tfrac{1}{3}(a^\dagger a^2 + a a^\dagger a + a^2 a^\dagger). \end{aligned} \tag{7.63}$$

Next, we note that

$$\int d^2\alpha \, e^{\zeta\alpha^* - \zeta^*\alpha} W(\alpha) = \mathrm{Tr}(\rho e^{\zeta a^\dagger - \zeta^* a}), \tag{7.64}$$

where we have made use of the identity

$$\delta^{(2)}(\alpha) = \frac{1}{\pi^2} \int d^2\beta \, e^{\alpha\beta^* - \alpha^*\beta}. \tag{7.65}$$

Now, we also have that

$$e^{\zeta a^\dagger - \zeta^* a} = \sum_{m,n=0}^{\infty} \frac{1}{m!n!} \zeta^m (-\zeta^*)^n \{(a^\dagger)^m a^n\}_s. \tag{7.66}$$

If we now expand both exponentials in Eq. (7.64) in powers of ζ and ζ^*, and equate the coefficients of $\zeta^m(-\zeta^*)^n$ on both sides, we find that

$$\langle :(a^\dagger)^n a^m :_0 \rangle = \int d^2\alpha \, W(\alpha)(\alpha^*)^m \alpha^n. \tag{7.67}$$

7.5.2 Quadrature components

As an example of this, the Wigner function can be used to easily evaluate moments of the quadrature components of the field. Remember from Eq. (5.25) that $X(\phi) = e^{i\phi}a^\dagger + e^{-i\phi}a$. We then have that

$$X^n(\phi) = \sum_{m=0}^{n} e^{i(n-m)\phi} e^{-im\phi} \frac{n!}{m!(n-m)!} \{(a^\dagger)^{n-m} a^m\}_s. \tag{7.68}$$

This implies that

$$\langle X^n(\phi)\rangle = \int d^2\alpha \, W(\alpha)(e^{i\phi}\alpha^* + e^{-i\phi}\alpha)^n. \tag{7.69}$$

Because homodyne measurements measure field quadrature components, the Wigner function is very useful in determining the results of homodyne measurements and their fluctuations.

The Wigner function can be expressed in terms of the Glauber P-function, $P(\alpha)$, as

$$W(\alpha) = \frac{2}{\pi} \int d^2\beta \, P(\beta) e^{-2|\alpha-\beta|^2}. \tag{7.70}$$

This formula reflects the fact that the Wigner function is usually less singular than the corresponding positive P-function. For example, if the field is initially in the coherent state $|\alpha_c\rangle$, i.e. $\rho = |\alpha_c\rangle\langle\alpha_c|$, the two distributions are

$$P(\alpha) = \delta^2(\alpha - \alpha_c) \tag{7.71}$$

and

$$W(\alpha) = \frac{2}{\pi} e^{-2|\alpha-\alpha_c|^2}. \tag{7.72}$$

This relationship is easily proved by Fourier-transforming the equation relating characteristic functions of different orderings, given in Eq. (7.57).

For nonlinear Hamiltonians, the corresponding Wigner evolution equation typically does not contain any second-order derivatives, but rather third-order derivatives, which invalidates any interpretation in terms of a Fokker–Planck equation. As it is expected that the third-order terms will only be significant when quantum effects are large, i.e. when the characteristic photon number $\bar{n}$ is small, or in the limit of large evolution times, it is possible to recover a stochastic interpretation by neglecting (or 'truncating') the third-order derivative terms, in an approximation valid at large particle number. Thus an approximate stochastic equation can be derived for the field.

The chief advantages of the truncated Wigner method are its classical appearance (after truncation), and direct correspondence to homodyne measurements. The chief disadvantage is its approximate nature, since it involves a truncation of quantum correlations higher than second order. Its lack of correspondence to normally ordered measurements like direct photo-detection is less of a problem, as this can be corrected for.

7.5.3 Q-representation

The Husimi Q-function is defined by coherent-state diagonal matrix elements of the density operator; in the single-mode case it is given by

$$Q(\alpha) = \frac{1}{\pi}\langle\alpha|\rho|\alpha\rangle. \tag{7.73}$$

The Q-function is explicitly always positive semi-definite, since the density matrix is a positive semi-definite operator, that is, it only has positive or zero eigenvalues. This facilitates the calculation of expectation values of anti-normally ordered operator products in terms of a probability distribution.

We next wish to show that this corresponds to the definition in terms of characteristic functions, given in Eq. (7.59). This follows from the definition of the anti-normally ordered characteristic function, which in the one-mode case is

$$C_{-1}(\zeta) = \mathrm{Tr}[\rho\exp(-\zeta a)\exp(\zeta a^{\dagger})]. \tag{7.74}$$

Next, the expansion of the identity operator in coherent states from Eq. (5.37) is inserted between the exponential terms. Using the cyclic properties of the trace, this becomes

$$C_{-1}(\zeta) = \frac{1}{\pi}\int d^2\alpha\,\langle\alpha|\exp(\zeta a^{\dagger})\rho\exp(-\zeta a)\,|\alpha\rangle. \tag{7.75}$$

Finally, using the eigenvalue equation for coherent states, and the definition of $Q(\alpha)$ given above, this becomes

$$C_{-1}(\zeta) = \int d^2\alpha\,\exp(\zeta\alpha^* - \zeta^*\alpha)Q(\alpha), \tag{7.76}$$

which is simply the defining equation of an anti-normally ordered phase-space representation, from Eq. (7.59).

A product of creation and annihilation operators is anti-normally ordered if all of the annihilation operators are to the left, and all of the creation operators are to the right. We can show that the Q-function can be used to evaluate anti-normally ordered products as follows:

$$\begin{aligned}\langle a^n(a^{\dagger})^m\rangle &= \mathrm{Tr}(\rho a^n(a^{\dagger})^m)\\ &= \frac{1}{\pi}\int d^2\alpha\,\langle\alpha|(a^{\dagger})^m\rho a^n|\alpha\rangle\\ &= \int d^2\alpha\,\alpha^n(\alpha^*)^m Q(\alpha).\end{aligned} \tag{7.77}$$

The Q-function is always nonnegative and in this sense has properties similar to an actual probability density. The reason why it is not a true probability density, but is a quasi-probability distribution function, is due to the fact that the quantities $\Re(\alpha)$ and $\Im(\alpha)$, which are the mean values of two noncommuting operators, i.e. $(a+a^{\dagger})/2$ and $(a-a^{\dagger})/2i$, respectively, are not simultaneously measurable with precise values. This means that the localized Q-function $\delta^2(\alpha-\alpha_c)$ cannot exist physically.

7.5.4 Husimi, Glauber and Wigner representations

Using the relationship between the characteristic functions as in the case of the Wigner function, the Q-function can be expressed in terms of the Glauber P-representation by a convolution,

$$Q(\alpha) = \frac{1}{\pi}\int d^2\beta\, P(\beta)e^{-|\alpha-\beta|^2}, \tag{7.78}$$

while its relation to the Wigner function is

$$Q(\alpha) = \frac{2}{\pi}\int d^2\beta\, W(\beta)e^{-2|\alpha-\beta|^2}. \tag{7.79}$$

Surprisingly, except in trivial linear cases, the Q-function usually does not satisfy a normal Fokker–Planck equation with positive definite diffusion even though the distribution is itself always positive definite. Thus, it is less useful for quantum dynamical calculations than one might expect at first.

7.6 Nonclassical representations

We shall now treat the generalized P-representations of the density matrix, ρ, which exist on a phase space of double the classical dimension. We term these 'nonclassical representations', as the phase space has no classical analog. First, it is useful to understand the Glauber R-representation, which was historically the first approach of this type, and has a useful existence theorem. We then move on to the generalized P-representations, which have more practical application in calculations.

7.6.1 Glauber R-representation

Glauber introduced a coherent-state representation that used a doubled phase-space dimension. This is called the R-representation, and is defined on a phase space of four real dimensions in the case of a single mode. This method relies for its existence on the coherent-state resolution of the identity in Eq. (5.37), which leads to

$$\begin{aligned}\rho &= \hat{I}\rho\hat{I} \\ &= \frac{1}{\pi^2}\iint d^2\alpha\, d^2\beta\, |\alpha\rangle\langle\alpha|\rho|\beta\rangle\langle\beta|.\end{aligned} \tag{7.80}$$

This expansion can be rewritten in Glauber's notation as

$$\rho = \frac{1}{\pi^2}\iint d^2\alpha\, d^2\beta\, R(\alpha^*,\beta)e^{-(|\alpha|^2+|\beta|^2)/2}|\alpha\rangle\langle\beta|, \tag{7.81}$$

where, clearly,

$$R(\alpha^*,\beta) = \langle\alpha|\rho|\beta\rangle e^{(|\alpha|^2+|\beta|^2)/2}. \tag{7.82}$$

Since we have an explicit construction, this representation always exists, albeit on a higher-dimensional space. It is nonsingular, and also analytic in α^* and β, which means that it generically has complex values. The analytic property is best seen from expanding the coherent states in Eq. (7.82) using the number-state expansion defined in Eq. (5.30), which leads to

$$R(\alpha^*, \beta) = \sum_{n,m} \frac{\langle n|\rho|m\rangle}{\sqrt{n!m!}} (\alpha^*)^n \beta^m. \tag{7.83}$$

This representation, being complex-valued, has no probabilistic interpretation. We will see in the next section that there are other forms of phase-space representation that are positive-valued and have a probabilistic interpretation. These are called positive P-representations.

The R-representation can be used to obtain the normally ordered characteristic function for any density matrix. We first use the eigenvalue equation for a coherent state, Eq. (5.31), and its conjugate, together with the cyclic properties of trace and the inner-product result of (5.35), to give

$$\mathrm{Tr}(e^{\zeta a^\dagger} e^{-\zeta^* a} |\alpha\rangle\langle\beta|) = e^{\zeta\beta^* - \zeta^*\alpha - (|\alpha|^2 + |\beta|^2)/2 + \beta^*\alpha}. \tag{7.84}$$

Next, we insert this result into the definition of the characteristic function, with ρ expressed using the R-representation, to obtain

$$\begin{aligned} C_1(\zeta) &= \mathrm{Tr}(\rho\, e^{\zeta a^\dagger} e^{-\zeta^* a}) \\ &= \frac{1}{\pi^2} \iint d^2\alpha\, d^2\beta\, R(\alpha^*, \beta) e^{(\zeta+\alpha)\beta^* - \zeta^*\alpha - |\alpha|^2 - |\beta|^2}. \end{aligned} \tag{7.85}$$

We can now use the complex function identity that, for any entire analytic function $f(\alpha)$,

$$f(\alpha') = \frac{1}{\pi} \int f(\alpha) \exp(-|\alpha - \alpha'|^2)\, d^2\alpha', \tag{7.86}$$

to carry out the integral over the β variable. This gives us a more elegant expression of the quantum characteristic function:

$$C_1(\zeta) = \frac{1}{\pi} \int d^2\alpha\, R(\alpha^*, \zeta + \alpha) e^{-\zeta^*\alpha - |\alpha|^2}. \tag{7.87}$$

While the R-representation is always well-defined, the fact that it has complex values means that it is not very useful for probabilistic sampling methods. Nevertheless, it has many useful properties that allow it to be employed to prove other results. We will see this in the next section.

7.6.2 Positive *P*-representation

We shall now treat the generalized P-representations of the density matrix, ρ, which, like the Glauber R-representation, exist on a phase space of double the classical dimension. Since the applications of this are typically for complex, multi-mode problems, we shall move immediately to a general multi-mode quantum state. These are defined using an M-mode coherent state $|\boldsymbol{\alpha}\rangle$, which generalizes Eq. (5.30) to a Hilbert space of M bosonic modes. Here the state $|\boldsymbol{\alpha}\rangle$ is a multi-mode coherent state, defined as

$$|\boldsymbol{\alpha}\rangle = \exp\left(\sum_{n=1}^{M} \alpha_n a_n^\dagger - |\boldsymbol{\alpha}|^2/2\right)|0\rangle, \tag{7.88}$$

where $|0\rangle$ is the vacuum state, and $a_n^\dagger$ is the creation operator for the nth boson mode of the system.

Generalized P-representations are defined as follows:

$$\rho = \iint d\mu(\boldsymbol{\alpha}, \boldsymbol{\beta})\, P(\boldsymbol{\alpha}, \boldsymbol{\beta})\hat{\lambda}_+(\boldsymbol{\alpha}, \boldsymbol{\beta}). \tag{7.89}$$

Here $d\mu(\boldsymbol{\alpha}, \boldsymbol{\beta})$ is an integration measure, the choice of which allows one to define different possible P-representations. The operator term or kernel $\hat{\lambda}_+(\boldsymbol{\alpha}, \boldsymbol{\beta})$ at the heart of the generalized P-representation is a projection operator onto an off-diagonal pair of coherent states:

$$\hat{\lambda}_+(\boldsymbol{\alpha}, \boldsymbol{\beta}) \equiv \frac{|\boldsymbol{\alpha}\rangle\langle\boldsymbol{\beta}^*|}{\langle\boldsymbol{\beta}^*|\boldsymbol{\alpha}\rangle}. \tag{7.90}$$

The complex vector $\boldsymbol{\alpha} = \boldsymbol{\alpha}^x + i\boldsymbol{\alpha}^y$ corresponds to one classical complex mode amplitude for each of the M modes present, while $\boldsymbol{\beta} = \boldsymbol{\beta}^x + i\boldsymbol{\beta}^y$ corresponds to another set of mode amplitudes, with the physical state being expressed in terms of both. This is the most general coherent-state expansion possible.

7.6.3 Types of generalized P-function

Glauber–Sudarshan P-function

The earliest P-representation is the Glauber–Sudarshan P-function, which we have already met. This corresponds to the density operator expansion in diagonal coherent-state projection operators, and can be obtained by choosing the integration measure as

$$d\mu(\boldsymbol{\alpha}, \boldsymbol{\beta}) = \delta^{2M}(\boldsymbol{\alpha} - \boldsymbol{\beta}^*)\, d^{2N}\alpha\, d^{2N}\beta, \tag{7.91}$$

so that we can immediately obtain the multi-dimensional P-function as

$$\rho = \int d^{2M}\alpha\, P(\boldsymbol{\alpha})|\boldsymbol{\alpha}\rangle\langle\boldsymbol{\alpha}|. \tag{7.92}$$

The Glauber–Sudarshan P-function exists and is most fruitfully employed for describing the quantum statistical properties of thermal light fields and coherent fields. However, for fields with nonclassical photon statistics, $P(\boldsymbol{\alpha})$ only exists as a highly singular generalized function, similar to a derivative of a delta-function. Alternative P-representation functions, the positive P-function and the complex P-function, avoid these problems. These can be employed successfully in cases where nonclassical photon statistics arise.

Complex P-representation

Other types of generalized P-representation are obtained by choosing an appropriate integration measure. The complex P-representation is like other generalized P-representations in the way it handles the doubling of the number of phase-space variables. In this case, the measure is chosen as a complex, multi-dimensional line integral or contour measure, as follows:

$$d\mu(\boldsymbol{\alpha}, \boldsymbol{\beta}) = d^M\boldsymbol{\alpha}\, d^M\boldsymbol{\beta}. \tag{7.93}$$

The corresponding integrals are contour integrals, with the integrations over each component of the independent complex vectors $\boldsymbol{\alpha}$ and $\boldsymbol{\beta}$ to be carried out on individual contours in complex planes. In this representation, complex values of $P(\boldsymbol{\alpha}, \boldsymbol{\beta})$ may occur. This method is mainly used for obtaining exact analytic solutions, where use can be made of the analytic properties of contour integrals.

Positive P-representation

The positive P-representation is defined by choosing the integration measure as a volume measure in the extended phase space

$$d\mu(\boldsymbol{\alpha}, \boldsymbol{\beta}) = d^{2M}\boldsymbol{\alpha}\, d^{2M}\boldsymbol{\beta}. \tag{7.94}$$

This leads to the following expansion of the density operator ρ using coherent-state projectors:

$$\rho = \iint P(\boldsymbol{\alpha}, \boldsymbol{\beta})\hat{\lambda}_+(\boldsymbol{\alpha}, \boldsymbol{\beta})\, d^{2M}\boldsymbol{\alpha}\, d^{2M}\boldsymbol{\beta}. \tag{7.95}$$

The modes represented by $\boldsymbol{\alpha}$ and $\boldsymbol{\beta}$ can correspond to all possible modes of excitation of the system, whether they correspond to the photon field or the damping reservoirs; however, it is most usual to eliminate the variables associated with the reservoirs. The positive nature of $P(\boldsymbol{\alpha}, \boldsymbol{\beta})$ means that, whenever the corresponding evolution equation is of the Fokker–Planck type, then an equivalent set of Ito stochastic differential equations can be written down which describe the motion of the coordinates $(\boldsymbol{\alpha}, \boldsymbol{\beta})$ of a fictitious particle whose movement in the phase space is governed by the stochastic equation.

7.6.4 Observables and measurements

The next property of this representation that is needed is the property of measurement, that is, how to extract information about observables from the distribution function. In order to calculate an operator expectation value, we can use the fact that there is a direct correspondence between the moments of the distribution and the normally ordered operator products. This correspondence comes directly from the fact that coherent states are eigenstates of the annihilation operator.

We first note that any normally ordered observable results in a trace over the kernel function of the representation. We follow a similar technique to that used for the diagonal

Glauber–Sudarshan distribution, in Eq. (5.41). We again suppose that $O_N(\mathbf{a}, \mathbf{a}^\dagger)$ is an operator, which is a function of $\mathbf{a}$ and $\mathbf{a}^\dagger$, expressed in normally ordered form, so that

$$O_N(\mathbf{a}, \mathbf{a}^\dagger) = \sum_{\mathbf{m},\mathbf{n}} c_{\mathbf{mn}} \prod_{j=1}^{M} (a_j^\dagger)^{m_j} \prod_{j=1}^{M} a_j^{n_j}. \tag{7.96}$$

The expectation value of O_N is then given by

$$\begin{aligned}\langle O_N(\mathbf{a}, \mathbf{a}^\dagger)\rangle &= \mathrm{Tr}(O_N(\mathbf{a}, \mathbf{a}^\dagger)\rho) \\ &= \iint d^{2M}\boldsymbol{\alpha}\, d^{2M}\boldsymbol{\beta}\, P(\boldsymbol{\alpha}, \boldsymbol{\beta}) \\ &\quad \times \sum_{\mathbf{m},\mathbf{n}} c_{\mathbf{mn}} \mathrm{Tr}\left[\prod_{j=1}^{M} \alpha_j^{n_j} \frac{|\boldsymbol{\alpha}\rangle\langle\boldsymbol{\beta}^*|}{\langle\boldsymbol{\beta}^*|\boldsymbol{\alpha}\rangle} \prod_{j=1}^{M} \beta_j^{m_j} \cdots\right] \\ &= \langle O_N(\boldsymbol{\alpha}, \boldsymbol{\beta})\rangle_P. \end{aligned} \tag{7.97}$$

Here we have used the eigenvalue property for coherent states, Eq. (5.31), together with its conjugate. The trace is eliminated by using the fact that the coherent-state projection operator has unit trace, which follows from the cyclic properties of the trace itself,

$$\mathrm{Tr}\left[\frac{|\boldsymbol{\alpha}\rangle\langle\boldsymbol{\beta}^*|}{\langle\boldsymbol{\beta}^*|\boldsymbol{\alpha}\rangle}\right] = \mathrm{Tr}\left[\frac{\langle\boldsymbol{\beta}^*|\boldsymbol{\alpha}\rangle}{\langle\boldsymbol{\beta}^*|\boldsymbol{\alpha}\rangle}\right] = 1. \tag{7.98}$$

What this procedure has done is similar to the result for the Glauber–Sudarshan P-representation. We can express a quantum expectation value in terms of c-number quantities: a probability related to the density matrix $P(\boldsymbol{\alpha}, \boldsymbol{\beta})$, and a complex function for the operator $O_N(\boldsymbol{\alpha}, \boldsymbol{\beta})$. The advantage of this approach, as we will show in the next section, is that $P(\boldsymbol{\alpha}, \boldsymbol{\beta})$ always exists as a positive probability. For this reason, we use the notation

$$\langle O_N(\boldsymbol{\alpha}, \boldsymbol{\beta})\rangle_P \equiv \int d^{2M}\boldsymbol{\alpha}\, d^{2M}\boldsymbol{\beta}\, P(\boldsymbol{\alpha}, \boldsymbol{\beta}) O_N(\boldsymbol{\alpha}, \boldsymbol{\beta}) \tag{7.99}$$

to indicate a probabilistic average over the P-function, as distinct from the usual quantum average.

A particular example is the overall integral of the total probability, which must be unity, since

$$1 = \langle 1\rangle_P = \iint d^{2M}\boldsymbol{\alpha}\, d^{2M}\boldsymbol{\beta}\, P(\boldsymbol{\alpha}, \boldsymbol{\beta}). \tag{7.100}$$

Similar results hold for time-ordered, normally ordered multi-time correlation functions, as found in spectral measurements. The important property here is that there is a direct relationship between the moments of the representation, and the usual ensemble averages obtained from a photo-detector measurement, which gives normally ordered moments. An example of this is the quantum characteristic function in the multi-mode case, which from Eq. (7.97) one can immediately obtain to be

$$\begin{aligned} C_1(\boldsymbol{\zeta}) &\equiv \mathrm{Tr}(\rho e^{\boldsymbol{\zeta}\cdot\mathbf{a}^\dagger} e^{-\boldsymbol{\zeta}^*\cdot\mathbf{a}}) \\ &= \langle e^{\boldsymbol{\zeta}\cdot\boldsymbol{\beta}-\boldsymbol{\zeta}^*\cdot\boldsymbol{\alpha}}\rangle_P. \end{aligned} \tag{7.101}$$

If other types of moment are needed, then the operator commutators must be used to calculate them.

7.6.5 Existence theorem

We will show that the distribution function $P(\boldsymbol{\alpha}, \boldsymbol{\beta})$ can always be chosen to be a positive function defined on the $4M$-dimensional phase space spanned by the complex coordinates $\boldsymbol{\alpha}$ and $\boldsymbol{\beta}$, which are to be regarded as distinct phase-space variables. This includes all diagonal and off-diagonal coherent-state components of the density matrix. A nonclassical field necessarily corresponds to a superposition of coherent states. These are represented by the off-diagonal terms in the coherent-state expansion, in which $\boldsymbol{\alpha} \neq \boldsymbol{\beta}^*$.

Since the coherent states comprise a complete basis set for the quantum states of a radiation field, it is not surprising that a representation exists for some distribution function P. It is less obvious that a positive function exists in all cases, but this can always be constructed, even for highly nonclassical fields.

Theorem 7.1 *A positive P-distribution exists for any density matrix.*

Consider the following ansatz for the positive P-distribution:

$$P_+(\boldsymbol{\alpha}, \boldsymbol{\beta}) = \frac{1}{(2\pi)^{2M}} e^{-|\boldsymbol{\alpha}-\boldsymbol{\beta}^*|^2/4} \left\langle \frac{\boldsymbol{\alpha} + \boldsymbol{\beta}^*}{2} \right| \rho \left| \frac{\boldsymbol{\alpha} + \boldsymbol{\beta}^*}{2} \right\rangle. \tag{7.102}$$

The above distribution is clearly positive, although it is not unique, owing to the nonorthogonal basis used. We will now give a proof that this is a positive P-distribution corresponding to the density matrix ρ, using the well-known result that the characteristic function uniquely defines any density matrix. The Glauber–Sudarshan R-distribution exists as a well-defined function for any density matrix. From Eq. (7.87), the characteristic function for a multi-mode system is

$$C_1(\boldsymbol{\zeta}) = \frac{1}{\pi^M} \int d^{2M}\boldsymbol{\alpha}\, R(\boldsymbol{\alpha}^*, \boldsymbol{\zeta} + \boldsymbol{\alpha}) e^{-\boldsymbol{\zeta}^* \cdot \boldsymbol{\alpha} - |\boldsymbol{\alpha}|^2}. \tag{7.103}$$

We now evaluate the characteristic function using the ansatz of Eq. (7.102), giving the result that

$$C_{P+}(\boldsymbol{\zeta}) = \int d^{2M}\boldsymbol{\alpha}\, d^{2M}\boldsymbol{\beta}\, e^{\boldsymbol{\zeta} \cdot \boldsymbol{\beta} - \boldsymbol{\zeta}^* \cdot \boldsymbol{\alpha}} P_+(\boldsymbol{\alpha}, \boldsymbol{\beta}). \tag{7.104}$$

Provided the two characteristic function results agree with each other, this shows that the ansatz gives a positive P-distribution that generates the density matrix ρ, as required. To achieve this, we must expand the ansatz for P_+ in terms of the density matrix ρ. This, in turn, must be expressed using the R-distribution so that we can compare the two results.

The proof is like peeling an onion, as three successive $2M$-dimensional integrals must be carried out to uncover the final result. We hope no tears will result! First, we introduce a variable change, through the definitions $\boldsymbol{\gamma} = (\boldsymbol{\alpha} + \boldsymbol{\beta}^*)/2$ and $\boldsymbol{\delta} = (\boldsymbol{\alpha} - \boldsymbol{\beta}^*)/2$. The Jacobian for the transformation gives us that $d^{2M}\boldsymbol{\alpha}\, d^{2M}\boldsymbol{\beta} = 4^M\, d^{2M}\boldsymbol{\gamma}\, d^{2M}\boldsymbol{\delta}$. Inserting the ansatz for

P_+ into Eq. (7.104), we obtain the characteristic function of the ansatz in the form

$$C_{P+}(\boldsymbol{\zeta}) = \frac{1}{\pi^{2M}} \int d^{2M}\boldsymbol{\gamma}\, d^{2M}\boldsymbol{\delta}\, e^{\boldsymbol{\zeta}\cdot\boldsymbol{\beta}-\boldsymbol{\zeta}^*\cdot\boldsymbol{\alpha}-|\boldsymbol{\delta}|^2} \langle\boldsymbol{\gamma}|\rho|\boldsymbol{\gamma}\rangle. \tag{7.105}$$

Expanding the original quantum density matrix ρ with the Glauber R-distribution, using Eq. (7.81) and the coherent-state inner product, Eq. (5.35), gives

$$C_{P+}(\boldsymbol{\zeta}) = \frac{1}{\pi^{4M}} \int d^{2M}\boldsymbol{\gamma}\, d^{2M}\boldsymbol{\delta}\, d^{2M}\boldsymbol{\alpha}'\, d^{2M}\boldsymbol{\beta}' \\ \times e^{\boldsymbol{\zeta}\cdot\boldsymbol{\beta}-\boldsymbol{\zeta}^*\cdot\boldsymbol{\alpha}-|\boldsymbol{\delta}|^2-|\boldsymbol{\gamma}|^2-|\boldsymbol{\alpha}'|^2-|\boldsymbol{\beta}'|^2+\boldsymbol{\gamma}^*\cdot\boldsymbol{\alpha}'+\boldsymbol{\gamma}\cdot\boldsymbol{\beta}'^*} R(\boldsymbol{\alpha}'^*, \boldsymbol{\beta}). \tag{7.106}$$

Next, we must carry out three successive complex integrations over the phase space, to reduce this expression to the required form. We use the complex function identity that, for any entire analytic function $f(\boldsymbol{\alpha})$,

$$f(\boldsymbol{\alpha}) = \frac{1}{\pi^M} \int f(\boldsymbol{\beta}) \exp(\boldsymbol{\alpha}\boldsymbol{\beta}^* - |\boldsymbol{\beta}|^2)\, d^2\boldsymbol{\beta}. \tag{7.107}$$

Since $R(\boldsymbol{\alpha}'^*, \boldsymbol{\beta})$ is analytic in $\boldsymbol{\beta}$, we can utilize this result to integrate over $\boldsymbol{\beta}'$. At the same time, we expand $\boldsymbol{\alpha} = \boldsymbol{\gamma} + \boldsymbol{\delta}$ and $\boldsymbol{\beta} = (\boldsymbol{\gamma} - \boldsymbol{\delta})^*$, so that

$$C_{P+}(\boldsymbol{\zeta}) = \frac{1}{\pi^{3M}} \int d^{2M}\boldsymbol{\gamma}\, d^{2M}\boldsymbol{\delta}\, d^{2M}\boldsymbol{\alpha}' \\ \times e^{\boldsymbol{\zeta}\cdot(\boldsymbol{\gamma}-\boldsymbol{\delta})^*-\boldsymbol{\zeta}^*\cdot(\boldsymbol{\gamma}+\boldsymbol{\delta})-|\boldsymbol{\delta}|^2-|\boldsymbol{\gamma}|^2-|\boldsymbol{\alpha}'|^2+\boldsymbol{\gamma}^*\cdot\boldsymbol{\alpha}'} R(\boldsymbol{\alpha}'^*, \boldsymbol{\gamma}). \tag{7.108}$$

Using the same method, where $\exp(-\boldsymbol{\zeta}^* \cdot (\boldsymbol{\gamma} + \boldsymbol{\delta}))$ is the analytic function, we can now integrate over $\boldsymbol{\delta}$ to obtain

$$C_{P+}(\boldsymbol{\zeta}) = \frac{1}{\pi^{2M}} \int d^{2M}\boldsymbol{\gamma}\, d^{2M}\boldsymbol{\alpha}'\, e^{|\boldsymbol{\zeta}|^2+(\boldsymbol{\zeta}+\boldsymbol{\alpha}')\cdot\boldsymbol{\gamma}^*-\boldsymbol{\zeta}^*\cdot\boldsymbol{\gamma}-|\boldsymbol{\gamma}|^2-|\boldsymbol{\alpha}'|^2} R(\boldsymbol{\alpha}'^*, \boldsymbol{\gamma}). \tag{7.109}$$

Since $R(\boldsymbol{\alpha}'^*, \boldsymbol{\gamma}) \exp(-\boldsymbol{\zeta}^* \cdot \boldsymbol{\gamma})$ is analytic in $\boldsymbol{\gamma}$, we can finally integrate over $\boldsymbol{\gamma}$. This in turn replaces $\boldsymbol{\gamma}$ by $\boldsymbol{\zeta} + \boldsymbol{\alpha}'$, to give

$$C_{P+}(\boldsymbol{\zeta}) = \frac{1}{\pi^M} \int d^{2M}\boldsymbol{\alpha}'\, e^{-\boldsymbol{\zeta}^*\cdot\boldsymbol{\alpha}'-|\boldsymbol{\alpha}'|^2} R(\boldsymbol{\alpha}'^*, \boldsymbol{\zeta} + \boldsymbol{\alpha}') \\ = C_1(\boldsymbol{\zeta}). \tag{7.110}$$

This is sufficient to prove the required existence theorem.

7.7 Operator identities and quantum dynamics

The time evolution or quantum dynamics of the positive P-representation can be calculated, as we shall see, by applying differential operator correspondence rules to the corresponding density matrix quantum evolution equations. These result in a Fokker–Planck equation for the motion of the distribution function, provided the distribution asymptotically vanishes sufficiently rapidly as $|\boldsymbol{\alpha}|, |\boldsymbol{\beta}| \to \infty$. The next step is to transform this into the stochastic formulation of a diffusion process, which solves a Fokker–Planck equation as an ensemble average over a set of stochastic trajectories. These classical-like equations are usually simpler to treat, and can always be numerically simulated if necessary – although, as in all Monte Carlo type methods, there is a sampling error in any finite set of trajectories. Since the variable $\boldsymbol{\beta}$ plays the role of a type of Hermitian conjugate to $\boldsymbol{\alpha}$, it is sometimes useful to use the notation of $\boldsymbol{\alpha}^+$ instead of $\boldsymbol{\beta}$.

7.7.1 Elementary identities

In order to derive the identities that allow us to treat time evolution for quasi-distribution functions, it is useful to start by establishing elementary identities for coherent states. We can write the expression of the positive P-kernel, Eq. (7.90), more explicitly using the operator form of the definition of coherent states, Eq. (5.34), and the inner product result of Eq. (5.35), so that

$$\hat{\lambda}_+(\boldsymbol{\alpha}, \boldsymbol{\beta}) = e^{(\mathbf{a}^\dagger - \boldsymbol{\beta})\cdot\boldsymbol{\alpha}}|\mathbf{0}\rangle\langle\mathbf{0}|\, e^{\mathbf{a}\cdot\boldsymbol{\beta}}. \tag{7.111}$$

This demonstrates that it is an analytic (i.e. holomorphic) function over the entire $2M$-dimensional complex domain of $\boldsymbol{\alpha},\ \boldsymbol{\beta}$.

We first note that, on differentiating the operator form of the kernel given in Eq. (7.111) above with respect to α_n, we get the following result:

$$\frac{\partial}{\partial \alpha_n}\hat{\lambda}_+(\boldsymbol{\alpha}, \boldsymbol{\beta}) = [a_n^\dagger - \beta_n]\hat{\lambda}_+(\boldsymbol{\alpha}, \boldsymbol{\beta}). \tag{7.112}$$

Combining this result and its conjugate with the standard eigenvalue equation for coherent states gives the following four identities:

$$\begin{aligned} a_n^\dagger\hat{\lambda}_+ &= \left[\frac{\partial}{\partial \alpha_n} + \beta_n\right]\hat{\lambda}_+, \\ a_n\hat{\lambda}_+ &= \alpha_n\hat{\lambda}_+, \\ \hat{\lambda}_+ a_n &= \left[\frac{\partial}{\partial \beta_n} + \alpha_n\right]\hat{\lambda}_+, \\ \hat{\lambda}_+ a_n^\dagger &= \beta_n\hat{\lambda}_+. \end{aligned} \tag{7.113}$$

Since the projector is an analytic function of both α_n and β_n, we can obtain alternative identities by replacing $\partial/\partial\alpha_n$ by either $\partial/\partial\alpha_n^x$ or $\partial/i\partial\alpha_n^y$, where $\alpha_n^x = \Re(\alpha_n)$ and $\alpha_n^y = \Im(\alpha_n)$, and similarly for β_n. Thus we have that

$$\begin{aligned} a_n^\dagger\hat{\lambda}_+ &= \left[\frac{\partial}{\partial \alpha_n^x} + \beta_n\right]\hat{\lambda}_+ = \left[-i\frac{\partial}{\partial \alpha_n^y} + \beta_n\right]\hat{\lambda}_+, \\ \hat{\lambda}_+ a &= \left[\frac{\partial}{\partial \beta_n^x} + \alpha_n\right]\hat{\lambda}_+ = \left[-i\frac{\partial}{\partial \beta_n^y} + \alpha_n\right]\hat{\lambda}_+. \end{aligned} \tag{7.114}$$

This equivalence is, of course, one of the fundamental properties of a complex analytic function, and it is the flexibility that this equivalence makes possible that will allow us always to be able to have a positive diffusion matrix in the time-evolution equation for the positive P-function.

7.7.2 Time-evolution identities

Next, consider a typical time-evolution master equation of the form given in Eq. (2.6):

$$\begin{aligned} i\hbar\frac{\partial}{\partial t}\rho &= [H,\rho] \\ &= \iint d\mu(\boldsymbol{\alpha},\boldsymbol{\beta})\,P(\boldsymbol{\alpha},\boldsymbol{\beta})[H,\hat{\Lambda}_+(\boldsymbol{\alpha},\boldsymbol{\beta})]. \end{aligned} \tag{7.115}$$

Typical terms appearing in H include multinomials in annihilation and creation operators. Suppose one of the terms was the creation operator $a_n^\dagger$. Then from the identities in Eq. (7.113):

$$\begin{aligned} i\hbar\frac{\partial}{\partial t}\rho &= [a_n^\dagger,\rho] \\ &= \iint d\mu(\boldsymbol{\alpha},\boldsymbol{\beta})\,P(\boldsymbol{\alpha},\boldsymbol{\beta})[a_n^\dagger,\hat{\Lambda}_+(\boldsymbol{\alpha},\boldsymbol{\beta})] \\ &= \iint d\mu(\boldsymbol{\alpha},\boldsymbol{\beta})\,P(\boldsymbol{\alpha},\boldsymbol{\beta})\left[\frac{\partial}{\partial\alpha_n}\hat{\Lambda}_+(\boldsymbol{\alpha},\boldsymbol{\beta})\right]. \end{aligned} \tag{7.116}$$

Assuming that $P(\boldsymbol{\alpha},\boldsymbol{\beta})$ vanishes sufficiently quickly at the boundaries (faster than any power of $\boldsymbol{\alpha},\ \boldsymbol{\beta}$), it is permissible to integrate by parts with vanishing boundary terms. We can then write this as

$$i\hbar\frac{\partial}{\partial t}\rho = \iint d\mu(\boldsymbol{\alpha},\boldsymbol{\beta})\,\hat{\Lambda}_+(\boldsymbol{\alpha},\boldsymbol{\beta})\left[\frac{-\partial}{\partial\alpha_n}P(\boldsymbol{\alpha},\boldsymbol{\beta})\right]. \tag{7.117}$$

We notice here that there is a sign change in the differential term, due to the integration by parts. A characteristic property of the final equation is that it is of the form of a differential operator acting on the distribution function, which means that the total probability (integrated over the phase space) is conserved.

7.7.3 Identities for general distributions

Similar identities can be obtained for the other distribution functions as well. Thus, transformation of the master equation into an evolution equation for all the different quasi-probability functions (P, W or Q) is carried out according to the following operator identities, after integrating by parts inside the integrals over phase space and assuming vanishing boundary terms:

$$\begin{aligned} a\rho &\rightarrow \left(\alpha - \frac{s-1}{2}\frac{\partial}{\partial\beta}\right)P_s, \\ a^\dagger\rho &\rightarrow \left(\beta - \frac{s+1}{2}\frac{\partial}{\partial\alpha}\right)P_s, \\ \rho a &\rightarrow \left(\alpha - \frac{s+1}{2}\frac{\partial}{\partial\beta}\right)P_s, \\ \rho a^\dagger &\rightarrow \left(\beta - \frac{s-1}{2}\frac{\partial}{\partial\alpha}\right)P_s. \end{aligned} \tag{7.118}$$

Here P_s with $s = 1$, $s = 0$ or $s = -1$ corresponds to the Glauber–Sudarshan P-function, the Wigner function and the Q-function, respectively. For these classical phase-space representations, we define $\beta = \alpha^*$.

To describe the cases where positive or complex P-representations are employed, we take $s = 1$ and $\beta \neq \alpha^*$. In these cases where $\alpha \neq \beta^*$, we can, as we have seen, obtain multiple identities by replacing $\partial/\partial\alpha$ by either $\partial/\partial\alpha_x$ or $\partial/i\partial\alpha_y$. In the other cases, where the phase space is two-dimensional, we must define $\partial/\partial\alpha = \frac{1}{2}[\partial/\partial\alpha_x + \partial/i\partial\alpha_y]$. This greatly reduces the freedom of choice of possible evolution equations.

In cases where multiple operators are involved, the above operator identities apply in sequential order, so that the nearest operator to ρ generates the nearest differential operator to W. Thus, the order of operations is reversed if the creation and annihilation operators act on ρ from the right. For example,

$$a^\dagger a\rho \to \left(\beta - \frac{s+1}{2}\frac{\partial}{\partial\alpha}\right)\left(\alpha - \frac{s-1}{2}\frac{\partial}{\partial\beta}\right) P_s,$$
$$\rho a^\dagger a \to \left(\alpha - \frac{s+1}{2}\frac{\partial}{\partial\beta}\right)\left(\beta - \frac{s-1}{2}\frac{\partial}{\partial\alpha}\right) P_s. \tag{7.119}$$

If there are operators on either side of ρ, the order of using the identities is immaterial. This can be seen by considering the two possibilities below:

$$a^\dagger \rho a \to \left(\beta - \frac{s+1}{2}\frac{\partial}{\partial\alpha}\right)\left(\alpha - \frac{s+1}{2}\frac{\partial}{\partial\beta}\right) P_s,$$
$$a^\dagger \rho a \to \left(\alpha - \frac{s+1}{2}\frac{\partial}{\partial\beta}\right)\left(\beta - \frac{s+1}{2}\frac{\partial}{\partial\alpha}\right) P_s. \tag{7.120}$$

Note that, in either case, the final result is the same, even though the identity rules were applied in a different order.

7.7.4 Example: the harmonic oscillator

Consider the harmonic oscillator Hamiltonian, which, from Eq. (2.191), also describes a single-mode electromagnetic field:

$$H = \hbar\omega a^\dagger a. \tag{7.121}$$

Using Eq. (2.6), this leads to the following time-evolution equation for the density matrix:

$$\frac{\partial\rho}{\partial t} = -i\omega[a^\dagger a\rho - \rho a^\dagger a]. \tag{7.122}$$

We now apply the operator mappings from Eq. (7.119), which gives

$$a^\dagger a\rho - \rho a^\dagger a \to \left(-\frac{s+1}{2}\left[\frac{\partial}{\partial\alpha}\alpha - \frac{\partial}{\partial\beta}\beta\right] - \frac{s-1}{2}\left[\beta\frac{\partial}{\partial\beta} - \alpha\frac{\partial}{\partial\alpha}\right]\right) P_s. \tag{7.123}$$

Simplifying this, we see that, once applied to the quasi-probability P_s, we obtain

$$\frac{\partial P_s}{\partial t} = i\omega\left(\frac{\partial}{\partial\alpha}\alpha - \frac{\partial}{\partial\beta}\beta\right) P_s. \tag{7.124}$$

This is a general result for the harmonic oscillator. The motion in phase space for this type of quadratic Hamiltonian is exactly the same as the corresponding classical trajectory, regardless of the operator ordering of the representation.

The solution to this first-order problem is by the method of characteristics. For an initial delta-function, corresponding to a coherent state in a Glauber P-representation, the solution remains a delta-function at all times:

$$P_1(\alpha, t) = \delta(\alpha - \alpha(t)). \tag{7.125}$$

Here the trajectory in the corresponding stochastic differential equation, (7.28), has no noise term, so that

$$\frac{d\alpha}{dt} = -i\omega\alpha. \tag{7.126}$$

This has the expected solution of an oscillatory motion, which in phase space gives a circular path around the origin, so that a coherent state becomes a coherent state with a rotated amplitude:

$$\alpha(t) = \alpha(0)e^{-i\omega t}. \tag{7.127}$$

This is perhaps the simplest case possible. More examples including damping and non-linear terms are given in the next chapter. We note that a solution exists provided the initial distribution exists. However, in some cases the initial distribution may be nonpositive or singular, depending on the ordering.

7.8 Quasi-probability Fokker–Planck equation

These rules result in a Fokker–Planck (diffusive) type of equation for the motion of the distribution function, provided the distribution asymptotically vanishes sufficiently rapidly as $\boldsymbol{\alpha}, \boldsymbol{\beta} \to \infty$. The next step is to use the stochastic formulation of a diffusion process, which transforms a Fokker–Planck equation into an ensemble average over a set of stochastic trajectories. These classical-like equations are usually simpler to treat, and can always be numerically simulated if necessary, although, as in all Monte Carlo type methods, there is a sampling error in any finite set of trajectories.

The Wigner evolution equation typically does not contain second-order derivatives but does contain third- or higher-order derivatives, which invalidates any interpretation in terms of a Fokker–Planck equation. It is expected that the third-order terms will only be significant when quantum effects are large, which occurs when the characteristic photon number per mode, $\bar{n}$, is small, or in the limit of large evolution times. Therefore, it is possible to recover a stochastic interpretation by neglecting (or 'truncating') the third-order derivative terms. This can be formally justified by means of defining a scaled variable, $\boldsymbol{\delta} = \boldsymbol{\alpha}/\sqrt{\bar{n}}$, so that $\boldsymbol{\delta} = O(1)$. The Fokker–Planck equation is then expressed in the new variables, so that each higher-order derivative is a term in a series expansion in $\bar{n}^{-1/2}$. The truncation then corresponds to omitting higher-order terms in this expansion.

The Q-function has the disadvantage that it normally produces non-positive definite diffusion terms in the Fokker–Planck type evolution equations, so that there are no equivalent stochastic differential equations that one can try to solve. In some cases, it can produce even higher-order (higher than second) derivative terms in the evolution equation, while using the P-representation instead produces a true Fokker–Planck equation.

Thus, in many quantum optics problems, the result of applying operator identities (7.119) is that the master equation for the system density operator ρ can be transformed into a Fokker–Planck equation describing the evolution of the quasi-probability distribution function. The usual form of the Fokker–Planck equation is a complex Fokker–Planck equation of the form given in Eq. (7.44):

$$\frac{\partial P(\boldsymbol{\alpha},t)}{\partial t} = \left[-\frac{\partial}{\partial \alpha_\mu} A_\mu(\boldsymbol{\alpha}) + \frac{1}{2}\frac{\partial^2}{\partial \alpha_\mu \partial \alpha_\nu} D_{\mu\nu}(\boldsymbol{\alpha}) \right] P(\boldsymbol{\alpha},t), \tag{7.128}$$

where $A_\mu(\boldsymbol{\alpha})$ are drift terms, $D_{\mu\nu}(\boldsymbol{\alpha})$ are the elements of the diffusion matrix, $\boldsymbol{\alpha} = \{\alpha_1, \ldots, \alpha_\mu, \ldots\}$ is a vector over all the complex coordinates, and summation over repeated indices (corresponding to the number of components of $\boldsymbol{\alpha}$) is assumed.

In the case of the Wigner or P-representation, the number of components of $\boldsymbol{\alpha}$ would be twice the number of degrees of freedom, M, usually the number of field modes. These variables we have previously referred to as $\{\alpha_n \mid n = 1, \ldots, M\}$, so in this case $\boldsymbol{\alpha} = (\alpha_1, \ldots, \alpha_M, \alpha_1^*, \ldots, \alpha_M^*)$. For complex P-representations, $\boldsymbol{\alpha}$ contains $2M$ independent complex components, and $\boldsymbol{\alpha} = (\alpha_1, \ldots, \alpha_M, \beta_1, \ldots, \beta_M)$. In this case, the diffusion matrix is not generally positive definite, and the complex conjugate derivatives are not utilized. To obtain a positive definite diffusion, a different procedure is needed. This will be treated in Section 7.8.1.

7.8.1 Potential solutions

In some cases, it is possible to find exact steady-state solutions to the Fokker–Planck equations. This is obtained using the method of potential solutions.

A steady-state solution $P_{ss}(\boldsymbol{\alpha})$, of the form

$$P_{ss}(\boldsymbol{\alpha}) = N_0 \exp\left(-\int_0^{\boldsymbol{\alpha}} d\boldsymbol{\alpha}' \cdot \mathbf{V}(\boldsymbol{\alpha}') \right), \tag{7.129}$$

to the Fokker–Planck equation (7.128) exists provided the potential conditions

$$\frac{\partial V_\nu}{\partial \alpha_\mu} = \frac{\partial V_\mu}{\partial \alpha_\nu} \tag{7.130}$$

are satisfied, where the potentials are vectors such that

$$V_\nu = (D^{-1})_{\nu\mu}\left(-2A_\mu + \frac{\partial D_{\mu\sigma}}{\partial \alpha_\sigma} \right), \tag{7.131}$$

and N_0 is the normalization constant. This follows from setting $\partial P(\boldsymbol{\alpha},t)/\partial t = 0$ and then noting that

$$\frac{\partial}{\partial \alpha_\mu}\left(-A_\mu P_{ss} + \frac{1}{2}\frac{\partial}{\partial \alpha_\nu} D_{\mu\nu} P_{ss} \right) = 0. \tag{7.132}$$

We next set the expression in parentheses in the above equation equal to zero, and try a solution of the form $P_{ss}(\boldsymbol{\alpha}) = \exp[-\phi(\boldsymbol{\alpha})]$. This gives us

$$D_{\mu\nu}\frac{\partial \phi}{\partial \alpha_\nu} = -2A_\mu + \frac{\partial D_{\mu\nu}}{\partial \alpha_\nu}, \tag{7.133}$$

which implies that

$$\frac{\partial \phi}{\partial \alpha_\nu} = V_\nu. \tag{7.134}$$

These equations can be integrated to yield ϕ if the potential conditions are satisfied. There are more general potential conditions possible than this as well, which include both reversible and irreversible terms; the results given above are for the simplest case.

For the Glauber–Sudarshan, Wigner and Q representations, there are, of course, only M complex variables or $2M$ real variables. However, the corresponding evolution equations generally do not have exact solutions, except for linear cases – due to various technical reasons like higher-order derivatives or simply a lack of detailed balance. Despite this, an approximate solution can often be found, especially in cases where nonlinear and nonclassical photon statistics do not arise.

7.8.2 Dimension doubling and positive diffusion

With the positive P-representation, it is possible to transform the second-derivative terms into a positive semi-definite diffusion operator, by choosing the appropriate derivative operator, out of the alternatives discussed previously. We will now show this explicitly. For these positive P-representations, we use the dimension-doubling method to obtain a $4M$-dimensional vector, with

$$\boldsymbol{\alpha} \to (\alpha_1, \ldots, \alpha_M, \beta_1, \ldots, \beta_M, \alpha_1^*, \ldots, \alpha_M^*, \beta_1^*, \ldots, \beta_M^*).$$

It is sometimes convenient to use the notation $\beta_j \equiv \alpha_j^+$, to indicate that this is a c-number variable which corresponds stochastically to the conjugate operator, even though it is independent.

We first write the equation (using the Einstein summation convention) in the form

$$\frac{\partial}{\partial t}P(\boldsymbol{\alpha}, t) = \left[-\partial_\mu A_\mu(\boldsymbol{\alpha}) + \frac{1}{2}\partial_\mu \partial_\nu D_{\mu\nu}(\boldsymbol{\alpha})\right] P(\boldsymbol{\alpha}, t). \tag{7.135}$$

Next, the complex matrix square root of $D_{\mu\nu}$, which is defined so that $D = BB^{\mathrm{T}}$, is expanded into its real and imaginary parts – written as $B = B^x + iB^y$. Note that, because B is, in general, a complex matrix, BB^{T} is not positive. If B were real, it would be. A similar procedure is followed for $A_\mu = A_\mu^x + iA_\mu^y$. This implies that the time-evolution equation can be written as

$$\frac{\partial}{\partial t}P(\boldsymbol{\alpha}, t) = \left[-\partial_\mu(A_\mu^x + iA_\mu^y) + \frac{1}{2}\partial_\mu\partial_\nu(B_{\mu\sigma}^x + iB_{\mu\sigma}^y)(B_{\nu\sigma}^x + iB_{\nu\sigma}^y)\right] P(\boldsymbol{\alpha}, t). \tag{7.136}$$

So far, we have been deliberately general about the choice of partial derivatives in the above equation, but now we shall make the choice explicit. The exact choice used will depend on the factor to the right of the partial derivative. In particular, the following choices for the drift terms are indicated,

$$\partial_\mu A_\mu \to \partial^x_\mu A^x_\mu + \partial^y_\mu A^y_\mu, \tag{7.137}$$

where $\partial^x_\mu = \partial/\partial\zeta^x_\mu$ and $\partial^y_\mu = \partial/\partial\zeta^y_\mu$, and a choice for the diffusion terms, so that

$$\partial_\mu\partial_\nu D_{\mu\nu} \to \partial^x_\mu\partial^x_\nu B^x_{\mu\sigma}B^x_{\nu\sigma} + \partial^y_\mu\partial^x_\nu B^y_{\mu\sigma}B^x_{\nu\sigma} + \partial^x_\mu\partial^y_\nu B^x_{\mu\sigma}B^y_{\nu\sigma} + \partial^y_\mu\partial^y_\nu B^y_{\mu\sigma}B^y_{\nu\sigma}. \tag{7.138}$$

In order to show what this implies, in real variables we can express the final equation as

$$\frac{\partial}{\partial t}P(\boldsymbol{\alpha}, t) = \left[\partial_m \widetilde{A}_m + \frac{1}{2}\partial_m\partial_n \widetilde{D}_{mn}\right] P(\boldsymbol{\alpha}, t), \tag{7.139}$$

where the indices m and n range over $[1, 4M]$, indicating that they cover the $4M$-dimensional phase space of real and imaginary parts of $\boldsymbol{\alpha}$. In particular, the range $1 \le m, n \le 2M$ corresponds to the real parts, and the range $2N + 1 \le m, n \le 4M$ corresponds to the imaginary parts.

The new drift and diffusion functions are real, with the $4M$-dimensional drift $\widetilde{A}$ defined so that

$$\widetilde{A} = \begin{bmatrix} \mathbf{A}^x \\ \mathbf{A}^y \end{bmatrix}. \tag{7.140}$$

Similarly, the diffusion matrix, $\widetilde{D} = \widetilde{B}\widetilde{B}^{\mathrm{T}}$ is positive semi-definite, because the matrix

$$\widetilde{B} = \begin{pmatrix} B^x & 0 \\ B^y & 0 \end{pmatrix} \tag{7.141}$$

is real. We have, explicitly, for $\widetilde{D}$,

$$\widetilde{D} = \begin{bmatrix} B^x(B^x)^{\mathrm{T}} & B^x(B^y)^{\mathrm{T}} \\ B^y(B^x)^{\mathrm{T}} & B^y(B^y)^{\mathrm{T}} \end{bmatrix}. \tag{7.142}$$

Therefore, it is always possible to construct a Fokker–Planck equation with a positive semi-definite diffusion matrix for the positive P-function.

Another way to understand this is that we are simply adding additional conjugate derivatives of the form $\partial/\partial\alpha^*$ to the complex identities that generate the Fokker–Planck equation, in order that it have the final expected form. This has no effect on the operator identities, as the conjugate derivatives vanish when acting on the analytic kernel operator $\hat{\lambda}_+$. However, after partial integration – provided boundary terms vanish – the result is a positive definite Fokker–Planck equation, of the form given in Eq. (7.44).

7.8.3 Stochastic differential equations in phase space

As explained in Section 7.3, in cases where direct solutions to the Fokker–Planck equation (FPE) are not available, one can alternatively employ the stochastic differential equation (SDE) approach. This approach is based on the fact that, for a positive definite FPE, there exists a set of equivalent SDEs. As we have seen, we can always find an operator mapping that will map a master equation into an FPE with a positive definite diffusion matrix. If the diffusion matrix $\mathbf{D}(\boldsymbol{\alpha})$, which is always symmetric, has nonnegative eigenvalues (when expressed in terms of real variables), then it can be factorized into the form

$$\mathbf{D}(\boldsymbol{\alpha}) = \mathbf{B}(\boldsymbol{\alpha})\mathbf{B}^{\mathrm{T}}(\boldsymbol{\alpha}), \tag{7.143}$$

and a set of SDEs equivalent to the FPE (7.128) can be written in the following (Ito) form:

$$\frac{d\boldsymbol{\alpha}(t)}{dt} = \mathbf{A}(\boldsymbol{\alpha}) + \mathbf{B}(\boldsymbol{\alpha}) \cdot \boldsymbol{\xi}(t). \tag{7.144}$$

Here $\xi_\mu(t)$ $(\mu = 1, 2, 3, \dots)$ are real, independent Gaussian white-noise terms with zero mean values, $\langle \xi_\mu(t) \rangle = 0$, and the following nonzero correlations:

$$\langle \xi_\mu(t)\xi_\nu(t') \rangle = \delta_{\mu\nu}\delta(t - t'). \tag{7.145}$$

The above SDEs can be treated using direct numerical simulation techniques or analytically.

When treating an SDE, it is necessary to realize that there is a choice of the Ito or Stratonovich form of SDE. Each of these assumes a specific definition of stochastic integration and calculus rules, and each has some advantages and disadvantages. For example, only in the Stratonovich form are the rules of stochastic calculus those of ordinary calculus. The Ito form has advantages arising from the increments being independent of the integration variable; the rules of calculus, however, are different from ordinary rules. Without going into the details of stochastic calculus, we give here Ito's formula for variable change. This is required when transforming from $\boldsymbol{\alpha}(t)$ to a new set of variables. For an arbitrary function of $\boldsymbol{\alpha}$, $f(\boldsymbol{\alpha})$, we may often need to know what SDE function f obeys, given the fact that $\boldsymbol{\alpha}(t)$ obeys Eq. (7.144). The SDE for the new variable f is given, according to Ito's variable-change formula (7.29), by

$$\frac{df}{dt} = \sum_\mu \frac{\partial f}{\partial \alpha_\mu}\left[A_\mu + \sum_\nu B_{\mu\nu}\xi_\nu(t)\right] + \frac{1}{2}\sum_{\mu,\nu} D_{\mu\nu}\frac{\partial^2 f}{\partial \alpha_\mu \partial \alpha_\nu}. \tag{7.146}$$

Here $D_{\mu\nu} = \left(BB^{\mathrm{T}}\right)_{\mu\nu}$, and the first terms are due to the ordinary rules of variable-change calculus, while the last term is Ito's correction term due to Ito's rule.

Of all the above representations, it is only the positive P-function that can generally be used in a stochastic form for nonlinear quantum systems. The diffusion matrix can always be obtained in a positive definite form in this representation, by means of using the freedom of choice in the operator identities to guarantee positivity. In some cases, the Wigner representation can be used to obtain an approximate (truncated) stochastic equation, after neglecting the higher-order terms.

We note that the positive P Fokker–Planck equation always has a positive initial condition, since we can choose the form of the initial distribution to correspond to Eq. (7.102), for any initial density matrix. As a result, the initial distribution can always be randomly sampled. Similarly, because the diffusion can be chosen positive definite, there is always a corresponding SDE. However, the transformation to a Fokker–Planck form relies on an integration by parts without additional boundary terms. This in turn requires the distribution to vanish sufficiently rapidly at infinity in the phase space.

In some cases, the double-dimensional equations are unstable. This is especially found under conditions of low damping and large nonlinearity. The resulting dynamics can lead to distribution tails that only vanish as fast as a power law. In other words, the stochastic equations themselves do not always preserve the exact structure of Eq. (7.102). Under these conditions, sampling errors become large, and systematic errors can occur due to the neglected boundary terms. For this reason, it is desirable to carefully check that boundary terms can in fact be neglected, when using this technique. For most cases in quantum optics, this is not an issue.

7.9 Linearized fluctuation theory

In many cases, we cannot solve the Fokker–Planck equation for the quasi-probability function, so we have to find a way to find an approximate solution. One of the most useful methods is linearization around a semiclassical solution. This allows us to find the quantum fluctuations about a stable steady state.

The semiclassical theory starts with the SDEs coming from the FPE, neglects quantum fluctuations (noise terms) and assumes the semiclassical approximation, i.e. it assumes that $\langle a_i \rangle = \alpha_i^{(0)} = \beta_i^{(0)}$ (in the dimension-doubled case) and that higher-order correlation functions factorize. Then the semiclassical steady states $\alpha_i^{(0)}$ are found from $d\boldsymbol{\alpha}(t)/dt = 0$, that is, from $A(\boldsymbol{\alpha}) = 0$. We are only interested in stable semiclassical solutions, because any kind of noise would drive the system away from an unstable solution.

The linearized fluctuation theory assumes small fluctuations around stable semiclassical steady states,

$$\begin{aligned} \alpha_i(t) &= \alpha_i^{(0)} + \delta\alpha_i(t), \\ \beta_i(t) &= \beta_i^{(0)} + \delta\beta_i(t), \end{aligned} \tag{7.147}$$

where $\delta\beta_i \neq \delta\alpha_i$, that is, $\delta\alpha_i$ and $\delta\beta_i$ can vary independently. Substituting this into Eq. (7.144) one can obtain the linearized SDEs, which can always be written in the matrix form

$$\frac{d\delta\alpha_\mu(t)}{dt} = -A_{\mu\nu}^{(0)}\delta\alpha_\nu(t) + B_{\mu\nu}^{(0)}\xi_\nu(t), \tag{7.148}$$

where $\delta\boldsymbol{\alpha} \equiv (\delta\alpha_1, \ldots, \delta\alpha_M, \delta\beta_1, \ldots, \delta\beta_M)$ and $B^{(0)} \equiv B(\boldsymbol{\alpha}^{(0)})$, so that $D^{(0)} \equiv D(\boldsymbol{\alpha}^{(0)}) = B^{(0)} \cdot (B^{(0)})^{\mathrm{T}}$, and

$$A_{\mu\nu}^{(0)} = \frac{\partial A_\mu(\boldsymbol{\alpha}^{(0)})}{\partial \alpha_\nu}. \tag{7.149}$$

The stability of the semiclassical steady states against small fluctuations is checked by means of the eigenvalue spectrum of the steady-state drift matrix $A^{(0)}$. All the eigenvalues of $A^{(0)}$ must simultaneously have positive real parts for stability of the corresponding semiclassical steady state. This guarantees that any fluctuation will decay rather than grow. This process can be iterated to obtain higher-order corrections, in order to treat large fluctuations that may occur near the points of phase transitions.

We are interested in the steady-state solution of these differential equations, that is, the solution after all of the transient effects have died out. This can be found by taking our initial time to be $-\infty$, so that the solution becomes

$$\delta\boldsymbol{\alpha}(t) = \int_{-\infty}^{t} dt'\, e^{-A^{(0)}(t-t')} B^{(0)} \boldsymbol{\xi}(t'). \tag{7.150}$$

This solution can also be found by Fourier-transforming the differential equations. In order to see this, we first define the Fourier transforms

$$\begin{aligned}\delta\boldsymbol{\alpha}(\omega) &= \frac{1}{\sqrt{2\pi}} \int_{-\infty}^{\infty} dt\, e^{-i\omega t} \delta\boldsymbol{\alpha}(t), \\ \delta\boldsymbol{\xi}(\omega) &= \frac{1}{\sqrt{2\pi}} \int_{-\infty}^{\infty} dt\, e^{-i\omega t} \delta\boldsymbol{\xi}(t),\end{aligned} \tag{7.151}$$

and note that the noise terms are correlated according to $\langle \xi_\mu(\omega)\xi_\nu(\omega')\rangle = \delta_{\mu\nu}\delta(\omega+\omega')$. Fourier-transforming Eq. (7.150) we find

$$\begin{aligned}\delta\boldsymbol{\alpha}(\omega) &= \frac{1}{\sqrt{2\pi}} \int_{-\infty}^{\infty} dt \int_{-\infty}^{t} dt'\, e^{-i\omega t} e^{-A^{(0)}(t-t')} B^{(0)} \delta\boldsymbol{\xi}(t') \\ &= \frac{1}{\sqrt{2\pi}} \int_{-\infty}^{\infty} dt' \int_{t'}^{\infty} dt\, e^{-i\omega t} e^{-A^{(0)}(t-t')} B^{(0)} \delta\boldsymbol{\xi}(t') \\ &= (A^{(0)} + i\omega)^{-1} B^{(0)} \delta\boldsymbol{\xi}(\omega).\end{aligned} \tag{7.152}$$

This implies that

$$(A^{(0)} + i\omega)\delta\boldsymbol{\alpha}(\omega) = B^{(0)} \delta\boldsymbol{\xi}(\omega), \tag{7.153}$$

which is just the Fourier transform of Eq. (7.148). Therefore, we can also find the fluctuations about the semiclassical solution by Fourier-transforming the SDEs for the fluctuations and solving the resulting algebraic equations.

7.10 Functional phase-space representations

In this section for the advanced student, we introduce functional phase-space representations for bosonic quantum fields. The s-ordering approach of previous sections is utilized here, as before. The results are obtained by translation of mode sums to integrals. We usually treat a single, complex bosonic field, but the formalism can include spin indices, $i = 1, \ldots, S$.

The phase-space representation for operator ordering s is defined using a distribution $P[\boldsymbol{\alpha}]$ over a nonorthogonal operator basis:

$$\rho = \int P[\boldsymbol{\alpha}] \hat{\lambda}_s[\boldsymbol{\alpha}]\, D[\boldsymbol{\alpha}], \tag{7.154}$$

Here $\boldsymbol{\alpha} = \boldsymbol{\alpha}(\mathbf{r})$ is a complex c-number field over a d-dimensional real coordinate $\mathbf{r}$, and $\hat{\lambda}_s[\boldsymbol{\alpha}]$ is a corresponding basis functional element in the space of field operators acting on $\mathcal{H}$. The functional integral measure is a product over all coordinates:

$$D[\boldsymbol{\alpha}] = \prod_{\mathbf{r},\ell} d\alpha_\ell(\mathbf{r})\, d\alpha_\ell^*(\mathbf{r}).$$

In practical calculations, this infinite product is usually replaced by a finite product over modes, with a momentum cutoff. The corresponding bosonic quantum characteristic functional for arbitrary orderings is a c-number functional given by

$$C_s[\boldsymbol{\zeta}] \equiv \mathrm{Tr}[\rho \hat{C}_s[\boldsymbol{\zeta}]]. \tag{7.155}$$

Here $\boldsymbol{\zeta} = \boldsymbol{\zeta}(\mathbf{r})$ is another complex function, and the operators $\hat{C}_s[\boldsymbol{\zeta}]$ are s-ordered characteristic functional generators,

$$\hat{C}_s[\boldsymbol{\zeta}] = :\exp\left[\int d\mathbf{r}\,(\boldsymbol{\zeta}(\mathbf{r})\cdot\boldsymbol{\psi}^\dagger(\mathbf{r}) - \boldsymbol{\zeta}^*(\mathbf{r})\cdot\boldsymbol{\psi}(\mathbf{r}))\right]:_s . \tag{7.156}$$

The notation $:\quad:_s$ indicates an s-ordered product, where $s = 1,\ 0,\ -1$ for normal, symmetric and anti-normal ordering, as previously, and $d\mathbf{r} = dr_1 \ldots dr_d$. The vector products indicate a sum over any spin indices. From the Baker–Hausdorff theorem, all these characteristic functions can be transformed into each other using

$$\hat{C}_s[\boldsymbol{\zeta}] = \exp\left[\frac{s}{2}\int d\mathbf{r}\,\boldsymbol{\zeta}(\mathbf{r})\cdot\boldsymbol{\zeta}^*(\mathbf{r})\right]\hat{C}_0[\boldsymbol{\zeta}]. \tag{7.157}$$

The characteristic functions of phase-space basis elements are

$$\mathrm{Tr}[\hat{\Lambda}_s[\boldsymbol{\alpha}]\hat{C}_s[\boldsymbol{\zeta}]] = \exp\left[\int d\mathbf{r}\,(\boldsymbol{\zeta}(\mathbf{r})\cdot\boldsymbol{\alpha}^*(\mathbf{r}) - \boldsymbol{\zeta}^*(\mathbf{r})\cdot\boldsymbol{\alpha}(\mathbf{r}))\right]. \tag{7.158}$$

It follows that the corresponding functional distributions $P[\boldsymbol{\alpha}]$ are different for the three different orderings, and the characteristic functionals can be calculated in each case from the phase-space distributions as

$$C_s[\xi] = \int D[\boldsymbol{\alpha}]P[\boldsymbol{\alpha}]\exp\left[\int d\mathbf{r}\,(\boldsymbol{\zeta}(\mathbf{r})\cdot\boldsymbol{\alpha}^*(\mathbf{r}) - \boldsymbol{\zeta}^*(\mathbf{r})\cdot\boldsymbol{\alpha}(\mathbf{r}))\right]. \tag{7.159}$$

In the normally ordered case of $s = 1$, this is simply the functional version of the Glauber P-function, and one similarly obtains functional versions of the Wigner and Q-functions. More general, dimension-doubled functional representations exist. Any ordering can be used to obtain a positive functional distribution $P_+[\boldsymbol{\alpha}, \boldsymbol{\beta}]$ with the s-ordered characteristic function

$$C_S[\boldsymbol{\zeta}] = \int P_+[\boldsymbol{\alpha}, \boldsymbol{\beta}]\exp\left[\int d\mathbf{r}\,(\boldsymbol{\zeta}(\mathbf{r})\cdot\boldsymbol{\beta}(\mathbf{r}) - \boldsymbol{\zeta}^*(\mathbf{r})\cdot\boldsymbol{\alpha}(\mathbf{r}))\right] D[\boldsymbol{\alpha}]\,D[\boldsymbol{\beta}]. \tag{7.160}$$

The operator identities that follow from this are essentially the same as in the discrete mode case, except that derivatives are replaced by functional derivatives:

$$\begin{aligned}
\psi_i\rho &\to \left(\alpha_i - \frac{s-1}{2}\frac{\delta}{\delta\beta_i}\right)P_s,\\
\psi_i^\dagger\rho &\to \left(\beta_i - \frac{s+1}{2}\frac{\delta}{\delta\alpha_i}\right)P_s,\\
\rho\psi_i &\to \left(\alpha_i - \frac{s+1}{2}\frac{\delta}{\delta\beta_i}\right)P_s,\\
\rho\psi^\dagger &\to \left(\beta_i - \frac{s-1}{2}\frac{\delta}{\delta\alpha_i}\right)P_s.
\end{aligned} \tag{7.161}$$

Here the meaning of β_i is just the same as with discrete modes: $\beta_i = \alpha_i^*$ in a classical phase space. Otherwise, it is an additional independent coordinate in a nonclassical phase space. Using these methods, we find almost identical stochastic equations to the previous ones, except that the resulting noise fields are delta-correlated in space as well as in time.

Note that this formalism needs to be interpreted judiciously. It is a useful way to simplify the resulting equations, but nonlinear quantum field theories are notoriously singular in the continuum limit. In quantum field theory, it is generally necessary to provide a momentum cutoff, and renormalize if necessary, in a practical calculation. The reader is referred to texts on quantum field theory for these details.

Assuming local interactions, this procedure leads to a Fokker–Planck equation of the form

$$\frac{\partial}{\partial t}P[\boldsymbol{\alpha}, t] = \int d^3\mathbf{r}\left[-\frac{\delta}{\delta\alpha_\mu(\mathbf{r})}A_\mu[\boldsymbol{\alpha}] + \frac{1}{2}\frac{\delta^2}{\delta\alpha_\mu(\mathbf{r})\delta\alpha_\nu(\mathbf{r})}D_{\mu\nu}[\boldsymbol{\alpha}]\right]P[\boldsymbol{\alpha}, t], \tag{7.162}$$

where the sums over the functional derivatives extend over the complex and conjugate derivatives of all the independent phase-space fields. Just as in the earlier cases, we define $\mathbf{D} = \mathbf{B}\mathbf{B}^{\mathrm{T}}$, and the sums over μ include all the independent complex fields and their conjugates. Transforming back to stochastic variables, the complex-valued partial stochastic differential equation (PSDE) that results has the form one might expect:

$$\frac{d\boldsymbol{\alpha}(\mathbf{r}, t)}{dt} = \mathbf{A}(\boldsymbol{\alpha}) + \mathbf{B}(\boldsymbol{\alpha}) \cdot \boldsymbol{\xi}(\mathbf{r}, t). \tag{7.163}$$

Here $\xi_\mu(\mathbf{r}, t)$ ($\mu = 1, 2, 3, \dots$) are real, independent Gaussian white-noise terms with zero mean values, $\langle\xi_\mu(\mathbf{r}, t)\rangle = 0$, and the following nonzero correlations:

$$\langle\xi_\mu(\mathbf{r}, t)\xi_\nu(\mathbf{r}', t')\rangle = \delta_{\mu\nu}\delta(t - t')\delta(\mathbf{r} - \mathbf{r}'), \tag{7.164}$$

where, in three dimensions, $\delta(\mathbf{r} - \mathbf{r}') = \delta(x - x')\delta(y - y')\delta(z - z')$. In numerical calculations, this must be replaced by a discrete delta-function on a lattice $r_j(\ell)$, with a lattice index ℓ and cell size Δr_j. In one dimension, one simply makes the substitution $\delta(x - x') \to \delta_{\ell\ell'}/\Delta x$. The momentum cutoff that is implied by a lattice of this type is then $k_{max} = 2\pi/\Delta x$.

Additional reading

Books

- Several books on stochastic processes, Fokker–Planck equations and stochastic differential equations are:

C. W. Gardiner, *Handbook of Stochastic Methods* (Springer, Berlin, 1985).

H. Risken and T. Frank, *The Fokker–Planck Equation: Methods of Solution and Applications* (Springer, Berlin, 1996).

N. G. Van Kampen, *Stochastic Processes in Physics and Chemistry* (Elsevier, Amsterdam, 2007).

The book by Gardiner also discusses c-number representations of operators, with particular emphasis on complex and positive P-representations. The standard P-representation, the Wigner function and the Q-function are discussed in most of the quantum optics books that have been mentioned in previous chapters.

Articles

- A review paper on quasi-distribution functions is:

M. Hillery, R. F. O'Connell, M. O. Scully, and E. P. Wigner, *Phys. Rep.* **106**, 121 (1984).

- Some original papers are:

G. S. Agarwal and E. Wolf, *Phys. Rev.* D **2**, 2161, 2187, 2206 (1970).
K. E. Cahill and R. J. Glauber, *Phys. Rev.* **177**, 1857, 1883 (1969).
P. D. Drummond and C. W. Gardiner, *J. Phys.* A **13**, 2353 (1980).

- A discussion of numerical methods for stochastic differential equations is given in:

P. D. Drummond and I. K. Mortimer, *J. Comput. Phys.* **93**, 144 (1991).
M. Werner and P. D. Drummond, *J. Comput. Phys.* **132**, 312 (1997).

Problems

7.1 Repeat the calculation of Hilbert space dimension in Section 2.10, for the case of fermions. Note that the occupation per mode is now restricted to zero or one. Here m is to be regarded as the total number of modes, including both the spin and spatial quantum numbers.

7.2 Consider a Gaussian stochastic variable x with a zero mean ($\langle x \rangle = 0$) and $\langle x^2 \rangle \neq 0$. For such a variable, the odd moments vanish ($\langle x^{2n+1} \rangle = 0$, where $n = 1, 2, 3, \ldots$), while the even moments factorize according to

$$\langle x^{2n} \rangle = \frac{(2n)!}{2^n n!} \langle x^2 \rangle^n .$$

Use this to show that

$$\langle e^x \rangle = e^{\langle x^2 \rangle /2} .$$

7.3 Show that the Wigner function can be expressed in terms of the P-representation as

$$W(\alpha) = \frac{2}{\pi} \int d^2\beta \, P(\beta) e^{-2|\alpha-\beta|^2} .$$

7.4 Find the Wigner function for the single-mode squeezed vacuum state $S(z)|0\rangle$, where $S(z)$ is the squeeze operator.

7.5 Find a positive P-function for the single-mode number state $\rho = |n_0\rangle\langle n_0|$.

7.6 The driven harmonic oscillator has the Hamiltonian $H = \omega a^\dagger a + \lambda a + \lambda^* a^\dagger$, and, if damping is added, obeys the master equation

$$\frac{d\rho}{dt} = \frac{-i}{\hbar}[H, \rho] + \gamma(2a\rho a^\dagger - a^\dagger a\rho - \rho a^\dagger a).$$

(a) For the Wigner and Glauber–Sudarshan P-representation, find Fokker–Planck equations equivalent to the above master equation.

(b) In the case of the Wigner function with no damping ($\gamma = 0$), find a solution to the Fokker–Planck equation of the form

$$W(t, \alpha, \alpha^*) = W(0, g(t)\alpha + f(t), g^*(t)\alpha^* + f^*(t)).$$

8 Single-mode devices

The simplest open, nonlinear optical system consists of a single bosonic mode with nonlinear self-interactions coupled to driving fields and reservoirs. In this chapter, we will examine a number of systems of this type using the methods that have been developed in the preceding chapters. Systems of this type can exhibit a variety of interesting physical effects. For example, in the case of a driven nonlinear oscillator, there can be more than one steady state (bistability), and quantum statistical effects can manifest themselves by affecting the stability of these states. Other quantum statistical effects have also been predicted in a variety of related systems – such as photon anti-bunching, squeezing and changes to spectra. In general, the size of these effects scales inversely with the size of the system. This 'system size' refers to a threshold photon number, a number of atoms, or some similar quantity.

The nature of the steady states themselves can become less than straightforward. Nonequilibrium dissipative systems, of which the systems considered in this chapter are examples, have parameters that describe the energy input to the system, and their steady states depend on these parameters. When driven far from equilibrium, such systems can exhibit bifurcations in their steady states. In more complex cases, this eventually leads to periodic oscillations and chaos. Devices like this can be realized with multiple types of nonlinearity, ranging from nonlinear dielectrics through to cavity QED (with near-resonant atoms), cavity optomechanics (with nanomechanical oscillators) and circuit QED (with superconducting Josephson junctions).

We will start with cases where all the interactions are with the reservoirs, which may be linear or nonlinear, while the cavity itself is harmonic in its response.

8.1 Linear cavity

Our first example is one we have encountered before, a driven, passive cavity. In the Heisenberg picture, the total system Hamiltonian is, including coupling to an external coherent field at frequency ω_L,

$$H_{sys} = \hbar\omega a^\dagger a + i\hbar(\mathcal{E}e^{-i\omega_L t}a^\dagger - \mathcal{E}^* e^{i\omega_L t}a), \tag{8.1}$$

where $\mathcal{E}$ is the amplitude of the external laser. By comparison to previous input–output theory results from Eq. (6.74), we see that this is related to the cavity input flux operator

b^{in} as follows:

$$\mathcal{E} = \sqrt{2\gamma}\,\langle b^{in}(t)\rangle e^{i\omega_L t}. \tag{8.2}$$

The resulting master equation represents the combined effect of the cavity resonant Hamiltonian, with frequency ω, together with an external reservoir almost entirely in the vacuum state at zero temperature. This reservoir includes so many modes that it is not changed by the special treatment of the input laser mode, which has a finite excitation. The master equation is

$$\frac{d\rho}{dt} = \frac{1}{i\hbar}[H_{sys}, \rho] + \gamma(2a\rho a^\dagger - a^\dagger a\rho - \rho a^\dagger a). \tag{8.3}$$

While it is straightforward to solve this master equation, it is often simpler to work in an interaction picture in which the density matrix only evolves according to the interactions. For convenience, we can define the free part of the Hamiltonian in a slightly unorthodox way, so that it includes the laser frequency rather than the cavity frequency ω:

$$H_0 = \hbar\omega_L a^\dagger a. \tag{8.4}$$

In other words, the operators will evolve according to the laser frequency, while the states evolve according to the rest of the system Hamiltonian – including part of the cavity energy term if the laser is not on-resonance. Having done this, one finds that the equation simplifies considerably. Since the operators all have a known time evolution of $a_I(t) = ae^{-i\omega_L t}$, all of the explicitly time-dependent terms disappear. This gives an interaction-picture master equation with operators having no intrinsic time dependence:

$$\frac{d\rho_I}{dt} = [-i\Delta\omega a^\dagger a + \mathcal{E}a^\dagger - \mathcal{E}^* a, \rho] + \gamma(2a\rho a^\dagger - a^\dagger a\rho - \rho a^\dagger a). \tag{8.5}$$

Here $\Delta\omega = \omega - \omega_L$ represents the detuning between the cavity-mode resonance and the laser frequency. This must be much less than the cavity-mode spacing for the single-mode approximation to be applicable. We call this the rotating-frame picture, and will denote expectation values in this picture by $\langle O \rangle_R$.

8.2 Phase-space representation methods

8.2.1 Diagonal P-representation

For the linear cavity system, it is often acceptable to use the diagonal (Glauber–Sudarshan) P-representation, provided that the initial condition has a diagonal P-representation. This is nonsingular as time evolves because, if the state of the cavity mode is initially classical, it will remain so. Applying the rules from the previous chapter to obtain a Fokker–Planck equation, we find that

$$\frac{\partial P(\alpha, t)}{\partial t} = \left\{ \frac{\partial}{\partial\alpha}[(\gamma + i\Delta\omega)\alpha - \mathcal{E}] + \frac{\partial}{\partial\alpha^*}[(\gamma - i\Delta\omega)\alpha^* - \mathcal{E}^*] \right\} P(\alpha, t). \tag{8.6}$$

This can be transformed using the standard rules to obtain the corresponding stochastic equation in the P-representation. In this case, since there is no diffusion, there is no noise term. Defining, for convenience, a complex relaxation constant,

$$\tilde{\gamma} = \gamma + i\Delta\omega, \tag{8.7}$$

we find that

$$\begin{aligned} \frac{d\alpha}{dt} &= -\tilde{\gamma}\alpha + \mathcal{E}, \\ \frac{d\alpha^*}{dt} &= -\tilde{\gamma}^*\alpha^* + \mathcal{E}^*. \end{aligned} \tag{8.8}$$

Such a linear system of equations is easily solved exactly, giving the result that, if $\alpha(0) = \alpha_0$, then

$$\begin{aligned} \alpha(t) &= \alpha_0 e^{-\tilde{\gamma}t} + \mathcal{E}\int_0^t dt'\, e^{-\tilde{\gamma}(t-t')}, \\ &= \alpha_0 e^{-\tilde{\gamma}t} + \frac{\mathcal{E}}{\tilde{\gamma}}(1 - e^{-\tilde{\gamma}t}). \end{aligned} \tag{8.9}$$

This represents a solution of the original Fokker–Planck equation by the method of characteristics. In particular, the solution trajectory is a deterministic quantity if there is a deterministic initial condition. In other words, suppose the initial density matrix was that of a pure coherent state,

$$\rho(0) = |\alpha_0\rangle\langle\alpha_0|,$$

then the initial P-representation would be

$$P(\alpha, 0) = \delta^2(\alpha - \alpha_0).$$

The P-representation at time t would then be

$$P(\alpha, t) = \delta^2(\alpha - \alpha(t)).$$

This implies that $\langle a^n\rangle_R = \langle\alpha^n(t)\rangle_R = \alpha^n(t)$, and furthermore that operator moments simply factorize in any normally ordered operator product, so that

$$\langle(a^\dagger)^m a^n\rangle_R = \langle(\alpha^*(t))^m(\alpha(t))^n\rangle = (\alpha^*(t))^m(\alpha(t))^n = \langle a^\dagger\rangle_R^m\langle a\rangle_R^n. \tag{8.10}$$

The relation $\langle(a^\dagger)^m a^n\rangle_R = \langle a^\dagger\rangle_R^m\langle a\rangle_R^n$ is only true for coherent states, and is sometimes used as an alternative definition of the coherent state. Therefore, the coherently driven and damped cavity preserves its initial coherence, so that the cavity field is always in a coherent state if it starts out that way. This preservation of coherence under damping is one of the remarkable properties of coherent states, which has made them a universal and basic entity in laser physics. In other cases, for example, if the initial state represented by $P(\alpha_0, 0)$ were a chaotic or thermal state, then we would have to integrate over all possible initial points of the trajectories, so that

$$P(\alpha, t) = \int d^2\alpha_0\, \delta^2(\alpha - \alpha(t))P(\alpha_0, 0).$$

8.2.2 Positive *P*-representation

A general quantum initial state may not have a nonsingular diagonal P-representation, but it must have a positive P-representation, since that always exists. In this representation, the Fokker–Planck equation for the damped cavity is then

$$\frac{\partial P(\boldsymbol{\alpha})}{\partial t} = \left\{ \frac{\partial}{\partial \alpha}[\widetilde{\gamma}\alpha - \mathcal{E}] + \frac{\partial}{\partial \beta}[\widetilde{\gamma}^*\beta - \mathcal{E}^*] \right\} P(\boldsymbol{\alpha}). \tag{8.11}$$

This can be transformed to stochastic differential equations in the positive P-representation. Once again, there is no diffusion, and this implies that there is no noise term in the stochastic equations. We have that

$$\frac{d\alpha}{dt} = -\widetilde{\gamma}\alpha + \mathcal{E}, \tag{8.12}$$

$$\frac{d\beta}{dt} = -\widetilde{\gamma}^*\beta + \mathcal{E}, \tag{8.13}$$

Such a linear system of equations is again exactly soluble, giving the result

$$\alpha(t) = \alpha_0 e^{-\widetilde{\gamma}t} + \frac{\mathcal{E}}{\widetilde{\gamma}}(1 - e^{-\widetilde{\gamma}t}), \tag{8.14}$$

$$\beta(t) = \beta_0 e^{-\widetilde{\gamma}^*t} + \frac{\mathcal{E}}{\widetilde{\gamma}^*}(1 - e^{-\widetilde{\gamma}^*t}). \tag{8.15}$$

Once again, this is a solution by characteristics, except in a four-dimensional phase space. The resulting distribution at time t is therefore

$$P(\alpha, \beta, t) = \iint d^2\alpha_0\, d^2\beta_0\, \delta^2(\alpha - \alpha(t))\delta^2(\beta - \beta(t))P(\alpha_0, \beta_0, 0).$$

Even if the representation is nondiagonal initially, with terms having $\alpha_0 \neq \beta_0$, it will evolve toward a coherent state in which $\alpha = \beta = E/\widetilde{\gamma}$. Any operator moment can be calculated directly from the initial positive P-representation, since

$$\begin{aligned} \langle (a^\dagger)^m a^n \rangle_R &= \langle (\beta(t))^m (\alpha(t))^n \rangle \\ &= \iint d^2\alpha_0\, d^2\beta_0 \left(\beta_0 e^{-\widetilde{\gamma}^*t} + \frac{\mathcal{E}^*}{\widetilde{\gamma}^*}(1 - e^{-\widetilde{\gamma}^*t}) \right)^m \\ &\quad \times \left(\alpha_0 e^{-\widetilde{\gamma}t} + \frac{\mathcal{E}}{\widetilde{\gamma}}(1 - e^{-\widetilde{\gamma}t}) \right)^n P(\alpha_0, \beta_0, 0). \end{aligned} \tag{8.16}$$

This shows very clearly that any nonclassical correlations present in the initial state will decay to leave behind the same coherent-state correlations that would have occurred had the initial state been a coherent state.

8.2.3 Wigner representation

The Wigner representation is positive for a somewhat larger class of initial density matrices than the Glauber–Sudarshan diagonal P-representation. The price that is paid for this is that a symmetrically ordered reservoir correlation is noisy, even in the vacuum state. Thus, for the linear cavity, one obtains for the stochastic differential equations

$$\frac{d\alpha}{dt} = -\tilde{\gamma}\alpha + \mathcal{E} + \sqrt{\gamma}\,\xi_c(t),$$
$$\frac{d\alpha^*}{dt} = -\tilde{\gamma}^*\alpha^* + \mathcal{E}^* + \sqrt{\gamma}\,\xi_c^*(t). \tag{8.17}$$

Here $\xi_c(t) = (\xi_1(t) + i\xi_2(t))/\sqrt{2}$ is a complex function with $\langle \xi_c(t)\xi_c^*(t')\rangle = \delta(t - t')$. The solution is more complicated than before, since

$$\alpha(t) = \alpha(0)e^{-\tilde{\gamma}t} + \frac{\mathcal{E}}{\tilde{\gamma}}(1 - e^{-\tilde{\gamma}t}) + \sqrt{\gamma}\int_0^t dt'\, e^{-\tilde{\gamma}(t-t')}\xi_c(t'). \tag{8.18}$$

8.2.4 Q-representation

The Q-representation, like the positive P-representation, is positive for all initial density matrices. The corresponding anti-normally ordered reservoir correlation is even noisier than in the case of the Wigner function. In this case, one obtains

$$\frac{d\alpha}{dt} = -\tilde{\gamma}\alpha + \mathcal{E} + \sqrt{2\gamma}\,\xi_c(t),$$
$$\frac{d\alpha^*}{dt} = -\tilde{\gamma}^*\alpha^* + \mathcal{E}^* + \sqrt{2\gamma}\,\xi_c^*(t). \tag{8.19}$$

The solution is similar to the case of the Wigner function:

$$\alpha(t) = \alpha(0)e^{-\tilde{\gamma}t} + \frac{\mathcal{E}}{\tilde{\gamma}}(1 - e^{-\tilde{\gamma}t}) + \sqrt{2\gamma}\int_0^t dt'\, e^{-\tilde{\gamma}(t-t')}\xi_c(t'). \tag{8.20}$$

We notice that the noise terms have twice the variance that they have for the Wigner function. This is a natural result of the quantum fluctuation–dissipation theorem. That is, the noise terms must compensate for the losses in order that the distribution function maintains an overall variance that corresponds to anti-normally ordered operator products. This can be thought of as a contribution from vacuum modes entering the cavity. In an anti-normally ordered representation, these must bring vacuum fluctuations with them.

8.2.5 Steady-state results

In either form of normally ordered P-representation, the steady-state solution is diagonal, with

$$\rho_I = |\alpha_s\rangle\langle\alpha_s|, \tag{8.21}$$

so that the density matrix in the usual Schrödinger picture, ρ, is time-dependent,

$$\rho = |\alpha_s e^{-i\omega_L t}\rangle\langle\alpha_s e^{-i\omega_L t}|, \tag{8.22}$$

where

$$\alpha_s = \mathcal{E}/\tilde{\gamma}. \tag{8.23}$$

Using the operator correspondences, this results in

$$\langle a \rangle = \mathcal{E}/\tilde{\gamma}. \tag{8.24}$$

As a consequence, the intra-cavity photon number is given by a well-known result, which is also found classically:

$$n = \langle a^\dagger a \rangle = |\mathcal{E}/\tilde{\gamma}|^2 = \frac{|\mathcal{E}|^2}{\gamma^2 + \Delta\omega^2}. \tag{8.25}$$

In the Wigner representation or the operator Heisenberg picture, similar results are obtained, but require the inclusion of a stochastic process or the input fields, respectively.

This demonstrates the usual, expected classical behavior of an empty cavity, which is that the intra-cavity intensity (and hence the transmitted intensity of a two-port device) has a peak at the cavity resonance frequency, with a Lorentzian lineshape. In fact, there is essentially no difference between the classical and quantum results. This is a result of the fact that there is no difference between the classical and quantum versions of Maxwell's equations for linear systems, provided the operators are normally ordered!

We can also use these results to discuss what happens when the cavity has two mirrors with nonzero transmissivity. That arrangement is known as a Fabry–Perot interferometer. We will consider the case of a symmetric Fabry–Perot interferometer, where the mirrors, one of which we shall call the input mirror and the other the output mirror, have equal reflectivities. The external field is incident on the input mirror, and we want to find the flux emerging from the output mirror. Each mirror contributes an equal loss, so $\gamma^{in} = \gamma^{out} = \gamma/2$. If the mean value of the input flux amplitude is b^{in}, then from Eq. (6.74) $\mathcal{E} = \sqrt{2\gamma^{in}}b^{in}$, so the intra-cavity photon number is

$$n = \langle a^\dagger a \rangle = \frac{\gamma \langle b^{in\dagger} b^{in} \rangle}{\gamma^2 + \Delta\omega^2}. \tag{8.26}$$

Note that on-resonance $\Delta\omega = 0$, and the cavity occupation number is just $n = \langle b^{in\dagger} b^{in} \rangle / \gamma$, which is the number of photons that arrive in a relaxation time $1/\gamma$. There are two output fields coming from each mirror – one reflected from the first mirror, and one transmitted through the second mirror. The transmitted flux at resonance from the second, output, mirror is, from Eq. (6.77),

$$\langle b^{out\dagger} b^{out} \rangle = 2\gamma^{out} \langle a^\dagger a \rangle = \langle b^{in\dagger} b^{in} \rangle. \tag{8.27}$$

This means that all the input photons are transmitted on-resonance, and none is reflected. Just as in the classical theory, one finds that the reflected field at the first mirror is completely canceled by the transmitted field through the first mirror, and all the input flux ends up being transmitted through the output mirror.

Similarly, it is possible to calculate the spectrum of the transmitted field. However, this has a rather trivial property. Since the input coherent state evolves at just one frequency, the spectrum is a delta-function centered at the laser frequency. This represents a characteristic property of all driven, linear systems – the only spectral feature is at the input frequency, a phenomenon called entrainment. In the case of the other representations, there are additional fluctuation terms, which represent the vacuum fluctuations. These are not affected by the input laser field – nor are they detected by any normally ordered detection device, like a photo-detector.

In summary, identical results can be readily obtained using any of the methods outlined here. However, as the system remains in a coherent state, the P-representation methods are clearly the simplest in this case.

8.3 Driven nonlinear absorber

The model we consider in this section is a single cavity mode that is driven by external coherent radiation, damped by one-photon losses, and, in addition, is subject to two-photon losses (e.g. due to a two-photon absorber). The resulting equations can also be applied to describe second-harmonic generation with the harmonic mode adiabatically eliminated.

8.3.1 Master equation, Fokker–Planck equation and stochastic differential equations

Using the results from previous chapters, we see that this system can be described by the following interaction-picture master equation:

$$\frac{d\rho}{dt} = [-i\Delta\omega a^\dagger a + \mathcal{E}a^\dagger - \mathcal{E}^* a, \rho] + \gamma(2a\rho a^\dagger - a^\dagger a\rho - \rho a^\dagger a) + \frac{\gamma^{(2)}}{2}(2a^2\rho a^{\dagger 2} - a^{\dagger 2}a^2\rho - \rho a^{\dagger 2}a^2). \tag{8.28}$$

Here $\Delta\omega = \omega - \omega_L$ is the detuning between the cavity resonance frequency ω and the driving field frequency ω_L, γ is the one-photon loss coefficient, $\gamma^{(2)}$ is the two-photon loss coefficient and $\mathcal{E}$ is the driving field amplitude. Applying the positive P-representation procedure and discarding the boundary terms results in the Fokker–Planck equation:

$$\frac{\partial P(\alpha,\beta)}{\partial t} = \left\{\frac{\partial}{\partial\alpha}[\tilde{\gamma}\alpha + \gamma^{(2)}\alpha^2\beta - \mathcal{E}] + \frac{\partial}{\partial\beta}[\tilde{\gamma}^*\beta + \gamma^{(2)}\beta^2\alpha - \mathcal{E}^*] + \frac{1}{2}\frac{\partial^2}{\partial\alpha^2}(-\gamma^{(2)}\alpha^2) + \frac{1}{2}\frac{\partial^2}{\partial\beta^2}(-\gamma^{(2)}\beta^2)\right\}P(\alpha,\beta). \tag{8.29}$$

This, in turn, is equivalent to the following Ito stochastic differential equations:

$$\frac{d\alpha}{dt} = -\tilde{\gamma}\alpha - \gamma^{(2)}\alpha^2\beta + \mathcal{E} + i\sqrt{\gamma^{(2)}}\alpha\xi_1(t),$$

$$\frac{d\beta}{dt} = -\tilde{\gamma}^*\beta - \gamma^{(2)}\beta^2\alpha + \mathcal{E}^* - i\sqrt{\gamma^{(2)}}\beta\xi_2(t). \tag{8.30}$$

Here $\tilde{\gamma} \equiv \gamma + i\Delta\omega$, and the $\xi_i(t)$ are real independent Gaussian noise terms with zero means and the following nonzero correlations:

$$\langle\xi_1(t)\xi_1(t')\rangle = \delta(t-t'), \tag{8.31}$$

$$\langle\xi_2(t)\xi_2(t')\rangle = \delta(t-t'). \tag{8.32}$$

If we write the above stochastic equations, Eq. (8.30), in matrix form, then the corresponding diffusion matrix D and the matrix B (such that $B \cdot B^{\mathrm{T}} = D$) are

$$D = \begin{pmatrix} -\gamma^{(2)}\alpha^2 & 0 \\ 0 & -\gamma^{(2)}\beta^2 \end{pmatrix}, \qquad B = \begin{pmatrix} i\sqrt{\gamma^{(2)}}\alpha & 0 \\ 0 & -i\sqrt{\gamma^{(2)}}\beta \end{pmatrix}. \tag{8.33}$$

The explicit form of the stochastic equations (8.30) allows us to understand why we needed to employ the positive P-representation in this problem. If we employed the Glauber–Sudarshan P-representation and naively converted the corresponding Fokker–Planck equation (FPE) into stochastic differential equations (SDEs), disregarding the fact that the diffusion matrix is not positive definite, then the SDEs would still have the form of Eq. (8.30). This would lead to β being replaced by α^*, the complex conjugate of α, while the noise increments ξ_1 and ξ_2 would still be independent. This, however, would be mathematically inconsistent. Since the noise terms are independent, the equations would not allow α and α^* to remain complex conjugate. Employing the positive P-representation (in which the diffusion matrix is positive definite) resolves this problem, since β is independent of α.

8.3.2 Linearization and stability

In what follows, we assume, for simplicity, an exact resonance of the driving field frequency with the cavity-mode frequency, i.e. we assume $\Delta\omega = 0$, so that $\tilde{\gamma} = \gamma$.

The semiclassical steady state α_0 (with $\beta_0 = \alpha_0^*$) is found from Eq. (8.30) by ignoring the noise terms and setting $d\alpha/dt = d\beta/dt = 0$. Introducing the amplitudes and phases for $\mathcal{E}$ and α_0, according to $\mathcal{E} = |\mathcal{E}| \exp(i\Phi)$ and $\alpha_0 = |\alpha_0| \exp(i\varphi_0)$, we then find that the steady-state phase φ_0 and the amplitude $|\alpha_0|$ of the cavity mode are determined by

$$\begin{aligned} \varphi_0 &= \Phi, \\ |\mathcal{E}| &= \gamma^{(2)}|\alpha_0|^3 + \gamma|\alpha_0|. \end{aligned} \tag{8.34}$$

Thus the semiclassical steady-state solution has a definite phase and definite amplitude.

Stability of the steady state is checked by means of the linearization technique. That is, we treat fluctuations as small deviations around the steady state, so that

$$\begin{aligned} \alpha(t) &= \alpha_0 + \delta\alpha(t), \\ \beta(t) &= \alpha_0^* + \delta\beta(t). \end{aligned} \tag{8.35}$$

Substituting this into Eq. (8.30) and keeping only linear terms in $\delta\alpha(t)$ and $\delta\beta^*(t)$, we arrive at the following matrix form of the linearized equations:

$$\frac{d}{dt}\begin{pmatrix} \delta\alpha \\ \delta\beta \end{pmatrix} = -A^{(0)}\begin{pmatrix} \delta\alpha \\ \delta\beta \end{pmatrix} + B^{(0)}\begin{pmatrix} \xi_1(t) \\ \xi_2(t) \end{pmatrix}, \tag{8.36}$$

where the steady-state matrices $A^{(0)}$ and $B^{(0)}$ are given by

$$A^{(0)} = \begin{pmatrix} 2\gamma^{(2)}|\alpha_0|^2 + \gamma & \gamma^{(2)}\alpha_0^2 \\ \gamma^{(2)}(\alpha_0^*)^2 & 2\gamma^{(2)}|\alpha_0|^2 + \gamma \end{pmatrix}, \qquad B^{(0)} = \begin{pmatrix} i\sqrt{\gamma^{(2)}}\alpha_0 & 0 \\ 0 & -i\sqrt{\gamma^{(2)}}\alpha_0^* \end{pmatrix}. \tag{8.37}$$

Steady states are stable (and hence the linearization around them is valid) if all the eigenvalues of $A^{(0)}$ have positive real parts, ensuring that fluctuations will decay in time, rather than grow. Solving the eigenvalue problem for the matrix $A^{(0)}$, we find that both the eigenvalues,

$$\lambda_1 = \gamma + 3\gamma^{(2)}|\alpha_0|^2,$$
$$\lambda_2 = \gamma + \gamma^{(2)}|\alpha_0|^2,$$

are positive, and hence the semiclassical steady states (8.34) are stable.

In order to calculate several quantities of interest, we need to evaluate the two-time correlation functions, for example, $\langle\delta\alpha(t)\delta\alpha(s)\rangle$, that follow from the preceding stochastic differential equations. Let us now demonstrate how to find one of them. We are interested in the steady-state value of the correlation functions, that is, their value after the effect of any initial transients has died out. We can accomplish this by putting our initial conditions at $t = -\infty$. The solution to the stochastic differential equations is then

$$\begin{pmatrix} \delta\alpha(t) \\ \delta\beta^*(t) \end{pmatrix} = \int_{-\infty}^{t} dt'\, e^{-A^0(t-t')} B^{(0)} \begin{pmatrix} \xi_1(t') \\ \xi_2(t') \end{pmatrix}. \tag{8.38}$$

Noting that the eigenstates of $A^{(0)}$ are

$$\frac{1}{\sqrt{2}}\begin{pmatrix} e^{2i\varphi_0} \\ 1 \end{pmatrix}, \qquad \frac{1}{\sqrt{2}}\begin{pmatrix} -e^{2i\varphi_0} \\ 1 \end{pmatrix}, \tag{8.39}$$

we have that

$$e^{-A^{(0)}t} = \frac{1}{2}e^{-\lambda_1 t}\begin{pmatrix} 1 & e^{2i\varphi_0} \\ e^{-2i\varphi_0} & 1 \end{pmatrix} + \frac{1}{2}e^{-\lambda_2 t}\begin{pmatrix} 1 & -e^{2i\varphi_0} \\ -e^{-2i\varphi_0} & 1 \end{pmatrix}. \tag{8.40}$$

It is now straightforward to find $\delta\alpha$ and $\delta\beta^*$ and then find their correlation functions. For example, we find that

$$\langle\delta\alpha(t)\delta\alpha(s)\rangle = \frac{-\gamma^{(2)}\alpha_0^2}{4}\left(\frac{1}{\lambda_1}e^{-\lambda_1|t-s|} + \frac{1}{\lambda_2}e^{-\lambda_2|t-s|}\right). \tag{8.41}$$

8.4 Squeezing and photon anti-bunching

In Section 5.4.2, we discussed the squeezing of a single mode of the free electromagnetic field. We now want to consider the case in which the field has time variation beyond its simple frequency dependence, that is, when the mode operator $a(t)$ has a time dependence more complicated than just $e^{-i\omega t}$. In this case, the additional time dependence is due to the source that emitted the field. We now want to discuss what we mean by squeezing in this situation. Here we will calculate the squeezing in the case of the nonlinear absorber, as a simple example of this process.

In order to do so, we look at what we do when we measure squeezing. As we have seen, squeezing is measured by a process known as homodyne detection. The signal to be measured is mixed at a beam-splitter with a strong local oscillator, which is just a mode in a large-amplitude coherent state. Then photon counting measurements are performed on the resulting combined signal. The quantity that is then measured is directly related to the spectrum of fluctuations (or squeezing spectrum, if fluctuations are suppressed below the shot noise level) of the quadrature-phase amplitude of the field incident on the homodyne detector. This incident field, in turn, can be an output from a cavity where a particular nonlinear optical process takes place, in which case we use the definitions given in the input–output formalism section to relate the field in the cavity to the field incident on the detector. The quadrature-phase amplitude operator is then defined as

$$X^{out\theta}(t) = b^{out}(t)e^{-i\theta+i\omega_L t} + b(t)^{out\dagger}(t)e^{i\theta-i\omega_L t}, \tag{8.42}$$

where θ is the phase (defined relative to the phase of the measured field) of the local oscillator field and ω_L is its carrier frequency, chosen to coincide with the carrier frequency of the measured field. The spectrum or variance of the quadrature fluctuations is defined as

$$V(\omega,\theta) = \int_{-\infty}^{+\infty} d\tau\, e^{-i\omega\tau}\langle \Delta X^{out\theta}(t+\tau)\Delta X^{out\theta}(t)\rangle, \tag{8.43}$$

where $\Delta X^{out\theta} = X^{out\theta} - \langle X^{out\theta}\rangle$. We want to express the expectation value in normally ordered form so we can use our c-number representations of the field to evaluate them. First, using the fact that the two-time correlation function for the steady state is independent of t, we can simply set $t = 0$. Next, using the commutation relations for the out operators, $[b^{out}(t), b^{out\dagger}(t')] = \delta(t-t')$, we have that

$$\langle \Delta X^{out\theta}(\tau)\Delta X^{out\theta}(0)\rangle = \delta(\tau) + \langle : \Delta X^{out\theta}(\tau)\Delta X^{out\theta}(0) : \rangle, \tag{8.44}$$

where

$$\begin{aligned}
&\langle : \Delta X^{out\theta}(0)\Delta X^{out\theta}(\tau) : \rangle \\
&= \langle e^{-2i\theta} b^{out}(\tau)b^{out}(0) + b^{out\dagger}(\tau)b^{out}(0) + b^{out\dagger}(0)b^{out}(\tau) + e^{2i\theta} b^{out\dagger}(\tau)b^{\dagger}(0)\rangle \\
&- \langle X^{out\theta}(\tau)\rangle\langle X^{out\theta}(0)\rangle.
\end{aligned} \tag{8.45}$$

This then gives us that

$$V(\omega,\theta) = 1 + \int_{-\infty}^{+\infty} d\tau\, e^{-i\omega\tau}\langle : \Delta X^{out\theta}(\tau)\Delta X^{out\theta}(0) : \rangle. \tag{8.46}$$

The 1 on the right-hand side of Eq. (8.46) corresponds to the so-called shot noise level, which is characteristic of a vacuum field or a light field in a coherent state. The nonclassical effect of the reduction of quadrature fluctuations below the shot noise limit, or squeezing, occurs if $V(\omega,\theta) < 1$, and maximal, or perfect (100%), squeezing corresponds to $V(\omega,\theta) = 0$. This can occur in a particular spectral range, and for a specific value of θ. This θ value will then specify the squeezed quadrature. The conjugate quadrature ($\theta \rightarrow \theta + \pi/2$) will be the non-squeezed quadrature, which will have increased fluctuations according to the Heisenberg uncertainty principle.

We now want to use the input–output formalism to express the out operators in terms of the cavity operators. We recall from Eqs (6.113) and (6.114) that, for coherent or vacuum inputs,

$$\begin{aligned} \langle b^{out\dagger}(t), b^{out}(t')\rangle &= 2\gamma\langle a^\dagger(t), a(t')\rangle, \\ \langle b^{out}(t), b^{out}(t')\rangle &= 2\gamma\langle a(\max(t,t')), a(\min(t,t'))\rangle. \end{aligned} \tag{8.47}$$

Let us first consider $\tau = t - t' > 0$. In that case, we have

$$\begin{aligned} \langle b^{out\dagger}(\tau), b^{out}(0)\rangle &= 2\gamma\langle a^\dagger(\tau), a(0)\rangle, \\ \langle b^{out\dagger}(0), b^{out}(\tau)\rangle &= 2\gamma\langle a^\dagger(0), a(\tau)\rangle. \end{aligned} \tag{8.48}$$

This shows us that there is a direct correspondence between internal and external operator orderings for products of annihilation and creation operators. Since photodetectors detect normally ordered operator products of the *out* fields, we simply have to calculate normally ordered products of the internal fields and scale by the appropriate transmission factors. However, while the external operators are not interacting, so that annihilation operators commute at different times, this is not the case for internal operators. Therefore, we can only have a direct correspondence between internal and external fields if we use both normal ordering and time ordering. The general combination of internal operators measured is therefore a time-ordered, normally ordered, product. For coherent inputs with amplitude $\alpha^{in}(t)$, this means that

$$\begin{aligned} &\langle (b^{\dagger out}(t_1) - \alpha^*(t_1)) \cdots (b^{\dagger out}(t_n) - \alpha^*(t_n)) \\ &\quad \times (b^{out}(t_{n+1}) - \alpha^{in}(t_{n+1})) \cdots (b^{out}(t_{n+m}) - \alpha^{in}(t_{n+m}))\rangle \\ &\quad = (2\gamma)^{[n+m]/2}\langle a^\dagger(t_1) \cdots a^\dagger(t_n) a(t_{n+1}) \cdots a(t_{n+m})\rangle, \end{aligned} \tag{8.49}$$

where $t_1 \le t_2 \le \cdots \le t_n$ and $t_{n+1} \ge t_{n+2} \ge \cdots \ge t_{n+m}$. We note the convention that annihilation operators in correlation functions are time-ordered, so that the largest time is written first. Similarly, creation operators in correlation functions are time-anti-ordered, so that the smallest time is written first. This means that all of the correlation functions necessary for calculating the squeezing spectrum for $\tau > 0$ can be directly evaluated either by using the positive P-function at $t = 0$ or from the corresponding correlation functions resulting from the stochastic differential equations.

In the case that $\tau < 0$, we find that the correlation functions for the cavity mode are normally ordered with respect to the creation and annihilation operators evaluated at time τ. These can then be evaluated by using the positive P-function at time τ, but, since we are looking at the steady state, this is the same as the positive P-function at time 0, which also allows us to use the correlation functions derived from the stochastic differential equations. Consequently, we have that

$$\begin{aligned} V(\omega,\theta) &= 1 + 2\gamma\int_{-\infty}^{+\infty} d\tau\, e^{-i\omega\tau}\langle :\Delta X^\theta(\tau)\Delta X^\theta(0):\rangle, \\ &= 1 + 2\gamma\int_{-\infty}^{+\infty} d\tau\, e^{-i\omega\tau}[\langle\beta(\tau),\alpha(0)\rangle + \langle\alpha(\tau),\beta(0)\rangle \\ &\quad + \langle\alpha(\tau),\alpha(0)\rangle e^{-2i\theta} + \langle\beta(\tau),\beta(0)\rangle e^{2i\theta}], \end{aligned} \tag{8.50}$$

where X^θ defines the quadrature-phase amplitude operator for the intra-cavity field,

$$X^\theta = ae^{-i\theta} + a^\dagger e^{i\theta}, \tag{8.51}$$

with a and $a^\dagger$ being slowly varying operators, which means that $\omega = 0$ corresponds to cavity resonance.

The calculation of $V(\omega, \theta)$ and its analysis for squeezing is now straightforward. For the two-photon damping problem under consideration, the squeezed quadrature is the one that determines the amplitude fluctuations ($\theta = 0$) of the cavity output field. The resulting expression for the corresponding squeezing spectrum is

$$V(\omega, \theta = 0) = 1 - \frac{16\gamma^{(2)}\gamma|\alpha_0|^2}{\omega^2 + [2\gamma + 6\gamma^{(2)}|\alpha_0|^2]^2}. \tag{8.52}$$

As we can see, the maximum squeezing is obtained in the zero frequency range ($\omega = 0$) and, maximizing the squeezing with respect to $|\alpha_0|^2$, at $3\gamma^{(2)}|\alpha_0|^2 = \gamma$, giving

$$V_{\text{min}}(\omega = 0, \theta = 0) = \frac{2}{3}. \tag{8.53}$$

This corresponds to $\sim 33\%$ noise reduction below the shot noise level.

The nonclassical photon statistics occurring due to the two-photon absorption are also manifested via the second-order correlation function, which is defined in terms of stochastic averages as

$$g^{(2)} = \frac{\langle a^\dagger a^\dagger a a\rangle}{\langle a^\dagger a\rangle^2} = \frac{\langle \beta^*\beta^*\alpha\alpha\rangle}{\langle \beta^*\alpha\rangle^2}. \tag{8.54}$$

If $g^{(2)} < 1$, the photon statistics are sub-Poissonian. Within the linearized treatment of fluctuations, this can be written in terms of equal-time correlations of $\delta\alpha$ and $\delta\beta^*$, yielding

$$g^{(2)} \simeq 1 + \frac{2\langle\delta\beta^*\delta\alpha\rangle + e^{-2i\phi_0}\langle\delta\alpha^2\rangle + e^{2i\phi_0}\langle\delta\beta^{*2}\rangle}{|\alpha_0|^2} = 1 - \frac{(\gamma^{(2)}/\gamma)}{1 + 3n_0(\gamma^{(2)}/\gamma)} < 1, \tag{8.55}$$

where $n_0 = |\alpha_0|^2$. As we see, $g^{(2)} < 1$, i.e. the second-order correlation function here is smaller than that of a coherent light, so that the photon statistics are, in fact, sub-Poissonian, which implies that the light is nonclassical. This effect is also called photon anti-bunching, which here is due to the anticorrelation between photon pairs. This is easy to understand as the two-photon absorber effectively removes the photon pairs that were present in the incident coherent driving field, thus making the probability of a joint photo-detection of the remaining photon pairs (described by $g^{(2)}$) smaller. The maximum anti-bunching is obtained in the limit $n_0 \to \infty$, yielding $g^{(2)} = 1 - 1/(3n_0)$. For n_0 large, however, this is a small effect, because the deviation of the value of $g^{(2)}$ from 1 is small.

A larger-scale manifestation of the nonclassical photon statistics can be seen via photon number fluctuations, corresponding to a direct photo-detection scheme. This is usually quantified in terms of the relative variance in the photon number fluctuations $\langle(\Delta n)^2\rangle/\langle n\rangle$, where $n = a^\dagger a$ and $\Delta n = n - \langle n\rangle$. For coherent light, $\langle(\Delta n)^2\rangle/\langle n\rangle = 1$; while for nonclassical light, the photon number fluctuations can be suppressed below the coherent level,

$\langle(\Delta n)^2\rangle/\langle n\rangle < 1$. In the present problem of two-photon absorption, the relative variance is

$$\frac{\langle(\Delta n)^2\rangle}{\langle n\rangle} \simeq 1 - \frac{n_0(\gamma^{(2)}/\gamma)}{1+3n_0(\gamma^{(2)}/\gamma)}, \tag{8.56}$$

which in the limit $n_0 \to \infty$ goes to $1 - \frac{1}{3}$. This implies that the photon number fluctuations are reduced compared to the case of coherent light.

8.5 High-Q laser

Here we consider the simplest model of a laser that includes all the essential features of any practical laser. These are linear or one-photon loss (due to the output coupler), linear gain (due to inverted electronic transitions) and saturation (due to nonlinear loss), which are experienced by a single cavity mode. All lasers must have these three elements to operate as a laser.

If there is only a linear gain, then, of course, there can be no output, since an output causes a loss. With both gain and loss, but no saturation, the laser intensity is either nearly zero (below threshold), or else it rises infinitely (above threshold). Neither case is very useful as a model of a real laser, which has a finite output above threshold. This is why a saturation mechanism – an intensity-dependent nonlinear loss – is needed. In detailed laser models, one can include the precise level structure in the gain medium.

8.5.1 Master equation including laser reservoir

Here the laser saturation will be modeled by the two-photon absorber term. The simple model used here is also obtainable in the limit of a high-Q laser operating with conventional gain saturation, not too far above threshold. Thus, combining linear loss, gain and nonlinear loss, we can model the laser by the following master equation in the rotating-frame picture, with the frame defined relative to the cavity resonance frequency:

$$\begin{aligned}\frac{d\rho}{dt} &= \gamma(2a\rho a^\dagger - a^\dagger a\rho - \rho a^\dagger a) + g(2a^\dagger\rho a - aa^\dagger\rho - \rho aa^\dagger) \\ &\quad + \frac{\gamma^{(2)}}{2}(2a^2\rho a^{\dagger 2} - a^{\dagger 2}a^2\rho - \rho a^{\dagger 2}a^2).\end{aligned} \tag{8.57}$$

Here the meanings of the terms are:

- γ, linear loss (units photons/s),
- g, linear gain (units photons/s),
- $\gamma^{(2)}$, nonlinear loss (units photons/s).

We have omitted any nonlinear refractive index terms – these are also present in most lasers. Another realistic factor that we are ignoring is the presence of competing modes. Finally, we note that the two-photon loss term here represents a non-saturable two-photon absorber, that is, a two-photon absorber that can handle arbitrarily large light intensity.

In reality, of course, the nonlinear medium consists of a finite number of atoms that can undergo a two-photon transition. If the light intensity is large enough to cause all the atoms to absorb pairs of photons, then further increase of the intensity will not be seen by the nonlinear medium – the absorber will saturate in an even more nonlinear way, and this regime is not described by the nonlinear loss term that we included in Eq. (8.57).

8.5.2 Fokker–Planck equation and stochastic differential equations

Transforming the above master equation into the FPE in P-representation, and neglecting the two-photon absorption term in the diffusion matrix, allows us to employ the Glauber–Sudarshan P-representation. This is a good approximation provided the operating intensity is not too far above the threshold. The corresponding Fokker–Planck equation then has the following form:

$$\frac{\partial P(\alpha,\alpha^*)}{\partial t} = \left\{ \frac{\partial}{\partial \alpha}[(\gamma - g)\alpha + \gamma^{(2)}|\alpha|^2\alpha] + \frac{\partial}{\partial \alpha^*}[(\gamma - g)\alpha^* + \gamma^{(2)}|\alpha|^2\alpha^*] + \frac{\partial^2}{\partial\alpha\partial\alpha^*}(2g)\right\} P(\alpha,\alpha^*), \tag{8.58}$$

The diffusion matrix $D(\boldsymbol{\alpha})$ and the matrix $B(\boldsymbol{\alpha})$ (where $B(\boldsymbol{\alpha})B^{\mathrm{T}}(\boldsymbol{\alpha}) = D(\boldsymbol{\alpha})$, and $\boldsymbol{\alpha} = (\alpha, \alpha^*)$) corresponding to this FPE are given by

$$D = \begin{pmatrix} 0 & 2g \\ 2g & 0 \end{pmatrix}, \qquad B = \sqrt{g}\begin{pmatrix} 1 & i \\ 1 & -i \end{pmatrix}. \tag{8.59}$$

The SDEs corresponding to the above FPE (in Ito form) are

$$\frac{d\alpha}{dt} = -(\gamma - g)\alpha + \gamma^{(2)}|\alpha|^2\alpha + F(t),$$
$$\frac{d\alpha^*}{dt} = -(\gamma - g)\alpha^* + \gamma^{(2)}|\alpha|^2\alpha^* + F^*(t). \tag{8.60}$$

Here we have used $F(t)$ and $F^*(t)$ to define the noise terms corresponding to the complex conjugate variables α and α^*, respectively. These are given by

$$F(t) = \sqrt{g}(\xi_1(t) + i\xi_2(t)),$$
$$F^*(t) = \sqrt{g}(\xi_1(t) - i\xi_2(t)),$$

where $\xi_1(t)$ and $\xi_2(t)$ are the independent noise increments. We see that F^* is indeed the complex conjugate of the noise term F, and therefore α and α^* remain complex conjugate, and no paradox arises (relating to the validity of employing the Glauber–Sudarshan representation). This is only true because we have neglected, in the FPE, the diffusion coefficient due to the nonlinear loss. With the use of the usual correlation properties of the independent increments $\xi_1(t)$ and $\xi_2(t)$, we find that the only nonzero correlations of the noise terms $F(t)$ and $F^*(t)$ are

$$\langle F(t)F^*(t')\rangle = \langle F^*(t)F(t')\rangle = g\delta(t - t'). \tag{8.61}$$

The semiclassical steady-state solutions and their stability properties follow immediately from the above SDE and standard linearization techniques. We find that there are two types of solutions:

- $\alpha_0 = 0$. This is stable if $g - \gamma < 0$, i.e. *below threshold*.

- $|\alpha_0|^2 = (g-\gamma)/\gamma^{(2)}$. This is physically meaningful for $g-\gamma > 0$, i.e. *above threshold*, but contains no information about the steady-state phase φ_0 of the complex field amplitude $\alpha_0 = |\alpha_0|\exp(i\varphi_0)$. The solution is not stable, since the linearization reveals a zero eigenvalue.

Thus, in the below-threshold regime, we can treat the system using the linearization technique around the trivial solution $\alpha_0 = 0$. This implies that $\alpha(t) = \alpha_0 + \delta\alpha(t) = \delta\alpha(t)$, and neglects (as long as $\langle\delta\alpha^*(t)\delta\alpha(t)\rangle \ll 1$) nonlinear terms in $\delta\alpha(t)$, in Eq. (8.60), so that the linearized SDEs are

$$\frac{d\delta\alpha}{dt} = -(\gamma - g)\delta\alpha + F(t),$$
$$\frac{d\alpha^*}{dt} = -(\gamma - g)\delta\alpha^* + F^*(t). \tag{8.62}$$

In the above-threshold region, the system is not stable (the linearized equations have a zero eigenvalue) and cannot be analyzed correctly by the assumption of small fluctuations and using the linearization techniques. As we will see, this instability is physically due to the phenomenon of phase diffusion.

8.5.3 Exact steady-state analysis

To analyze the problem more carefully, we transform to new variables, the intensity and the phase,

$$I = \alpha^*\alpha, \qquad \varphi = \frac{1}{2i}\ln\frac{\alpha}{\alpha^*}. \tag{8.63}$$

Using the usual rules of differentiation, we have

$$\frac{\partial}{\partial\alpha}\alpha = \frac{\partial}{\partial I}I + \frac{1}{2i}\frac{\partial}{\partial\varphi} = \left(\frac{\partial}{\partial\alpha^*}\alpha^*\right)^*,$$
$$\frac{\partial^2}{\partial\alpha\partial\alpha^*} = I\frac{\partial^2}{\partial I^2} + \frac{\partial}{\partial I} + \frac{1}{4I}\frac{\partial^2}{\partial\varphi^2} = \frac{\partial^2}{\partial I^2}I - \frac{\partial}{\partial I} + \frac{\partial^2}{\partial\varphi^2}\frac{1}{4I}, \tag{8.64}$$

and may transform the original FPE (8.58) into the following form:

$$\frac{\partial P(I,\varphi)}{\partial t} = \left\{2\frac{\partial}{\partial I}[(\gamma - g)I + \gamma^{(2)}I^2 - g] + \frac{1}{2}\frac{\partial^2}{\partial I^2}(4gI) + \frac{1}{2}\frac{\partial^2}{\partial\varphi^2}\left(\frac{g}{I}\right)\right\}P(I,\varphi). \tag{8.65}$$

The normalization condition for the distribution function

$$P(I,\varphi) \equiv P(\alpha = \sqrt{I}e^{i\varphi}, \alpha^* = \sqrt{I}e^{-i\varphi})$$

is given by

$$\int d^2\alpha\, P(\alpha,\alpha^*) = \int_0^{2\pi} d\varphi \int_0^{\infty} r\, dr\, P(re^{i\varphi}, re^{-i\varphi}) = \frac{1}{2}\int_0^{2\pi} d\varphi \int_0^{\infty} dI\, P(I,\varphi) = 1, \tag{8.66}$$

where we have let $r = \sqrt{I}$.

The FPE (8.65) is still exact, and can readily be solved for the steady state, using the method of potential equations (see Eqs (7.129)–(7.131)). This yields the following steady-state distribution:

$$P_{ss}(I,\varphi) = N\exp\left[-\frac{\gamma^{(2)}}{g}I^2 + \frac{g-\gamma}{g}I\right], \tag{8.67}$$

where N is the normalization constant. As we see, the distribution function $P_{ss}(I,\varphi)$ depends only on the intensity I, but not on the phase φ. The steady-state phase distribution function alone, which is defined as

$$\Phi(\varphi) = \int_0^\infty r\,dr\,P(re^{i\varphi}, re^{-i\varphi}) = \frac{1}{2}\int_0^\infty dI\,P(I,\varphi), \tag{8.68}$$

can therefore be readily found from the normalization condition (8.66). Defining $P_1(I) = P_{ss}(I,\varphi)/2$ to reflect the fact that the (joint) distribution $P_{ss}(I,\varphi)$ does not actually depend on φ, we have, from Eq. (8.66),

$$1 = \frac{1}{2}\int_0^{2\pi} d\varphi \int_0^\infty dI\,P_{ss}(I,\varphi) = \int_0^{2\pi} d\varphi \int_0^\infty dI\,P_1(I) = 2\pi\Phi_{ss}(\varphi), \tag{8.69}$$

so that the steady-state phase distribution is

$$\Phi_{ss}(\varphi) = \frac{1}{2\pi}. \tag{8.70}$$

This implies that the steady-state phase of the cavity field is uniformly distributed and does not have a well-defined value, which is due to phase diffusion.

In contrast to this, the distribution function $P_1(I)$ demonstrates a peak, corresponding to the most probable value of the intensity. The peak is located at zero in the below-threshold regime and at a nonzero value in the above-threshold region. The width of the distribution function, corresponding to the uncertainty in the intensity fluctuations, becomes smaller well above the threshold, so that the laser intensity acquires a well-defined value. This distribution function can be used to calculate equal-time intensity correlation functions, including the laser operating intensity (in photon number units)

$$\langle a^\dagger a\rangle = \langle \alpha^* \alpha\rangle = \frac{1}{2}\int_0^{2\pi} d\varphi \int_0^\infty dI\,I P_{ss}(I,\varphi) = 2\pi\int_0^\infty dI\,I P_1(I). \tag{8.71}$$

In the far-above-threshold region ($g - \gamma \gg \sqrt{\gamma^{(2)}g}$), the steady-state intensity can be approximated by

$$\langle a^\dagger a\rangle \simeq \frac{g-\gamma}{\gamma^{(2)}} \quad \text{(far above threshold).} \tag{8.72}$$

In the far-below-threshold regime ($\gamma - g \gg \sqrt{\gamma^{(2)}g}$), the distribution function is approximated by $P_{ss}(\alpha,\alpha^*) \simeq N\exp[-(\gamma-g)|\alpha|^2/g]$, i.e. by a simple Gaussian centered at $|\alpha| = 0$, giving the average intensity

$$\langle a^\dagger a\rangle \simeq \frac{g}{\gamma - g} \quad \text{(far below threshold).} \tag{8.73}$$

8.6 Laser linewidth

While the above steady-state analysis can provide exact results for equal-time intensity correlations (or moments), it does not solve time-domain problems, such as two-time correlation functions or the problem of practical importance – the laser linewidth.

8.6.1 Below-threshold regime

In the below-threshold regime, problems involving two-time correlation functions can be solved within the linearized treatment of fluctuations, using the solutions to Eq. (8.62), because the system is stable. In the long-time limit we have

$$\delta\alpha(t) = \int_{-\infty}^{t} dt' \exp[-(\gamma - g)(t - t')]F(t'), \tag{8.74}$$

so that $\langle a(t)\rangle = \langle \delta\alpha(t)\rangle = 0$, while the two-time correlation function of the amplitudes is given by

$$\langle a^\dagger(t)a(t+\tau)\rangle = \langle \delta\alpha^*(t)\delta\alpha(t+\tau)\rangle = \frac{g}{\gamma - g}\exp[-(\gamma - g)|\tau|]. \tag{8.75}$$

The calculation of the laser intensity spectrum

$$I(\omega) = \int_{-\infty}^{+\infty} d\tau\, e^{i(\omega-\omega_c)\tau}\langle a^\dagger(t)a(t+\tau)\rangle, \tag{8.76}$$

where ω_c is the light carrier frequency determined by the cavity resonance frequency, is now straightforward. Substituting (8.75) into (8.76) gives

$$I(\omega) = \frac{2g}{(\omega - \omega_c)^2 + (\gamma - g)^2}, \tag{8.77}$$

i.e. the laser spectrum below threshold is given by a Lorentzian with half-width $(\gamma - g)$, and is broad far below threshold (the width is only limited by the cavity linewidth γ). Together with the fact that the higher-order correlation functions have the same statistical properties as the Gaussian noise term $F(t)$ (so that, for example, the second-order correlation function $g^{(2)} = \langle a^\dagger a^\dagger aa\rangle / \langle a^\dagger a\rangle^2 \simeq 2$), this means that the laser light below threshold is essentially the same as thermal light.

8.6.2 Above-threshold regime

To analyze the problem in the above-threshold region, we use the (Ito) SDEs equivalent to the FPE (8.65):

$$\begin{aligned}
\frac{dI}{dt} &= -2(\gamma - g)I - 2\gamma^{(2)}I^2 + 2g + \sqrt{4gI}\,\xi_I(t), \\
\frac{d\varphi}{dt} &= \sqrt{\frac{g}{I}}\,\xi_\varphi(t),
\end{aligned} \tag{8.78}$$

where the noise sources $\xi_I(t)$ and $\xi_\varphi(t)$ are not correlated with each other, and have the usual delta-function autocorrelations

$$\langle \xi_I(t)\xi_I(t')\rangle = \langle \xi_\varphi(t)\xi_\varphi(t')\rangle = \delta(t-t'). \tag{8.79}$$

Equations (8.78) can also be derived from Eqs (8.60), using Ito's variable-change formula (7.146).

As we see, the equation for the intensity variable I is decoupled from the equation for the phase φ. Furthermore, its steady-state solution I_0 (which is found from $\gamma^{(2)}I_0^2 - (g - \gamma)I_0 - g = 0$) is stable against small fluctuations above threshold, since the corresponding eigenvalue of the linearized equation is positive for $g - \gamma > 0$. The instability in the system is due to the phase variable. These properties of the factorized equations can be used to treat the intensity subsystem within the linearized theory of fluctuations, while the equation for the phase variable is to be treated exactly. Within the linearized theory, i.e. assuming smallness of intensity fluctuations $\delta I(t) = I - I_0$ (which is valid far above threshold), the intensity I in the equation for φ can be replaced by I_0. The evolution of the phase variable is then governed purely by a Gaussian noise term

$$\frac{d\varphi}{dt} = F_\varphi^0(t), \tag{8.80}$$

where $F_\varphi^0(t) = \sqrt{g/I_0}\,\xi_\varphi(t) \equiv \sqrt{D_\varphi}\,\xi_\varphi(t)$. This in turn corresponds to an FPE for a phase distribution function, having the form of the diffusion equation

$$\frac{\partial \phi(\varphi, t)}{\partial t} = \frac{D_\varphi}{2}\frac{\partial^2 \phi}{\partial \varphi^2}, \tag{8.81}$$

where the diffusion coefficient is

$$D_\varphi = \frac{g}{I_0}. \tag{8.82}$$

The steady-state solution to the diffusion equation, such that $\phi_{ss}(\varphi + 2\pi) = \phi_{ss}(\varphi)$, is $\phi_{ss}(\varphi) = 1/(2\pi)$, which agrees with the earlier exact result, Eq. (8.70).

Using the above approximations, and Eq. (8.80), we can now calculate the two-time amplitude correlation function $\langle a^\dagger(t)a(t+\tau)\rangle$ and find the intensity spectrum of the laser operating well above threshold. Replacing I by I_0, we have for the correlation function

$$\langle a^\dagger(t)a(t+\tau)\rangle = \left\langle \sqrt{I(t)I(t+\tau)}\, e^{-i[\varphi(t+\tau)-\varphi(t)]}\right\rangle \simeq I_0\left\langle e^{i[\varphi(t+\tau)-\varphi(t)]}\right\rangle. \tag{8.83}$$

To evaluate the average $\langle e^{i[\varphi(t+\tau)-\varphi(t)]}\rangle$, we write the solution to Eq. (8.80) as

$$\varphi(t) = \varphi(t_0) + \int_{t_0}^{t} F_\varphi^0(t')\,dt', \tag{8.84}$$

and note that $\varphi(t+\tau) - \varphi(t)$ has the statistical properties of a Gaussian variable with zero mean. For a Gaussian variable x such that $\langle x\rangle = 0$, higher-order moments factorize according to $\langle x^{2n}\rangle = \langle x^2\rangle^n (2n)!/(2^n n!)$ and $\langle x^{2n+1}\rangle = 0$, and therefore $\langle \exp(x)\rangle = \exp(\langle x^2\rangle/2))$. This results in

$$\langle a^\dagger(t)a(t+\tau)\rangle \simeq I_0\left\langle e^{i[\varphi(t+\tau)-\varphi(t)]}\right\rangle = I_0 e^{-\langle[\varphi(t+\tau)-\varphi(t)]^2\rangle/2} = I_0 e^{-D_\varphi|\tau|/2}. \tag{8.85}$$

The laser intensity spectrum above threshold is therefore

$$I(\omega) = \int_{-\infty}^{+\infty} d\tau\, e^{i(\omega-\omega_c)\tau} \langle a^\dagger(t) a(t+\tau)\rangle \simeq \frac{I_0 D_\varphi}{(\omega-\omega_c)^2 + (D_\varphi/2)^2}, \tag{8.86}$$

where the diffusion coefficient is $D_\varphi = g/I_0$. Thus the spectrum has the form of a Lorentzian centered at ω_c with half-width $D_\varphi/2$. The width is essentially due to the phase diffusion, and becomes narrow with the increase of I_0. From Eq. (8.85), one can introduce the characteristic phase correlation time $t_c = 2/D_\varphi = 2I_0/g$, which grows with I_0 and can be very large. In other words, well above threshold, the laser can maintain the value of its phase within a sufficiently long time period, since the phase diffusion is slow (D_φ is small). The validity of the linearized treatment of the intensity fluctuations, on the other hand, implies that one can apply coherent-state factorization properties to higher-order intensity correlation functions (giving, for example, $g^{(2)} \simeq 1$). This means that the laser operating well above threshold produces essentially coherent light, with coherence time determined by $t_c \sim I_0/g$.

8.7 Laser quantum state: number or coherent?

As long as lasers have been around, there has been a long-standing debate about their quantum state. Is it a coherent state or a number state? This is a subtle point, because the current state of quantum mechanics is such that we can only really talk about whether or not a theory predicts what is measured. If two quantum systems have pure states, then a distinction is possible, as there are different predictions.

However, as we can see from the calculations given above, the laser is not in a pure state. For mixed states, it is possible to have different probabilistic mixtures of states that predict the same outcomes. Under these circumstances, we cannot use experiment to distinguish whether a laser is in a coherent state or a number state. The two descriptions are completely equivalent. There may be an advantage to one or the other description in terms of ease of calculation or philosophical interpretation. However, in terms of science, these criteria do not lead to distinct and falsifiable predictions.

To demonstrate this equivalence, we first note that our coherent-state representation predicts a random phase for the laser in the steady state. The simplest random phase P-function is one with fixed amplitude:

$$P(\alpha) = \frac{1}{2\pi} \int d\varphi\, \delta^2(\alpha - \sqrt{I} e^{i\phi}). \tag{8.87}$$

The corresponding density matrix is then a probabilistic mixture of coherent states with random phases:

$$\rho = \frac{1}{2\pi} \int d\varphi\, |\sqrt{I} e^{i\phi}\rangle\langle\sqrt{I} e^{i\phi}|. \tag{8.88}$$

We now expand the density matrix using Eq. (5.30), to give

$$\rho = \frac{1}{2\pi} \int d\varphi\, e^{-I} \sum_{n,m=0}^{\infty} \frac{(\sqrt{I})^{n+m} (e^{i\varphi})^{n-m}}{\sqrt{n!}\sqrt{m!}} |n\rangle\langle m|. \tag{8.89}$$

The terms with $m \neq n$ all vanish after phase-integration, so that

$$\rho = e^{-I} \sum_{n=0}^{\infty} \frac{I^n}{n!} |n\rangle\langle n|. \tag{8.90}$$

This is just a probabilistic distribution over number states, with a Poissonian weighting.

Accordingly, the best answer to the question about which is the quantum state of a laser is: *both.*

We note that a laser has other sources of number fluctuation. In the laser theory given above, one obtains a slightly broader than Poissonian number distribution, because of this. It is also possible, by controlling the pump fluctuations, to achieve a narrower than Poissonian distribution, but this is usually only possible in a restricted frequency range of the output field.

In general, the important point is that the number distribution is much narrower than for thermal light. This leads to Nobel Laureate Glauber's important observation, which is that the only way to reliably distinguish a laser from a light-bulb is to measure the second-order coherence, $g^{(2)}$. As far as the first-order coherence is concerned, one can filter a light-bulb's thermal radiation output to any desired bandwidth, until it matches a laser bandwidth. Of course, this is very inefficient, and not useful, but we are talking about questions of principle here.

For a laser, which we call a 'coherent' source, one obtains the second-order coherence of

$$g_c^{(2)}(0) \approx 1. \tag{8.91}$$

For a light-bulb, which we call a 'thermal' source, a single-mode measurement of number fluctuations gives

$$g_{th}^{(2)}(0) \approx 2. \tag{8.92}$$

Note that this measurement requires spectral and spatial filtering to achieve a bandwidth less than the photo-detector inverse response time, so that only a single mode is being measured.

We often model the output of a laser as a coherent state for simplicity. The cautious reader will ask, with good reason, why are phase fluctuations being ignored?

In fact, the overall phase uncertainty is not an observable property. There is no Hermitian observable that corresponds to the phase. One can only measure relative phase in practice, and indeed the outputs of two independent lasers are observed to have a random relative phase, as expected.

What is more important, when it comes to practical measurements, is the time-scale of the phase fluctuations. If these drift on much slower time-scales than the interferometer or atoms that are driven, then the coherent-state approach works well, although it may have to be supplemented by the inclusion of relevant phase noise and intensity noise to obtain more accurate predictions.

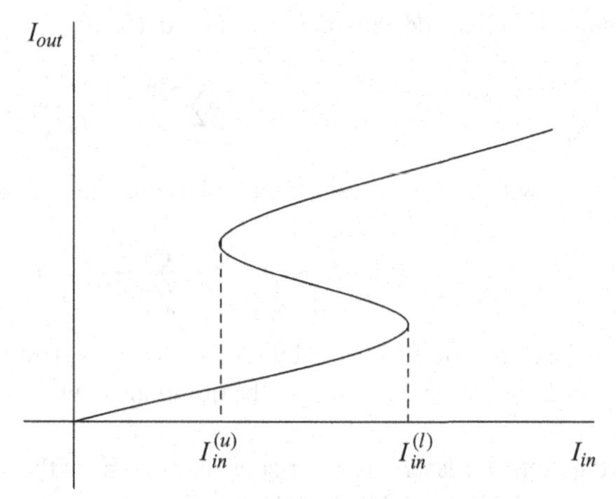

Fig. 8.1 A plot of output intensity, I^{out}, versus input intensity, I^{in}, for a bistable system. For input intensities between $I^{in(u)}$ and $I^{in(l)}$, the system can have two possible output intensities (the states on the backward-sloping part of the curve are unstable, and will, consequently, not occur). Which output intensity actually occurs depends on the history of the system.

8.8 Open nonlinear interferometer

We have already met the Hamiltonian for a single-mode field in a Kerr medium and seen some of the effects to which it can give rise in Section 5.8. Here we shall express the interaction part as

$$H_{int} = \frac{\kappa}{2}(a^{\dagger}a)^2. \tag{8.93}$$

What we now want to do is place this medium in a cavity, which will turn the system into a nonlinear interferometer.

Before going into the quantum mechanical treatment, it is useful to see what happens with classical fields. Our system consists of a cavity containing a medium with a intensity-dependent refractive index, and an input field that drives the cavity. What occurs is a phenomenon known as optical bistability. For a range of values of the input intensity, I^{in}, there are two possible values of the intensity of the cavity field (or the output field, see Figure 8.1). A plot of the cavity intensity versus the input intensity has two branches: a lower branch that starts at the origin and only extends up to a value we shall call $I^{in(l)}$; and an upper branch that starts at a point we shall call $I^{in(u)}$, where $I^{in(l)} > I^{in(u)}$. The values of the cavity field on the upper branch are always greater than those on the lower branch.

Which value the cavity field takes for a given value of I^{in} in the range $I^{in(l)} \geq I^{in} \geq I^{in(u)}$ is determined by the past history of the system, a phenomenon known as hysteresis. If we start the input intensity at small values and increase it, the system will stay on the lower branch until the input intensity reaches $I^{in(l)}$, at which point it will jump to the upper branch. If we start the input intensity at large values and decrease it, the system will stay on the upper branch until the input intensity reaches $I^{in(u)}$ and then jump to the lower branch.

In the quantum case, the nonlinear single-mode interferometer we consider will have an external input, which will take the form of an external driving field, and both linear and nonlinear losses. To allow for this, we include the usual one-photon losses for the cavity mode, and, in addition, we incorporate two-photon absorption. The combined effect of the two-photon absorption and the $\chi^{(3)}$ nonlinearity can be described by a complex nonlinear coupling constant in the FPE and SDE, thus allowing us to incorporate both types of nonlinear effects into a single model. We find that complex P-representation gives an exact analytical steady-state solution to this quantum nonlinear dissipative system, while the positive P-representation allows us to find approximate solutions (valid at large photon numbers) by using the SDE approach and the linearization technique.

Our nonlinear system can now be modeled, in the rotating-wave approximation, by the following Hamiltonian:

$$H = \hbar\omega a^\dagger a + \frac{\hbar\kappa}{2} a^{\dagger 2} a^2 + i\hbar(\mathcal{E}e^{-i\omega_L t} a^\dagger - \mathcal{E}^* e^{i\omega_L t} a) + a\Gamma_1^\dagger + a^\dagger\Gamma_1 + a^2\Gamma_2^\dagger + a^{\dagger 2}\Gamma_2. \tag{8.94}$$

Here $a^\dagger$ (a) is the creation (annihilation) operator for the cavity mode with the frequency ω, κ is the nonlinear coupling constant for the self-phase modulation process, proportional to the third-order susceptibility $\chi^{(3)}$, while $\mathcal{E}$ is the amplitude of the coherent driving field with the frequency ω_L. In addition, Γ_1, $\Gamma_1^\dagger$ and Γ_2, $\Gamma_2^\dagger$ are reservoir operators describing one-photon and two-photon losses, which will give rise to the damping rates γ and $\gamma^{(2)}$, respectively.

Utilizing standard techniques to eliminate the reservoir operators, one can obtain the following rotating-frame-picture master equation for the density matrix ρ of the cavity mode:

$$\frac{d\rho}{dt} = [-i\Delta\omega a^\dagger a + Ea^\dagger - E^* a, \rho] - i\frac{\kappa}{2}[a^{\dagger 2}a^2, \rho] + \gamma(2a\rho a^\dagger - \rho a^\dagger a - a^\dagger a\rho) + \frac{\gamma^{(2)}}{2}(2a^2\rho a^{\dagger 2} - \rho a^{\dagger 2}a^2 - a^{\dagger 2}a^2\rho), \tag{8.95}$$

where $\Delta\omega = \omega - \omega_L$ is the cavity detuning, as before.

The master equation is next transformed into a Fokker–Planck equation. Assuming vanishing boundary terms, we obtain the following FPE (in either the complex or positive P-representation):

$$\frac{\partial P(\alpha,\beta)}{\partial t} = \left\{ \frac{\partial}{\partial\alpha}(\tilde{\gamma}\alpha + \tilde{\kappa}\alpha^2\beta^* - \mathcal{E}) + \frac{\partial}{\partial\beta}(\tilde{\gamma}^*\beta + \tilde{\kappa}^*\beta^2\alpha - \mathcal{E}^*) + \frac{1}{2}\frac{\partial^2}{\partial\alpha^2}(-\tilde{\kappa}\alpha^2) + \frac{1}{2}\frac{\partial^2}{\partial\beta^2}(-\tilde{\kappa}^*\beta^2) \right\} P(\alpha,\beta), \tag{8.96}$$

where $\tilde{\gamma}$ and $\tilde{\kappa}$ are defined as follows: $\tilde{\gamma} = \gamma + i\Delta\omega$ and $\tilde{\kappa} = \gamma^{(2)} + i\kappa$.

8.8.1 Semiclassical analysis and bistability

If the FPE (8.96) is treated in the positive P-representation, then the equivalent set of (Ito) SDEs has the form

$$\frac{d\alpha}{dt} = -\tilde{\gamma}\alpha - \tilde{\kappa}\alpha^2\beta + \mathcal{E} + i\sqrt{\tilde{\kappa}}\,\alpha\xi_1(t),$$
$$\frac{d\beta}{dt} = -\tilde{\gamma}^*\beta - \tilde{\kappa}^*\beta^2\alpha + \mathcal{E}^* - i\sqrt{\tilde{\kappa}^*}\,\beta\xi_2(t), \tag{8.97}$$

where the $\xi_i(t)$ are independent Gaussian noise terms with zero means and the following nonzero correlations:

$$\langle \xi_i(t)\xi_j(t')\rangle = \delta_{ij}\delta(t-t'). \tag{8.98}$$

The semiclassical steady states ($\alpha_0 = |\alpha_0|\exp(i\varphi_0)$) are determined by

$$0 = -\tilde{\gamma}\alpha_0 - \tilde{\kappa}\alpha_0|\alpha_0|^2 + \mathcal{E}, \tag{8.99}$$

which yields the equations

$$|\mathcal{E}|^2 = |\alpha_0|^2[(\gamma^{(2)}|\alpha_0|^2 + \gamma)^2 + (\kappa|\alpha_0|^2 + \Delta\omega)^2],$$
$$\varphi_0 = \varphi - \tan^{-1}\left(\frac{\kappa|\alpha_0|^2 + \Delta\omega}{\gamma^{(2)}|\alpha_0|^2 + \gamma}\right), \tag{8.100}$$

where φ is the phase of the driving field, and $\mathcal{E} = |\mathcal{E}|\exp(i\varphi)$. Defining the dimensionless parameters ε and λ as

$$\varepsilon = \frac{2\mathcal{E}}{\tilde{\kappa}}, \qquad \lambda = \frac{2\tilde{\gamma}}{\tilde{\kappa}}, \tag{8.101}$$

the above state equation for the intensity $n_0 = |\alpha_0|^2$ can be rewritten as

$$|\varepsilon|^2 = 4(n_0)^3 + 4(\Re(\lambda))(n_0)^2 + |\lambda|^2 n_0 \equiv F(n_0). \tag{8.102}$$

The right-hand side of this equation is cubic in n_0 and it has a local maximum at n_0^+ and a local minimum at n_0^-, where

$$n_0^\pm = [-2(\Re(\lambda)) \pm \sqrt{4(\Re(\lambda))^2 - 3|\lambda|^2}\,]/6. \tag{8.103}$$

These points give the region in which the system is bistable (for bistability to occur, we must have $\Re(\lambda) < 0$ for the solution to make sense, since we must have $n_0 > 0$), because for $F(n_0^-) < |\varepsilon|^2 < F(n_0^+)$ there are actually three possible values of n_0 corresponding to each value of $|\varepsilon|^2$, but one of these values is always unstable. Therefore, for $|\varepsilon|^2$ in this range, there are two physically possible values of n_0. Note that, besides the condition $\Re(\lambda) \sim \gamma\kappa + \Delta\omega\kappa'' < 0$, the existence of bistability also requires that $(\Re(\lambda))^2 > 3|\lambda|^2$. It is easily seen that, if the cavity is only filled by a two-photon absorber (i.e. $\kappa = 0$), then we obtain $\Re(\lambda) > 0$, and, therefore, bistability cannot be realized.

8.8.2 Exact steady-state solution

At this point, we could simply calculate fluctuations about the semiclassical solutions we have just found. However, it turns out that in this case we can actually find exact steady-state solutions of the Fokker–Planck equation if we switch from the positive P-representation to the complex P-representation. The Fokker–Planck equation is the same for both, and it will shortly become apparent why we have to switch P-representations. We first focus on the exact steady-state solution to the FPE (8.96) that is available in the complex P-representation. In the complex P method, the amplitudes α and β are independent complex c-number variables (corresponding to the operators a and $a^\dagger$), whose integration domains are to be chosen as contours in the individual complex planes.

The Fokker–Planck equation (8.96) can be solved exactly, in the steady-state regime, using the method of potential equations. The generalized forces are

$$\begin{aligned} F_\alpha &= -\frac{\partial \phi}{\partial \alpha} = \frac{2}{\tilde{\kappa}\alpha^2}[(\tilde{\gamma} - \tilde{\kappa})\alpha + \tilde{\kappa}\alpha^2\beta - \mathcal{E}], \\ F_{\beta^*} &= -\frac{\partial \phi}{\partial \beta} = \frac{2}{\tilde{\kappa}^*\beta^2}[(\tilde{\gamma}^* - \tilde{\kappa}^*)\beta + \tilde{\kappa}^*\beta^2\alpha - \mathcal{E}^*]. \end{aligned} \tag{8.104}$$

These obey the potential condition, so the potential $\phi(\alpha, \beta^*)$ is well defined. This results in the following steady-state P-function:

$$P_{ss}(\alpha, \beta) = \mathcal{N}\alpha^{\lambda-2}\beta^{\lambda^*-2}\exp\left(\frac{\varepsilon}{\alpha} + \frac{\varepsilon^*}{\beta} + 2\alpha\beta\right), \tag{8.105}$$

where $\mathcal{N}$ is the normalization constant. If we try to interpret this as a positive P-function, we are immediately in trouble, because it is not normalizable if the domains of α and β are the entire complex plane. This does not mean that there is no positive P-function, as this always exists in some form, although it is never unique. Rather, it simply means that the positive P-function solution given using this procedure is not normalizable.

Instead, it is possible to obtain an exact solution using the complex P-function method. In order to see what contours we should use for α and β^*, let us look at the normalization integral for this distribution. We change variables to $\xi = 1/\alpha$ and $\zeta = 1/\beta$, giving us

$$I(\lambda) = \int_C d\xi \int_C d\zeta \sum_{n=0}^{\infty} \frac{2^n}{n!}\xi^{-n-\lambda}\zeta^{-n-\lambda^*}e^{\varepsilon\xi + \varepsilon^*\zeta}, \tag{8.106}$$

where $\mathcal{N}I(\lambda) = 1$. The integrand has a branch cut, which we will take to be along the negative real axis. If we then take the contour C to start from below the negative real axis at $-\infty$, go to and around the origin, and then go back out to $-\infty$, but this time above the axis, we can make use of a representation of the gamma function:

$$\frac{1}{\Gamma(z)} = \frac{1}{2\pi i}\int_C d\tau\, \tau^{-z}e^{\tau}. \tag{8.107}$$

This then gives us

$$I(\lambda) = -4\pi^2 \sum_{n=0}^{\infty} \frac{2^n \varepsilon^{\lambda+n-1}(\varepsilon^*)^{\lambda^*+n-1}}{n!\,\Gamma(\lambda+n)\Gamma(\lambda^*+n)}. \tag{8.108}$$

Finally, making use of the hypergeometric function,

$${}_0F_2(a, b, z) = \sum_{k=0}^{\infty} \frac{z^k\Gamma(a)\Gamma(b)}{k!\,\Gamma(k+a)\Gamma(k+b)}, \tag{8.109}$$

this can be expressed as

$$I(\lambda) = -4\pi^2 \frac{\varepsilon^{\lambda-1}(\varepsilon^*)^{\lambda^*-1}}{\Gamma(\lambda)\Gamma(\lambda^*)} \, {}_0F_2(\lambda, \lambda^*, 2|\varepsilon|^2). \tag{8.110}$$

In order to evaluate normally ordered moments of the form $\langle a^{\dagger m} a^m \rangle$, it is necessary to evaluate an integral very similar to $I(\lambda)$; in fact, the only difference is to replace λ by $\lambda + m$. We, therefore, find

$$\langle a^{\dagger m} a^m \rangle = \frac{|\varepsilon|^{2m} \Gamma(\lambda)\Gamma(\lambda^*) {}_0F_2(m+\lambda,\, m+\lambda^*,\, 2|\varepsilon|^2)}{\Gamma(m+\lambda)\Gamma(m+\lambda^*) {}_0F_2(\lambda, \lambda^*, 2|\varepsilon|^2)}. \tag{8.111}$$

Additional reading

Books

- Treatments of the quantum properties of the laser field can be found in the books listed at the end of Chapter 6. The nonlinear interferometer is discussed in the following quantum optics book:

D. F. Walls and G. J. Milburn, *Quantum Optics* (Springer, Heidelberg, 1994).

Articles

- The original treatment of the nonlinear interferometer is given in:

P. Drummond and D. Walls, *J. Phys.* A **13**, 725 (1980).

Problems

8.1 Consider the laser spectrum below threshold:

$$I(\omega) = \frac{2g}{(\omega - \omega_c)^2 + (\gamma - g)^2}.$$

Use this result to calculate the corresponding total intensity (in photon number units)

$$\langle a^\dagger a \rangle = \int_{-\infty}^{+\infty} d\omega \, I(\omega)/(2\pi).$$

8.2 In semiclassical laser theory, that is, laser theory in which the field is classical and not quantized, one finds the following equation for the laser intensity,

$$\frac{dI}{dt} = 2I(g - \gamma - \beta I),$$

where g is the linear gain, γ is the linear loss, and β is a parameter that describes gain saturation.

(a) For $g < \gamma$, find the steady-state solution (or solutions) and examine their stability.

(b) For $g > \gamma$, find the steady-state solution (or solutions) and examine their stability.

(c) Solve the above equation for $I(t)$ in terms of $I(0)$, and show that, in both of the above cases, $I(t)$ goes to its steady-state value as $t \to \infty$.

8.3 We looked at the case of what is called dispersive optical bistability, but optical bistability can also be caused by a saturable absorber. In that case, the absorption per unit length of the medium, α, is given by

$$\alpha = \frac{\alpha_0}{1 + (I/I_s)},$$

where I is the light intensity, α_0 is the unsaturated absorption per unit length, and I_s is the saturation intensity. The relation between the input intensity to the cavity, I^{in}, and the output intensity, I^{out}, is given by

$$I^{in} = I^{out}\left(1 + \frac{C_0}{1 + 2I^{out}/I_0}\right)^2,$$

where C_0 depends on α_0, the length of the cavity and the properties of the cavity mirror, and I_0 is just I_s times the transmissivity of the mirror. Define the normalized intensities as $X = 2I^{in}/I_0$ and $Y = 2I^{out}/I_0$. Find a relation between X and Y of the form $F(Y) = 0$, where F is a polynomial in Y whose coefficients depend on X. Show that there is always at least one solution of $F(Y) = 0$ for $Y > 0$, and that a necessary condition for there to be more than one solution is that

$$X^2 + 4(5 - C_0)X - 8(1 + C_0)^2 > 0.$$

8.4 Consider the following (Ito) stochastic differential equations, being treated according to the positive P-method,

$$\frac{d\alpha}{dt} = -\gamma\alpha - i\kappa\alpha^2\beta^* + \mathcal{E} + \sqrt{-i\kappa}\,\alpha\xi_1(t),$$
$$\frac{d\beta^*}{dt} = -\gamma\beta^* + i\kappa\beta^2\alpha + \mathcal{E} + \sqrt{i\kappa}\,\beta\xi_2(t),$$

where γ, κ and $\mathcal{E}$ are real and positive, and the $\xi_i(t)$ are independent Gaussian noise terms with zero means and $\langle\xi_i(t)\xi_j(t')\rangle = \delta_{ij}\delta(t - t')$. These equations are known to describe a nonlinear (anharmonic) oscillator, i.e. a damped (only one-photon loss) and driven cavity mode that undergoes the process of self-phase modulation (or Kerr interaction) in a medium with cubic nonlinearity $\chi^{(3)}$ (i.e. $\kappa \sim \chi^{(3)}$, and the effective interaction Hamiltonian is $H_{eff} = (\hbar\kappa/2)a^{\dagger 2}a^2$). Assume the semiclassical approximation ($\beta = \alpha^*$ and no noise terms), and that $\gamma = 0$ and $\mathcal{E} = 0$. What are the solutions for the intensity (in photon number units) $|\alpha(t)|^2$ and for the amplitude $\alpha(t)$, in this case?

8.5 Find the semiclassical steady-state solution α_0 ($\beta_0 = \alpha_0^*$) to the stochastic differential equations in problem 8.4 (in the general case of nonzero and positive γ, E and χ''), in terms of the amplitude $|\alpha_0|$ and the phase φ_0 of $\alpha_0 = |\alpha_0|\exp(i\varphi_0)$.

8.6 The degenerate parametric oscillator in the parametric approximation has the interaction Hamiltonian

$$H_{int} = \frac{i\hbar\varepsilon}{2}(a^2 - a^{\dagger 2}),$$

where ε is real. The interaction-picture master equation for this system is then

$$\frac{d\rho}{dt} = \frac{-i}{\hbar}[H_{int}, \rho] + \gamma(2a\rho a^\dagger - a^\dagger a\rho - \rho a^\dagger a).$$

Find the corresponding Fokker–Planck equation for the Glauber–Sudarshan P-representation, and show that the resulting diffusion matrix is not positive definite (be sure to use real variables when finding the diffusion matrix).

8.7 If we interpret the Fokker–Planck equation in problem 8.6 as one for the positive P-representation, by letting $\alpha^* \to \beta$, and then use this equation to obtain stochastic differential equations, we get

$$\frac{d\alpha}{dt} = -\varepsilon\beta - \gamma\alpha + i\sqrt{\varepsilon}\xi_1,$$

$$\frac{d\beta}{dt} = -\varepsilon\alpha - \gamma\beta + i\sqrt{\varepsilon}\xi_2,$$

where $\xi_1(t)$ and $\xi_2(t)$ are independent Gaussian noise sources satisfying

$$\langle\xi_j(t)\xi_j(t')\rangle = \delta(t - t').$$

(a) Solve these equations for $\alpha(t)$ and $\beta(t)$ in terms of $\alpha(0)$ and $\beta(0)$. Assuming the system is initially in the vacuum state, so that $\langle\alpha(0)\rangle = \langle\beta(0)\rangle = 0$ and $\langle\alpha^2(0)\rangle = \langle\beta^2(0)\rangle = \langle\beta(0)\alpha(0)\rangle = 0$, find $\langle\alpha(t)\rangle = \langle a(t)\rangle$, $\langle\alpha^2(t)\rangle = \langle a^2(t)\rangle$ and $\langle\beta(t)\alpha(t)\rangle = \langle a^\dagger(t)a(t)\rangle$. Show that this gives the resulting quadrature variance expected for a squeezed vacuum state as $\gamma \to 0$.

(b) The problem of the degenerate parametric oscillator coupled to a damping reservoir can also be solved by using operator Langevin equations. The equation for $a(t)$ is

$$\frac{da}{dt} = -\varepsilon a^\dagger - \gamma a + \Gamma,$$

where Γ is the operator noise term satisfying $\langle\Gamma(t)\Gamma^\dagger(t')\rangle = 2\gamma\delta(t - t')$, with all other correlation functions involving one or two noise operators being equal to zero. Solve the above equation, and the corresponding one for $a^\dagger$, for $a(t)$ and $a^\dagger(t)$ in terms of $a(0)$ and $a^\dagger(0)$. Assuming that the system is in the vacuum state at $t = 0$, find $\langle a(t)\rangle$ and $\langle a^\dagger(t)a(t)\rangle$. Your answers should agree with what you found in part (a).

9 Degenerate parametric oscillator

The optical parametric oscillator is one of the most well-characterized devices in nonlinear optics, with many useful applications, especially as a frequency converter and as a wide-band amplifier. Novel discoveries made with these devices in the quantum domain include demonstrations of large amounts of both single-mode and two-mode squeezing, and, in the two-mode case, significant quantum intensity correlations between the two modes and quadrature correlations, which provided the first experimental demonstration of the original Einstein–Podolsky–Rosen (EPR) paradox. In this chapter, a systematic theory is developed of quantum fluctuations in the degenerate parametric oscillator.

The quantum statistical effects produced by the degenerate parametric amplifier and oscillator have been intensely studied. As we have seen, the amount of transient squeezing produced in an ideal lossless parametric amplifier was calculated using asymptotic methods by Crouch and Braunstein. It was found to scale inversely with the square root of the initial pump photon number. When the system is put into a cavity, and driving and dissipation are introduced, we find that the steady state of a driven degenerate parametric oscillator bifurcates. There is a single steady state at low driving strengths, but, as the driving field increases, a threshold value is reached and the single state splits and forms two stable states. Below the threshold, a linearized analysis of squeezing in the quantum quadratures is possible; while above this threshold, or critical point, the tunneling time between the two states due to quantum fluctuations can be calculated.

We will first study the behavior of the degenerate parametric oscillator using a linearized positive P-function approach. In the case that the pump mode can be adiabatically eliminated, we can obtain an exact solution of the Fokker–Planck equation by using the complex P-representation. We will also look at some explicitly multi-mode treatments. The first is an analysis of the down-conversion process in free space, and the second considers the same process but in a cavity with a classical pump.

9.1 Hamiltonian and stochastic equations

The model considered here is the degenerate parametric oscillator. The system of interest is an idealized cavity, which is resonant at two frequencies, ω_1 and $\omega_2 \approx 2\omega_1$. It is externally driven at the larger of the two frequencies. Both frequencies are damped due to cavity losses. Down-conversion of the pump photons to resonant subharmonic mode photons occurs due to a $\chi^{(2)}$ nonlinearity present inside the cavity. The Heisenberg-picture Hamiltonian that describes this open system is

$$H = \sum_{j=1,2} \hbar\omega_j a_j^\dagger a_j + H_{sys} + \sum_{j=1,2} \hbar(a_j\Gamma_j^\dagger + a_j^\dagger\Gamma_j), \tag{9.1}$$

where

$$H_{sys} = i\hbar\frac{\chi}{2}(a_1^{\dagger 2}a_2 - a_1^2 a_2^\dagger) + i\hbar\sum_{j=1,2}(\mathcal{E}_j e^{-i\omega_{Lj}t}a_j^\dagger - \mathcal{E}^* e^{i\omega_{Lj}t}a_j). \tag{9.2}$$

Here $\mathcal{E}_j$ represents the jth external driving field at frequency ω_{Lj}, where the driving laser(s) are assumed to be frequency-locked with $\omega_{L2} = 2\omega_{L1}$. The coupling χ is the internal nonlinearity due to a $\chi^{(2)}$ nonlinear medium internal to the cavity, and Γ_j are reservoir operators that describe the losses of the internal mode of frequency ω_j.

Next, we wish to consider an interaction picture, obtained with the definition that

$$H_0 = \sum_{j=1,2} \hbar\omega_{Lj}a_j^\dagger a_j. \tag{9.3}$$

In other words, the operators will evolve according to the laser frequency, while the states evolve according to the rest of the system Hamiltonian, including part of the cavity energy term if the laser is not on-resonance. Having said this, one then makes a further simplification, just as in the case of a single-mode cavity. Since (in this picture) the operators all have a known time evolution of $a_j(t) = a_j e^{-i\omega_{Lj}t}$, it is possible to cancel all of the explicit time-dependent terms, giving an interaction-picture master equation with operators a_j that do not change in time. We call this, as before, the rotating-frame picture,

$$\frac{d\rho}{dt} = \frac{1}{i\hbar}[H_{int}, \rho] + \sum_{j=1,2}\gamma_j(2a_j\rho a_j^\dagger - a_j^\dagger a_j\rho - \rho a_j^\dagger a_j), \tag{9.4}$$

where the interaction Hamiltonian is given, in this picture, by

$$\frac{H_{int}}{\hbar} = \sum_{j=1,2}\Delta_j a_j^\dagger a_j + i\frac{\chi}{2}(a_1^{\dagger 2}a_2 - a_1^2 a_2^\dagger) + i\sum_{j=1,2}(\mathcal{E}_j a_j^\dagger - \mathcal{E}_j^* a_j). \tag{9.5}$$

Here $\Delta_j = \omega_j - \omega_{Lj}$ represents the detuning between the cavity-mode resonance and the jth laser (angular or carrier) frequency. In practical terms, this must be much less than the cavity-mode spacing for the above approximations to be applicable. A much more serious issue is the possibility that pairs of neighboring cavity modes may be coupled as well, and the down-conversion process could, indeed, be resonant to a whole 'comb' of modes. It is possible to arrange cavity dispersion parameters so that this does not occur, especially when the mode spacings are large, as they would be in a small cavity. We will assume, for the sake of simplicity, that no other modes are excited

In the case in which only the higher-frequency field is driven, and all interactions are on-resonance, the Hamiltonian used reduces to the standard one for a nondegenerate, single-mode parametric amplifier or oscillator,

$$\frac{H_{int}}{\hbar} = i\mathcal{E}[a_2^\dagger - a_2] + i\frac{\chi}{2}[a_2 a_1^{\dagger 2} - a_2^\dagger a_1^2]. \tag{9.6}$$

Here $\mathcal{E} = \mathcal{E}_2$ is proportional to the coherent input or driving field at the second-harmonic frequency, assumed to be at exact resonance with the cavity mode, and we have assumed it to be real. The term χ is a coupling parameter for the $\chi^{(2)}$ nonlinearity of the medium. For simplicity, we have chosen the field mode functions so that $\mathcal{E}$ and χ are real.

9.2 Classical results

As a starting point, the operator equations can be readily calculated in the Heisenberg picture. Suppose there is just a second-harmonic driving field at frequency $\omega_{L2} = 2\omega_L$, then

$$\begin{aligned}\frac{da_1}{dt} &= -i\omega_1 a_1 + \chi a_1^\dagger a_2, \\ \frac{da_2}{dt} &= \mathcal{E}e^{-i\omega_{L2}t} - i\omega_2 a_2 - \frac{1}{2}\chi a_1^2.\end{aligned} \tag{9.7}$$

Of course, it is certainly possible to include damping in these equations, giving quantum Langevin equations – but these must also involve quantum noise terms as well, giving

$$\begin{aligned}\frac{da_1}{dt} &= -(\gamma_1 + i\omega_1)a_1 + \chi a_1^\dagger a_2 + \Gamma_1(t), \\ \frac{da_2}{dt} &= \mathcal{E}e^{-2i\omega_L t} - (\gamma_2 + i\omega_2)a_2 - \frac{1}{2}\chi a_1^2 + \Gamma_2(t).\end{aligned} \tag{9.8}$$

These quantum Langevin equations, as they stand, are not easily solved. Similar results are obtained if the rotating-frame-picture master equation is used as a starting point. In this case, one obtains the following results for the expectation values of the operators a_j, which were defined as time-independent:

$$\begin{aligned}\frac{d\langle a_1\rangle_R}{dt} &= -\tilde{\gamma}_1\langle a_1\rangle_R + \chi\langle a_1^\dagger a_2\rangle_R, \\ \frac{d\langle a_2\rangle_R}{dt} &= \mathcal{E} - \tilde{\gamma}_2\langle a_2\rangle_R - \frac{1}{2}\chi\langle a_1^2\rangle_R.\end{aligned} \tag{9.9}$$

Here the complex decay rate $\tilde{\gamma}_j$ is defined through

$$\tilde{\gamma}_j = i(\omega_j - j\omega_L) + \gamma_j'.$$

These equations eliminate the time dependence of the driving field term. However, it is important to recall that the full operator time dependence of the expectation values requires use of the mapping to interaction-picture time-dependent operators, as follows:

$$\langle a_j(t)\rangle = \langle a_j\rangle_R e^{-ij\omega_L t}$$

We shall use a fully explicit time dependence of the operator to denote that all relevant time dependences are included, even the carrier frequency terms themselves. Otherwise, when the time dependence of the operator is not indicated, its time dependence is only due to interactions, damping and detuning effects, which come from the time evolution of the quantum state.

In the classical limit, all noise and commutation terms are neglected, and all operator expectation values are assumed to factorize. If classical expectation values are defined through the relation

$$\langle a_j(t)\rangle = \langle a_j\rangle e^{-ij\omega_L t} = \alpha_j e^{-ij\omega_L t},$$

then the usual results of classical nonlinear optics are regained, giving

$$\begin{aligned} \frac{d\alpha_1}{dt} &= -\tilde{\gamma}_1\alpha_1 + \chi\alpha_1^*\alpha_2, \\ \frac{d\alpha_2}{dt} &= \mathcal{E} - \tilde{\gamma}_2\alpha_2 - \frac{1}{2}\chi\alpha_1^2. \end{aligned} \tag{9.10}$$

The steady state is approached with a time-scale of γ_1^{-1} or γ_2^{-1}, depending on which is the slower time-scale. The steady-state solutions are given by

$$\begin{aligned} \alpha_1 &= \chi\alpha_1^*\alpha_2/\tilde{\gamma}_1, \\ \alpha_2 &= [\mathcal{E} - \tfrac{1}{2}\chi\alpha_1^2]/\tilde{\gamma}_2. \end{aligned} \tag{9.11}$$

In general, there is always one trivial solution, which is called the below-threshold solution,

$$\begin{aligned} \alpha_1 &= 0, \\ \alpha_2 &= \mathcal{E}/\tilde{\gamma}_2. \end{aligned} \tag{9.12}$$

In addition, there can be a second solution in which $\alpha_1 \neq 0$. In this solution, clearly

$$\alpha_1 = \chi\alpha_1^*[\mathcal{E} - \tfrac{1}{2}\chi\alpha_1^2]/\gamma_2\gamma_1.$$

Next, suppose that $\alpha_1 = |\alpha_1|\exp(i\theta)$ and $\tilde{\gamma}_2\tilde{\gamma}_1 = |\tilde{\gamma}_2\tilde{\gamma}_1|^2\exp(i\phi)$, so that

$$|\tilde{\gamma}_2\tilde{\gamma}_1|\exp(i\phi) = \chi\left[\mathcal{E}\exp(-2i\theta) - \frac{|\alpha_1|^2\chi}{2}\right].$$

Taking the modulus of both sides, it is clear that

$$\frac{|\tilde{\gamma}_2\tilde{\gamma}_1|^2}{\chi^2} = \mathcal{E}^2\sin^2(2\theta) + \left[\mathcal{E}\cos(2\theta) - \frac{|\alpha_1|^2\chi}{2}\right]^2.$$

This is generally rather complex, except in the simplest case where the cavity is doubly resonant (or symmetrically off-resonant) with the driving field, so that $\phi = 0$, and

$$\begin{aligned} \alpha_1 &= \pm\sqrt{(2/\chi)[\mathcal{E} - |\tilde{\gamma}_2\tilde{\gamma}_1|/\chi]}, \\ \alpha_2 &= \mathcal{E}_c/\tilde{\gamma}_2. \end{aligned} \tag{9.13}$$

Here $\mathcal{E}_c = |\tilde{\gamma}_2\tilde{\gamma}_1|/\chi$ is the critical or threshold point, and clearly it is necessary that $\mathcal{E}$ must be larger than this to obtain a solution of this type, which is called the above-threshold solution. At this point, the below-threshold solutions themselves become unstable. The above-threshold solutions are stable themselves up to a second threshold, where an oscillatory behavior occurs. Above this point, there are no stable steady-state solutions, although limit-cycle behavior (self-pulsing) occurs.

9.3 Fokker–Planck and stochastic equations

Next, we shall make use of operator representations to solve for the quantum fluctuations. In particular, we shall have occasion to use both the positive and complex P-representations. Following the standard procedures developed in Chapter 7, and the assumption of vanishing boundary terms, the master equation can be rewritten as a Fokker–Planck equation in the generalized P-representation, which can then be turned into stochastic equations with real noise. The assumption of vanishing boundary terms is critical to this procedure. This occurs when the ratio of nonlinearity to damping is small (i.e. $|\chi/\gamma_k| \ll 1$, where γ_k are the relevant damping rates). The stochastic positive P-representation is therefore only valid for small $|\chi/\gamma_k|$, in which case the boundary terms are exponentially suppressed. The required ratio of nonlinearity to damping is extremely well satisfied in all current experiments, where the ratio is typically 10^{-6} or less.

The generalized P-function, $P(\boldsymbol{\alpha}, \boldsymbol{\beta})$, is now a function of four variables, two for each mode. The pump mode is described by the complex variables α_2 and β_2, and the signal mode, which is at half the pump frequency, is described by α_1 and β_1. The generalized P-function obeys the Fokker–Planck equation

$$\begin{aligned}\frac{\partial P(\boldsymbol{\alpha}, \boldsymbol{\beta})}{\partial t} = \Bigg\{ & \frac{\partial}{\partial \alpha_1}[-\mathcal{E}_1 + \tilde{\gamma}_1\alpha_1 - \chi\alpha_2\beta_1] + \frac{\partial}{\partial \beta_1}[-\mathcal{E}_1^* + \tilde{\gamma}_1^*\beta_1 - \chi\beta_2\alpha_1] \\ & + \frac{1}{2}\frac{\partial^2}{\partial \alpha_1^2}\chi\alpha_2 + \frac{1}{2}\frac{\partial^2}{\partial \beta_1^2}\chi\beta_2 \\ & + \frac{\partial}{\partial \alpha_2}\left[-\mathcal{E}_2 + \tilde{\gamma}_2\alpha_2 + \frac{\chi}{2}\alpha_1^2\right] + \frac{\partial}{\partial \beta_2}\left[-\mathcal{E}_2^* + \tilde{\gamma}_2^*\beta_2 + \frac{\chi}{2}\beta_1^2\right]\Bigg\} P(\boldsymbol{\alpha}, \boldsymbol{\beta}).\end{aligned} \tag{9.14}$$

For the sake of completeness, we have put in driving fields for both modes. We now specialize to the case in which $\mathcal{E}_2 = \mathcal{E}$, where $\mathcal{E}$ is real, the driving field for the pump mode is on-resonance and $\mathcal{E}_1 = 0$. The Ito stochastic equations implied by the above Fokker–Planck equation, after doubling the dimension to include conjugate derivatives in a positive definite form, are

$$\begin{aligned}\frac{d\alpha_1}{dt} &= -\gamma_1\alpha_1 + \chi\beta_1\alpha_2 + \sqrt{\chi\alpha_2}\xi_1(t), \\ \frac{d\beta_1}{dt} &= -\gamma_1\beta_1 + \chi\alpha_1\beta_2 + \sqrt{\chi\beta_2}\xi_2(t), \\ \frac{d\alpha_2}{dt} &= -\gamma_2\alpha_2 + \mathcal{E} - \frac{1}{2}\chi\alpha_1^2, \\ \frac{d\beta_2}{dt} &= -\gamma_2\beta_2 + \mathcal{E} - \frac{1}{2}\chi\beta_1^2.\end{aligned} \tag{9.15}$$

Here the terms γ_k represent the amplitude damping rates, and are real and positive. The stochastic correlations are given by

$$\begin{aligned}\langle\xi_k(t)\rangle &= 0, \\ \langle\xi_k(t)\xi_l(t')\rangle &= \delta_{kl}\delta(t - t'),\end{aligned} \tag{9.16}$$

so that $\xi_k(t)$ represent two real Gaussian and uncorrelated stochastic processes.

Let us now use a linearized treatment to find the output squeezing spectrum above and below threshold. If $\alpha_j^{(0)}$ and $\beta_j^{(0)}$, for $j = 1, 2$, are the steady-state values for α_j and β_j, then define $\delta\alpha_j = \alpha_j - \alpha_j^{(0)}$, and similarly for $\delta\beta_j$. Below threshold we have $\alpha_1^{(0)} = \beta_1^{(0)} = 0$ and $\alpha_2^{(0)} = \beta_2^{(0)} = \mathcal{E}/\gamma_2$. The linearized equations for the fluctuations are

$$\frac{d}{dt}\begin{pmatrix} \delta\alpha_1 \\ \delta\beta_1 \\ \delta\alpha_2 \\ \delta\beta_2 \end{pmatrix} = \begin{pmatrix} -\gamma_1 & \chi\alpha_2^{(0)} & 0 & 0 \\ \chi\beta_2^{(0)} & -\gamma_1 & 0 & 0 \\ 0 & 0 & -\gamma_2 & 0 \\ 0 & 0 & 0 & -\gamma_2 \end{pmatrix} + \begin{pmatrix} \sqrt{\chi\alpha_2^{(0)}}\xi_1 \\ \sqrt{\chi\beta_2^{(0)}}\xi_2 \\ 0 \\ 0 \end{pmatrix}. \tag{9.17}$$

Solving these equations for $\delta\alpha_1$ and $\delta\beta_1$, we find

$$\delta\alpha_1 = \sqrt{\chi\alpha_2^{(0)}} \int_{-\infty}^{t} dt' \, e^{-\gamma_1(t-t')} \times \left\{ \xi_1(t') \cosh\left[\sqrt{\chi\alpha_2^{(0)}}(t-t')\right] + \xi_2(t') \sinh\left[\sqrt{\chi\alpha_2^{(0)}}(t-t')\right] \right\}, \tag{9.18}$$

with a similar expression for $\delta\beta_1$ in which cosh and sinh are interchanged. For the correlation function $\langle\delta\alpha_1(t)\delta\alpha_1(s)\rangle$, we find

$$\langle\delta\alpha_1(t)\delta\alpha_1(s)\rangle = \frac{1}{4}\chi\alpha_2^{(0)} e^{-\gamma_1|t-s|}\left[\frac{1}{\gamma_1 - \chi\alpha_2^{(0)}} e^{\chi\alpha_2^{(0)}|t-s|} + \frac{1}{\gamma_1 + \chi\alpha_2^{(0)}} e^{-\chi\alpha_2^{(0)}|t-s|}\right], \tag{9.19}$$

and for $\langle\delta\beta_1(t)\delta\alpha_1(s)\rangle$ we find a very similar expression, the only difference being that the last term in brackets is preceded by a minus instead of a plus sign.

Before proceeding to find the squeezing spectrum, let us first find the squeezing of the cavity mode. Defining $X(\theta) = e^{-i\theta}a_1 + e^{i\theta}a^\dagger$, we have that

$$\begin{aligned} \Delta X(\theta)^2 &= 1 + \langle : (X(\theta) - \langle X(\theta)\rangle)^2 : \rangle \\ &= 1 + \langle\delta\alpha_1^2(0)\rangle(e^{2i\theta} + e^{-2i\theta}) + 2\langle\delta\beta_1(0)\delta\alpha_1(0)\rangle \\ &= 1 + \chi\alpha_2^{(0)} \frac{\gamma_1\cos(2\theta) + \chi\alpha_2^{(0)}}{\gamma_1^2 - (\chi\alpha_2^{(0)})^2}. \end{aligned} \tag{9.20}$$

This achieves its minimum value at $\theta = \pi/2$, where we find

$$\Delta X(\pi/2)^2 = \frac{\gamma_1}{\gamma_1 + \chi\alpha_2^{(0)}}. \tag{9.21}$$

Now, the condition for the below-threshold solution to be valid is $\chi\alpha_2^{(0)} < \gamma_1$, so, as threshold is approached, this value goes to $1/2$. The linearized theory will not be valid too close to threshold, but it is valid throughout most of the below-threshold region. Therefore, we see that, in this region, the amount of squeezing we can obtain inside the cavity is severely limited.

This was originally thought to be a serious limitation on squeezing, but it was soon realized by Yurke that the squeezing outside the cavity can be much larger than the squeezing inside the cavity. Let us now go ahead and show this. As we have seen, the squeezing spectrum outside the cavity can be expressed as

$$V(\omega,\theta) = 1 + 2\gamma_1 \int_{-\infty}^{\infty} dt\, e^{-i\omega t} \langle : \Delta X(0,\theta)\Delta X(t,\theta) : \rangle. \tag{9.22}$$

In our case, this becomes

$$\begin{aligned} V(\omega,\theta) &= 1 + 2\gamma_1 \int_{-\infty}^{\infty} dt\, e^{-i\omega t} [\langle \delta\alpha_1(0)\delta\alpha_1(t)\rangle(e^{2i\theta} + e^{-2i\theta}) + 2\langle \delta\beta_1(t)\delta\alpha_1(0)\rangle] \\ &= 1 + 2\gamma_1 \chi \alpha_2^{(0)} \left[\frac{\cos(2\theta)+1}{\omega^2 + (\gamma_1 - \chi\alpha_2^{(0)})^2} + \frac{\cos(2\theta)-1}{\omega^2 + (\gamma_1 + \chi\alpha_2^{(0)})^2} \right]. \end{aligned} \tag{9.23}$$

The minimum value occurs when $\theta = \pi/2$, yielding

$$V(\omega,\pi/2) = 1 - \frac{4\gamma_1\chi\alpha_2^{(0)}}{\omega^2 + (\gamma_1 + \chi\alpha_2^{(0)})^2}. \tag{9.24}$$

Recalling that $\chi\alpha_2^{(0)} < \gamma_1$ for this solution to be valid, we see that, as we approach threshold at $\omega = 0$ (remember that $\omega = 0$ corresponds to the frequency of the cavity mode), $V(0,\pi/2) \to 0$. This implies that, though the squeezing inside the cavity is quite limited, we can obtain very large squeezing outside the cavity.

Now let us look at the above-threshold case. We now have

$$\begin{aligned} \alpha_1^{(0)} &= \beta_1^{(0)} = \pm\xi = \pm\sqrt{\frac{2}{\chi}(\mathcal{E} - \mathcal{E}_c)}, \\ \alpha_2^{(0)} &= \beta_2^{(0)} = \frac{\gamma_1}{\chi}, \end{aligned} \tag{9.25}$$

where $\mathcal{E}_c = \gamma_1\gamma_2/\chi$ is the critical field. These are the stable solutions when $\mathcal{E} > \mathcal{E}_c$. The equations for the fluctuations about the classical solutions are now

$$\frac{d}{dt}\begin{pmatrix} \delta\alpha_1 \\ \delta\beta_1 \\ \delta\alpha_2 \\ \delta\beta_2 \end{pmatrix} = \begin{pmatrix} -\gamma_1 & \chi\alpha_2^{(0)} & \pm\chi\xi & 0 \\ \chi\beta_2^{(0)} & -\gamma_1 & 0 & \pm\chi\xi \\ \mp\chi\xi & 0 & -\gamma_2 & 0 \\ 0 & \mp\chi\xi & 0 & -\gamma_2 \end{pmatrix} + \begin{pmatrix} \sqrt{\chi\alpha_2^{(0)}}\xi_1 \\ \sqrt{\chi\beta_2^{(0)}}\xi_2 \\ 0 \\ 0 \end{pmatrix}. \tag{9.26}$$

The eigenvalues of this matrix are independent of the choice of sign for the $\chi\xi$ terms, and have negative real parts for $\mathcal{E} > \mathcal{E}_c$, so that both solutions are stable in this region. The calculation of the squeezing spectrum is similar to that in the below-threshold case, so we will only give the results. We find that

$$V(\omega,\pi/2) = 1 - \frac{4\gamma_1^2(\gamma_2^2 + \omega^2)}{[2\gamma_1\gamma_2 + 2\chi(\mathcal{E} - \mathcal{E}_c) - \omega^2] + \omega^2(2\gamma_1 + \gamma_2)^2}. \tag{9.27}$$

For

$$2\chi(\mathcal{E} - \mathcal{E}_c) < \gamma_2\{[\gamma_2^2 + (2\gamma_1 + \gamma_2)^2]^{1/2} - (2\gamma_1 + \gamma_2\},\tag{9.28}$$

the spectrum has a minimum at $\omega = 0$ given by

$$V(0,\pi/2) = 1 - \frac{(\gamma_1\gamma_2)^2}{[\gamma_1\gamma_2 + \chi(\mathcal{E} - \mathcal{E}_c)]^2}. \tag{9.29}$$

In this region, we see that the maximum squeezing decreases as we get further above threshold. When the condition in Eq. (9.28) is not satisfied, i.e. if the driving field is sufficiently large, then $\omega = 0$ becomes a local maximum, and the minimum value of squeezing is obtained at nonzero values of ω located symmetrically about $\omega = 0$.

9.4 Adiabatic approximation

So far, we have relied on linearization methods to study the degenerate parametric oscillator. If we try to find an exact solution to the Fokker–Planck equation describing this system, we find that the equation does not satisfy the potential condition. However, an exact solution to the quantum operator moments can be found in the adiabatic limit of a rapidly decaying pump. In this situation, we can find a Fokker–Planck equation that does satisfy the potential condition.

In the limit of $\gamma_2 \gg \gamma_1$, the pump mode can be eliminated adiabatically. The idea is that, in this limit, the pump mode relaxes quickly to its steady-state value. The formal technique for accomplishing the adiabatic elimination is simply to set $d\alpha_2/dt = d\beta_2/dt = 0$, and then to solve for α_2 and β_2 in terms of the slowly varying α_1 and β_1 variables. Doing so, and then substituting the results back into the stochastic differential equations for α_1 and β_1, yields

$$\frac{d\alpha_1}{dt} = -\gamma_1\alpha_1 + \frac{\chi}{\gamma_2}\beta_1\left(\mathcal{E} - \frac{1}{2}\chi\alpha_1^2\right) + \left[\frac{\chi}{\gamma_2}\left(\mathcal{E} - \frac{1}{2}\chi\alpha_1^2\right)\right]^{1/2}\xi_1(t),$$
$$\frac{d\beta_1}{dt} = -\gamma_1\beta_1 + \frac{\chi}{\gamma_2}\alpha_1\left(\mathcal{E} - \frac{1}{2}\chi\beta_1^2\right) + \left[\frac{\chi}{\gamma_2}\left(\mathcal{E} - \frac{1}{2}\chi\beta_1^2\right)\right]^{1/2}\xi_2(t).$$

Note that we are assuming that $\mathcal{E}$ is real and that we are on-resonance. We wish to obtain a Fokker–Planck equation corresponding to these stochastic differential equations for the remaining signal mode. Hence, we simply reverse the procedure outlined in Section 7.8.2. Thus, we transform the stochastic equation to a positive definite Fokker–Planck equation and hence to a physically equivalent complex Fokker–Planck equation, to obtain

$$\begin{aligned}\frac{\partial P(\alpha_1,\beta_1)}{\partial t} = &\left\{\frac{\partial}{\partial\alpha_1}\left[\gamma_1\alpha_1 - \frac{\chi}{\gamma_2}\left(\mathcal{E} - \frac{\chi}{2}\alpha_1^2\right)\beta_1\right]\right.\\ &+ \frac{\partial}{\partial\beta_1}\left[\gamma_1\beta_1 - \frac{\chi}{\gamma_2}\left(\mathcal{E} - \frac{\chi}{2}\beta_1^2\right)\alpha_1\right]\\ &\left.+ \frac{1}{2}\frac{\partial^2}{\partial\alpha_1^2}\frac{\chi}{\gamma_2}\left(\mathcal{E} - \frac{\chi}{2}\alpha_1^2\right) + \frac{1}{2}\frac{\partial^2}{\partial\beta_1}\frac{\chi}{\gamma_2}\left(\mathcal{E} - \frac{\chi}{2}\beta_1^2\right)\right\}P(\alpha_1,\beta_1). \end{aligned} \tag{9.30}$$

This equation does obey the potential condition, so we can find an exact solution. It is given by

$$P_{ss}(\alpha,\beta_1) = \mathcal{N}(\alpha_1^2 - \varepsilon)^{\lambda-1}(\beta_1^2 - \varepsilon)^{\lambda-1}e^{2\alpha_1\beta_1}, \tag{9.31}$$

where we have introduced

$$\varepsilon = \frac{2\mathcal{E}}{\chi}, \qquad \lambda = \frac{2\gamma_1\gamma_2}{\chi^2}. \tag{9.32}$$

We will assume that $\lambda \gg 1$, i.e. that the nonlinearity is small. This function diverges in the direction $\beta_1 = \alpha_1^*$, so we will have to restrict its domain of applicability to specific contours in the α_1 and β_1 planes.

We can express expectation values in terms of this complex P-function as

$$\langle (a^\dagger)^n a^{n'} \rangle_R = \int d\alpha_1 \, d\beta_1 \, \beta_1^n \alpha_1^{n'} P_{ss}(\alpha_1, \beta_1). \tag{9.33}$$

Expanding the exponential in P_{ss} and making the change of variables

$$z_1 = \frac{1}{2}\left(1 + \frac{\chi \alpha_1}{v}\right), \qquad z_2 = \frac{1}{2}\left(1 + \frac{\chi \beta_1}{v}\right), \tag{9.34}$$

where $v = \sqrt{2\chi \mathcal{E}}$, this can be expressed as

$$\begin{aligned} \langle (a^\dagger)^n a^{n'} \rangle_R = \mathcal{N}' \left(\frac{-v}{\chi}\right)^{n+n'} \sum_{m=0}^{\infty} \frac{(2\varepsilon)^m}{m!} \int dz_1 \, z_1^{\lambda-1} (1 - z_1)^{\lambda-1} ((1 - 2z_1)^{m+n'} \\ \times \int dz_2 \, z_2^{\lambda-1} (1 - z_2)^{\lambda-1} ((1 - 2z_2)^{m+n}, \end{aligned} \tag{9.35}$$

where some constants have been absorbed into the new normalization constant, $\mathcal{N}'$. If we now take the contours for both z_1 and z_2 to be the interval on the real axis between 0 and 1 in the complex plane with a branch cut extending from 1 to ∞, we can express the result in terms of the hypergeometric function. The hypergeometric function, $F(a, b, c; z)$, where $\Re(c) > \Re(b) > 0$, can be expressed as

$$F(a, b, c; z) = \frac{\Gamma(c)}{\Gamma(b)\Gamma(c - b)} \int_0^1 dt \, t^{b-1} (1 - t)^{c-b-1} (1 - tz)^{-a}. \tag{9.36}$$

This gives us

$$\langle (a^\dagger)^n a^{n'} \rangle_R = \mathcal{N}'' \left(\frac{-v}{\chi}\right)^{n+n'} \sum_{m=0}^{\infty} \frac{(2\varepsilon)^m}{m!} F(-m - n', \lambda, 2\lambda; 2) F(-m - n, \lambda, 2\lambda; 2). \tag{9.37}$$

The new normalization constant can be obtained by setting $n = n' = 0$ on the right-hand side and setting the left-hand side equal to one.

9.5 Multi-mode treatment of parametric down-conversion in a cavity

So far, we have neglected the spatial dependence of the fields generated by degenerate parametric down-conversion. Let us now look at a simple model for a nonlinear medium in a cavity that will allow us to study the spatial dependence of the down-converted field. We will use the treatment of input–output theory from Chapter 6, in particular, the version that incorporates the spatial dependence of the modes. We will make one small change from the situation in Chapter 6, however. There, light hitting the mirror from the left had no phase-shift on reflection, while light incident on the mirror from the right had a π phase-shift on reflection. We want to reverse this situation: now light incident on the mirror from the right will have a π phase-shift on reflection, and light incident from the left will have no phase-shift on reflection. This has the effect of replacing $\sqrt{R}$ by $-\sqrt{R}$. The reason for making this replacement is that now the cavity resonances occur when kL/π is an integer, and, in the limit that the reflectivity goes to one, we would have the usual standing-wave modes for the cavity, with the mode functions vanishing at the mirrors at $x = 0$ and $x = L$.

With the medium filling the entire cavity, the interaction Hamiltonian for this system is given by

$$\begin{aligned} H_{int} &= \frac{\eta^{(2)}}{3} \int_0^L dx\, (\partial_x \Lambda)^3 \\ &= \frac{\eta^{(2)}}{3} \left(\frac{-i\sqrt{\hbar}}{\sqrt{\pi\mu_0}} \right)^3 \int_0^L dx \prod_{j=1}^{3} \int dk_j \, \frac{1}{\sqrt{\omega_{k_j}}} (v(k_j) a_{k_j} - v^*(k_j) a_{k_j}^\dagger) k_j \cos(k_j x), \end{aligned} \tag{9.38}$$

where

$$v(k) = \frac{\sqrt{T}}{1 - \sqrt{R}\, e^{2ikL}}.$$

We are now going to make a number of approximations. We shall assume that one of modes, in particular, mode 3, is highly excited in a coherent state with amplitude α_0 and wavevector k_0, so that we shall replace a_{k_3} by $\alpha_0 \exp(-i\nu_0 t)\delta(k_1 - k_0)$. We will also only want to consider the down-conversion process, in which photons with wavevectors k_1 and k_2 are created, where $k_1 + k_2 = k_0$, so we will only keep terms that correspond to this process. For the integral over x, we have

$$\begin{aligned} &\int_0^L dx\, \cos(k_0 x) \cos(k_2 x) \cos(k_3 x) \\ &= \frac{\sin(k_0 - k_1 - k_2)L}{4(k_0 - k_1 - k_2)} + \frac{1}{4} \int_0^L dx\, \{\cos[(k_0 + k_1 - k_2)L] \\ &\quad + \cos[(k_0 - k_1 + k_2)L] + \cos[(k_0 + k_1 + k_2)L]\}. \end{aligned} \tag{9.39}$$

Because we are only interested in the range $k_1, k_2 < k_0$, and, in particular, we will mainly be interested in the case when k_1 and k_2 are close to $k_0/2$, the first term in the above expression will be the dominant term. In addition, if we restrict our attention to the case when both k_1 and k_2 are in a range Δ, where Δ is an interval of size approximately $1/L$ centered about $k_0/2$, then we can replace this term with $L/4$. We are assuming that $k_0/2$ is a cavity resonance, and we wish only to consider modes in the vicinity of that resonance. Making all of these approximations, we have that

$$\begin{aligned} H_{int} &= \frac{\eta^{(2)}}{3} \left(\frac{-i\sqrt{\hbar}}{\sqrt{\pi\mu_0}} \right)^3 \left(\frac{L}{4} \right) \frac{\nu_0^{3/2}}{2c^3} \int_{k_1 \in \Delta} dk_1 \int_{k_2 \in \Delta} dk_2 \\ &\times [\alpha_0^* e^{i\nu_0 t} v^*(k_0) v(k_1) v(k_2) a_{k_1} a_{k_2} - \alpha_0 e^{-i\nu_0 t} v(k_0) v^*(k_1) v^*(k_2) a_{k_1}^\dagger a_{k_2}^\dagger], \end{aligned} \tag{9.40}$$

where we have set $\nu_0 = \omega_{k_0}$.

Now let us define the slowly varying operators $b_{k_j} = e^{i\omega_{k_j} t} a_{k_j}$, and the operator

$$\xi = \int_{k \in \Delta} dk\, v(k) e^{-i[k - (k_0/2)]ct} b_k. \tag{9.41}$$

These operators obey the commutation relations

$$[\xi, \xi^\dagger] = \int_{k\in\Delta} dk\, |v(k)|^2 \simeq \int_{k\in\Delta} dk \frac{T}{(1-\sqrt{R})^2 + 4RL^2(k-k_0/2)^2}$$
$$\simeq \int_{-\infty}^{\infty} dk' \frac{T}{(1-\sqrt{R})^2 + 4RL^2(k')^2} = \frac{\pi}{2(1-\sqrt{R})\sqrt{R}L}. \tag{9.42}$$

In order to obtain this result, we expanded the exponential around $k_0/2$, which is assumed to be a resonance, so that $e^{2i(k_0/2)L} = 1$. We then note that the resulting function is highly peaked about $k_0/2$, and the peak is much narrower than the width of the interval Δ, so we can extend the limits of integration to plus and minus infinity. We also have that

$$[b_k, \xi^\dagger] = v^*(k) e^{i(k-k_0/2)ct}. \tag{9.43}$$

The interaction Hamiltonian can now be expressed as

$$H_{int} = \frac{\eta^{(2)}}{3}\left(\frac{-i\sqrt{\hbar}}{\sqrt{\pi\mu_0}}\right)^3 \left(\frac{L}{4}\right) \frac{v_0^{3/2}}{2c^3}[\alpha_0^* v^*(k_0)\xi^2 - \alpha_0 v(k_0)\xi^{\dagger 2}]. \tag{9.44}$$

We can use the equation of motion for a_k and the definition of ξ to find the equation of motion for ξ as

$$\frac{d\xi}{dt} = \frac{i}{\hbar}[H_{int}, \xi] = \kappa \xi^\dagger, \tag{9.45}$$

where

$$\kappa = -\frac{\eta^{(2)}}{3} \frac{\sqrt{\hbar}}{(\pi\mu_0)^{3/2}} \left(\frac{L}{4}\right) \frac{v_0^{3/2}}{2c^3} \frac{2\pi}{(1-\sqrt{R})(2\sqrt{R}L)}.$$

This differential equation, along with the corresponding one for $\xi^\dagger$, is easily solved to give

$$\xi(t) = \xi(0)\cosh(|\kappa|t) + e^{i\phi}\sinh(|\kappa|t)\xi^\dagger(0), \tag{9.46}$$

where $\kappa = |\kappa|e^{i\phi}$. This result can now be used to find $b_k(t)$. We have that

$$\frac{db_k}{dt} = \frac{i}{\hbar}[H_{int}, b_k] = g v^*(k) e^{i(k-k_0/2)ct}\xi^\dagger(t), \tag{9.47}$$

where

$$\kappa = g\frac{\pi}{(1-\sqrt{R})(2\sqrt{R}L)}.$$

Integrating this equation gives us

$$b_k(t) = b_k(0) + g v^*(k) \int_0^t dt'\, e^{i(k-k_0/2)ct'}[\xi^\dagger(0)\cosh(|\kappa|t') + \xi(0)e^{-i\phi}\sinh(|\kappa|t')]. \tag{9.48}$$

It is convenient to define the functions

$$f_c(k,t) = g v^*(k) \int_0^t dt'\, e^{i(k-k_0/2)ct'}\cosh(|\kappa|t'),$$
$$f_s(k,t) = g v^*(k) \int_0^t dt'\, e^{i(k-k_0/2)ct'}\sinh(|\kappa|t'), \tag{9.49}$$

so that

$$b_k(t) = b_k(0) + f_c(k,t)\xi^\dagger(0) + f_c(k,t)\xi(0). \tag{9.50}$$

Now that we have the operators as a function of time, we can use them to calculate something. The probability that a detector atom located at position x will be excited between time t and $t + \Delta t$ is

$$p(x,t) = \int_t^{t+\Delta t} dt' \int_t^{t+\Delta t} dt''\, S(t'-t'')\langle D^{(-)}(x,t')D^{(+)}(x,t'')\rangle, \tag{9.51}$$

where $S(t)$ describes the frequency response of the detector. We saw an expression similar to this one in Eq. (5.10), where $S(t)$ was just an exponential, $e^{-i\nu t}$. This represents the case in which the detector is sensitive primarily to a particular frequency, in that case, ν, with a bandwidth determined by the integration interval, Δt. The operators for which we have found the time evolution, $\xi(t)$ and $b_k(t)$, only contain frequencies in an interval of size $c\Delta \sim c/L$ centered around $\nu_0/2$. A detector with $S(t) \propto e^{-i\nu_0 t/2}$ and a measurement interval of $\Delta t \sim L/c$ would only be sensitive to these frequencies. We shall go ahead and find the correlation function in the above equation using $\xi(t)$ and $b_k(t)$, but it should be kept in mind that it only contains contributions from a limited frequency range, and should only be used in Eq. (9.51) with an appropriate choice of $S(t)$ and Δt, so that only that frequency range is measured by the detector.

In order to find the correlation function, we note that

$$D(x,t) = \int_0^\infty dk \sqrt{\frac{\hbar}{2\mu_0\omega}}\,[b_k\partial_x u_k(x)e^{-ikct} + b_k^\dagger \partial_x u^*(x)e^{ikct}], \tag{9.52}$$

so that

$$\begin{aligned}&\langle D^{(-)}(x,t')D^{(+)}(x,t'')\rangle \\ &= \int_0^\infty dk \int_0^\infty dk' \left(\frac{\hbar}{2\mu c\sqrt{kk'}}\right)\partial_x u_k^*(x)\partial u_{k'}(x)e^{i(kt'-k't'')c}\langle b_k^\dagger(t')b_{k'}(t'')\rangle.\end{aligned} \tag{9.53}$$

We will now assume that the field is initially in the vacuum state. This implies that, for $k, k' \in \Delta$,

$$\langle b_k^\dagger(t')b_{k'}(t'')\rangle = f_c^*(k,t')f_c(k',t''), \tag{9.54}$$

which implies that

$$\langle D^{(-)}(x,t')D^{(+)}(x,t'')\rangle = \frac{\hbar}{\mu_0\nu_0}F^*(x,t')F(x,t''), \tag{9.55}$$

where

$$F(x,t) = \int_{k\in\Delta} dk\, e^{-ikct}\partial_x u_k(x) f_c(k,t). \tag{9.56}$$

We will now concentrate on calculating $F(x,t)$ in the case that $x > L$, that is, our detector is outside the medium. Let us look at the individual pieces. First, we have that

$$f_c(k,t) = \frac{g}{2} v^*(k) \left[\frac{1}{i(k - k_0/2) + |\kappa|} (e^{i(k-k_0/2)+|\kappa|t} - 1) + \frac{1}{i(k - k_0/2) - |\kappa|} (e^{i(k-k_0/2)-|\kappa|t} - 1) \right]. \tag{9.57}$$

Before moving on to the next piece, let us take a closer look at $v^*(k)$, which, for $k \in \Delta$, can be approximated as

$$v^*(k) \cong \frac{\sqrt{T}}{(1 - \sqrt{R}) + 2i\sqrt{R}(k - k_0/2)L}. \tag{9.58}$$

This function is highly peaked at $k = k_0/2$, and the width of the peak is given by $(1 - \sqrt{R})/(2\sqrt{R}L)$, which, for R close to one, is much less than $1/L$. Next, we have that

$$\partial_x u_k(x) = -\frac{ik}{\sqrt{2\pi}} \left[e^{-ikx} + \left(\frac{1 - \sqrt{R}e^{-2ikL}}{1 - \sqrt{R}e^{2ikL}} \right) e^{ikx} \right]. \tag{9.59}$$

In evaluating the k integral, this expression is multiplied by $v^*(k)$, which, as we noted, is peaked at $k_0/2$, and the peak has a width that is much narrower than $1/L$. That means that, to good approximation, we can replace k by $k_0/2$ in the term multiplying e^{ikx}. If we also note that we are assuming that $k_0/2$ is a cavity resonance, then $e^{ik_0L} = 1$, which means that this factor can be set equal to one. The remaining integral can be evaluated as follows. We first define a new integration variable, $k' = k - k_0/2$, which implies that the range of integration is now $-(1/L) \leq k' \leq (1/L)$. Since the integrand is now highly peaked around $k' = 0$ with a width that is much less than $1/L$, we can extend the lower and upper limits of integration to minus and plus infinity, respectively. The integral can then be evaluated by contour methods. Doing so, we find that the incoming wave part of the mode function makes no contribution. The final result, for $x < ct$, is

$$\begin{aligned} F(x,t) = &-\left(\frac{ik_0 g\sqrt{2\pi T}}{4} \right) e^{ik_0(x-ct)/2} \\ &\times \left[\frac{1}{2\sqrt{R}L|\kappa| - (1 - \sqrt{R})c} e^{|\kappa|t} (e^{-x(1-\sqrt{R})/2\sqrt{R}L} - e^{-|\kappa|x/c}) \right. \\ &\left. + \frac{1}{2\sqrt{R}L|\kappa| + (1 - \sqrt{R})c} e^{-|\kappa|t} (e^{|\kappa|x/c} - e^{-x(1-\sqrt{R})/2\sqrt{R}L}) \right]. \end{aligned} \tag{9.60}$$

Note the exponential increase in time in the first term. There is no gain saturation here due to making the parametric approximation, which implies that this expression would only be valid while the field in the cavity is building up to its steady-state value. The above expression also contains two length-scales. The first, $c/|\kappa|$, is associated with the gain, and the second, $2\sqrt{R}L/(1 - \sqrt{R})$, is associated with the width of the cavity resonance. The exponentially growing term is proportional to one over the difference of these two lengths. There is, however, no divergence when they are equal. This is because the reciprocal of the difference of the lengths is multiplied by the difference of exponentials in x, which also vanishes when the lengths are equal, so the result remains finite.

Additional reading

Articles

- The treatment of the degenerate parametric oscillator using the positive and complex P-representations comes from the papers:

P. D. Drummond, K. J. McNeil, and D. F. Walls, *J. Mod. Opt.* **27**, 321 (1980).
P. D. Drummond, K. J. McNeil, and D. F. Walls, *J. Mod. Opt.* **28**, 211 (1981).

Problems

9.1 Let us look at the phase properties of the light produced by the degenerate parametric amplifier. We will look at the simple system without damping or driving described by the Hamiltonian

$$H = \hbar\omega a_1^\dagger a_1 + 2\hbar\omega a_2^\dagger a_2 + \frac{i\hbar\chi}{2}(a_2 a_1^{\dagger 2} - a_2^\dagger a_1^2).$$

A phase distribution for a single-mode state, $|\psi\rangle$, can be defined in terms of the nonnormalizable phase states

$$|\theta\rangle = \frac{1}{\sqrt{2\pi}}\sum_{n=0}^{\infty} e^{in\theta}|n\rangle,$$

which form a resolution of the identity, $\int_0^{2\pi} d\theta\, |\theta\rangle\langle\theta| = I$. The phase distribution for $|\psi\rangle$ is just $P(\theta) = |\langle\theta|\psi\rangle|^2$.

(a) Show that

$$|\langle\psi|e^{i\phi a^\dagger a}|\psi\rangle| \leq \int_0^{2\pi} d\theta\, [P(\theta)P(\theta-\phi)]^{1/2}.$$

This inequality relates the overlap of a state with a rotated version of itself (rotated in phase space) to the overlap of the phase distribution of the original state with that of the rotated state.

(b) We can find a phase distribution for two-mode states by using two-mode phase states $|\theta_1,\theta_2\rangle = |\theta_1\rangle_1 \otimes |\theta_2\rangle_2$ and defining the phase distribution of the two-mode state $|\Psi\rangle$ to be $P(\theta_1,\theta_2) = |\langle\theta_1,\theta_2|\Psi\rangle|^2$. Using the fact that $M = a_1^\dagger a_1 + 2a_2^\dagger a_2$ commutes with the above Hamiltonian, show that

$$|\langle\Psi(0)|e^{i\phi M}|\Psi(0)\rangle| \leq \int_0^{2\pi} d\theta_1\, [P_1(\theta_1,t)P_1(\theta_1-\phi,t)]^{1/2},$$

where $P_1(\theta_1,t) = \int_0^{2\pi} d\theta_2\, P(\theta_1,\theta_2,t)$, and $P(\theta_1,\theta_2,t)$ is the phase distribution for the state $|\Psi(t)\rangle$. This inequality can be used to obtain information about the signal-mode phase distribution at time t in terms of the properties of the initial state.

(c) Find the left-hand side of this inequality in the case that the initial state is the vacuum in mode 1 and a coherent state $|\alpha_2\rangle$ in mode 2.

9.2 We found the Fokker–Planck equation for the degenerate parametric oscillator using the generalized P-representation (see Eq. (9.14)). Find the corresponding equation for the Wigner function.

9.3 Consider the master equation for the degenerate parametric oscillator in the interaction picture with everything on-resonance,

$$\frac{d\rho}{dt} = \frac{1}{i\hbar}[H_{int}, \rho] + \sum_{j=1,2} \gamma_j (2a_j \rho a_j^\dagger - a_j^\dagger a_j \rho - \rho a_j^\dagger a_j),$$

where

$$\frac{H_{int}}{\hbar} = i\frac{\chi}{2}(a_1^{\dagger 2} a_2 - a_1^2 a_2^\dagger) + i(\mathcal{E}_2 a_j^\dagger - \mathcal{E}_2^* a_j).$$

(a) In the case that $\mathcal{E}_2 = 0$ and $\gamma_1 = \gamma_2$, find $\langle M(t)\rangle$ in terms of $\langle M(0)\rangle$, where $M = a_1^\dagger a_1 + 2a_2^\dagger a_2$.

(b) Now set $\mathcal{E}_2 = |\mathcal{E}_2| e^{i\phi}$, and let us look at the general case of the above master equation. If $\rho(t;\phi)$ is the solution of the master equation corresponding to the phase of the driving field, ϕ, show that $\rho(t;\phi+\theta) = U\rho(t;\phi)U^{-1}$, where $U = e^{i\theta M}$.

10 Quantum field dynamics

In a realistic treatment of a three-dimensional nonlinear optical experiment, the complete Maxwell equations in $(3+1)$ space-time dimensions should be employed. It is then necessary to utilize a multi-mode Hamiltonian that correctly describes the propagating modes. There is an important difference between these experiments and traditional particle scattering. Quantum field dynamics in nonlinear media is dominated by multiple scattering, which is the reason why perturbation theory is less useful.

Nevertheless, it is interesting to make a link to conventional perturbation theory. Accordingly, we start by considering a perturbative theory of propagation in a one-dimensional nonlinear optical system in a $\chi^{(3)}$ medium. While this calculation cannot treat long interaction times, it does give a qualitative understanding of the important features.

This problem is the 'hydrogen atom' of quantum field theory: it has fully interacting fields with exact solutions for their energy levels. This is because a photon in a waveguide is an elementary boson in one dimension. The interactions between these bosons are mediated by the Kerr effect, which in quantum field theory is a quartic potential, equivalent to a delta-function interaction.

The quantum field theory involved is the simplest model of a quantum field that has an exact solution. This elementary model is still nontrivial in terms of its dynamics, as the calculation of quantum dynamics using standard eigenfunction techniques would require exponentially complex sums over multi-dimensional overlap integrals. This is not practicable, and accordingly we use other methods including quantum phase-space representations to solve this problem.

Next, we consider a $\chi^{(2)}$ nonlinear medium, this time in three dimensions. We will treat the down-conversion of light in a nonlinear crystal, which is one of the most well-known and famous cases of nonlinear quantum optics. The resulting creation of photon pairs is a workhorse of experiments in quantum information. Here we use perturbation theory to give an easily soluble first approximation, using a simplified nondispersive, homogeneous model.

Finally, we treat a microscopic model – the interaction of radiation with a medium of near-resonant two-level atoms, leading to the Maxwell–Bloch equations. These equations also have exact solutions in one dimension, although it is not trivial to create such a one-dimensional system. For this reason, we include the full three-dimensional equations here, which in general can only be treated numerically, especially at the quantum level.

10.1 Kerr medium

We will start this chapter with a simple calculation based on the in and out fields that were discussed in Chapter 2. Suppose we have a medium with a finite region of Kerr-type nonlinearity located between $x = -L$ and $x = L$. We assume that the fields are confined in an otherwise homogeneous waveguide with an effective area A in the y and z directions, and a single transverse mode. Our object is to find the out field in terms of the in field using first-order perturbation theory. As we will see in more detail in the next chapter, the effective area of a real waveguide depends on the mode function, which depends on the transverse coordinates, $\mathbf{r}_\perp = (y, z)$. Thus, if $\mathbf{u}(\mathbf{r}_\perp)$ is the transverse mode function for the displacement field, normalized so that $\int |\mathbf{u}(\mathbf{r}_\perp)|^2\, d^2\mathbf{r}_\perp = 1$, then

$$A = \left[\int |\mathbf{u}(\mathbf{r}_\perp)|^4\, d^2\mathbf{r}_\perp\right]^{-1}. \tag{10.1}$$

We can take the dual potential to point in the y direction, $\mathbf{\Lambda}(x,t) = \Lambda(x,t)\hat{\mathbf{y}}$, which implies that $\mathbf{D}(x,t) = \partial_x \Lambda(x,t)\hat{\mathbf{z}}$, where $\partial_x \equiv \partial/\partial_x$. Using a continuum quantization, and assuming for simplicity that the transverse mode function is uniform, we have the fields on the infinite line:

$$\Lambda(x,t) = \int_{-\infty}^{\infty} dk \left(\frac{\hbar}{4\pi A\mu\omega}\right)^{1/2} (e^{ikx}a(k) + e^{-ikx}a^\dagger(k)),$$
$$B(x,t) = i\int_{-\infty}^{\infty} dk \left(\frac{\hbar\omega\mu}{4\pi A}\right)^{1/2} (e^{-ikx}a^\dagger(k) - e^{ikx}a(k)), \tag{10.2}$$

where $[a(k), a^\dagger(k')] = \delta(k - k')$. These fields satisfy the canonical commutation relations

$$[\Lambda(x,t), B(x',t)] = \frac{i\hbar}{A}\delta(x - x'). \tag{10.3}$$

Defining the displacement field as $D(x,t) = \partial_x \Lambda(x,t)$, we find that

$$[\Lambda(x,t), D(x',t)] = 0. \tag{10.4}$$

For the Hamiltonian, we have, including an area factor of A that is needed for dimensional reasons,

$$H = A\int_{-\infty}^{\infty} dx \left(\frac{B^2}{2\mu} + \frac{D^2}{2\epsilon}\right) + \frac{A}{4}\eta^{(3)}\int_{-L}^{L} dx\, D^4. \tag{10.5}$$

The equation of motion for $B(x,t)$ is then

$$\frac{\partial B}{\partial t} = \frac{1}{\epsilon}\frac{\partial D}{\partial x} + \eta^{(3)}(x)\frac{\partial}{\partial x}D^3, \tag{10.6}$$

where $\eta^{(3)}(x) = 0$ for $x < -L$ and $x > L$. If we write this equation in terms of $\Lambda(x,t)$, using $B = \mu(\partial\Lambda/\partial t)$, we find

$$\frac{\partial^2\Lambda}{\partial t^2} - v^2\frac{\partial^2\Lambda}{\partial x^2} = J(x,t), \tag{10.7}$$

where $J = \eta^{(3)}(x)\partial_x(\partial_x\Lambda)^3/\mu$, and $v = 1/\sqrt{\mu\epsilon}$ is the phase velocity in the dielectric, which is also the group velocity in a nondispersive medium.

For the in and out fields, we now have

$$\Lambda(x,t) = \Lambda^{in}(x,t) + \int_{-\infty}^{\infty} dt' \int_{-\infty}^{\infty} dx'\, \Delta_{ret}(x-x', t-t')J(x',t';\Lambda)$$
$$\Lambda(x,t) = \Lambda^{out}(x,t) + \int_{-\infty}^{\infty} dt' \int_{-\infty}^{\infty} dx'\, \Delta_{adv}(x-x', t-t')J(x',t';\Lambda), \tag{10.8}$$

where we have indicated the dependence of $J(x,t;\Lambda)$ on space, time and the field, Λ. We want to solve these equations to lowest order in J for Λ^{out} in terms of Λ^{in}. To lowest order, the first integral equation becomes

$$\Lambda(x,t) = \Lambda^{in}(x,t) + \int_{-\infty}^{\infty} dt' \int_{-\infty}^{\infty} dx'\, \Delta_{ret}(x-x', t-t')J(x',t';\Lambda^{in}), \tag{10.9}$$

and this is then substituted into the second equation to give

$$\Lambda^{out}(x,t) = \Lambda^{in}(x,t) + \int_{-\infty}^{\infty} dt' \int_{-\infty}^{\infty} dx' \times [\Delta_{ret}(x-x', t-t') - \Delta_{adv}(x-x', t-t')]J(x',t';\Lambda^{in}). \tag{10.10}$$

The difference of the Green's functions is given, as in Chapter 2, by

$$\Delta_{ret}(x,t) - \Delta_{adv}(x,t) = \frac{1}{4v}[\varepsilon(x+vt) - \varepsilon(x-vt)], \tag{10.11}$$

where

$$\varepsilon(s) = \begin{cases} +1, & s \geq 0, \\ -1, & s < 0. \end{cases} \tag{10.12}$$

We can now use this equation to see how the expectation value of the field changes after it passes through the nonlinear medium. Let us assume that the state of the system is given by the coherent state

$$|\Psi\rangle = e^{a^{in\dagger}[f] - a^{in}[f]}|0\rangle, \tag{10.13}$$

where $a^{in}[f] = \int dk\, f^*(k)a^{in}(k)$. We shall assume that $f(k)$, which describes the shape of the photon wavepackets, is sharply peaked around some value $k_0 = \omega_0/v$. In order to find $\langle\Psi|\Lambda^{out}(x,t)|\Psi\rangle$, we need to find $\langle\Psi| : \partial_x[\partial_x\Lambda^{in}(x,t)]^3 : |\Psi\rangle$, where we are assuming normal ordering. We find that

$$\begin{aligned}\langle\Psi| &: \partial_x[\partial_x\Lambda^{in}(x,t)]^3 : |\Psi\rangle \\ &\simeq \left(\frac{\hbar k_0}{4\pi A\mu v}\right)^{3/2} \int_{-\infty}^{\infty} dk_1 \cdots \int_{-\infty}^{\infty} dk_3 \\ &\times [3k_0 e^{i(k_1+k_2+k_3)(x-ct)} f(k_1)f(k_2)f(k_3) \\ &- 3k_0 e^{i(k_1+k_2-k_3)(x-ct)} f(k_1)f(k_2)f^*(k_3) + c.c.].\end{aligned} \tag{10.14}$$

We now need to insert this expression into the equation for the out field. Defining $\xi(x,t) = \varepsilon(x+vt) - \varepsilon(x-vt)$, we see that we have to evaluate integrals of the form

$$\int_{-\infty}^{\infty} dt' \int_{-L}^{L} dx' \, \xi(x-x', t-t') e^{ik(x'-vt')}$$
$$= e^{i\kappa(x-vt)} \int_{-\infty}^{\infty} dt' \int_{x-L}^{x+L} dx'' \, \xi(x'', t'') e^{ik(x''-vt'')-\delta|t''|}, \tag{10.15}$$

where we have inserted a convergence factor δ into the time integration. We take $\delta \to 0^+$ at the end of the calculation. Keeping in mind that we will be interested in the cases when k is close to either k_0 or $3k_0$, and that $k_0 L \gg 1$, we find that

$$\int_{-\infty}^{\infty} dt' \int_{-L}^{L} dx' \, \xi(x-x', t-t') e^{ik(x'-vt')} \simeq \frac{4iL}{kv} e^{ik(x-vt)}. \tag{10.16}$$

This, finally, gives us that

$$\langle\Psi|\Lambda^{out}(x,t)|\Psi\rangle = \langle\Psi|\Lambda^{in}(x,t)|\Psi\rangle + \frac{iL\eta^{(3)}}{2v^2\mu}\left(\frac{\hbar k_0}{\pi\mu v A}\right)^{3/2}$$
$$\times \int_{-\infty}^{\infty} dk_1 \cdots \int_{-\infty}^{\infty} dk_3 \, [e^{i(k_1+k_2+k_3)(x-vt)} f(k_1) f(k_2) f(k_3)$$
$$- 3e^{i(k_1+k_2-k_3)(x-vt)} f(k_1) f(k_2) f^*(k_3) - c.c.]. \tag{10.17}$$

Looking at the above equation, we see that there are two effects of passing through the medium. First, a component at frequency $3\omega_0$ was generated, that is to say, there has been third-harmonic generation. Next, we see that the original component at frequency ω_0 has been modified. This modification is due to the intensity-dependent refractive effect.

10.2 Quantum solitons

We now wish to treat in detail the Hamiltonian for light propagating in a $\chi^{(3)}$ medium that was derived at the end of Chapter 3. As we shall shortly see, that Hamiltonian gives rise to quantum solitons. A classical soliton in a dispersive, nonlinear medium is a pulse that propagates without changing shape. They were first observed in water waves and, more recently, and of more relevance to us, they have been observed in optical fibers. Besides their fundamental interest, they could be of use in optical communications. When these waves are quantized, new effects emerge. We will use the formalism we have developed to study these quantized solitons.

Solitons themselves are persistent 'solitary waves' or pulses that are resistant to disruptions that would ordinarily distort a pulse as it propagates. A short pulse is made up of a range of oscillation frequencies; the shorter the pulse, the broader its spectrum. In any physical medium, the propagation velocity depends on frequency, an effect known as group velocity dispersion. By itself, this causes different portions of the pulse spectrum to travel with different velocities, leading to pulse broadening. Solitons exist in nonlinear media where the velocity also depends on the amplitude of the pulse. If the signs of these two effects are opposite, a soliton can classically propagate without distortion by either effect.

In the case of solitons in optical fibers, the balance is between the wavelength dependence and the intensity dependence of the refractive index of the fiber waveguide. The group velocity dispersion causes the shorter wavelengths to travel faster in the fiber (for the case of interest here) than the longer wavelengths. This would cause the leading edge of the pulse to become blue-shifted and the trailing edge red-shifted. At the same time, the intensity dependence of the refractive index causes a time dependence of the pulse phase as it passes along the fiber. This results in an opposite frequency shift, toward red on the leading edge and blue on the trailing edge, tending to cancel the effect of group velocity dispersion. When these opposing effects balance, an optical soliton is formed. In this case, the classical nonlinear wave equation is called the nonlinear Schrödinger equation, which is an example of a classically soluble or 'integrable' nonlinear wave equation in one dimension.

Classical arguments predict that a soliton is invariant as it propagates. However, the classical picture of solitons is *not* correct quantum mechanically – a coherent soliton undergoes phase diffusion and wavepacket spreading. Coherent solitons consist of linear superpositions of quantum solitons with different photon numbers and momentum eigenstates. Each component soliton has a different phase velocity and group velocity, so the phase and the wavepacket spread out as they propagate.

10.2.1 Quantum theory of optical fibers

We now know that solitons in optical fibers are superpositions of macroscopic quantum states that correspond to clusters of 10^8 or more photons, bound together. This makes them even larger than macromolecules or metallic clusters. Intrinsically, quantum mechanical effects emerge and can be detected, even for such macroscopic objects. The quantitative implications have been explored in the quantum theory of optical solitons.

These effects of quantum phase diffusion were first experimentally observed at IBM Almaden Research Center, in one of the few experimental tests of a simple, exactly soluble quantum field theory. The experiments give direct evidence for the quantum nature of soliton propagation. Quantum phase diffusion also provides a promising way for generating 'squeezed' states of light.

The theory of quantum solitons in optical fibers predicts that quantum solitons are squeezed, which was subsequently confirmed experimentally. This theory and its subsequent elaborations are comprehensive. It takes into account losses, Brillouin scattering and Raman processes, and its agreement with experiment is excellent. In this chapter, we only wish to present some of the basic features of quantum solitons, and to do so we will make use of an approximate theory due to Lai and Haus (see the additional reading). It is based on the time-dependent Hartree approximation. More precise stochastic methods will be covered in the next section, and in the subsequent chapter.

Our starting point is the Hamiltonian derived at the end of Chapter 3:

$$H = \hbar \int dx \left[\omega \psi^\dagger \psi + \frac{i}{2} v \left(\frac{\partial \psi^\dagger}{\partial x} \psi - \psi^\dagger \frac{\partial \psi}{\partial x} \right) + \frac{1}{2} \omega'' \frac{\partial \psi^\dagger}{\partial x} \frac{\partial \psi}{\partial x} \right] + \frac{3\eta^{(3)}}{8A} (\hbar k_0 v \epsilon)^2 \int dx \, (\psi^\dagger)^2 \psi^2. \tag{10.18}$$

As a first step, we shall go into a kind of interaction picture. Let

$$H_0 = \hbar \int dx\, \omega \psi^\dagger \psi, \tag{10.19}$$

and define operators

$$\psi_I = e^{-iH_0t/\hbar} \psi e^{iH_0t/\hbar}, \qquad H_I = e^{-iH_0t/\hbar} H e^{iH_0t/\hbar}. \tag{10.20}$$

We then find that

$$H_I = \frac{\hbar}{2} \int dx \left[iv \left(\frac{\partial \psi_I^\dagger}{\partial x} \psi_I - \psi_I^\dagger \frac{\partial \psi_I}{\partial x} \right) + \omega'' \frac{\partial \psi_I^\dagger}{\partial x} \frac{\partial \psi_I}{\partial x} + \kappa (\psi_I^\dagger)^2 \psi_I^2 \right], \tag{10.21}$$

where we have set

$$\kappa = \frac{3\hbar \eta^{(3)} (k_0 v \epsilon)^2}{4A}. \tag{10.22}$$

Nonlinear couplings are often quoted in different units, so for completeness, we note that

$$\kappa = \frac{-3\hbar \chi^{(3)} (\omega v)^2}{4\epsilon A c^2} = \frac{-\hbar n_2 (\omega v)^2}{Ac}, \tag{10.23}$$

where n_2 is the nonlinear refractive index per unit intensity, and $\chi^{(3)}$ is the Bloembergen nonlinear coefficient in S.I. units.

This gives us the equation of motion

$$i \left(\frac{\partial}{\partial t} + v \frac{\partial}{\partial x} \right) \psi_I = -\frac{\omega''}{2} \frac{\partial^2 \psi_I}{\partial x^2} + \kappa \psi_I^\dagger \psi_I^2. \tag{10.24}$$

From here on, we shall drop the subscript I with the understanding that all of the operators are in our interaction picture. Next, we go into a moving frame by considering ψ to be a function of $x_v = x - vt$ and t rather than x and t. In terms of the new coordinates, the equation of motion for ψ becomes

$$i \frac{\partial \psi}{\partial t} = -\frac{\omega''}{2} \frac{\partial^2 \psi}{\partial x_v^2} + \kappa \psi^\dagger \psi^2. \tag{10.25}$$

The above equation is an operator version of a nonlinear Schrödinger equation. The classical version of this equation gives rise to solitons. The operator version describes particles interacting by means of a delta-function potential, and the solutions to this problem are known. Let us make a short detour to describe the interacting-particle system, and then use the results to examine the properties of our solitons in a nonlinear medium.

10.2.2 Bosons on a wire

For comparison purposes, suppose that we have 'bosons on a wire': particles of mass m in one dimension interacting via a potential $V(x - x')$. This can be created via a strong transverse confinement of ultra-cold bosonic atoms. The second-quantized Hamiltonian for this system is

$$H = \int dx \frac{\hbar}{2m} \frac{\partial \psi^\dagger}{\partial x} \frac{\partial \psi}{\partial x} + \int dx \int dx'\, V(x - x') \psi^\dagger(x,t) \psi^\dagger(x',t) \psi(x',t) \psi(x,t), \tag{10.26}$$

where $\psi(x,t)^\dagger$ creates a particle at position x, and

$$[\psi(x,t), \psi(x',t)^\dagger] = \delta(x-x'). \tag{10.27}$$

If the potential is given by $V(x-x') = \hbar\kappa\delta(x-x')/2$, then we obtain a stationary-frame Hamiltonian that is identical to the moving-frame Hamiltonian describing the propagation of the field in a nonlinear fiber. The fact that the number operator commutes with the Hamiltonian implies that the eigenstates will be states of well-defined particle number, and hence be of the form

$$|\psi_n\rangle = \frac{1}{\sqrt{n!}} \int dx_1 \ldots dx_n\, f_n(x_1,\ldots,x_n)\psi(x_1,0)^\dagger \cdots \psi(x_n,0)^\dagger|0\rangle. \tag{10.28}$$

The function f_n satisfies the equation

$$E_n f_n = \left[-\frac{\hbar^2}{2m}\sum_{j=1}^{n}\frac{\partial^2}{\partial x_j^2} + \hbar\kappa \sum_{1\leq j<k\leq n} \delta(x_k - x_j)\right] f_n. \tag{10.29}$$

If $\kappa < 0$, then there are bound states, and for these f_n is proportional to

$$f_n(x_1,\ldots,x_n) \propto \exp\left[ip\sum_{j=1}^{n} x_j + \frac{m\kappa}{2\hbar}\sum_{1\leq j<k\leq n}|x_k - x_j|\right]. \tag{10.30}$$

In the case of photon–polaritons (which we will abbreviate as photons) in a nonlinear fiber, this solution would be a bound state of n photons, and, therefore, be a state of definite photon number. Because the solitons that propagate in nonlinear fibers are a result of light from a laser propagating through the fiber, and laser light, being close to a coherent state, does not have a definite photon number, the states we are looking for are superpositions of many of these states with different values of n. We also note that the above solutions hold for attractive interactions between photons. There are also exact solutions for the repulsive case.

10.3 Time-dependent Hartree approximation

Rather than trying to find these soliton solutions by superposing the exact solutions, we shall make use of an approximate method, the time-dependent Hartree approximation. We start from the time-dependent equation for f_n, which is an n-particle Schrödinger equation,

$$i\hbar\frac{\partial f_n}{\partial t} = \left[-\frac{\hbar^2}{2m}\sum_{j=1}^{n}\frac{\partial^2}{\partial x_j^2} + \hbar\kappa \sum_{1\leq j<k\leq n} \delta(x_k - x_j)\right] f_n, \tag{10.31}$$

and we assume that $f_n(x_1,\ldots,x_n) = \prod_{j=1}^{n} h_n(x_j,t)$. The effective potential felt by one of the particles is the n-particle potential averaged over all of the other particles:

$$V_{eff}(x,t) = \hbar\kappa\sum_{j=2}^{n}\int dx_j\, |h_n(x_j,t)|^2\delta(x-x_j) = \hbar\kappa(n-1)|h_n(x,t)|^2. \tag{10.32}$$

We assume that $h_n(x,t)$ satisfies the one-particle Schrödinger equation with the effective potential

$$i\hbar\frac{\partial h_n}{\partial t} = \frac{\hbar^2}{2m}\frac{\partial^2 h_n}{\partial x^2} + \hbar\kappa(n-1)|h_n(x,t)|^2 h_n(x,t). \tag{10.33}$$

For $g < 0$ this equation has the soliton solution

$$h_n(x,t) = \frac{2\sqrt{2}z}{|\kappa(n-1)|^{1/2}}\exp\left[-4i(\xi^2 - z^2)t - 2i\xi\sqrt{\frac{2m}{\hbar}}(x - x_0)\right] \times \operatorname{sech}\left[2z\left(\sqrt{\frac{2m}{\hbar}}(x - x_0) + 4\xi t\right)\right], \tag{10.34}$$

where ξ, z and x_0 are arbitrary parameters. The normalization condition,

$$\int dx\,|h_n(x,t)|^2 = 1, \tag{10.35}$$

however, fixes the value of z at

$$z = \frac{1}{8}|\kappa|(n-1)\sqrt{\frac{2m}{\hbar}}. \tag{10.36}$$

10.3.1 Nonlinear propagation in the Hartree approximation

We now want to apply what we have learned about interacting particles to a field propagating in a nonlinear fiber. In order to do so, it is useful to look at the field propagation problem in the Schrödinger picture. For the state $|\Phi\rangle$, we have

$$i\hbar\frac{\partial}{\partial t}|\Phi\rangle = (H_{lin} + H_{nlin})|\Phi\rangle. \tag{10.37}$$

Because the Hamiltonian conserves photon number, we can assume that

$$|\Psi\rangle = \frac{1}{\sqrt{n!}}\int dx_1 \ldots dx_n\, f_n(x_1,\ldots,x_n,t)\psi^\dagger(x_1,0)\cdots\psi^\dagger(x_n,0)|0\rangle, \tag{10.38}$$

where

$$\int dx_1 \ldots dx_n\,|f_n(x_1,\ldots,x_n)|^2 = 1. \tag{10.39}$$

Inserting $|\Phi\rangle$ into the Schrödinger equation, we find that f_n satisfies

$$i\frac{\partial f_n}{\partial t} = n\omega f_n - iv\sum_{j=1}^{n}\frac{\partial f_n}{\partial x_j} - \frac{1}{2}\omega''\sum_{j=1}^{n}\frac{\partial^2 f_n}{\partial x_j^2} + \kappa\sum_{j=1}^{n}\sum_{k=1}^{j-1}\delta(x_j - x_k)f_n. \tag{10.40}$$

If we now define $\tilde{f}_n = \exp(in\omega t)f_n$ and we consider $\tilde{f}_n$ to be a function of $\{x_{vj} = x_j - vt \mid j = 1,\ldots,n\}$ and t rather than of $\{x_j \mid j = 1,\ldots,n\}$ and t, then we have that

$$i\frac{\partial \tilde{f}_n}{\partial t} = -\frac{1}{2}\omega''\sum_{j=1}^{n}\frac{\partial^2 \tilde{f}_n}{\partial x_{vj}^2} + \kappa\sum_{j=1}^{n}\sum_{k=1}^{j-1}\delta(x_{vj} - x_{vk})\tilde{f}_n, \tag{10.41}$$

which has the same form as the Schrödinger equation for n particles interacting via a delta-function potential. If we assume that $\omega'' > 0$ and $\kappa < 0$, we can make use of the solutions we obtained for the delta-function-interacting system.

In particular, we can use the solutions we obtained from the Hartree approximation and set

$$f_n(x_1, \ldots, x_n, t) = \prod_{j=1}^{n} \tilde{h}_n(x_j, t), \tag{10.42}$$

where

$$\tilde{h}_n(x_j, t) = e^{i\omega_1 t} h_n(x_j - v_1 t, t), \tag{10.43}$$

and in the explicit expression for h_n we have made the replacement

$$\frac{\hbar}{2m} \to \frac{\omega''}{2}. \tag{10.44}$$

As was mentioned, we are interested in solutions that are superpositions of states with different photon numbers, and, in particular, solutions that resemble coherent states. Defining

$$\psi^\dagger[\tilde{h}_n] = \int dx\, \tilde{h}_n(x, t)\psi^\dagger(x, 0), \tag{10.45}$$

we note that $[\psi(x, 0), \psi^\dagger[\tilde{h}_n]] = \tilde{h}_n(x, t)$. Now consider the superposition of approximate solutions,

$$|\Psi\rangle = e^{-|\alpha|^2/2} \sum_{n=0}^{\infty} \frac{\alpha^n}{n!} (\psi^\dagger[\tilde{h}_n])^n |0\rangle. \tag{10.46}$$

This state has the property that

$$\langle\Psi|\psi(x, 0)|\Psi\rangle = \alpha e^{-|\alpha|^2} \sum_{n=0}^{\infty} \frac{|\alpha|^{2n}}{n!} \tilde{h}_{n+1}(x, t)(\langle\tilde{h}_n|\tilde{h}_{n+1}\rangle)^n. \tag{10.47}$$

Now let us assume that $n_0 = |\alpha|^2 \gg 1$, which implies that the average photon number in the state is large. The terms in the sum that will contribute most are those for which $n_0 - \sqrt{n_0} \leq n \leq n_0 + \sqrt{n_0}$. In this range we have that

$$(\langle\tilde{h}_n|\tilde{h}_{n+1}\rangle)^n \cong e^{it\kappa^2 n(2n-1)/8\omega_1''} \tag{10.48}$$

and that the dominant n dependence in $\tilde{h}_{n+1}(x, t)$ is the exponential factor $\exp(itn^2\kappa^2/4\omega_1'')$.

Therefore, if $tn_0\sqrt{n_0}\,\kappa^2/4\omega_1'' \ll 1$, the n dependence of the terms in the sum in the important range will be weak, and we can replace them by their values at n_0, so that

$$\langle\Psi|\psi(x, 0)|\Psi\rangle = \alpha\tilde{h}_{n_0}(x, t). \tag{10.49}$$

This is just a propagating soliton, so that, for sufficiently short times, the average value of the field is just that of a classical soliton. For longer times, however, the n dependence of the different terms in the sum comes into play, and these terms are no longer in phase. The phase of what had been a coherent state starts to diffuse, and this will cause the average value of the field to decay. This phase diffusion is a quantum effect; classically, the soliton would propagate with no change to its shape (ignoring losses). This phase diffusion can lead to squeezing, and this has been observed experimentally.

10.4 Quantum solitons in phase space

The techniques described above give some insight, but they have drawbacks in describing real experiments. Exact solutions have the complexity problem that one needs to carry out an exponentially large number of overlap integrals to obtain the expansion of an initial state in terms of the eigenstates, which is not generally feasible for large numbers of modes and particles. The Hartree method introduces uncontrolled approximations, and is unable to handle the realistic noise and loss processes of real experiments.

Instead, we can obtain quantitative results using phase-space techniques. Assuming vanishing boundary terms as usual, and using the methods of Chapter 7, the equivalent Ito stochastic equations in the positive P-representation in the cubic nonlinearity case are

$$\frac{\partial}{\partial t}\alpha(x,t_v) = \left[\frac{i\omega''}{2}\frac{\partial^2}{\partial x_v^2} - i\kappa\beta\alpha + \sqrt{-i\kappa}\xi_1(x,t_v)\right]\alpha(x,t_v),$$

$$\frac{\partial}{\partial t}\beta(x,t_v) = \left[-\frac{i\omega''}{2}\frac{\partial^2}{\partial x_v^2} + i\kappa\beta\alpha + \sqrt{i\kappa}\xi_2(x,t_v)\right]\beta(x,t_v), \qquad (10.50)$$

where, following the techniques of functional representations outlined earlier, the noises are delta-correlated in both space and time:

$$\langle \xi_i(x,t_v)\xi_j(x',t'_v)\rangle = \delta_{ij}\delta(t-t')\delta(x_v - x'_v). \qquad (10.51)$$

These equations are equivalent to the original operator Heisenberg equations, but involve only c-number equations, rather than noncommuting operators. To solve these equations, either computer simulation techniques can be used, in which case the space must be subdivided into a lattice with a momentum cutoff, or methods utilizing solutions with linearization around the classical soliton can be employed. Both approaches were able to successfully predict phase diffusion and squeezing in propagating solitons or other pulses, which was subsequently observed experimentally.

The main drawback to this procedure is that, while the method is exact, the sampling errors can grow large during time evolution. This is more readily controlled with the truncated Wigner method, although the representation is only applicable at large photon number owing to the truncation approximation. With photon numbers of around 10^7–10^9 in most typical fiber experiments, these conditions are easily met.

The approximate Wigner equations, suitable for calculating symmetrically ordered correlations, are identical to the semiclassical mean-field equations except with stochastic initial conditions:

$$\frac{\partial}{\partial t}\alpha(x,t_v) = \left[\frac{i\omega''}{2}\frac{\partial^2}{\partial x_v^2} - i\kappa|\alpha|^2\right]\alpha(x,t_v). \qquad (10.52)$$

The initial fields entering the fiber at $t = 0$ are assumed to be in a coherent state, and have vacuum fluctuations correlated according to

$$\langle \Delta\alpha(x,0)\Delta\alpha(x',0)\rangle = 0,$$
$$\langle \Delta\alpha(x,0)\Delta\alpha^*(x',0)\rangle = \tfrac{1}{2}\delta(x-x'). \tag{10.53}$$

The photon density (in photons/m) is represented by

$$\left\langle |\alpha(x,t)|^2 - \frac{1}{2\Delta x}\right\rangle,$$

where Δx^{-1} is the fundamental spatial frequency cutoff in the theory. This correction factor is necessary because the Wigner function represents symmetrically ordered operators, which have a diverging vacuum noise term as the cutoff is taken to infinity.

Computer simulations of these equations reveal that there is good agreement between this technique and the corresponding technique using the positive P-representation at large photon number.

10.5 Parametric down-conversion

So far, we have considered descriptions of parametric down-conversion in which only a small number of modes are present or in which the nonlinear medium is in a cavity. Now let us look at a multi-mode theory for down-conversion in the absence of a cavity. With the advent of quantum information, this has become a process that is extensively employed and intensively studied. A multi-mode quantum treatment due to Ou, Wang and Mandel has been a model for many later treatments, and we shall pattern our treatment after theirs (see the additional reading). Their approach does not take into account the issues with field quantization, which have concerned us in this book.

As we have seen, down-conversion takes place when a photon is incident on a material with a $\chi^{(2)}$ nonlinearity, and two photons, the sum of whose energies is equal to the energy of the incident photon, are produced. We shall assume that the medium is centered on the origin, and is of extent l_x, l_y and l_z in the x, y and z directions, respectively. In addition, we shall assume that the medium is nondispersive and that there are no free charges. In reality, dispersion in the medium is a problem. It typically requires a serious experimental effort to eliminate this, using birefringence or temperature tuning to match the refractive index at the two frequencies involved. For simplicity, we do not treat this here, although it is an important practical issue.

10.5.1 Nondispersive Hamiltonian

We then find for the Hamiltonian

$$H = H_0 + H', \tag{10.54}$$

where the free and interacting parts are given by

$$H_0 = \int_V d^3\mathbf{r}\,\frac{1}{2}\left(\frac{\mathbf{D}^2}{\epsilon} + \frac{\mathbf{B}^2}{\mu}\right),$$
$$H' = \int_V d^3\mathbf{r}\,\frac{1}{3}\,\mathbf{D}\cdot\eta^{(2)}\mathbf{D}\mathbf{D}. \tag{10.55}$$

Here V is the quantization volume and V_m is the volume of the medium. By taking the curl of the first of Eqs (3.90), we obtain an expression for $\mathbf{D}$ in terms of plane-wave creation and annihilation operators as

$$\mathbf{D}(\mathbf{r}) = \sum_{\mathbf{k},\alpha} \frac{i\hbar}{\sqrt{2\mu\omega V}}(\mathbf{k}\times\hat{\mathbf{e}}_{\mathbf{k},\alpha})(a_{\mathbf{k},\alpha}e^{i\mathbf{k}\cdot\mathbf{r}} - a^{\dagger}_{\mathbf{k},\alpha}e^{-i\mathbf{k}\cdot\mathbf{r}}), \tag{10.56}$$

where $\alpha = 1, 2$ and $\mathbf{k}\cdot\hat{\mathbf{e}}_{\mathbf{k},\alpha} = 0$. This expression and the corresponding one for $\mathbf{B}$ (see Eq. (3.90)) can now be inserted into the Hamiltonian. For H_0 we find, as before,

$$H_0 = \sum_{\mathbf{k},\alpha}\hbar\omega a^{\dagger}_{\mathbf{k},\alpha}a_{\mathbf{k},\alpha}. \tag{10.57}$$

For H' we shall make several assumptions. We shall assume that $\eta^{(2)}$ is symmetric, and that the operators are normally ordered, and, since we are interested only in down-conversion, we shall drop the two terms that do not contribute to that process. The result is

$$H' = i\sum_{\mathbf{k}_1,\alpha_1}\sum_{\mathbf{k}_2,\alpha_2}\sum_{\mathbf{k}_3,\alpha_3}\frac{1}{V^{3/2}}F(\mathbf{k}_1,\mathbf{k}_2,\mathbf{k}_3;\alpha_1,\alpha_2,\alpha_3)$$
$$\times\, h(\mathbf{k}_3 - \mathbf{k}_1 - \mathbf{k}_2)(a^{\dagger}_{\mathbf{k}_3,\alpha_3}a_{\mathbf{k}_1,\alpha_1}a_{\mathbf{k}_2,\alpha_2} - a^{\dagger}_{\mathbf{k}_2,\alpha_2}a^{\dagger}_{\mathbf{k}_1,\alpha_1}a_{\mathbf{k}_3,\alpha_3}), \tag{10.58}$$

where

$$F(\mathbf{k}_1,\mathbf{k}_2,\mathbf{k}_3;\alpha_1,\alpha_2,\alpha_3) = \left(\frac{\hbar}{\sqrt{2\mu}}\right)^3 \sum_{j,k,l=1}^{3}\frac{\eta^{(2)}_{jkl}}{\sqrt{\omega_{\mathbf{k}_1}\omega_{\mathbf{k}_2}\omega_{\mathbf{k}_3}}}$$
$$\times\,(\mathbf{k}_1\times\hat{\mathbf{e}}_{\mathbf{k}_1,\alpha_1})_j(\mathbf{k}_2\times\hat{\mathbf{e}}_{\mathbf{k}_2,\alpha_2})_k(\mathbf{k}_3\times\hat{\mathbf{e}}_{\mathbf{k}_3,\alpha_3})_l \tag{10.59}$$

and

$$h(\mathbf{k}) = \prod_{j=x,y,z}\frac{2\sin(k_j l_j/2)}{k_j}. \tag{10.60}$$

Note that F depends only on the direction of the wavevectors and not on their magnitudes.

10.5.2 Two-point correlation function

We would now like to use this Hamiltonian to calculate a two-point correlation function that is related to the probability of detecting a single photon at each point. Suppose we have single-atom detectors at points $\mathbf{r}_1$ and $\mathbf{r}_2$. The detector at $\mathbf{r}_1$ is turned on between times t_1 and $t_1 + \Delta t$, and the detector at $\mathbf{r}_2$ is turned on between t_2 and $t_2 + \Delta t$, where $t_2 > t_1 + \Delta t$. The initial state of the system is a product of coherent states,

$$|\psi^{in}\rangle = \prod_{\mathbf{k}}|\alpha_{\mathbf{k}}\rangle, \tag{10.61}$$

where $\alpha_{\mathbf{k}}$ is only appreciable for $\mathbf{k}$ in a neighborhood of $k_0\hat{\mathbf{z}}$. The nonzero values of $\alpha_{\mathbf{k}}$ represent the initial state of the pump; the signal and idler are assumed to be initially in the vacuum state. Let us assume that the detector atoms have ground states $|g_1\rangle$ and $|g_2\rangle$ and excited states $\{|q_{1j}\rangle\}$ and $\{|q_{2j}\rangle\}$ for some range of j, and we shall assume that these atoms are identical. Let ω_{jg} be the frequency difference between the levels $|q_{1j}\rangle$ and $|g_1\rangle$, and let the dipole matrix element between these two levels be $\mathbf{d}_{jg}$. We shall assume that the frequencies ω_{jg} are less than the frequency of the pump mode, and are sufficiently far from the pump frequency that the chance of a pump photon being absorbed by one of the detector atoms is negligible. The part of the atom–field interaction governing the absorption of a photon from the ground state for the atom at $\mathbf{r}_1$ is (the expression for the atom at $\mathbf{r}_2$ is similar)

$$\sum_j (\mathbf{d}_{jg} \cdot \mathbf{D}^{(+)}(\mathbf{r}_1)|q_{1j}\rangle\langle g_1| + h.c.),$$

where

$$\mathbf{D}^{(+)}(\mathbf{r}, t) = \sum_{\mathbf{k},\alpha} \frac{i\hbar}{\sqrt{2\mu\omega V}}(\mathbf{k} \times \hat{\mathbf{e}}_{\mathbf{k},\alpha})a_{\mathbf{k},\alpha}e^{i(\mathbf{k}\cdot\mathbf{r}-\omega t)}. \tag{10.62}$$

The process in which we are interested is that where H_{int} creates two photons in modes (signal and idler), which were initially in the vacuum state, some time between times 0 and t_1, and each of these photons is then absorbed at a later time by a detector atom. This implies that the final state of the field will be the same as its initial state, $|\psi^{in}\rangle$. To lowest order in perturbation theory, where both the detector atoms and H_{int} are treated as perturbations, the probability that each atom will have absorbed a photon, so that atom 1 is in $|q_{1j}\rangle$ and atom 2 is in $|q_{2j'}\rangle$ after time $t_2 + \Delta t$ (we assume $t_2 \geq t_1$), is

$$P(\mathbf{r}_1, t_1; \mathbf{r}_2, t_2) = \left| \int_{t_2}^{t_2+\Delta t} dt_2' \int_{t_1}^{t_1+\Delta t} dt_1' \int_0^{t_1} dt\, e^{i\omega_{jg}t_1'}e^{i\omega_{j'g}t_2'} \right. \\ \left. \times \langle\psi^{in}|\mathbf{d}_{j'g} \cdot \mathbf{D}^{(+)}(\mathbf{r}_2, t_2')\mathbf{d}_{jg} \cdot \mathbf{D}^{(+)}(\mathbf{r}_1, t_1')H_{int}^{(I)}(t)|\psi^{in}\rangle \right|^2. \tag{10.63}$$

In this equation, $H_{int}^{(I)}(t)$ is the interaction Hamiltonian in the interaction picture. It differs from H_{int} only in the fact that the creation and annihilation operators now have their free-field time dependence, i.e. $a_{k_j,\alpha_l} \rightarrow a_{k_j,\alpha_l}e^{-i\omega_{\mathbf{k}_j}t}$ and $a_{k_j,\alpha_l}^{\dagger} \rightarrow a_{k_j,\alpha_l}^{\dagger}e^{i\omega_{\mathbf{k}_j}t}$. The above probability can be understood as describing the series of events in which the medium creates two photons between time 0 and time t_1, one of these photons is absorbed by the atom at $\mathbf{r}_1$ between t_1 and $t_1 + \Delta t$, sending it to the state $|q_{1j}\rangle$, and the other is absorbed by the atom at $\mathbf{r}_2$ between times $t_2 + \Delta t$, sending it into the state $|q_{2j'}\rangle$. If we define

$$C_{\mu\nu}(\mathbf{r}_1, t_1'; \mathbf{r}_2, t_2') = \int_0^{t_1} dt\, \langle\psi^{in}|D_\mu^{(+)}(\mathbf{r}_2, t_2')D_\nu^{(+)}(\mathbf{r}_1, t_1')H_{int}^{(I)}(t)|\psi^{in}\rangle \tag{10.64}$$

and

$$S_{\mu\mu'}(t) = \sum_j e^{i\omega_{jg}t}d_{jg,\mu}d_{jg,\mu'}^*, \tag{10.65}$$

then we have that

$$P(\mathbf{r}_1, t_1; \mathbf{r}_2, t_2) = \int_{t_2}^{t_2+\Delta t} dt_2' \int_{t_2}^{t_2+\Delta t} dt_2'' \int_{t_1}^{t_1+\Delta t} dt_1' \int_{t_1}^{t_1+\Delta t} dt_1''$$
$$\times \sum_{\mu,\mu'=1}^{3} \sum_{\nu,\nu'=1}^{3} S_{\nu\nu'}(t_1' - t_1'') S_{\mu\mu'}(t_2' - t_2'')$$
$$\times C_{\mu'\nu'}(\mathbf{r}_1, t_1''; \mathbf{r}_2, t_2'')^* C_{\mu\nu}(\mathbf{r}_1, t_1'; \mathbf{r}_2, t_2'). \tag{10.66}$$

Note that the function $S_{\mu\mu'}(t)$ characterizes the frequency bandwidth of the detector.

Finally, in order to obtain an idea of how this probability behaves, we will find an expression for $C_{\mu\nu}(\mathbf{r}_1, t_1'; \mathbf{r}_2, t_2')$. Converting the sums over momentum into integrals, we find that

$$C_{\mu\nu}(\mathbf{r}_1, t_1'; \mathbf{r}_2, t_2') = \frac{\hbar^2 \sqrt{V}}{2\mu_0 (2\pi)^9} \sum_{\alpha_1,\alpha_2,\alpha_3} \int d^3k_1 \int d^3k_2 \int d^3k_3 \frac{F}{\sqrt{\omega_1\omega_2}}$$
$$\times M_{\mu\nu}(\mathbf{r}_1, t_1'; \mathbf{r}_2, t_2') h(\mathbf{k}_3 - \mathbf{k}_1 - \mathbf{k}_2)$$
$$\times \frac{e^{i(\omega_1+\omega_2-\omega_3)t_1} - 1}{\omega_1 + \omega_2 - \omega_3} \alpha_{\mathbf{k}_3,\alpha_3}, \tag{10.67}$$

where we have set $\omega_j = \omega_{k_j}$ and $\hat{\mathbf{e}}_{j,\alpha_j} = \hat{\mathbf{e}}_{\mathbf{k}_j,\alpha_j}$ for $j = 1, 2, 3$. We also have that

$$M_{\mu\nu}(\mathbf{r}_1, t_1'; \mathbf{r}_2, t_2') = (\mathbf{k}_2 \times \hat{\mathbf{e}}_{2,\alpha_2})_\mu (\mathbf{k}_1 \times \hat{\mathbf{e}}_{1,\alpha_1})_\nu e^{i\mathbf{k}_2\cdot\mathbf{r}_1} e^{i\mathbf{k}_1\cdot\mathbf{r}_2} e^{-i(\omega_2 t_1' + \omega_1 t_2')}$$
$$+ (\mathbf{k}_1 \times \hat{\mathbf{e}}_{1,\alpha_1})_\mu (\mathbf{k}_2 \times \hat{\mathbf{e}}_{2,\alpha_2})_\nu e^{i\mathbf{k}_1\cdot\mathbf{r}_1} e^{i\mathbf{k}_2\cdot\mathbf{r}_2} e^{-i(\omega_1 t_1' + \omega_2 t_2')}. \tag{10.68}$$

We can now make some rather rough approximations in order to make some sense of the above expression. If we assume that the nonlinear crystal is large and that $|\mathbf{r}_1|$ and $|\mathbf{r}_2|$ are comparable in size to l_z, then $h(\mathbf{k}_3 - \mathbf{k}_1 - \mathbf{k}_2) \to \delta^{(3)}(\mathbf{k}_3 - \mathbf{k}_1 - \mathbf{k}_2)$. We shall also assume that the interaction time t_1 satisfies $ct_1 \gg |\mathbf{r}_1|, |\mathbf{r}_2|$, and in this case we can replace $(e^{i(\omega_1+\omega_2-\omega_3)t_1} - 1)/(\omega_1 + \omega_2 - \omega_3)$ by $\delta(\omega_1 + \omega_2 - \omega_3)$. We shall also assume that the pump mode is specified by $\alpha_{\mathbf{k}_3,\alpha_3} = \alpha_0 \delta_{\alpha_3,1} \delta^{(3)}(\mathbf{k}_3 - k_0\hat{\mathbf{z}})$. Finally, because the detectors are not sensitive to frequencies at the pump frequency and above, we shall cut the $\mathbf{k}_1$ and $\mathbf{k}_2$ integrals off at k_0. This gives us that

$$C_{\mu\nu}(\mathbf{r}_1, t_1'; \mathbf{r}_2, t_2') \sim \int_{|\mathbf{k}_1|<k_0} d^3k_1 \int_{|\mathbf{k}_2|<k_0} d^3k_2\, \delta(\omega_1 + \omega_2 - \omega_3) \delta^{(3)}(\mathbf{k}_3 - \mathbf{k}_1 - \mathbf{k}_2)$$
$$\times [e^{i\mathbf{k}_2\cdot\mathbf{r}_1} e^{i\mathbf{k}_1\cdot\mathbf{r}_2} e^{-i(\omega_2 t_1' + \omega_1 t_2')} + e^{i\mathbf{k}_1\cdot\mathbf{r}_1} e^{i\mathbf{k}_2\cdot\mathbf{r}_2} e^{-i(\omega_1 t_1' + \omega_2 t_2')}], \tag{10.69}$$

where, now, $\mathbf{k}_3 = k_0\hat{\mathbf{z}}$.

It is helpful to look at one of these integrals in some detail. Let

$$I_1 = \int_{|\mathbf{k}_1|<k_0} d^3k_1 \int_{|\mathbf{k}_2|<k_0} d^3k_2\, \delta(\omega_1 + \omega_2 - \omega_3) \delta^{(3)}(\mathbf{k}_3 - \mathbf{k}_1 - \mathbf{k}_2)$$
$$\times e^{i\mathbf{k}_1\cdot\mathbf{r}_1} e^{i\mathbf{k}_2\cdot\mathbf{r}_2} e^{-i(\omega_1 t_1' + \omega_2 t_2')}. \tag{10.70}$$

Doing the $\mathbf{k}_1$ integral, we get

$$I_1 = \frac{1}{c}\int_{|\mathbf{k}_2|<k_0} d^3k_2\,\delta(k_0 - k_2 - |k_0\hat{\mathbf{z}} - \mathbf{k}_2|) \times e^{i(k_0\hat{\mathbf{z}}-\mathbf{k}_2)\cdot\mathbf{r}_1} e^{i\mathbf{k}_2\cdot\mathbf{r}_2} e^{-ic|k_0\hat{\mathbf{z}}-\mathbf{k}_2|t_1'} e^{-ick_2t_2'}. \quad (10.71)$$

Now, the argument of the delta-function in the above integral can only be zero when $\mathbf{k}_2$ points along the positive z axis. Noting this and using polar coordinates, with $\mu = \cos\theta = \hat{\mathbf{k}}_2\cdot\hat{\mathbf{z}}$, we have

$$\begin{aligned} I_1 &= \frac{1}{c}\int_0^{k_0} dk_2\,k_2^2 \int_0^{2\pi} d\phi \int_{-1}^{1} d\mu\,\delta(k_0 - k_2 - [k_0^2 + k_2^2 - 2\mu k_0 k_2]^{1/2}) \\ &\quad\times e^{i(k_0\hat{\mathbf{z}}-\mathbf{k}_2)\cdot\mathbf{r}_1} e^{i\mathbf{k}_2\cdot\mathbf{r}_2} e^{-ic|k_0\hat{\mathbf{z}}-\mathbf{k}_2|t_1'} e^{-ick_2t_2'} \\ &= \frac{\pi}{c} e^{ik_0(z_1-ct_1')} \int_0^{k_0} dk_2 \frac{k_2(k_0-k_2)}{k_0} e^{ik_2[z_2-z_1-c(t_2'-t_1')]}. \end{aligned} \quad (10.72)$$

Performing the remaining integral and incorporating the result into the expression for $C_{\mu\nu}(\mathbf{r}_1, t_1'; \mathbf{r}_2, t_2')$, we have, setting $\Delta Z = (z_2 - z_1) - c(t_2' - t_1')$,

$$\begin{aligned} C_{\mu\nu}(\mathbf{r}_1, t_1'; \mathbf{r}_2, t_2') &\sim \frac{\pi}{ck_0} e^{i(z_1-ct_1')} \left[\frac{2i}{(\Delta Z)^3}(1 - e^{ik_0\Delta Z}) - \frac{k_0}{(\Delta Z)^2}(1 + e^{ik_0\Delta Z})\right] \\ &\quad + (\mathbf{r}_1, t_1' \longleftrightarrow \mathbf{r}_2, t_2'). \end{aligned} \quad (10.73)$$

This expression tells us the following. Because the photons in down-conversion are emitted simultaneously, it is most likely to detect them both at points satisfying the condition $\Delta Z = 0$, and the likelihood decays as approximately $1/(\Delta Z)^2$ as ΔZ increases.

10.6 Maxwell–Bloch equations

We will conclude this chapter by deriving equations, known as the Maxwell–Bloch equations, that describe the propagation of near-resonant fields in a dilute medium consisting of two-level atoms. These equations describe a situation that is complementary to the one in the previous section, where the frequencies had to be sufficiently detuned from resonance that the medium could be characterized by a set of susceptibilities. The Maxwell–Bloch equations give excellent quantitative agreement with many experiments on propagating, near-resonant fields, in dilute media (typically atomic vapors), at both the semiclassical (classical fields and quantum atoms) and the fully quantum levels.

10.6.1 Inhomogeneous broadening

Our starting point is the rotating-wave approximation Hamiltonian, Eq. (4.104), which is valid near resonance. The frequencies ν_β are the transition frequencies of atomic transitions, with transition operator $\hat{\sigma}_\beta^-$. Allowing different atoms to have different resonance frequencies makes it possible for us to include the effect of inhomogeneous broadening. This is caused by the fact that, in many near-resonant systems, the resonance frequency is not the same for all constituents of the system. For example, Doppler broadening, due to the fact that different atoms in a gas have different velocities, leads to a range of resonant frequencies. The typical inhomogeneous environment for atomic impurities in solids has a similar effect.

The simplest case of inhomogeneous broadening to understand physically is a noninteracting gas with atoms or molecules of mass M, at finite temperature T above quantum degeneracy, with a resonant transition (at rest) of ν_0. In this case, there is a Maxwell–Boltzmann distribution of atomic velocities $\mathbf{v}$, given as usual by

$$P(\mathbf{v})\,d^3\mathbf{v} = \left(\frac{M}{2\pi k_b T}\right)^{3/2} \exp\left(-\frac{M\mathbf{v}^2}{2k_b T}\right) d^3\mathbf{v}. \tag{10.74}$$

This leads to a frequency shift in the effective resonance frequencies, due to Doppler broadening. The effective frequency ν for photons emitted and absorbed in (say) the z direction by atoms with velocity $\mathbf{v} = (v_x, v_y, v_z)$ is determined by the usual Doppler formula, which is

$$\nu = \nu_0\left(1 + \frac{v_z}{c}\right). \tag{10.75}$$

The result of this is that there is a Gaussian distribution of effective resonance frequencies,

$$P(\nu)\,d\nu = \sqrt{\frac{1}{2\pi\delta\nu}}\,\exp\left(-\frac{(\nu-\nu_0)^2}{2(\delta\nu)^2}\right) d\nu, \tag{10.76}$$

with a standard deviation in (angular) frequency of

$$\delta\nu = \sqrt{\frac{k_B T \nu_0^2}{Mc^2}}. \tag{10.77}$$

At room temperature, such broadening is typically of the order of 1 GHz, although it is obviously strongly dependent on temperature, atomic mass and resonant frequency. Doppler broadening is therefore a dominant effect in atomic gases at or above room temperatures. It is usually much larger than the natural linewidth determined by the spontaneous emission rate. A third type of line broadening can also occur, which is collisional broadening due to interatomic or intermolecular potentials at high density.

10.6.2 Lattice operators

To derive the Maxwell–Bloch equations from first principles, we suppose that the sum over k is restricted to a single polarization, under the assumption that only a single polarization is excited. Consequently, we will drop the polarization index on the mode creation and annihilation operators. We will also assume that only modes with $k \simeq k_0$ are relevant. We divide the spatial and frequency regions of interest into a lattice of cells. There are $N_{\mathbf{n}}$ atoms in the $\mathbf{n}$th cell, where $\mathbf{n} = (\mathbf{j}, n)$, given that $\mathbf{j}$ labels the M spatial cells and n the frequency cell. The volume of each cell is $\Delta V = \Delta_1\Delta_2\Delta_3$, and the position of cell $\mathbf{j}$ is given by

$$\mathbf{R_j} = \mathbf{R}_0 + (j_1\Delta_1, j_2\Delta_2, j_3\Delta_3). \tag{10.78}$$

The set of all atoms in the $\mathbf{n}$th cell is labeled $s(\mathbf{n})$. There are typically several center frequencies ν_n, each with a width in frequency space of $\Delta\nu_n$.

We take the lattice approach, since this method of analysis helps to link the continuum theory with the cavity electrodynamics approach we gave in Section 5.9. Introducing a lattice is necessary so that discrete quantum objects like atoms can be integrated into a wave equation that is continuous in the classical limit. In addition, any computational approach to these problems will (almost inevitably) use a lattice decomposition in some form. Quantum noise treatments also require a cell decomposition. Therefore, we now continue the discrete treatment of atoms that was introduced in Section 4.7, as a more precise alternative to the continuous linear polarization model, for treating near-resonant, paraxial propagation. We will next take a formal continuum limit of this lattice theory, to obtain analytic solutions in certain limits.

We assume that the range of propagating wavenumbers near a carrier wavenumber of $\mathbf{k}_0 = (0, 0, k_0)$ in the z direction is relatively small, so that for atoms within the cell $s(\mathbf{n})$ the paraxial approximation implies that

$$|(\mathbf{k} - \mathbf{k}_0) \cdot (\mathbf{R}_\beta - \mathbf{R}_\mathbf{j})| \ll 1. \tag{10.79}$$

Here $ck_0 = \nu_0$, which is the atomic transition frequency. We can now use the atomic equivalence classes to define a set of inequivalent atomic spin operators, each defined as in the Tavis–Cummins model:

$$\begin{aligned} S_\mathbf{n}^{(\pm)} &= \sum_{\mu \in s(\mathbf{n})} \sigma_\mu^{(\pm)} e^{\pm i \mathbf{k}_0 \cdot \mathbf{R}_\mathbf{j}}, \\ S_\mathbf{n}^z &= \frac{1}{2} \sum_{\mu \in s(\mathbf{n})} \sigma_\mu^z. \end{aligned} \tag{10.80}$$

It should be noted that the Pauli matrix notation, with components like σ_μ^z, does not indicate a physical spin in the z direction. It is simply a useful way of representing the two-level atomic operators, without any direct link to angular momentum.

For a pure two-level system, the coupling term is now a function of momentum $\mathbf{k}$ and also of position. It can now be expressed as

$$g_{\mathbf{k},\mathbf{j}} = d\sqrt{\frac{\omega_0}{2\hbar\varepsilon_0 V}}\, e^{i(\mathbf{k}-\mathbf{k}_0)\cdot\mathbf{R}_\mathbf{j}} = \frac{g}{\sqrt{V}} e^{i(\mathbf{k}-\mathbf{k}_0)\cdot\mathbf{R}_\mathbf{j}}. \tag{10.81}$$

This creates a simplified interaction Hamiltonian, which is essentially a sum over collective spin Hamiltonians like those in Section 4.7, defined for every cell in space and frequency, but with a phase-dependent coupling:

$$H' \approx \frac{i\hbar g}{\sqrt{V}} \sum_{\mathbf{k},\mathbf{n}} (e^{-i(\mathbf{k}-\mathbf{k}_0)\cdot\mathbf{R}_\mathbf{j}} a_\mathbf{k}^\dagger S_\mathbf{n}^{(-)} - h.c.). \tag{10.82}$$

Next, we can remove the phase dependence of the coupling, by means of introducing discrete Fourier transforms of the mode operators. We define a lattice of momenta:

$$\mathbf{k}(\mathbf{m}) = \mathbf{k}_0 + \Delta\mathbf{k}(\mathbf{m}), \tag{10.83}$$

with

$$\Delta\mathbf{k}(\mathbf{m}) = (m_1 \Delta k_1, m_2 \Delta k_2, m_3 \Delta k_3), \tag{10.84}$$

where $m_i = -M_i, \ldots, 0, \ldots, +M_i$ labels the momentum cells in the ith direction. The total number of lattice points is $M = \prod_i (2M_i + 1)$. Here

$$\Delta k_i = 2\pi/(\Delta_i(2M_i + 1)),$$

and we can now introduce localized field operators $b_{\mathbf{j}}$ according to a discrete Fourier transform:

$$b_{\mathbf{j}} = \frac{1}{\sqrt{M}} \sum a_{\mathbf{k}} e^{i\mathbf{R}_{\mathbf{j}}\cdot\Delta\mathbf{k}(\mathbf{m})}. \tag{10.85}$$

This is a unitary transformation, so these field operators have the usual canonical commutators, and have the interpretation that they generate a localized photonic excitation. Noting that $V = M\Delta V$, and defining $\tilde{g} = g/\sqrt{\Delta V}$, the total Hamiltonian is now

$$H = \hbar \left[\sum_{\mathbf{j},\mathbf{j}'} \omega_{\mathbf{j},\mathbf{j}'} b_{\mathbf{j}}^{\dagger} b_{\mathbf{j}'} + \sum_{\mathbf{n}} \nu_n S_{\mathbf{n}}^{(z)} + i\hbar\tilde{g} \sum_{\mathbf{j},\mathbf{n}} (b_{\mathbf{j}}^{\dagger} S_{\mathbf{n}}^{(-)} - h.c.) \right]. \tag{10.86}$$

Here, the coupling terms are

$$\omega_{\mathbf{j},\mathbf{j}'} = \frac{1}{\sqrt{M}} \sum_{\mathbf{m}} \omega_{\mathbf{k}(\mathbf{m})} e^{i(\mathbf{R}_{\mathbf{j}}-\mathbf{R}_{\mathbf{j}'})\cdot\Delta\mathbf{k}(\mathbf{m})}. \tag{10.87}$$

The resulting Heisenberg-picture field and atomic equations in the rotating-wave and paraxial approximations give a discrete, lattice-based version of the operator Maxwell–Bloch equations, as follows:

$$\begin{aligned}
\frac{\partial}{\partial t} b_{\mathbf{j}} &= -i \sum_{\mathbf{j}'} \omega_{\mathbf{j},\mathbf{j}'} b_{\mathbf{j}'} + \tilde{g} \sum_{n} S_{\mathbf{j},n}^{(-)}, \\
\frac{\partial}{\partial t} S_{\mathbf{n}}^{(-)} &= -i\nu_{\mathbf{n}} S_{\mathbf{n}}^{(-)} + 2\tilde{g} S_{\mathbf{n}}^{z} b_{\mathbf{j}}, \\
\frac{\partial}{\partial t} S_{\mathbf{n}}^{z} &= -\tilde{g}[b_{\mathbf{j}}^{\dagger} S_{\mathbf{n}}^{(-)} + b_{\mathbf{j}} S_{\mathbf{n}}^{(+)}].
\end{aligned} \tag{10.88}$$

If mean values are taken, and we assume that all operator products factorize, the resulting mean-value equations are called the Maxwell–Bloch equations. They describe nonlinear propagation through a saturable, resonant medium. However, it is conventional to write these equations in a continuous form, which we treat next.

10.6.3 Paraxial wave equation

In the continuum limit, the terms involving $\omega_{\mathbf{j},\mathbf{j}'}$ simply generate the paraxial wave equation given in Eq. (1.14). In order to see how this is obtained, we simply consider the continuum limit of the operator equations for the field modes $b_{\mathbf{j}}$. The $\omega_{\mathbf{k}(\mathbf{m})}$ terms can be expanded in powers of $\Delta\mathbf{k}/k_0$, giving

$$\begin{aligned}
\omega_{\mathbf{k}(\mathbf{m})} &= c|\mathbf{k}_0 + \Delta\mathbf{k}| \\
&= ck_0 \left[1 + \frac{\Delta k_z}{k_0} + \frac{\mathbf{k}_{\perp}^2}{2k_0^2} + O\left(\frac{|\Delta\mathbf{k}|^3}{k_0^3} \right) \right].
\end{aligned} \tag{10.89}$$

Here $\mathbf{k}_\perp = (k_y, k_z)$ is the transverse component of momentum. In the continuum limit, we can define a slowly varying photonic operator field as a three-dimensional generalization of the one-dimensional photon field of Eq. (3.154):

$$\begin{aligned}\psi(\mathbf{R_j}) &= \frac{1}{\sqrt{\Delta V}} b_{\mathbf{j}} e^{i\omega_0 t} \\ &= \frac{1}{\sqrt{V}} \sum a_{\mathbf{k}} e^{i(\mathbf{R_j}\cdot\Delta\mathbf{k}(\mathbf{m})+\omega_0 t)}.\end{aligned} \tag{10.90}$$

This is also essentially identical to the standard definition of a scalar quantum field given in Eq. (2.34), except that a carrier frequency and wavenumber are introduced to give a slowly varying envelope definition. However, it is often more convenient to define a scaled version of this field as a quantum Rabi frequency field, Ω, similar to the classical Rabi frequency in Eq. (1.76), with dimensions of s^{-1}:

$$\Omega(\mathbf{r}) = 2g\psi(\mathbf{r}). \tag{10.91}$$

With these definitions, the Fourier-transformed kernel $\omega_{\mathbf{j},\mathbf{j}'}$ becomes equivalent to a differential operator, so that

$$-i\omega_{\mathbf{j},\mathbf{j}'} \to -ick_0\left[1 - \frac{i}{k_0}\frac{\partial}{\partial z} - \frac{\nabla_\perp^2}{2k_0^2}\right]. \tag{10.92}$$

Using these substitutions, the first equation in (10.89), for the quantum field, now becomes the quantum version of the paraxial wave equation, Eq. (1.14), that we encountered earlier, except with a dipole source term

$$\left[\frac{\partial}{\partial t} + \frac{c\partial}{\partial z} - \frac{ic\nabla_\perp^2}{2k_0}\right]\Omega = \frac{2g^2}{\Delta V}\sum_n S_{\mathbf{j},n}^{(-)} e^{i\omega_0 t}. \tag{10.93}$$

Next, we introduce corresponding slowly varying operators for the atomic fields, with $N_{\mathbf{j}}$ atoms at lattice point $\mathbf{j}$, which are given by

$$\begin{aligned}\sigma^{(\pm)}(\mathbf{R_j}, \nu_n) &= \left[\frac{1}{N_{\mathbf{j}}\Delta\nu_n}\right] S_{\mathbf{j},n}^{\pm} e^{\mp i\omega_0 t}, \\ \sigma^z(\mathbf{R_j}, \nu_n) &= \left[\frac{2}{N_{\mathbf{j}}\Delta\nu_n}\right] S_{\mathbf{j},n}^z.\end{aligned} \tag{10.94}$$

Since the equations describe an excitation traveling near light speed, it is also useful to introduce a moving frame of reference, with

$$\tau = t - z/c. \tag{10.95}$$

To treat inhomogeneous frequency broadening, we define a continuous detuning $\Delta = \nu_n - \omega_0$ as the detuning from the central frequency ω_0, and introduce $P(\Delta)$ as the probability density of Δ due to inhomogeneous frequency broadening. This is generally a Gaussian lineshape due to Doppler broadening at finite temperature. Using the chain rule of calculus to change the independent variables results in the following continuum equations:

$$\begin{aligned}
\left[\frac{\partial}{\partial z} - \frac{i\nabla_\perp^2}{2k_0}\right]\Omega &= G\int d\Delta P(\Delta)\sigma^{(-)}(\Delta), \\
\frac{\partial}{\partial \tau}\sigma^{(-)}(\Delta) &= -i\Delta\sigma^{(-)}(\Delta) + \Omega\sigma^z(\Delta), \qquad (10.96) \\
\frac{\partial}{\partial \tau}\sigma^z(\Delta) &= -\frac{1}{2}[\Omega\sigma^{(+)}(\Delta) + h.c.],
\end{aligned}$$

where the coupling constant at number density ρ_n is given by

$$G = \frac{\rho_n d^2 \omega_0}{c\hbar\varepsilon_0}. \qquad (10.97)$$

Here, G and $P(\Delta)$ are functions of space in general, owing to inhomogeneous atomic density and temperature. The space-time dependence of the propagating field means that $\sigma^{\pm}(\Delta)$ and $\sigma^z(\Delta)$ are also functions of their space-time coordinates. To solve these equations, it is common to use a mean-field approximation. This replaces operators by c-numbers, under the assumption that the expectation values of operator products all factorize. In order to include quantum noise, it is necessary to use phase-space representations, similar to techniques already described for single modes.

In this three-dimensional form, which includes inhomogeneous broadening and diffraction, these equations must be treated numerically. Even more generally, one can also include spontaneous emission and other forms of dissipation, which was treated in earlier chapters. However, there are a number of special cases, in one dimension, that correspond to exactly soluble models in both the quantum and semiclassical (mean-field) limits. These result in soliton propagation or 'self-induced transparency': pulsed solutions that propagate through a resonant medium without absorption.

Additional reading

Articles

- The theories of quantized optical solitons on which this chapter is based can be found in:

S. J. Carter and P. D. Drummond, *J. Opt. Soc. Am.* B **4**, 1565 (1987).
S. J. Carter and P. D. Drummond, *Phys. Rev. Lett.* **67**, 3757 (1991).
Y. Lai and H. A. Haus, *Phys. Rev.* A **40**, 844 (1989).

- The solution of a system of n particles in one dimension interacting via a delta-function potential is presented in:

E. H. Lieb and W. Liniger, *Phys. Rev.* **130**, 1605 (1963).
C. N. Yang, *Phys. Rev.* **130**, 1920 (1967).

- For the Maxwell–Bloch equations in the semiclassical and quantum cases, see:

P. D. Drummond and M. G. Raymer, *Phys. Rev.* A **44**, 2072 (1991).

S. L. McCall and E. L. Hahn, *Phys. Rev.* A **83**, 457 (1969).

- The original multi-mode treatment of the degenerate parametric amplifier in free space appeared in:

Z. Y. Ou, L. J. Wang, and L. Mandel, *Phys. Rev.* A **40**, 1428 (1989).

Problems

10.1 Show that Eq. (10.15) is correct.

10.2 Show that Eq. (10.30) is a solution of Eq. (10.29), and find E_n.

10.3 In going from Eq. (10.47) to Eq. (10.49), we made the simplest approximation, which was just to replace n by n_0. We can do somewhat better. We first set $n = n_0 + \delta n$, and, because the important range for n is between $n_0 - \sqrt{n_0}$ and $n_0 + \sqrt{n_0}$, we can assume that $\delta n / n_0 \ll 1$. We expand the terms in the sum in Eq. (10.47) up to second order in δn, making use of Stirling's formula, $n! \simeq \sqrt{2\pi n}\, n^n e^{-n}$ for the factorial. We then approximate the sum as an integral, and then perform the integral. Do this and find a more accurate expression for $\langle \Psi | \psi(x, 0) | \Psi \rangle$.

11 Quantum propagation in fibers and waveguides

Optical fibers are the backbone of the world's communication systems. They also represent the simplest example of a photonic device, where photons in waveguides replace electrons in wires. More sophisticated devices than this can be made, in which all the optical components – including multiple waveguided modes – are integrated on a chip. It is important to understand the quantum noise properties of these systems.

While in some respects these systems can be treated as one-dimensional, there are additional features. Dispersion is due not just to material properties, but also to the waveguide geometry. The nonlinearity comes from a combination of material and geometric properties as well. There is, of course, a quantum aspect to these photonic devices, in which quantum noise occurs due to random effects caused by nonlinearity and gain.

Optical devices based on waveguides are, of course, not just simple fibers. One can have multiple waveguides with various types of linear coupling. The intrinsic nonlinearity of the dielectric leads to four-wave mixing, which can be an important source of correlated photons, both for fundamental tests of quantum mechanics, and for applications in quantum information.

It is often essential to include effects due to anisotropies of various types, as well as material dispersion. In dielectrics, another effect that should be included is the coupling of vibrational modes of the (solid) dielectric to the propagating field, which gives rise to Raman and Brillouin scattering effects. This allows us to obtain a more detailed and correct theory of quantum noise than the simplified case treated in Chapter 3.

11.1 Order-of-magnitude estimates

We start by considering orders of magnitude. High intensities and hence nonlinear effects can occur in optical fibers, normally considered the domain of relatively low-power lasers. The ultimate limit of quantum optics is a single photon in a single cycle, which gives the highest communications bit rate for a given power input in a single transverse mode. With single-cycle modulation and single-photon signals, one can achieve a communication bit rate equal to the optical frequency. This is about one petabit per second, where 1 Pbit s^{-1} $= 10^{15}$ bit s^{-1}, at optical wavelengths of 0.5 μm. The corresponding power requirement is $P \sim hf^2 \sim 0.5$ mW. By comparison, the highest communication rate achieved

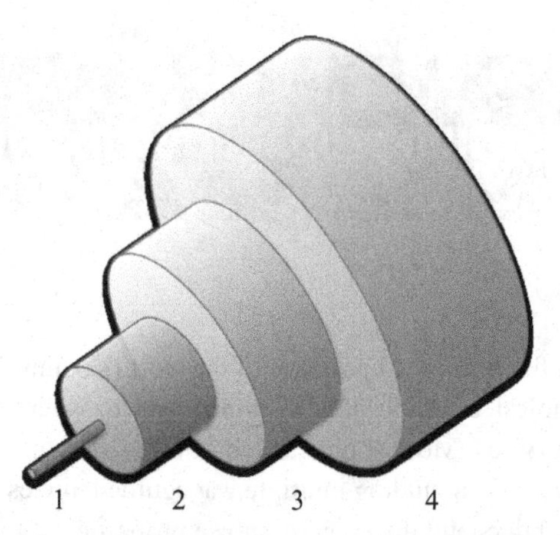

Fig. 11.1 Optical fiber. Regions 1, 2, 3 and 4, respectively, are the silica core ($\sim$ 8 μm diameter), silica cladding ($\sim$ 125 μm diameter), plastic transition coating ($\sim$ 250 μm diameter), and plastic jacket ($\sim$ 400 μm diameter).

in single-mode optical fiber systems is around a hundred terabits per second at a wavelength of about 1.55 μm, where 100 Tbit $s^{-1} = 10^{14}$ bit s^{-1}, limited by fiber-optical bandwidth restrictions and nonlinear effects. Here, one uses multiple channels of different wavelengths rather than short pulses, for technical reasons.

Even a single photon can cause nonlinear effects in principle. Consider a single photon per cycle pulse, focused to wavelength dimensions. Using the fact that $c = f\lambda$ in free space, one obtains a maximum single-photon intensity of

$$I_{max}(\lambda) = hf^2/\lambda^2 = hc^2/\lambda^4. \tag{11.1}$$

At optical wavelengths of $\lambda = 0.5$ μm, the intensity would reach $I \sim 0.2$ GW m^{-2}. With this approach, a single-photon pulse in the ultraviolet could have peak powers that easily exceed ~ 1 GW m^{-2}. While the nonlinear optics treated in this chapter focuses on larger numbers of photons, we see that extreme quantum effects are also possible.

In optical fibers, it is usual to have a less extreme focus, and more photons per pulse. To understand the numbers that are relevant, we must consider the typical parameter values as given for current optical fibers in Figure 11.1. The core is a region of increased refractive index. This acts by total internal reflection to confine the mode. We assume that this is a single mode in this chapter. Although multi-mode fibers also exist, and are widely used, single-mode fibers are optimal for communications systems, as dispersion is reduced. They also have the highest nonlinear effects, as a rule, due to the small mode diameter. However, while fibers have a single spatial mode, they generally support two orthogonal polarizations.

A typical physical system of interest here is illustrated in Figure 11.1, showing an optical waveguide in the form of a step-index fiber made of fused silica, or other material. The change in index (of order 0.01–0.05) is obtained by means of doping the transparent fiber material with other atomic species, either in the core or in the cladding. This results in an increased core index, or a depressed cladding index. However, the change may be gradual rather than abrupt, and the fiber is not necessarily rotationally symmetric. In addition, other materials – usually a plastic polymer – are often employed for jacketing the fiber to increase its mechanical strength. These may be applied in two layers, with different mechanical properties. This last step has little effect on the optical properties, but can change the low-frequency Brillouin response.

11.2 Waveguide modes

In obtaining the mode functions of a waveguide or cavity for quantization, there are two main requirements. The first is that each mode has a well-defined frequency, so that it corresponds to an eigenfunction of the linear Hamiltonian. The second is that the modes should be normalized to give the correct Hamiltonian when written in the usual form with annihilation and creation operators. This requires an orthonormality relationship.

The treatment of a realistic fiber or waveguide mode needs to combine the treatment of two types of dispersion: first, the mode structure is frequency-dependent; second, the material itself is dispersive. As a result, the combined dispersion relation of the waveguide can be very strongly modified by the mode structure, especially near the zero-dispersion point of the bulk material.

This requires a more general treatment than we gave in the early chapters. Previously, we treated waveguides that were either inhomogeneous or intrinsically dispersive due to material properties. Now we must unify these two sources of frequency dependence in the wave equation.

11.2.1 Mode expansion

The frequency eigenvalues (or dispersion relation) of a dispersive waveguide are (is) found using the standard linear wave equation of the fiber or waveguide. This is explained in Chapter 3, which treats modal dispersion and material dispersion separately. We now wish to treat the general problem where both appear together. In any case, the fields satisfy Maxwell's equations, so that $\mathbf{D} = \nabla \times \mathbf{\Lambda}$ and $\mathbf{B} = \mu \partial \mathbf{\Lambda}/\partial t$. The eigenvalue equation is then obtained from the remaining Maxwell's equations,

$$\begin{aligned} \nabla \times \mathbf{B} &= \mu \frac{\partial \mathbf{D}}{\partial t}, \\ \nabla \times \mathbf{E} &= -\frac{\partial \mathbf{B}}{\partial t}, \end{aligned} \tag{11.2}$$

together with the definitions of the linear response function $\eta(\mathbf{r}, \omega,) = \boldsymbol{\varepsilon}^{-1}(\mathbf{r}, \omega)$, at a frequency ω near the carrier frequency ν_0, so that complex envelope fields are given by

$$\mathcal{E}(\omega) = \eta(\mathbf{r}, \omega)\mathcal{D}(\omega). \tag{11.3}$$

This gives a wave equation, and hence an eigenvalue equation for mode functions $\boldsymbol{u}_n$ with eigenvalues ω_n. Apart from the frequency dependence of the inverse permittivity, the resulting equation is identical to the equation for a nondispersive waveguide treated in Eq. (3.96),

$$\nabla \times [\eta(\mathbf{r}, \omega_n)\nabla \times \boldsymbol{U}_n(\mathbf{r})] = \mu\omega_n^2 \boldsymbol{U}_n(\mathbf{r}), \tag{11.4}$$

where we have introduced mode functions, labeled by the subscript n, which are slowly varying envelopes,

$$\mathbf{A}(\mathbf{r}, t) = \sum_n a_n e^{-i\omega_n t} \boldsymbol{U}_n(\mathbf{r}) + h.c. \tag{11.5}$$

If the waveguide is considered as approximately homogeneous in the x direction, then periodic boundary conditions are typically imposed at $x = 0$ and $x = \ell$, with vanishing transverse boundary conditions as $|\mathbf{r}_\perp| \to \infty$, to give a set of modal solutions. If we take $\ell \to \infty$, the modes become continuous. Only modes that satisfy $\nabla \cdot \boldsymbol{U}_n(\mathbf{r}) = 0$ can be regarded as physical modes. Typically, a discrete set of modes with different transverse behavior exists. There is a transverse continuum set of modes as well, and they can be included in the current formalism by assuming a finite periodic boundary direction, then taking an infinite-area limit as $|\mathbf{r}_\perp| \to \infty$.

The resulting general modal expansion for the complex fields is written in terms of time-varying amplitudes near a carrier frequency ω_n, so that $a_n(t) = a_n \exp(-i\omega_n t)$, as

$$\begin{aligned} \mathcal{D}(\mathbf{r}, t) &= \sum_n \mathcal{D}_n(\mathbf{r}) a_n(t) = \sum_n \nabla \times \boldsymbol{U}_n(\mathbf{r}) a_n(t), \\ \mathcal{B}(\mathbf{r}, t) &= \sum_n \mathcal{B}_n(\mathbf{r}) a_n(t) = -i \sum_n \mu\omega_n \boldsymbol{U}_n(\mathbf{r}) a_n(t). \end{aligned} \tag{11.6}$$

The real fields are then obtained by combining $\mathbf{D} = \mathcal{D} + \mathcal{D}^*$, and similarly for the other electromagnetic fields.

11.2.2 Orthogonality relations

We now consider the mode equations for a general lossless dispersive dielectric, with no other restrictions. There is an orthogonality equation for modes of distinct frequencies ω_n and $\omega_{n'}$, which is obtained by multiplying the Maxwell equation for mode n' by the conjugate field $\boldsymbol{U}_n^*(\mathbf{x})$, and integrating, so that

$$\int d^3\mathbf{r} \, \{\boldsymbol{U}_{n'}^*(\mathbf{r}) \cdot \nabla \times [\eta(\mathbf{r}, \omega_n)\nabla \times \boldsymbol{U}_n(\mathbf{r})] - \mu\omega_n^2 \boldsymbol{U}_{n'}^*(\mathbf{r}) \cdot \boldsymbol{U}_n(\mathbf{r})\} = 0. \tag{11.7}$$

Next, we follow the general procedure outlined in Section 3.5.3. We use the standard vector identity, $\nabla \cdot [\mathbf{A} \times \mathbf{A}'] = \mathbf{A}' \cdot [\nabla \times \mathbf{A}] - \mathbf{A} \cdot [\nabla \times \mathbf{A}']$, where $\mathbf{A} = \boldsymbol{U}_n^*(\mathbf{x})$ and $\mathbf{A}' = \eta(\omega_n', \mathbf{x})\nabla \times \boldsymbol{U}_{n'}(\mathbf{x})$. After integrating by parts, assuming vanishing boundary terms, we obtain

$$\int d^3\mathbf{x} \left\{ \frac{1}{\omega_n} \nabla \times \boldsymbol{U}_{n'}^*(\mathbf{r}) \cdot [\eta(\mathbf{r}, \omega_n)\nabla \times \boldsymbol{U}_n(\mathbf{r})] - \mu\omega_n \boldsymbol{U}_{n'}^*(\mathbf{r}) \cdot \boldsymbol{U}_n(\mathbf{r}) \right\} = 0. \tag{11.8}$$

Assuming that $\eta(\omega_n)$ is Hermitian for a lossless medium, the identities obtained at two distinct mode frequencies are subtracted to give

$$\int \left[\mu(\omega_n - \omega_{n'}) \boldsymbol{U}^*_{n'}(\mathbf{r}) \cdot \boldsymbol{U}_n(\mathbf{r}) - \nabla \times \boldsymbol{U}^*_{n'}(\mathbf{r}) \cdot \left(\frac{\eta(\mathbf{r}, \omega_n)}{\omega_n} - \frac{\eta(\mathbf{r}, \omega_{n'})}{\omega_{n'}} \right) \nabla \times \boldsymbol{U}_n(\mathbf{r}) \right] d^3\mathbf{r} = 0. \tag{11.9}$$

We now multiply this equation by $\omega_n \omega_{n'}/(\omega_n - \omega_{n'})$, giving, for $n \neq n'$,

$$\int [\mu\omega_n\omega_{n'} \boldsymbol{U}^*_{n'}(\mathbf{r}) \cdot \boldsymbol{U}_n(\mathbf{r}) + \nabla \times \boldsymbol{U}^*_{n'}(\mathbf{r}) \cdot \tilde{\eta}(\omega_{n'}, \omega_n, \mathbf{r}) \cdot \nabla \times \boldsymbol{U}_n(\mathbf{r})] \, d^3\mathbf{r} = 0, \tag{11.10}$$

where the kernel of this integral identity is defined (for $\omega_n \neq \omega_{n'}$) as

$$\tilde{\eta}(\omega', \omega, \mathbf{r}) = \frac{\omega'\eta(\mathbf{r}, \omega) - \omega\eta(\mathbf{r}, \omega')}{\omega' - \omega}. \tag{11.11}$$

In the limit of $\omega' \to \omega$, we can reduce this to a convenient form by taking a continuous limit, to give a single expression for orthogonality and normalization. This allows us to *define* the orthogonality kernel, $\tilde{\eta}(\omega, \omega, \mathbf{r})$, for degenerate frequencies as

$$\tilde{\eta}(\omega, \omega, \mathbf{r}) = -\omega^2 \frac{\partial[\eta(\omega, \mathbf{r})/\omega]}{\partial\omega}. \tag{11.12}$$

Since the theory given here is intended for normalized modes, we make use of the methods of Chapter 3 to choose a frequency-dependent orthonormality condition. Hence, the general orthonormality requirement for any two modes $\boldsymbol{U}_n$ and $\boldsymbol{U}_{n'}$ can now be *defined* (even for $n = n'$) as

$$\int [\mu\omega_n\omega_{n'} \boldsymbol{U}^*_n(\mathbf{r}) \cdot \boldsymbol{U}_{n'}(\mathbf{r}) + \nabla \times \boldsymbol{U}^*_{n'}(\mathbf{r}) \cdot \tilde{\eta}(\omega_{n'}, \omega_n, \mathbf{r}) \cdot \nabla \times \boldsymbol{U}_n(\mathbf{r})] \, d^3\mathbf{r} = \hbar\omega_n\delta_{nn'}. \tag{11.13}$$

11.3 Dispersive energy

With the use of the wave equation, together with vanishing boundary terms, the orthonormality expression can be transformed to the classical dispersive energy term of Eq. (1.61). To see this, we use the general dispersive energy expression from Eq. (1.61) to define

$$H_1 = \frac{1}{2} \int \left[\int \int_{-\infty}^{\infty} d\omega \, d\omega' \boldsymbol{\mathcal{D}}^*(\mathbf{r}, \omega') \cdot \tilde{\eta}(\omega', \omega, \mathbf{r}) \cdot \boldsymbol{\mathcal{D}}(\mathbf{r}, \omega) + \frac{1}{\mu} |\mathbf{B}(\mathbf{r})|^2 \right] d^3\mathbf{r}. \tag{11.14}$$

Here, we assume a lossless medium, where $\eta(\mathbf{r}, \omega) = \eta^\dagger(\mathbf{r}, \omega)$, and include a convergence factor δ to ensure convergence of the integrals, with definition

$$\tilde{\eta}(\omega', \omega, \mathbf{r}) = \frac{\omega'\eta(\mathbf{r}, \omega) - \omega\eta(\mathbf{r}, \omega')}{\omega' - \omega + i\delta}. \tag{11.15}$$

The above choice of the normalization constant immediately gives the dispersive energy in the alternative form:

$$H_1 = \hbar \sum_n \omega_n a_n^* a_n. \tag{11.16}$$

This allows for a three-dimensional Lagrangian to be written down in the form already derived in Eq. (2.219) for a collection of harmonic oscillators:

$$\mathcal{L}_1 = \frac{i\hbar}{2} \sum_n [a_n^* \dot{a}_n - \dot{a}_n^* a_n] - \hbar \sum_n \omega_n a_n^* a_n. \tag{11.17}$$

This will generate both the correct Hamiltonian and the classical equations of motion in the form

$$\frac{\partial a_n}{\partial t} = -i\omega_n a_n. \tag{11.18}$$

Once a Lagrangian is available, quantization is possible following the usual rules of constrained Dirac–Jackiw quantization, as explained in Section 2.9. In this case, the a_n are simply replaced by the raising and lowering operators $\hat{a}_n$, with the usual harmonic oscillator commutation relations. So far, nothing here is particularly specific to either a waveguide or a fiber. In fact, the same expressions can be used to define the mode structure of a dielectric in a cavity. Now suppose that the waveguide structure is uniform in the x direction, apart from small inhomogeneities that can be treated perturbatively. In particular, periodic structures or bandgaps will not be treated, for simplicity. However, the extension to treat these is reasonably straightforward.

In the translationally invariant case, we set $\eta(\omega, \mathbf{r}) = \bar{\eta}(\omega, \mathbf{r}_\perp)$, where $\mathbf{r}_\perp = (y, z)$, in the wave equation, so that any inhomogeneities in the x direction are not included. Next, suppose that the solutions have an e^{ikx} type longitudinal behavior, with all mode functions being expressed as

$$\mathbf{U}_n = c_{k,s} \mathbf{u}_{k,s}(\mathbf{r}_\perp) e^{ikx}. \tag{11.19}$$

Here the subscript s indicates the transverse mode structure, which labels the two polarizations in the case of a single-mode optical fiber, and $c_{k,s}$ is a normalization constant. More generally, s could label all the higher-order transverse modes as well, in the case of multi-mode fibers or waveguides.

By labeling the mode frequencies with their longitudinal momentum k, a frequency dispersion relation is obtained, of the form $\omega_s(k_n) = \omega_n$, which includes both the material and the modal dispersion. The longitudinal wavevector k and polarization s are thereby related to the corresponding frequency $\omega_{ks} = \omega_s(k)$. There is a known relationship between the linear energy density W_s and the time-averaged propagating power P_s for a traveling-wave mode structure. This is that

$$v_{ks} = \frac{\partial \omega_s(k)}{\partial k} = \frac{P_{ks}}{W_{ks}}. \tag{11.20}$$

Thus, the normalization of the traveling-wave mode can be reduced to the form

$$\left| \int \Re[\bar{\eta}(\omega_{ks}, \mathbf{r}_\perp) \boldsymbol{\mathcal{D}}_{k,s}(\mathbf{r}_\perp) \times \boldsymbol{\mathcal{B}}^*_{k,s}(\mathbf{r}_\perp)]\, d^2\mathbf{r}_\perp \right| = \frac{v_{ks}\hbar\mu\omega_{ks}}{2}. \tag{11.21}$$

With this equation, the normalization of $\mathbf{u}_{k,s}$ is obtained for any choice of $c_{k,s}$, with a somewhat more straightforward equation than before. The situation is especially simple in the limit of a weakly guided wave. Here $\bar{\eta}(\omega, \mathbf{r}_\perp) \simeq \eta_{k,s}$, where $\eta_{k,s} \equiv \mu(\omega_{k,s}/k)^2$ is the power-averaged inverse permittivity, and

$$\nabla \times [\mathbf{u}_{k,s}(\mathbf{r}_\perp)e^{ikx}] \simeq ik\hat{\mathbf{x}} \times [\mathbf{u}_{k,s}(\mathbf{r}_\perp)e^{ikx}]. \tag{11.22}$$

Thus, a suitable choice of $c_{k,s}$ in the weakly guided wave limit is identical to that in the one-dimensional model:

$$c_{k,s} = \frac{1}{i}\sqrt{\frac{\hbar k v_{k,s}}{2\mu\omega^2_{k,s}}}. \tag{11.23}$$

The weakly guided mode function normalization then has the following approximate form:

$$\int \mathbf{u}^*_s(\mathbf{r}_\perp)\mathbf{u}_{s'}(\mathbf{r}_\perp)\, d^2\mathbf{r}_\perp \simeq \delta_{ss'}. \tag{11.24}$$

More generally, the exact expressions given above should be used. Although the approximate normalization for $\mathbf{u}$ given above is appropriate in most cases, the more precise form of the normalization equation is necessary if there is severe modal dispersion near a mode cutoff.

For a completely uniform waveguide, the modal expansion results in the following expression for the displacement field operator:

$$\begin{aligned} \mathbf{D}(\mathbf{r}) &= \sum_{ks} \sqrt{\frac{\hbar v_{ks} k}{2\mu\omega^2_{ks} L}}\, \nabla \times [\mathbf{u}_{ks}(\mathbf{r}_\perp)e^{ikx}]a_{ks} + h.c. \\ &= \frac{1}{\sqrt{L}} \sum_{ks} \boldsymbol{\mathcal{D}}_{k,s}(\mathbf{r})a_{ks} + h.c. \end{aligned} \tag{11.25}$$

The coordinates described by $\mathbf{r}$ are the longitudinal and radial pair $(x, \mathbf{r}_\perp)$. The fiber length is ℓ, which is taken to be large, and can be eventually taken to infinity to remove any effects due to the boundaries. The main difference between the dispersive expansion and that for a nondispersive medium is that ω is replaced under the square root sign by kv_k. This correction is needed in order to obtain equivalence to the correct form of Maxwell's equations for a dispersive waveguide. There are no restrictions as yet on the type of waveguide, apart from the isotropy of the medium. Thus, these results have a general applicability to waveguides of other types as well as those in fiber optics.

In weakly guided cases, typical of optical fibers, the changes in refractive index are small. For example, in a quadratic index profile, the mode functions are a normalized Gaussian in the paraxial approximation. These are parameterized by the beam waist size W, so that $\mathbf{u}_s(\mathbf{r}) \simeq \mathbf{e}_s\, e^{-r^2/W^2}/\sqrt{\pi W^2/2}$. More generally the solutions would be of Bessel function form in the case of a step-index fiber, although these can often be approximated by a suitable Gaussian. In this text, these mode functions are only used to calculate the nonlinear coupling coefficients. Since these coefficients involve nonlinear parameters, which are often not known to more than 1% accuracy, the use of Gaussian modes is a reasonable approximation in many cases, except for highly structured waveguides with modes that are highly non-Gaussian.

11.4 Nonlinear Hamiltonian

After quantization of a field with frequency near ω_0, we define $\mathbf{D}(t, \mathbf{r}) = \mathcal{D}(t, \mathbf{r}) + \mathcal{D}^\dagger(t, \mathbf{r})$, where $\mathcal{D}(t, \mathbf{r}) \sim e^{-i\omega_0 t}$. The dual potential of the electric displacement field is used here as a canonical variable for greater simplicity in quantization, as it results in much simpler transversality requirements. This is needed to define the Lagrangian and commutation relations, which give an operational meaning to the mode operators.

The nonlinear response is most readily described by expanding the polarization field – and hence the electric field – in terms of the displacement field, as explained in the nondispersive case in Chapter 1. This must now be extended to treat the full dispersive nonlinear response. Taylor expansion coefficients $\eta^{(n)}_{i,j_1,\ldots,j_n}(t_1, \ldots, t_n; \mathbf{x})$ are therefore introduced as $(n+1)$th-rank tensors, with

$$\mathbf{E}(t, \mathbf{r}) = \sum_n \int_0^\infty d^n \mathbf{t}\, \eta^{(n)}(\mathbf{t}, \mathbf{r}) : \mathbf{D}(t - t_1, \mathbf{r}) \otimes \cdots \otimes \mathbf{D}(t - t_n, \mathbf{r}). \qquad (11.26)$$

Next, coefficients $\eta^{(n)}(\omega_1, \ldots, \omega_n, \mathbf{r})$ are introduced as the Fourier transforms of the time-dependent dielectric response tensors, with the notation that

$$\eta^{(n)}_{i,j_1,\ldots,j_n}(\omega_1, \ldots, \omega_n, \mathbf{r}) = \int_0^\infty d^n \mathbf{t}\, \eta^{(n)}_{i,j_1,\ldots,j_n}(\mathbf{t}, \mathbf{r}) e^{i\omega \cdot \mathbf{t}}. \qquad (11.27)$$

An abbreviated notation will be introduced for nonlinear refractive index type nonlinearities, where $\eta^{(3)}(\omega_0, \mathbf{r}) \equiv \eta^{(3)}(\omega_0, -\omega_0, \omega_0, \mathbf{r})$. Not all the tensor components of $\eta^{(3)}$ are independent in isotropic media. This is clear from the nonlinear Hamiltonian, which must be a rotational scalar. Thus, we can represent the nonlinear response using two terms $\eta_a^{(3)}$ and $\eta_b^{(3)}$. This is because the most general scalar dielectric potential energy in the rotating-wave approximation is

$$\bar{U}^N = \{\eta_a^{(3)}(\omega_0, \mathbf{r})[\mathcal{D}(\mathbf{r}) \cdot \mathcal{D}^\dagger(\mathbf{r})]^2 + \tfrac{1}{2}\eta_b^{(3)}(\omega_0, \mathbf{r})|[\mathcal{D}(\mathbf{r}) \cdot \mathcal{D}(\mathbf{r})]|^2\}. \qquad (11.28)$$

Next, this is combined with Eq. (11.17) of the previous section, to generate the total Lagrangian

$$\mathcal{L} = \mathcal{L}_0 - \int_V \bar{U}^N[\mathcal{D}(\mathbf{r}), \mathcal{D}^\dagger(\mathbf{r}), \mathbf{r}]\, d^3\mathbf{r}. \qquad (11.29)$$

Only these terms can occur, as these correspond to all the possible ways of obtaining a slowly varying, rotationally invariant nonlinear scalar potential, from the transversely polarized fields.

For the case of a single-frequency field in an isotropic, lossless medium, the resulting Hamiltonian is $H = \sum_{j=1}^{4} : H_j :$, where $:\;:$ indicates normal ordering, and

$$H_2 = \int_V \{\eta_a^{(3)}(\omega_0, \mathbf{r})[\mathcal{D}(\mathbf{r}) \cdot \mathcal{D}^\dagger(\mathbf{r})]^2 + \tfrac{1}{2}\eta_b^{(3)}(\omega_0, \mathbf{r})|[\mathcal{D}(\mathbf{r}) \cdot \mathcal{D}(\mathbf{r})]|^2\}\, d^3\mathbf{r},$$
$$H_3 = \sum_n \frac{m_n}{2}\hat{\mathbf{v}}_n^2 + \frac{1}{2}\sum_n \sum_{n'} \kappa_{nn'} : \delta\hat{\mathbf{r}}_n \delta\hat{\mathbf{r}}_{n'},$$
$$H_4 = \sum_n \delta\hat{\mathbf{r}}_n \cdot \boldsymbol{\eta}^P(\nu_0, \mathbf{r}_n) : \mathcal{D}^\dagger(\mathbf{r})\mathcal{D}(\mathbf{r}). \tag{11.30}$$

The meaning of these terms is as follows:

1. H_1 is the energy of a linear dispersive medium. The Hamiltonian includes the effects of an inhomogeneous dielectric. If the linear dielectric response is also assumed to be isotropic, the linear tensor response functions are diagonal. Thus, for comparison with the usual dielectric permittivity, $\eta_{ij}(\omega_0, \mathbf{x}) = \eta(\omega_0, \mathbf{x})\delta_{ij}$. Here the subscripts indicate spatial coordinates as usual, and $\eta(\omega_0, \mathbf{x}) = 1/\epsilon(\omega_0, \mathbf{x})$, where $\epsilon(\nu_0, \mathbf{x})$ is the usual permittivity. If required, the assumption of linear isotropy can be relaxed.
2. H_2 is the energy due to an electronically induced nonlinear polarizability, neglecting the nonlinear dispersion in the electronic response. Differentiating the potential to obtain the response function, and taking account of combinatoric factors, gives a three-dimensional tensor,
$$\eta_{ijkl}^{(3)}(\omega_0, \mathbf{r}) = \tfrac{1}{3}\{\eta_a^{(3)}(\omega_0, \mathbf{r})[\delta_{ij}\delta_{kl} + \delta_{il}\delta_{jk}] + \eta_b^{(3)}(\omega_0, \mathbf{r})\delta_{ik}\delta_{jl}\}. \tag{11.31}$$
3. H_3 treats the nuclear displacement energy involved in thermal excitation of the medium. The interaction matrix $\kappa_{nn'}$ gives the potential energy for a nuclear displacement $\delta\mathbf{r}_n$, due to interatomic forces. Molecular structure in fused silica is a disordered structure, which results in a range of resonant phonon frequencies up to 10 THz, as well as guided acoustic wave Brillouin scattering (GAWBS) phonons at frequencies less than 10 GHz.
4. H_4 gives the term describing a coupling of the phonons to the radiation field. Here, the couplings to both the GAWBS phonons and the Raman phonons are written in terms of multipolar interactions. Thus, the interaction involves the electric displacement operator at each nuclear position $\mathbf{r}_n$, together with the nuclear displacements $\delta\mathbf{r}_n$. The appropriate transition matrix elements are all included here in the phonon coupling tensor $\boldsymbol{\eta}^P$. The coupling is to a lattice-averaged displacement field, which is assumed to have a slowly varying envelope. Any local-field corrections needed to transform to the microscopic, non-lattice-averaged displacement field is assumed to be included in the coupling tensor $\boldsymbol{\eta}^P$.

The present macroscopic procedure focuses on the slowly varying envelope part of the field, rather than on short-wavelength components of the field, which determine the displacement field variations on length-scales of the order of a lattice constant. To simplify the quantum theory, it is usual to expand the envelope functions in terms of creation and annihilation operators. These, in turn, are defined using mode functions that are eigenfunctions of the linear, dispersive wave equation, as discussed already. The resulting modes have an orthonormality property that is most simply expressed using the Poynting vector for energy transport along the waveguide or fiber. In the case of a translationally invariant waveguide, the properties of the linear modes are summarized in the previous subsection.

Optical fibers or waveguides are *not* translationally or temporally invariant in practice. To account for this, low-frequency and static inhomogeneities must be included via a permittivity fluctuation term $\Delta\eta(t, \mathbf{r})$. That is, $\eta(\mathbf{r}, \omega_0) = \bar{\eta}(\mathbf{r}, \omega_0) + \Delta\eta(t, \mathbf{r})$. The average, translationally invariant dielectric permittivity is $\bar{\eta}(\mathbf{r}, \omega_0)$, which determines the general modal behavior. In addition, there are local fluctuations $\Delta\eta(t, \mathbf{r})$, which vary slowly in time and space. In this approach, any small frequency dependence in the fluctuations of the waveguide refractive index are omitted. Like nonlinear dispersion, this additional effect is negligible in most cases.

Refractive index variations along the length of the fiber can cause the depolarization of an initially polarized input field. In addition, there are quasi-static inhomogeneities that have low-frequency temporal fluctuations, due to tunneling of defects in the amorphous silica crystal lattice. Defect tunneling gives rise to a characteristic $1/f$ phase noise spectrum. Both these effects are included by allowing $\Delta\eta$ to have a low-frequency classical time variation. Since the tunneling effects go beyond the usual quadratic approximation in the nuclear displacement Hamiltonian, they cannot be treated in the linear response formalism used for the Raman–Brillouin response. They have similar measurable effects to low-frequency Brillouin scattering, but a different temperature dependence.

While this chapter focuses on the third-order nonlinearity that is present in all dielectric media, it is also possible to include a second-order nonlinear response in addition to these terms, for cases with a non-reflection symmetric crystal structure.

11.5 Fiber optic Hamiltonian

The treatments given above are rather general. Next, we focus on a single transverse mode fiber with dispersion and nonlinearity. It is useful to take the infinite-volume limit, which replaces a summation over wavevectors with an integral. The effect of a transverse mode structure will also be included. The basic normally ordered, nonlinear Hamiltonian for the fiber in this case is $H_F = H_1 + H_2$, so that

$$H_F = \int dk\hbar\omega(k)a^\dagger(k)a(k) + \int d^3\mathbf{r}\,[\tfrac{1}{2}\Delta\eta^{(1)}(\mathbf{r}) : |\mathbf{D}|^2(\mathbf{r}) : +\tfrac{1}{4}\eta^{(3)}(\mathbf{r}) : |\mathbf{D}|^4(\mathbf{r}) :], \tag{11.32}$$

where $\omega(k)$ is the angular frequency of modes with wavevector k, describing the average linear response of the fiber, in the limit of a spatially uniform environment, including dispersion. Since optical fibers are not perfect crystals, it is necessary to add additional inhomogeneous terms to the Hamiltonian, of generic form $\Delta\eta^{(1)}(\mathbf{r})$. Here $a(k)$ is an annihilation operator defined so that

$$[a(k'), a^\dagger(k)] = \delta(k - k'). \tag{11.33}$$

Neglecting modal dispersion, the electric displacement field operator $\mathbf{D}(\mathbf{r})$ is

$$\mathbf{D}(\mathbf{r}) = i \int dk \sqrt{\frac{\hbar k \epsilon(\omega(k)) v(k)}{4\pi}}\, a(k)\mathbf{u}(\mathbf{r}_\perp) e^{ikx} + h.c., \tag{11.34}$$

where the mode function is defined as in the previous section, so that

$$\int d^2\mathbf{r}_\perp \, |\mathbf{u}(\mathbf{r}_\perp)|^2 = 1, \tag{11.35}$$

and $v(k) = \partial\omega(k)/\partial k$ is the usual group velocity.

A slowly varying polariton field in an interaction picture is defined by expanding

$$\psi(x,t) = \frac{1}{\sqrt{2\pi}} \int dk\, a(k,t) e^{i(k-k_0)x + i\omega_0 t}. \tag{11.36}$$

The field operator $\psi(x,t)$ corresponds to the linear quasi-particle excitations traveling through the fiber. The equal-time commutations relations are

$$[\psi(x,t), \psi^\dagger(x,t')] = \delta(x - x'). \tag{11.37}$$

The Hamiltonian is now

$$H_F = \hbar \int dx \int dx'\, \omega(x,x')\psi^\dagger(x,t)\psi(x,t') + \frac{\hbar}{2}\int dx\, \kappa_e \psi^{\dagger 2}(x,t)\psi^2(x,t). \tag{11.38}$$

Here we have defined the kernel $\omega(x,x')$, which is the linear dielectric component of the Hamiltonian. This is Taylor-expanded around $k = k_0$, and approximated to quadratic order in $(k - k_0)$, so that

$$\begin{aligned}\omega(x,x') &= \int \frac{dk}{2\pi}\, \omega(k) e^{i(k-k_0)(x-x')} + \frac{\epsilon}{2} k_0 v(k_0) \int d^2\mathbf{r}_\perp\, \Delta\eta^{(1)}(\mathbf{x})|\mathbf{u}(\mathbf{r})|^2 \delta(x - x') \\ &\simeq [\omega_0 + \Delta\omega(x)]\delta(x - x') + \int \frac{dk}{4\pi}\, [i\omega_0'(\partial_{x'} - \partial_x) + \omega_0''(\partial_x \partial_{x'}) + \cdots] e^{ik(x-x')}.\end{aligned} \tag{11.39}$$

In addition, a nonlinear coupling term was also introduced, which gives rise to an electronic contribution n_{2e} to the intensity-dependent refractive index, where $n = n_0 + In_2 = n_0 + I(n_{2e} + n_{2p})$. Thus, we define κ_e, in units of [m s^{-1}] as

$$\kappa_e \equiv \frac{3\hbar\eta^{(3)}(k_0 v \epsilon)^2}{4A} \equiv -\left[\frac{\hbar n_{2e}(x)\omega_0^2 v^2}{Ac}\right]. \tag{11.40}$$

Here $A = [\int d^2\mathbf{r}\, |\mathbf{u}(\mathbf{r})|^4]^{-1}$ is the effective modal cross-section, and n_{2e} is the refractive index change per unit field intensity due to fast electronic transitions. This is somewhat less than the observed value of n_2, since phonon contributions need to be included.

The resulting interaction Hamiltonian H_F' in the slowly varying envelope and rotating-wave approximations is

$$\begin{aligned} H_F' &= H_F - \int dk \hbar\omega_0 a^\dagger(k)a(k) \\ &= \frac{\hbar}{2}\int_{-\infty}^{\infty} dx \left[\Delta\omega(x)\psi^\dagger\psi + \frac{iv}{2}(\nabla\psi^\dagger\psi - \psi^\dagger\nabla\psi) + \frac{\omega''}{2}\nabla\psi^\dagger\nabla\psi + \frac{\kappa_e}{2}\psi^{\dagger 2}\psi^2\right]. \end{aligned} \tag{11.41}$$

For simplicity, only quadratic dispersion is included here. Higher-order dispersion can be included by treating the dispersion as part of the reservoir response function, or by expanding to higher order. The response function approach has the advantage that an arbitrary polarization response can be included.

11.5.1 Heisenberg equation

We obtain the Heisenberg equation of motion for the field operator propagating in the $+x$ direction as

$$\left[v\frac{\partial}{\partial x} + \frac{\partial}{\partial t}\right]\psi(x,t) = \left[-i\Delta\omega(x) + \frac{i\omega''}{2}\frac{\partial^2}{\partial x^2} - i\kappa_e\psi^\dagger\psi\right]\psi(x,t). \tag{11.42}$$

This describes a Bose gas of quasi-particles with an average velocity of v, for photons near to the carrier frequency. It includes spatial variation, through the term $\Delta\omega(x)$.

11.6 Raman Hamiltonian

There can also be couplings to linear gain, absorption and phonon reservoirs, as described in earlier chapters. The phonon field arises from the displacement of atoms from their mean locations in the dielectric lattice. As described earlier, the Raman interaction energy of a fiber, in terms of atomic displacements from their mean lattice positions, has the form

$$H_R = \frac{1}{2}\sum_j \eta_j^R : \mathbf{D}(\bar{\mathbf{r}}^j)\mathbf{D}(\bar{\mathbf{r}}^j)\delta\mathbf{r}^j + \frac{1}{2}\sum_{ij}\kappa_{ij} : \delta\mathbf{r}^i\delta\mathbf{r}^j. \tag{11.43}$$

Here $\mathbf{D}(\bar{\mathbf{r}}^j)$ is the electric displacement at the jth mean atomic location $\bar{\mathbf{r}}^j$, while $\delta\mathbf{x}^j$ is the atomic displacement operator, η_j^R is a Raman coupling tensor, and κ_{ij} represents the short-range atom–atom interactions.

In order to quantize the Raman interaction, we now take into account a corresponding set of phonon operators, $b(\omega, x)$ and $b^\dagger(\omega, x)$. These diagonalize the atomic displacement Hamiltonian near x, and have well-defined eigenfrequencies. The frequency spectrum and normal modes of vibration for vitreous silica may be obtained from the random network theory of disordered systems, or measured from observed Raman gain profiles. The resulting phonon–photon coupling induces Raman transitions and scattering from acoustic waves (the Brillouin effect), causing extra noise sources and an additional nonlinearity. The initial state of phonons is assumed to be thermal, with $n_{th}(\omega) = [\exp(\hbar\omega/kT) - 1]^{-1}$.

11.6.1 Hamiltonian and Heisenberg equations

In terms of the phonon operators, the fiber Hamiltonian in the interaction picture and within the rotating-wave approximation for a single polarization is $H' = H_R + H'_F$, where we now introduce a Raman interaction Hamiltonian, $H_R = H_3 + H_4$, so that for the present case of a single polarization we have

$$H_R = \hbar \int_{-\infty}^{\infty} dx \int_0^{\infty} d\omega \, \{\psi^\dagger(x)\psi(x)R(\omega, x)[b(\omega, x) + b^\dagger(\omega, x)] + \omega b^\dagger(\omega, x)b(\omega, x)\}. \tag{11.44}$$

Here, the atomic vibrations in the fiber are modeled as localized oscillators, and are coupled to the radiation modes by a frequency-dependent coupling $R(\omega, x)$. This coupling is determined empirically through measurements of the Raman gain spectrum. The atomic displacement is proportional to $b + b^\dagger$, where the phonon operators, b and $b^\dagger$, have the equal-time commutation relations

$$[b(\omega, x, t), b^\dagger(\omega', x', t)] = \delta(x - x')\delta(\omega - \omega'). \tag{11.45}$$

Thus, Raman excitations are treated here as an inhomogeneously broadened continuum of modes, localized at each longitudinal location x. GAWBS is a special case in the low-frequency limit. Since neither Raman nor Brillouin excitations are completely localized, this treatment requires a frequency and wavenumber cutoff.

Suppressing the time and space coordinates for simplicity, the corresponding nonlinear operator equations are

$$\begin{aligned} \left[v\frac{\partial}{\partial x} + \frac{\partial}{\partial t}\right]\psi &= i\left[-\Delta\omega(x) + \frac{\omega''}{2}\frac{\partial^2}{\partial x^2} - \kappa_e\psi^\dagger\psi\right]\psi \\ &\quad - i\left[\int_0^{\infty} R(\omega, x)[b(\omega) + b^\dagger(\omega)]\, d\omega\right]\psi, \\ \frac{\partial}{\partial t}b(\omega) &= -i\omega b(\omega) - iR(\omega, x)\psi^\dagger\psi. \end{aligned} \tag{11.46}$$

The nonlinear quantum field equations now include both the electronic and the Raman nonlinearity. The result is a delayed nonlinear response to the field due to the Raman coupling. On integrating the Raman reservoirs, one obtains

$$\begin{aligned} &\left[v\frac{\partial}{\partial x} + \frac{\partial}{\partial t}\right]\psi(x, t) \\ &= i\left[-\Delta\omega(x) + \frac{\omega''}{2}\frac{\partial^2}{\partial x^2} - \int_{0-}^{\infty} dt'\, \kappa(x, t')[\psi^\dagger\psi](t - t') - \Gamma^R(x, t)\right]\psi(x, t), \end{aligned} \tag{11.47}$$

where, defining $\Theta(t)$ as the step function,

$$\kappa(x,t) = \kappa_e \delta(t) - 2\Theta(t)\int_0^\infty R^2(\omega,x)\sin(\omega t)\,d\omega,$$

$$\Gamma^R(x,t) = \int_0^\infty R(\omega,x)[b(\omega,x,0) + e^{-i\omega t} + h.c.]\,d\omega. \tag{11.48}$$

The noise terms are stochastic operators defined so that

$$\Gamma^R(x,\omega) = \frac{1}{\sqrt{2\pi}}\int dt\ \exp(i\omega t)\Gamma^R(x,t),$$

$$\Gamma^{R\dagger}(x,\omega) = \frac{1}{\sqrt{2\pi}}\int dt\ \exp(-i\omega t)\Gamma^R(x,t). \tag{11.49}$$

Their frequency-space correlations are

$$\langle \Gamma^{R\dagger}(x',\omega')\,\Gamma^R(x,\omega)\rangle = -2\kappa''(x,|\omega|)[n^{th}(|\omega|) + \Theta(-\omega)]\delta(x-x')\delta(\omega-\omega'). \tag{11.50}$$

In this expression, we introduce a Raman amplitude gain of κ'', equal to the imaginary part of the Fourier transform of $\kappa(x,t)$, so that $\kappa''(x,|\omega|) = -\pi R^2(|\omega|,x)$. This uses the Bloembergen normalization for response function Fourier transforms,

$$\widetilde{\kappa}(x,\omega) = \int dt\ \exp(i\omega t)\kappa(x,t), \tag{11.51}$$

which does not include a $\sqrt{2\pi}$ factor.

We finally note that the use of a single polarization is an oversimplification in many cases. Typically, there are two transverse polarization modes in a waveguide. These are included in the complete Hamiltonian in the previous section. With both the polarization modes excited, there will be a response function and noise term for each polarization, as well as cross-coupling terms due to the tensor nature of the dielectric response. It is also possible to have multiple spatial modes.

11.6.2 Raman gain measurements

The measured intensity gain due to Raman effects, per unit photon flux $I_0 = v\langle\psi^\dagger\psi\rangle$, is

$$\frac{1}{I_0}\frac{\partial \ln I}{\partial x} = 2\frac{\kappa''(\omega,x)}{v^2}. \tag{11.52}$$

Here the gain is positive for red-shifted frequencies ($\omega < 0$), and negative for blue-shifted ($\omega > 0$). These are called Stokes and anti-Stokes lines for historical reasons. This allows the coupling to be estimated from measured Raman properties. The simplest way to do this is by expanding the Raman response function using a multiple-Lorentzian model, which is fitted to observed Raman fluorescence data. We therefore expand

$$\kappa(x,t) = \kappa_e\delta(t) + \kappa(x)\Theta(t)\sum_{j=0}^{n} F_j\delta_j e^{-\delta_j t}\sin(\omega_j t). \tag{11.53}$$

For normalization purposes, we have introduced $\kappa(x)$, which is defined as the *total* effective phase-shift coefficient per unit time and photon density (in units of rad m s^{-1}). This is given by integrating over time:

$$\kappa(x) = \kappa_e - 2\int_0^\infty \int_0^\infty R^2(\omega, x)\sin(\omega t)\, d\omega\, dt. \tag{11.54}$$

In the above expansion, F_j are a set of positive, dimensionless Lorentzian strengths, and ω_j and δ_j are the resonant frequencies and widths, respectively. The $j = 0$ Lorentzian models the Brillouin contribution to the response function. All of these parameters could be x-dependent, but, from now on, we assume that they are constant in space. The coefficient of the electronic nonlinearity is

$$\kappa_e = \frac{\hbar(f-1)n_2\omega_0^2 v^2}{Ac}, \tag{11.55}$$

where ω_0 is the carrier frequency, A is the effective cross-sectional area of the traveling mode, and f is the fraction of the nonlinearity due to the Raman gain, which has been estimated using the procedure outlined above as

$$\begin{aligned} f = \frac{\kappa_r}{\kappa} &= -\frac{2}{\kappa}\int_0^\infty dt \int_0^\infty d\omega\, R^2(\omega, x)\sin(\omega t) \\ &\simeq 0.2. \end{aligned} \tag{11.56}$$

A result of this model is that the phonon operators have a colored-noise property, which invalidates the usual Markovian and rotating-wave approximations. Finally, there is another important low-frequency effect. This is tunneling due to lattice defects. The effects of this $1/f$ type noise may be included approximately by modifying the refractive index perturbation term so that it becomes $\Delta\omega(x, t)$, and generates the measured low-frequency refractive index fluctuations.

11.7 Gain and absorption

In silica fibers, there is a flat absorption profile, with a minimum absorption coefficient of approximately 0.2 dB km^{-1} around $\lambda = 1.5$ μm. This effect can be compensated for by the use of fiber laser amplifiers, resulting in nearly zero net absorption over a total link that includes both normal and amplified fiber segments.

11.7.1 Absorbing reservoirs

The absorption reservoir is modeled by a coupling to a continuum of harmonic oscillators labeled with their resonant frequency ω. In the interaction picture used here, the Hamiltonian term is

$$H'_A = \hbar\int_{-\infty}^\infty dx \int_0^\infty d\omega\, \{[\Psi(x)a^\dagger(\omega, x)A(\omega, x) + h.c.] + (\omega - \omega_0)a^\dagger a(\omega, x)\}, \tag{11.57}$$

where the reservoir annihilation and creation operators, a and $a^\dagger$, have the commutation relations

$$[a(\omega, x), a^\dagger(\omega', x')] = \delta(x - x')\delta(\omega - \omega'). \tag{11.58}$$

The equations for the absorbing photon reservoirs can be integrated immediately, following procedures used in Chapter 6. The response function and reservoir terms are obtained so that

$$\begin{aligned} \gamma^A(x, t) &\approx \Theta(t) \int_{-\infty}^{+\infty} d\omega \, |A(\omega, x)|^2 \, e^{-i(\omega - \omega_0)t}, \\ \Gamma^A(x, t) &= -i \int_0^\infty d\omega \, A^*(\omega, x) e^{-i(\omega-\omega_0)(t-t_0)} a(\omega, x, t_0). \end{aligned} \tag{11.59}$$

The response function integral is a deterministic term, with a Fourier transform of

$$\widetilde{\gamma}^A(x, \omega) = \int \gamma^A(x, t) e^{i\omega t} \, dt = \gamma^A(x, \omega) + i\gamma^{A''}(x, \omega), \tag{11.60}$$

so that the amplitude loss rate is

$$\gamma^A(x, \omega) = \pi \, |A(\omega_0 + \omega, x)|^2. \tag{11.61}$$

In the case of a spatially uniform reservoir with a flat spectral density, we obtain a uniform Markovian loss term with

$$\gamma^A(t) \approx 2\gamma^A \delta(t), \tag{11.62}$$

where the average loss coefficient is

$$\begin{aligned} \gamma^A &= \int_0^\infty \int_{-\infty}^\infty dt \, d\omega \, |A(\omega)|^2 \, e^{-i(\omega-\omega_0)t} \\ &= \gamma^{A\prime} + i\gamma^{A\prime\prime}. \end{aligned} \tag{11.63}$$

This approximation is generally rather accurate for the absorbing reservoirs. The second quantity, $\Gamma^A(x, t)$, behaves like a stochastic term due to the random initial conditions. Neglecting the frequency dependence of the thermal photon number, the corresponding correlation functions are

$$\langle \Gamma^A(x, t) \, \Gamma^{A\dagger}(x', t') \rangle = \int_0^\infty d\omega \, |A(\omega)|^2 e^{-i(\omega-\omega_0)(t-t')} (n^{th}(\omega) + 1) \delta(x - x') \tag{11.64}$$

and

$$\langle \Gamma^{A\dagger}(x', t') \, \Gamma^A(x, t) \rangle = \int_0^\infty d\omega \, |A(\omega)|^2 e^{-i(\omega-\omega_0)(t-t')} n^{th}(\omega) \delta(x - x'). \tag{11.65}$$

At optical or infrared frequencies, it is a good approximation to set $n^{th}(\omega_0) = 0$. On Fourier-transforming the noise sources, one then obtains for homogeneous fibres

$$\langle\Gamma^A(x,\omega)\,\Gamma^{A\dagger}(x',\omega')\rangle = 2\gamma^A\delta(x-x')\delta(\omega-\omega'). \tag{11.66}$$

Here we have taken the simplifying case of a spatially uniform reservoir in the Wigner–Weiskopff limit, neglecting frequency shifts, so on Fourier transforming this becomes:

$$\begin{aligned}\langle\Gamma^A(x,t)\,\Gamma^{A\dagger}(x',t')\rangle &= 2\gamma^A\delta(t-t')\delta(x-x'),\\ \langle\Gamma^{A\dagger}(x,t)\,\Gamma^{A}(x',t')\rangle &= 0.\end{aligned} \tag{11.67}$$

Here $2\gamma^A/v$ corresponds to the linear absorption coefficient for fibers during propagation. A typical measured absorption figure in current fused silica communications fibers is 0.2 dB km^{-1} in the minimum region of absorption (near $\lambda = 1.5$ μm), so that $2\gamma^A/v \simeq 2.3\times 10^{-5}$ m^{-1}.

11.7.2 Waveguide gain

The equations for gain or laser reservoirs are generally more complex. In the case of silica fibers, a commonly used lasing transition uses erbium impurities. It is possible to develop a detailed theory of erbium laser amplifiers. Here we treat the simplest possible quantum theory of a traveling-wave quantum-limited laser amplifier. Assuming that the laser amplifier is polarization-insensitive, we omit the polarization index. The reservoir variable $\sigma_\mu = |1\rangle_\mu\langle 2|_\mu$ is an atomic transition operator, as in Chapter 6, except with spatial dependence and density $\rho(\omega,x)$.

The gain terms $\sigma^{\pm}(\omega,x)$ represent the raising and lowering Pauli field operators, for two-level lasing transitions at frequency ω and position x, so that

$$H'_G = \hbar\int_{-\infty}^{\infty}dx\int_0^{\infty}d\omega\left[\{\psi\sigma^+(\omega,x)G(\omega,x)+h.c.\}+\frac{\omega-\omega_0}{2}\sigma^z(\omega,x)\right], \tag{11.68}$$

where the atomic raising and lowering field operators, $\sigma^\pm$, are defined as

$$\begin{aligned}\sigma^+(\omega,x) &= \frac{1}{\sqrt{\rho(\omega,x)}}\sum_\mu |2\rangle\langle 1|_\mu e^{-i\omega_0 t}\delta(x-x_\mu)\delta(\omega-\omega_\mu),\\ \sigma^-(\omega,x) &= \frac{1}{\sqrt{\rho(\omega,x)}}\sum_\mu |1\rangle\langle 2|_\mu e^{i\omega_0 t}\delta(x-x_\mu)\delta(\omega-\omega_\mu),\\ \sigma^z(\omega,x) &= \frac{1}{\rho(\omega,x)}\sum_\mu [|2\rangle\langle 2|-|1\rangle\langle 1|]_\mu\delta(x-x_\mu)\delta(\omega-\omega_\mu).\end{aligned} \tag{11.69}$$

These have the commutation relations

$$[\sigma^+(\omega,x,t),\sigma^-(\omega',x',t')] = \sigma^z(\omega,x,t)\delta(x-x')\delta(\omega-\omega'). \tag{11.70}$$

If we follow the procedure in Chapter 6, Fourier-transforming the response function gives

$$\tilde{\gamma}^G(x,\omega) = \int\gamma^G(x,t)e^{i\omega t}\,dt = \gamma^{G'}(x,\omega)+i\gamma^{G''}(x,\omega), \tag{11.71}$$

and, dropping small imaginary terms, the (real) resonant amplitude gain coefficient is

$$\gamma^G(x,\omega) = \pi\,|G(\omega+\omega_0,x)|^2. \tag{11.72}$$

Here, $\Gamma^G(x,t)$ behaves like a stochastic term due to the random initial conditions, with correlation functions

$$\langle \Gamma^{G\dagger}(x',t')\,\Gamma^G(x,t)\rangle = \int_0^\infty d\omega\,|G(\omega,x)|^2 e^{i(\omega-\omega_0)(t-t')}\delta(x-x'). \tag{11.73}$$

Fourier-transforming gives

$$\langle \Gamma^{G\dagger}(x',\omega')\,\Gamma^G(x,\omega)\rangle = 2\gamma^G(x,\omega)\delta(x-x')\delta(\omega-\omega'). \tag{11.74}$$

Taking a uniform fiber in the Wigner–Weiskopff limit as before, so $\gamma^G = \gamma^G(x,0)$, this becomes

$$\begin{aligned}
\langle \Gamma^G(x,t)\,\Gamma^{G\dagger}(x',t')\rangle &= 0,\\
\langle \Gamma^{G\dagger}(x,t)\,\Gamma^G(x',t')\rangle &= 2\gamma^G\delta(t-t')\delta(x-x').
\end{aligned} \tag{11.75}$$

The results presented here are only valid in the linear gain regime. Finally, it is necessary to consider the result of incomplete inversion of an amplifier. Noninverted atoms give rise to absorption, not gain, and must be treated as in the previous section, including non-Markovian effects if the absorption line is narrow-band. An important consequence is that the measured gain only gives the difference between the gain and the loss. This causes difficulties in determining the amplifier quantum noise levels, which can only be uniquely determined through spontaneous fluorescence measurements.

11.8 Combined Heisenberg equations

Adding couplings to reservoirs leads to a combined quantum Langevin equation. In the case of a single polarization, the resulting field equations are

$$\begin{aligned}
\left[v\frac{\partial}{\partial x}+\frac{\partial}{\partial t}\right]\psi(x,t) &= \int_0^\infty dt'\,\gamma(x,t')\psi(t-t')] + \Gamma(x,t)\\
&\quad + i\left[\frac{\omega''}{2}\frac{\partial^2}{\partial x^2} - \int_{0-}^\infty dt'\,\kappa(x,t')[\psi^\dagger\psi](x,t-t') - \Gamma^R(x,t)\right]\psi.
\end{aligned} \tag{11.76}$$

In this equation,

$$\gamma(x,t) = \gamma^A(x,t) - \gamma^G(x,t) + i\Delta\omega(x)\delta(t) \tag{11.77}$$

is a net linear response function including the effects of a spatially varying refractive index. Similarly, $\Gamma(x,t)$ is the quantum noise due to gain and absorption. The measured intensity gain at frequency $\omega + \omega_0$ is given by

$$\frac{\partial \ln I}{\partial x} = 2(\gamma^G(x,\omega) - \gamma^A(x,\omega))/v. \tag{11.78}$$

The minimal linear quantum noise terms for the gain/absorption reservoirs are (neglecting thermal photons, since usually $\hbar\omega_0 \gg kT$):

$$\begin{aligned}
\langle \Gamma^{R\dagger}(x',\omega')\,\Gamma^R(x,\omega)\rangle &= -2\kappa''(|\omega|,x)[n^{th}(|\omega|) + \Theta(-\omega)]\delta(x-x')\delta(\omega-\omega'),\\
\langle \Gamma^{\dagger}(x',\omega')\,\Gamma(x,\omega)\rangle &= 2\gamma^G(x,\omega)\delta(x-x')\delta(\omega-\omega'),\\
\langle \Gamma(x,\omega)\,\Gamma^{\dagger}(x',\omega')\rangle &= 2\gamma^A(x,\omega)\delta(x-x')\delta(\omega-\omega').
\end{aligned} \tag{11.79}$$

This complete Heisenberg equation gives a consistent quantum theoretical description of optical fiber quantum dynamics, including dispersion, nonlinear refractive index, Raman/GAWBS scattering, linear gain and absorption. It is sometimes easier to do calculations in a moving frame. This can be carried out either using a standard moving frame ($x_v = x - vt$) or with a propagative time ($t_v = t - x/v$), with the space coordinate unchanged. For propagative calculations, it is most convenient to use photon flux operators:

$$\psi_f(x,t_v) = \sqrt{v}\,\psi(x, t - x/v). \tag{11.80}$$

For long pulses, assuming a uniform gain/loss response in the frequency domain, the propagative transformation gives the following approximate equations:

$$\begin{aligned}
\frac{\partial}{\partial x}\psi_f(x,t_v) &= -\int_0^\infty dt_v'\,\frac{\gamma(x,t_v')}{v}\psi_f(x,t_v - t_v') + \frac{\Gamma(t)}{\sqrt{v}}\\
&\quad - i\left[-\frac{k''}{2}\frac{\partial^2}{\partial t_v{}^2} + \int_{0-}^\infty dt'\,\frac{\kappa(t_v')}{v^2}[\psi_f{}^\dagger\psi_f](t_v - t_v') + \frac{1}{v}\Gamma^R\right]\psi_f(x,t_v).
\end{aligned} \tag{11.81}$$

Here $k'' = \partial^2 k/\partial\omega^2 = -\omega''/v^3$. In addition, if the pulses are narrow-band compared to the gain and loss bandwidths, and the reservoirs are uniform, then the gain and absorption reservoirs are nearly delta-correlated, with

$$\begin{aligned}
\langle \Gamma^{\dagger}(x,t_v)\,\Gamma(x',t_v')\rangle &= 2\gamma^G\delta(x_v - x_v{}')\delta(t-t'),\\
\langle \Gamma(x,t_v)\,\Gamma^{\dagger}(x,t_v')\rangle &= 2\gamma^A\delta(x_v - x_v{}')\delta(t-t').
\end{aligned} \tag{11.82}$$

This set of approximate equations can be used predict the soliton self-frequency shift and related effects in soliton propagation.

11.9 Phase-space methods

To generate more tractable equations, operator representation theory can be used, as explained in Chapter 7. The positive P-representation produces exact results, although often with large sampling error, while a truncated Wigner representation gives approximate results. The latter method represents symmetrically ordered rather than normally ordered operator products, and so has finite quantum noise terms even for a vacuum field. Applying the appropriate operator correspondences to the master equation for the reduced density operator ρ_a in which the reservoir modes have been traced over,

$$\dot{\rho}_\Psi = \mathrm{Tr}_R\,\dot{\rho} = \mathrm{Tr}_R\,\frac{1}{i\hbar}[H,\rho], \tag{11.83}$$

gives a functional equation for the corresponding operator representation.

In the positive P case, the equation is defined on a functional phase space of double the classical dimensions, so that a complete expansion in terms of a coherent-state basis $|\alpha(\mathbf{x})\rangle$ is obtained. The resulting Fokker–Planck equation has only second-order derivative terms. The equation for the Wigner function also contains third- and fourth-order derivative terms, which are negligible at large photon number. The resultant equation in either case can be converted into an equivalent Ito stochastic equation.

11.9.1 Modified nonlinear Schrödinger equation

It is simplest to use the propagative reference frame with the normalized variables $\tau = (t - x/v)/t_0$ and $\zeta = x/x_0$. Here t_0 is a typical pulse duration used for scaling purposes, and $x_0 = t_0^2/|k''|$. This change of variables is useful when slowly varying second-order derivatives involving ζ can be neglected, which is possible for $vt_0/x_0 \ll 1$. This inequality is well satisfied in many cases with $vt_0 < 10^{-4}$ m. We will make the same transformation for the stochastic equations, and scale the variables used in a dimensionless form.

In the spatially uniform case, the resultant truncated Wigner equation is a Raman-modified nonlinear Schrödinger (NLS) equation with stochastic noise terms:

$$\begin{aligned}\frac{\partial}{\partial\zeta}\alpha(\zeta,\tau) = &-\int_{-\infty}^{\infty} d\tau'\, g(\tau-\tau')\alpha(\zeta,\tau') + \Gamma(\zeta,\tau) + i\left[\pm\frac{1}{2}\frac{\partial^2\alpha}{\partial\tau^2}\right.\\ &\left.+\int_{-\infty}^{\infty} d\tau'\, h(\tau-\tau')\alpha^*(\zeta,\tau')\alpha(\zeta,\tau') - \Gamma^R(\zeta,\tau)\right]\alpha(\zeta,\tau),\end{aligned} \tag{11.84}$$

where $\alpha \sim \psi\sqrt{vt_0}$ is a dimensionless photon field amplitude. The photon flux is $|\alpha|^2/t_0$. The positive sign in front of the second-derivative term applies for anomalous dispersion ($k'' < 0$), and the negative sign applies for normal dispersion ($k'' > 0$).

The positive P-distribution leads to an equation of the same form, although with modified noise sources and initial conditions, described below. A second equation is also obtained in this case, except that α^* and $\Gamma^{R*}(\tau,\zeta)$ are replaced by independent fields denoted α^+ and $\Gamma^{R+}(\tau,\zeta)$.

The causal *linear* response function $g(\tau)$ is defined as

$$g(\tau) = \frac{\gamma(\tau t_0)x_0}{v}. \tag{11.85}$$

The causal nonlinear response function $h(\tau)$ includes both electronic and Raman nonlinearities, and is defined to have a positive sign in the usual case of $n_2 > 0$, so:

$$h(\tau) = h^E(\tau) + h^R(\tau) = -\frac{x_0 \kappa(\tau t_0)}{v^2}. \tag{11.86}$$

The Raman response function is $h^R(\tau)$. The Raman gain can be modeled as a sum of n Lorentzians, as explained above. This gives a response function of the form

$$h^R(t/t_0) = \Theta(t) \sum_{j=0}^{n} F_j \delta_j t_0 e^{-\delta_j t} \sin(\omega_j t). \tag{11.87}$$

It is most convenient to express these through the definitions $\Omega_j = \omega_j t_0$ and $\Delta_j = \delta_j t_0$, giving

$$h^R(\tau) = \Theta(\tau) \sum_{j=0}^{n} F_j \Delta_j e^{-\Delta_j \tau} \sin(\Omega_j \tau). \tag{11.88}$$

Here Δ_j are dimensionless widths, and the Ω_j are the dimensionless center frequencies, in normalized units. The dimensionless intensity gain function is $2|h''(\Omega)|$, where the response function Fourier transform is given by

$$\tilde{h}(\Omega) = \int d\tau \exp(i\Omega\tau) h(\tau) = h'(\Omega) + i h''(\Omega). \tag{11.89}$$

These stochastic partial differential equations can be discretized and numerically simulated using a split-step Fourier integration routine.

11.9.2 Initial conditions

A typical initial condition is the multi-mode coherent state, since this is the simplest model for the output of mode-locked lasers. To represent this in the positive P-distribution, one just takes

$$\alpha_P(0,\tau) = [\alpha_P^+(0,\tau)]^* = \sqrt{v t_0} \langle \psi(0,\tau) \rangle. \tag{11.90}$$

In the case of the Wigner equations, only a subset of possible states can be represented with a positive probability distribution. In this case, which corresponds to symmetric operator ordering, one must also include complex quantum vacuum fluctuations, in order to correctly represent operator fields. For coherent inputs, the Wigner vacuum fluctuations are Gaussian, and are correlated as

$$\begin{aligned} \langle \alpha_W(0,\tau) \rangle &= \sqrt{v t_0} \langle \psi(0,\tau) \rangle, \\ \langle \Delta\alpha_W(0,\tau) \Delta\alpha_W^*(0,\tau') \rangle &= \tfrac{1}{2}\delta(\tau - \tau'). \end{aligned} \tag{11.91}$$

These equations imply that an appropriate correction is made for any losses at the input interface.

11.9.3 Wigner noise

Both fiber loss and the gain medium contribute quantum noise to the equations. This symmetrically ordered noise source is present for both gain and loss reservoirs. In this case,

$$\langle \Gamma(\zeta,\Omega)\,\Gamma^*(\zeta',\Omega')\rangle = (g^A + g^G)\delta(\zeta-\zeta')\delta(\Omega-\Omega'), \tag{11.92}$$

where $\Gamma(\zeta,\Omega)$ is the Fourier transform of the noise source,

$$\begin{aligned}\Gamma(\zeta,\Omega) &= \frac{1}{\sqrt{2\pi}}\int_{-\infty}^{\infty} d\tau\,\Gamma(\tau,\zeta)e^{i\Omega\tau},\\ \Gamma^*(\zeta,\Omega) &= \frac{1}{\sqrt{2\pi}}\int_{-\infty}^{\infty} d\tau\,\Gamma^*(\tau,\zeta)e^{-i\Omega\tau},\end{aligned} \tag{11.93}$$

and $g^A = \gamma^A x_0/v t_0$, $g^G = \gamma^G x_0/(v t_0)$ are, respectively, the real parts of the response function Fourier transforms. Similarly, the real Raman noise, which appears as a multiplicative stochastic variable Γ^R, has correlations

$$\langle \Gamma^R(\zeta,\Omega)\,\Gamma^{R*}(\zeta',\Omega')\rangle = h''(|\Omega|)[2n_{th}(|\Omega|/t_0)+1]\delta(\zeta-\zeta')\delta(\Omega-\Omega'). \tag{11.94}$$

Here $h'' = -\kappa'' x_0/(t_0 v^2)$ is the dimensionless Raman gain term. Thus the Raman noise is strongly temperature-dependent, but it also contains a spontaneous component that provides vacuum fluctuations even at $T = 0$. Since the spontaneous component can occur through coupling to either a gain or a loss reservoir, in a symmetrically ordered representation, it is present for both positive and negative frequency detunings.

11.9.4 Positive P noise

The positive P-representation is a useful alternative strategy, because it does not require truncation of higher-order derivatives in a Fokker–Planck equation, and corresponds directly to observable normally ordered, time-ordered operator correlations. It has no vacuum fluctuation terms. Provided the phase-space boundary terms are negligible, one can then obtain a set of c-number stochastic differential equations in a phase space of double the usual classical dimensions. These are very similar to the classical equations. Here the additive stochastic term is as before, except that it *only* depends on the gain term. The conjugate term Γ^* is used in the α^+ equation:

$$\langle \Gamma(\zeta,\Omega)\,\Gamma^*(\zeta',\Omega')\rangle = 2g^G\delta(\zeta-\zeta')\delta(\Omega-\Omega'). \tag{11.95}$$

Since this representation is normally ordered, the only noise sources present are due to the gain reservoirs. There is no vacuum noise term for the absorbing reservoirs, because absorption simply maps a coherent state into another coherent state.

The complex terms Γ^R and Γ^{R+} include both Raman and electronic terms (through $h'(\Omega)$). As elsewhere in this book, we regard $\Gamma^{R+}(\zeta,\Omega)$ as a complex conjugate Fourier transform (with a negative frequency exponent, as in the Wigner noise conjugate term in

Eq. (11.93)):

$$\langle \Gamma^R(\zeta,\Omega)\,\Gamma^R(\zeta',\Omega')\rangle = \delta(\zeta-\zeta')\delta(\Omega+\Omega')\{[2n_{th}(|\Omega|/t_0)+1]h''(|\Omega|) - ih'(\Omega)\},$$
$$\langle \Gamma^{R+}(\zeta',\Omega')\,\Gamma^R(\zeta,\Omega)\rangle = 2\delta(\zeta-\zeta')\delta(\Omega-\Omega')[n_{th}(|\Omega|/t_0)+\Theta(-\Omega)]h''(|\Omega|). \tag{11.96}$$

This is an expected result, since it states that, when $\Omega < 0$, the spectral intensity of noise due to the Stokes process, in which a photon is down-shifted in frequency by an amount Ω with the production of a phonon of the same frequency, is proportional to $n_{th} + 1$. However, the anti-Stokes process in which a phonon is absorbed ($\Omega > 0$) is only proportional to n_{th}.

As one might expect, the two forms of the equation are identical at high phonon occupation numbers, when classical noise is so large that it obscures the differences due to the operator orderings of the two representations. Another, less obvious, result is that the two equations have identical additive noise sources, provided the gain and loss are balanced. To understand this, we can see that, in the absence of any net gain or loss, the difference in the operator correlations due to ordering is a constant, contained in the initial conditions. While the truncated Wigner method is simpler and faster to use, it has approximations that can cause systematic errors at low mode occupation numbers. For this reason, a cautious approach is to check the truncated Wigner predictions against the exact results of the positive P method.

In general, the Wigner and positive P reservoir correlations are obtainable simply by examining the expectation values of the Heisenberg reservoir terms, with symmetric and normal ordering, respectively. The additional term proportional to $h'(\Omega)$ in the positive P noise terms is due to dispersive nonlinear effects, and gives rise to a nonclassical noise source, which is responsible for the observed quantum squeezing effects.

We are now faced with solving nonlinear stochastic equations. Two basic approaches are as follows. The first approach is to linearize about a classical solution. One assumes that the field is given by the sum of a large classical part, which is a solution of the equations with the noise sources set equal to zero, and a stochastic part – only linear terms in the stochastic part are kept.

The second approach is to solve the c-number stochastic equations numerically. The results are in very good agreement with experiments that have been carried out. In particular, one finds that, despite the noise due to Raman scattering and absorption, it is possible to observe quantum effects in fields propagating in fibers. Rosenbluh and Shelby (see the additional reading), in a landmark experiment, were the first to observe squeezing in quantum solitons. They were able to obtain a noise level of 0.68 times the vacuum noise level in the squeezed quadrature using 5 m of fiber at liquid nitrogen temperature. These results have since been greatly improved with increased quantitative agreement between theory and experiment in more recent experiments by Leuchs' group at Erlangen.

11.10 Polarization squeezing

To conclude, we look at an example to which the techniques described in this chapter have been applied: quantum dynamical polarization squeezing in a single-mode optical fiber.

We shall start with a description involving only a single spatial mode, with two orthogonal polarization states, and then go on to discuss how polarization squeezing can be produced by propagation in a nonlinear fiber.

The polarization of a single quantized mode is described by using the Stokes operators

$$S_0 = a_x^\dagger a_x + a_y^\dagger a_y, \qquad S_1 = a_x^\dagger a_x - a_y^\dagger a_y,$$
$$S_2 = a_x^\dagger a_y + a_y^\dagger a_x, \qquad S_3 = i(a_y^\dagger a_x - a_x^\dagger a_y). \tag{11.97}$$

Here, x and y denote orthogonal polarization states. They could correspond to orthogonal linear polarizations or to right and left circular polarizations. These operators obey the commutation relations

$$[S_0, S_j] = 0, \qquad [S_j, S_k] = 2i\epsilon_{jjl}S_l, \tag{11.98}$$

where $j, k, l = 1, 2, 3$. This implies that

$$\Delta S_j \Delta S_k \geq |\langle S_l\rangle|. \tag{11.99}$$

The Stokes operators are identical to those appearing in the Schwinger representation of the angular momentum operators.

In order to define polarization squeezing, we rotate the coordinate system so that only one of the $\langle S_j\rangle$, for $j = 1, 2, 3$, is nonzero. Let us suppose for convenience that it is $\langle S_3\rangle$ that is nonzero. Then the only nontrivial uncertainty relation is $\Delta S_1 \Delta S_2 \geq |\langle S_3\rangle|$. If either $(\Delta S_1)^2$ or $(\Delta S_2)^2$ is less than $|\langle S_3\rangle|$, then the state is polarization squeezed. We can make this more general. If we define the operators

$$S(\theta) = S_1 \cos\theta + S_2 \sin\theta,$$
$$S_\perp(\theta) = S(\theta + \pi/2) = -S_1 \sin\theta + S_2 \cos\theta, \tag{11.100}$$

we find that $[S(\theta), S_\perp(\theta)] = iS_3$, so that $\Delta S(\theta)\Delta S_\perp(\theta) \geq |\langle S_3\rangle|$. Therefore, if there exists a θ such that $(\Delta S(\theta))^2 < |\langle S_3\rangle|$, then the state is polarization squeezed.

Next, by looking at a simple model, let us see how a fiber with a nonlinear index of refraction can produce polarization squeezing. Suppose the interaction-picture Hamiltonian is given by

$$H_I = \hbar g[(a_x^\dagger a_x)^2 + (a_y^\dagger a_y)^2], \tag{11.101}$$

and our initial state is a product of two coherent states, one for the x polarization and one for the y polarization, $|\psi\rangle = |\alpha/\sqrt{2}\rangle_x |i\alpha/\sqrt{2}\rangle_y$. We can calculate all of the expectation values that we need in order to check whether, at a time t, the state has become polarization squeezed. We first note that $[S_1, H] = 0$, so that $\langle S_1\rangle$ and $\langle S_1^2\rangle$ do not change with time. In fact, we have that $\langle S_1\rangle = 0$ and $\langle S_1^2\rangle = |\alpha|^2$. In order to calculate the other expectation values, the relations

$$e^{itH_I/\hbar} a_j e^{-itH_I/\hbar} = e^{-igt(2n_j+1)} a_j = a_j e^{-igt(2n_j-1)}, \tag{11.102}$$

where $n_j = a_j^\dagger a_j$ and $j = x, y$, are useful. A second useful relation is, for a coherent state β,

$$\langle\beta| e^{-2igtn} |\beta\rangle = \exp[|\beta|^2(e^{-2igt} - 1)]. \tag{11.103}$$

Using these relations we find that

$$\langle S_2\rangle = 0, \qquad \langle S_3\rangle = |\alpha|^2 e^{|\alpha|^2(\cos\tau - 1)}, \tag{11.104}$$

where we have set $\tau = gt$, and

$$\begin{aligned}\langle S_2^2\rangle &= |\alpha|^2 + \tfrac{1}{2}|\alpha|^2(1 - e^{|\alpha|^2(\cos 4\tau - 1)}),\\ \langle S_1S_2 + S_2S_1\rangle &= -2|\alpha|^4 \sin(2\tau) e^{|\alpha|^2(\cos 2\tau - 1)}.\end{aligned} \tag{11.105}$$

This yields

$$\begin{aligned}(\Delta S(\theta))^2 = |\alpha|^2 + |\alpha|^4[\tfrac{1}{2}(1 - e^{|\alpha|^2(\cos 4\tau - 1)})\sin^2\theta \\ - 2\sin(2\tau)e^{|\alpha|^2(\cos 2\tau - 1)}\sin\theta\cos\theta].\end{aligned} \tag{11.106}$$

Now, let us make some assumptions. In particular, we shall assume that $\tau|\alpha| \ll 1$, which implies that $\tau \ll 1$, and that $\tau|\alpha|^2 > 1$. Note that this implies that $\langle S_3\rangle = |\alpha|^2$, so that the state will exhibit polarization squeezing if the second term in the previous equation, the one proportional to $|\alpha|^4$, is negative. With the above assumptions, we have that

$$(\Delta S(\theta))^2 = |\alpha|^2 + 4|\alpha|^2[(\tau|\alpha|^2)^2\sin\theta - (\tau|\alpha|^2)\cos\theta\sin\theta]. \tag{11.107}$$

For a fixed value of $\tau|\alpha|^2$, we can minimize the above expression with respect to θ. We find $\tan(2\theta_{min}) = 1/(\tau|\alpha|^2)$, so that, setting $s = \tau|\alpha|^2$, we have

$$(\Delta S(\theta_{min}))^2 = |\alpha|^2[1 - 2s(s - \sqrt{1 + s^2})] < |\alpha|^2, \tag{11.108}$$

and the state is polarization squeezed. Polarization squeezing can be measured by rotating the polarization so that the squeezed Stokes parameter coincides with S_1, which can be measured by a simple number difference measurement.

The preceding treatment is highly simplified, and it completely ignores the propagation of the fields down the fiber. A much more sophisticated treatment including propagation was given by Corney *et al.* (see the additional reading). They defined two fields, $\psi_j(x,t)$ for $j = x, y$, one for each polarization. Each of these fields obeys a nonlinear Schrödinger equation and is subject to absorption and Raman noise. These equations are first turned into c-number equations by using either the Wigner or the positive P-function, and the resulting Fokker–Planck equations are used to find equivalent stochastic differential equations, as described in previous sections. These equations are then solved numerically. In the case of the Wigner function, terms in the Fokker–Planck equation involving derivatives of order higher than second, which result from the nonlinear terms, are dropped. In the same paper, results of experiments carried out at the Max Planck Institute for Optics, Information and Photonics were described. The agreement between theory and experiment was, as we shall see, excellent.

In order to describe the polarization squeezing in this setting, we first define the operators

$$N_{jj'}(t) = \int_0^L dx\, \psi_j^\dagger(x,t)\psi_{j'}(x,t), \tag{11.109}$$

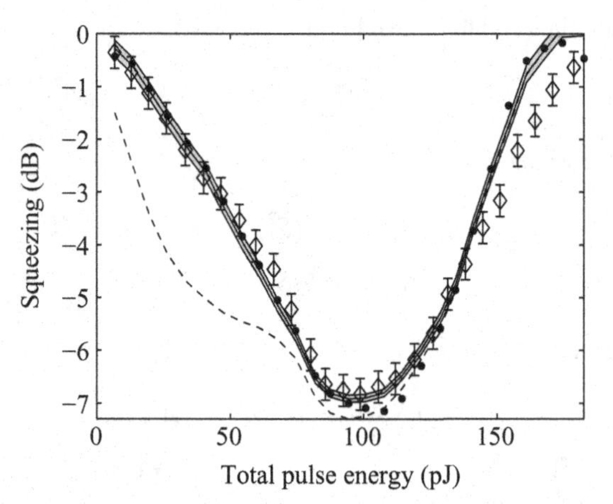

Fig. 11.2 Polarization squeezing as a function of pulse energy in a 31.2 m fiber. The diamonds are experimental results, and the theoretical results are given by the solid line. These include the effects of third-order linear dispersion, which arises from the rate of change of curvature of the dispersion. The theoretical prediction without this effect is given by the dotted line. Reproduced (with permission) from J. F. Corney *et al.*, *Phys. Rev.* A **78**, 023831 (2008).

where L is the length of the fiber, and we will choose t to be the time it takes the light to propagate down the fiber, which we shall denote by T. The Stokes operators are now

$$S_0 = N_{xx}(T) + N_{yy}(T), \quad S_1 = N_{xx}(T) - N_{yy}(T),$$
$$S_2 = N_{xy}(T) + N_{yx}(T), \quad S_3 = i[N_{yx}(T) - N_{xy}(t)]. \tag{11.110}$$

We define $S(\theta)$ as before, that is, $S(\theta) = S_1 \cos\theta + S_2 \sin\theta$. We express the squeezing in decibels, defined as

$$\text{squeezing (dB)} = \log_{10} \frac{(\Delta S(\theta))^2}{\langle S_0 \rangle}. \tag{11.111}$$

The angle θ is chosen to be the angle of optimum squeezing. As can be seen from Figure 11.2, significant polarization squeezing can be obtained, and the agreement between theory and experiment is very good. For the relatively high photon numbers in these experiments, both the approximate truncated Wigner and exact positive P methods gave identical polarization squeezing predictions.

Additional reading

Books

G. P. Agrawal, *Nonlinear Fiber Optics* (Academic Press, London, 2001).

Articles

- For a discussion of quantum and classical noise in waveguides, see:

P. D. Drummond and J. F. Corney, *J. Opt. Soc. Am.* B, **18**, 139–152 (2001).
S. H. Perlmutter, M. D. Levenson, and R. M. Shelby, *Phys. Rev.* B **42**, 5294–5305 (1990).
R. M. Shelby, M. D. Levenson, and P. W. Bayer, *Phys. Rev.* B **31**, 5244 (1985).

- For the first report of squeezing in quantum solitons, see:

M. Rosenbluh and R. M. Shelby, *Phys. Rev. Lett.* **66**, 153 (1991).

- Two papers on polarization squeezing are given below:

J. F. Corney, J. Heersink, R. Dong, V. Josse, P. D. Drummond, G. Leuchs, and U. L. Andersen, *Phys. Rev.* A **78**, 023831 (2008).
N. Korolkova, G. Leuchs, R. Loudon, T. C. Ralph, and C. Silberhorn, *Phys. Rev.* A **65**, 052306 (2002).

Our discussion in this chapter was based on the first.

Problems

11.1 Consider a step-index fiber waveguide. Calculate the lowest-order transverse mode function **u** for propagation at a vacuum wavelength of 1.5 μm in a cylindrical fiber with core index of refraction 1.55, a cladding index 1.50 and a core radius of 10 μm.

11.2 Prove Eq. (11.20), giving the propagated power in a waveguide.

11.3 Show that, if a state satisfies $(\Delta S_2)^2 < \langle S_0 \rangle$, it is nonclassical.

12 Quantum information

Quantum information, which studies the representation of information by quantum mechanical systems and the type of information processing this makes possible, is a relatively new field. Its roots, however, go back to early discussions of the interpretation of the quantum mechanical formalism. In 1935 Einstein, Podolsky and Rosen suggested that there were interpretational issues with quantum mechanics, having to do with local realism. Einstein was puzzled by an apparent lack of locality in quantum mechanical descriptions of reality, and suggested that quantum mechanics was not a complete theory. This stimulated a subsequent series of theoretical and experimental investigations. In particular, John Bell realized that Einstein's proposal for a 'completion' of quantum mechanics by the addition of more variables – called 'hidden' variables – was not consistent with quantum predictions, and could not be carried out.

While these interpretational issues set the stage for what followed, the modern field of quantum information arose in the 1980s stimulated by the ever decreasing size of the elements in information processing circuits. It was realized that at some point a threshold would be crossed and the devices would start to exhibit quantum mechanical behavior. It was first determined that, if they did, it would not present a problem, and, in addition, it could even be advantageous. This led to the idea of a quantum computer that stores information in quantum states, or 'qubits', rather than as binary digits, or 'bits', in a standard digital computer.

The field took off when, in 1994, Peter Shor found an efficient quantum algorithm for finding the prime factors of an integer. The best-known classical algorithms for accomplishing this require a number of steps that is exponential in the size (number of digits) of the integer, while Shor's algorithm requires a number of steps proportional to the cube of the size. This showed that, if we could build a quantum computer, it could do something useful – perhaps much faster than could a standard, classical computer. Since then, considerable progress has been made in understanding the information processing and transmission properties of quantum systems.

While the basic unit of classical information is a bit, which can be either 0 or 1, the basic unit of quantum information is the qubit, which is a two-level system, one level, denoted by $|0\rangle$, representing 0, and the other level, denoted by $|1\rangle$, representing 1. The key point about a qubit is that it can be in a superposition of the states $|0\rangle$ and $|1\rangle$. This is one of the things that makes quantum information processing different from classical information processing. Quantum information can, however, also be stored by multi-level systems or even in something like a field mode, which has continuous degrees of freedom. Nonlinear optical systems, in particular the parametric amplifier, can be useful at both the qubit and continuous-variable level.

In this chapter, we shall look at the original paradoxes of Einstein and Bell, as well as two modern aspects of quantum information. We will see how nonlinear quantum optics, in particular the parametric amplifier, can be applied to each. The first application is the copying, or cloning, of information. While this is something that is trivial on the classical level, things are very different when considering quantum systems. The second is teleportation, which involves transporting a quantum state from one system to another.

12.1 The Einstein–Podolsky–Rosen paradox

In 1935, Einstein, Podolsky and Rosen (EPR) published a paper claiming that quantum mechanics is incomplete, i.e. that there are aspects of reality that it does not include. This was the first work to emphasize the strange properties of entangled states, although it was Schrödinger who coined the term. They considered an ideal *gedanken* experiment. Nonlinear optics provides a way to turn their idealized experiment into a real one.

Einstein, Podolsky and Rosen started by considering a two-particle entangled state proportional to $\delta(x_a - x_b - x_0)$, where x_a is the position of particle A and x_b is the position of particle B. This state is, of course, not normalizable, but it can be well approximated by a normalizable state. In this state, the position of each particle is undetermined – it can be anywhere – but the difference in the positions of the particles is definitely x_0. The momentum-space state is proportional to $\delta(p_a + p_b)$, where p_a and p_b are the momenta of particles A and B, respectively. Note that this implies that, while the individual particle momenta are undetermined, we definitely have $p_a = -p_b$.

The next step in their argument was to state two postulates, which serve to define what they call an 'element of reality'.

1. 'If, without disturbing a system, we can predict with certainty the value of a physical quantity,' then 'there exists an element of physical reality corresponding to this physical quantity.' Note that this assumes that the specified quantity has a value before it is measured, that is, this statement assumes what is sometimes called *realism*.
2. If systems A and B are space-like separated, then a measurement at B will not disturb system A. This is what is generally called *locality*.

Consider our two-particle state, and assume that we perform measurements at space-like separated locations. By measuring the position of particle B, one can determine the position of particle A, since $x_a = x_b + x_0$. This will, according to the second postulate, not disturb system A, so the position of particle A must be an element of reality. One could also measure the momentum of particle B, which will tell us the momentum of particle A. Therefore, the momentum of particle A is also an element of reality.

The fact that both x_a and p_a are elements of reality is in disagreement with the fact that, according to quantum mechanics, both of these quantities cannot have well-defined values – one or the other can, but *not* both. Therefore, since these quantities are both elements of reality, but quantum mechanics does not allow them both to have well-defined values, quantum mechanics cannot be complete, i.e. there are aspects of reality that quantum mechanics cannot explain. Einstein therefore proposed that additional elements of reality – sometimes called 'hidden variables' – would be required for a complete physical theory.

12.1.1 EPR paradox in real systems

However, we note that there is problem of a practical nature with the above argument. The quantum state $\delta(x_a - x_b - x_0)$ is infinitely delocalized, and cannot be realistically prepared. Does the argument still hold in more realistic wavefunctions? In addition, it has proved difficult to prepare even approximate versions of this state for mechanical particles. This suggests that, in principle, one way that nature might avoid the paradox would be if the state proposed by Einstein, Podolsky and Rosen was not physically realizable. In other words, there could be a kind of super-selection rule or 'cosmic censor' that prevents these paradoxical states from ever occurring in the first place. Then the paradox would be easily resolved, since quantum mechanics is not obliged to describe elements of reality that never occur in practice. For this reason, it is fundamentally important to have an experimental route toward generating and observing the EPR paradox.

The EPR argument can also be used as a starting point to study quantum correlations in an entangled state. In order to understand how to apply the argument to real experiments, we have to sharpen the concepts that Einstein, Podolsky and Rosen used. We need to apply this logic to actual quantum states, which can violate a measurable inequality, instead of the unnormalizable states that we have been using so far. One method to do this was proposed by Reid, whose theoretical work obtained the first measurable signatures for the continuous-variable EPR paradox.

Let us begin by considering, as we have so far, a two-particle state. Suppose that we measure x_b and use the result to infer a result for x_a, which we shall call $x_a^{(est)}(x_b)$, the estimator for x_a given the measurement result x_b. We can define a conditional uncertainty for a subsequent measurement of x_a by

$$\Delta_{inf}^2(x_a \mid x_b) = \int_{-\infty}^{\infty} dx_a \, [x_a - x_a^{(est)}(x_b)]^2 P(x_a \mid x_b), \tag{12.1}$$

where $P(x_a \mid x_b)$ is the conditional probability for obtaining a measurement result x_a for particle A given that a result x_b was obtained for particle B. We can obtain a full uncertainty by averaging this result over x_b,

$$\begin{aligned}\Delta_{inf}^2(x_a) &= \int_{-\infty}^{\infty} dx_b \, \Delta_{inf}^2(x_a \mid x_b) P(x_b) \\ &= \int_{-\infty}^{\infty} dx_a \int_{-\infty}^{\infty} dx_b \, [x_a - x_a^{(est)}(x_b)]^2 P(x_a, x_b),\end{aligned} \tag{12.2}$$

where $P(x_a, x_b) = P(x_a \mid x_b)P(x_b)$ is the joint distribution for measurement results of x_a and x_b. Similarly, for the momenta we have

$$\Delta_{inf}^2(p_a) = \int_{-\infty}^{\infty} dp_a \int_{-\infty}^{\infty} dp_b \, [p_a - p_a^{(est)}(p_b)]^2 P(p_a, p_b), \tag{12.3}$$

where $P(p_a, p_b)$ is the joint distribution for obtaining momentum measurement results of p_a and p_b.

Now, for a single-particle state, the standard deviations of x and p, $\Delta(x)$ and $\Delta(p)$, respectively, satisfy $\Delta(x)\Delta(p) \geq \hbar/2$. An inferred estimate made from measuring x_b should not disturb x_a, which is at a distant location. If we now apply the assumption that elements of reality exist whenever we can measure something without disturbing it, we see that the inferred estimate is an element of reality.

But for quantum mechanics to be complete *and* to satisfy local realism, Reid has pointed out that these estimates *must* satisfy the Heisenberg uncertainty principle, since there should be a local wavefunction description of the measurements at a, i.e. we have an inferred uncertainty principle

$$\Delta_{inf}(x_a)\Delta_{inf}(p_a) \geq \frac{\hbar}{2}. \tag{12.4}$$

If this condition is violated, we will say that the state exhibits EPR correlations. That is, by performing measurements on particle B, one can infer values of the position and momentum of particle A whose accuracies are better than allowed for a single-particle state defined locally at A. In addition, violation of this condition guarantees that the state is entangled (this requires some work to show, but we will not give the proof here). However, in order to reach Einstein's ultimate conclusion that quantum mechanics is incomplete, we must accept the local realism postulate.

It is important to recall that the assumption of local realism was a common belief of most scientists of this time. As we will see in the next section, this is not compatible with what modern experiments tell us. This does not diminish the value of the proposal of Einstein, Podolsky and Rosen. However, we must re-interpret their argument. Logically, it does not tell us that quantum mechanics is incomplete, but rather it says that the two assumptions of local realism and completeness of quantum mechanics are *inconsistent* with each other. We cannot hold both viewpoints simultaneously.

We will treat this in more detail in the next section, where we will see that the modern resolution of the EPR paradox is to reject local realism, rather than to suggest that quantum mechanics is incomplete. Nevertheless, a demonstration of the EPR paradox is an important step toward understanding the nature of reality in the quantum world. It also has applications in areas like quantum cryptography, where a violation of an inferred uncertainty principle can be used for security proofs.

12.1.2 EPR correlations in two-mode squeezing

To actually obtain these strongly correlated states is another matter still. How does one create the required correlations? The first practical route, proposed by Reid and Drummond, used the technology of entangled electromagnetic field quadratures obtained using nonlinear quantum optics. This technique was not known in Einstein's time.

Let us consider generating the required correlations with a two-mode squeezed state, $S(r)|0\rangle$, where S is the two-mode squeezing operator and $r > 0$. We will consider field quadratures instead of position and momentum. This state can be produced, as we have noted, by parametric down-conversion. The quadrature operators for mode a, considered to be localized at position A, are X_a and Y_a, and the operators for mode b, localized at position B, are X_b and Y_b. Now, from Eq. (5.72), this state has the property that

$$\langle (X_a + X_b)^2 \rangle = 2e^{-2r}, \\ \langle (Y_a - Y_b)^2 \rangle = 2e^{-2r}. \tag{12.5}$$

This implies that, in the large-squeezing limit, a measurement of X_b yielding a value of x_b would imply that a measurement of X_a would yield approximately $-x_b$. Of course, a measurement of Y_b yielding y_b would similarly imply that a measurement of Y_a would yield a value of approximately y_b. Therefore, let us set

$$x_a^{(est)}(x_b) = -x_b, \qquad y_a^{(est)}(y_b) = y_b. \tag{12.6}$$

Doing so, we find that

$$\Delta_{inf}^2(x_a) = \int_{-\infty}^{\infty} dx_a \int_{-\infty}^{\infty} dx_b\, (x_a + x_b)^2 P(x_a, x_b) \\ = \langle (X_a + X_b)^2 \rangle = 2e^{-2r} \tag{12.7}$$

and

$$\Delta_{inf}^2(y_a) = \int_{-\infty}^{\infty} dy_a \int_{-\infty}^{\infty} dy_b\, (y_a - y_b)^2 P(y_a, y_b) \\ = \langle (Y_a - Y_b)^2 \rangle = 2e^{-2r}. \tag{12.8}$$

Now, because for a single-mode state $\Delta(X_a)\Delta(Y_a) \geq 1$, then, if

$$\Delta_{inf}(x_a)\Delta_{inf}(y_a) < 1, \tag{12.9}$$

we can conclude that the state exhibits EPR correlations and is entangled. This will clearly be the case if r is sufficiently large, in particular, if $e^{-2r} < 1/2$. The first experiment along these lines was carried out by Ou *et al.* They used an intra-cavity nondegenerate down-conversion scheme to generate the state and homodyne detection to measure the output quadratures. They were able to demonstrate EPR correlations. We note that this is a simplified version of the inference argument, and more sophisticated inference methods are described in the review by Reid *et al.* in the additional reading.

Although this is not as strong a demonstration of a quantum paradox as the Bell inequality, an advantage of this method is that quadrature detection using local oscillators is very efficient (close to 100%). It therefore does not have the efficiency loopholes that Bell tests of local realism often have, as we see in the next section. Owing to space restrictions in their optical table, the Ou *et al.* experiment was not able to reach full causal separation of the measurements at the A and B locations. This necessary improvement seems technically feasible.

12.2 Bell inequality

Nonlinear quantum optics, in particular, parametric down-conversion, gives us ways to study issues in the foundations of quantum mechanics that are not readily available using other technologies. As we have seen, one of these concerns the possibility that a class of classical probabilistic theories, known as local hidden-variable theories, might explain all of the experimental results that can be explained by quantum mechanics. The original motivation behind studying this question was to see if quantum mechanics, with all of its unusual features, is really necessary or whether something less drastic, in particular, a classical probabilistic theory, could produce the same results. The proposal of Einstein, Podolsky and Rosen suggested that this was needed to 'complete' quantum mechanics.

Bell's ground-breaking theoretical work proved that no local hidden-variable theory can duplicate all of the predictions of quantum mechanics. In other words, there is no consistent way to carry out Einstein's proposal of completion. This implies that experiments can be done in the regimes where hidden-variable theories must give different results from quantum mechanics, in order to test which is correct.

Ground-breaking experiments have been carried out by groups led by Clauser, Aspect and Zeilinger to test these quantum predictions, which have successfully closed off various objections to this work. Quantum mechanics has always been vindicated in these tests, although there are some remaining loopholes due to experimental inefficiencies. While these ever-shrinking loopholes could allow exotic hidden-variable theories to replicate quantum predictions, this is becoming increasingly unlikely as experiments improve.

One might wonder why tests of quantum mechanics are necessary, given the tremendous successes of quantum mechanics in the last 100 years. There are a number of reasons. First, it is always a good idea to check one's assumptions, especially in dealing with something as unusual as quantum mechanics. Quantum mechanics is highly counter-intuitive, and the Bell inequalities show that there is no classical theory that can produce the same results. Experimental Bell inequality violations show that these counter-intuitive features are really necessary.

Second, situations in which the Bell inequalities are violated are examples of true quantum behavior, and it is interesting to probe this quantum regime. For example, the violation of Bell inequalities requires entanglement, so Bell inequalities can be used to detect entanglement.

Finally, there are even practical reasons. Just as with EPR tests, there are quantum cryptography schemes that use Bell inequalities to detect the presence of an eavesdropper, and these security tests are the strongest known.

12.2.1 Local hidden-variable theories

In general terms, a Bell inequality is a constraint on the outcomes of measurements on a physical system that obeys a local hidden-variable (LHV) theory. This is a theory that has the property of local realism, as defined by Einstein; that is, properties that are well defined before a measurement at a causally separated location have objective reality.

There are a number of ways to formulate mathematically the conditions imposed by a local hidden-variable theory. In the case of particles emitted from a common source, the measurements of M spatially separated observers (say, Alice and Bob for the bipartite case) are obtained by taking random samples of a common parameter λ, to which the observers do not have access. This is the hidden variable. Measured values are then functions of some local detector settings and the hidden parameter λ, which could be (for example) a multicomponent real or complex variable. The value observed by Alice with detector setting a is $A(a, \lambda)$; for Bob, with detector setting b, it is $B(b, \lambda)$; and similarly for other observers. Mathematically, the correlations in a hidden-variable theory are obtained from a probabilistic calculation of the form, for the two-party case (if λ is discrete, the integral will be replaced by a sum),

$$\langle A(a)B(b)\rangle = \int d\lambda \, P(\lambda)A(a, \lambda)B(b, \lambda). \tag{12.10}$$

As shown by Fine (see the additional reading), a completely equivalent condition is that there is a joint distribution for all of the observables. Quantum mechanically, this will, in general, not be the case, because not all of the operators corresponding to different detector settings at a given location will commute. We will be making use of this second formulation.

To begin, we will make use of a slightly less formal description of local hidden-variable theories. In theories of this type, we can think of particles emitted from a common source as carrying instruction sets that specify the results of possible measurements. For example, a spin-1/2 particle on which we are going to measure one of two spin components, say S_x or S_y, might carry an instruction set of the following form:

1. If S_x is measured, the result will be $\hbar/2$.
2. If S_y is measured, the result will be $-\hbar/2$.

In this type of theory, unlike in quantum mechanics, observables have values before they are measured. In philosophy, this is called realism. The description can be probabilistic if we allow the particle to have different instruction sets with different probabilities. In addition, for sets consisting of more than one particle, the instructions for one particle should not depend on what is measured on another particle.

For example, for two spin-1/2 particles, which we shall call A and B, the instruction set for each particle can list what will happen for each particle, that is, it will be a list of four numbers specifying what the results will be for S_x and S_y for particle A and for S_x and S_y for particle B. However, instructions of the form 'if S_x is measured on particle A, the result will be $\hbar/2$ if S_x is measured on particle B, but will be $-\hbar/2$ if S_y is measured on particle B' are not allowed, as they would violate the causality and local realism postulates of an LHV theory.

The property of demanding that the instructions for particle A do not depend on what is measured on particle B is known as locality. It prevents instantaneous signaling. If Bob, holding particle B, can influence the measurement results on particle A, held by Alice, just by changing what he decides to measure, he could instantly send her messages, and that would violate special relativity. Hence, we require our hidden-variable theories to be local. Because the hidden-variable theory we are considering here satisfies both realism and locality, theories of this type are known as local realistic theories.

We see that, in a local realistic theory, we have instruction sets that list possible measurement outcomes, and each of these instruction sets can occur with some probability. That means what we have is a joint probability distribution for the variables being measured. For our two spin-1/2 particles, an LHV could have a joint distribution of the form

$P(S_x^{(A)}, S_y^{(A)}, S_x^{(B)}, S_y^{(B)})$, where each of the variables can take one of two values, $\pm\hbar/2$. That is, λ is a vector of four integer values, and the integration over λ is a summation. Therefore, a local realistic theory is one that can be described by a joint probability distribution. Quantum mechanically, no such joint distribution exists, because S_x and S_y do not commute.

12.2.2 Derivation of the Bell inequality

Let us now see what this implies. Suppose we have a two-particle system, and we shall again label the particles as A and B. On particle A we can measure A_1 or A_2, and on particle B we can measure B_1 or B_2. We shall assume that each of these observables has only two possible values, and for convenience we shall take them to be ± 1. If this system is described by a local realistic theory, we have a joint probability distribution $P(A_1, A_2, B_1, B_2)$. We now have that

$$
\begin{aligned}
P(A_2 = 1, B_1 = 1) - P(A_1 = 1, A_2 = 1, B_1 = 1, B_2 = 1) \\
\leq P(A_1 = -1, B_1 = 1) + P(A_2 = 1, B_2 = -1), \\
P(A_1 = 1, B_2 = 1) - P(A_1 = 1, A_2 = 1, B_1 = 1, B_2 = 1) \geq 0.
\end{aligned}
\tag{12.11}
$$

The first of these inequalities can be verified simply by expressing both sides in terms of the joint probability distribution. Subtracting the second inequality from the first, we get

$$
\begin{aligned}
P(A_2 = 1, B_1 = 1) - P(A_1 = 1, B_2 = 1) \\
\leq P(A_1 = -1, B_1 = 1) + P(A_2 = 1, B_2 = -1).
\end{aligned}
\tag{12.12}
$$

Noting that

$$
\begin{aligned}
P(A_1 = -1, B_1 = 1) = P(B_1 = 1) - P(A_1 = 1, B_1 = 1), \\
P(A_2 = 1, B_2 = -1) = P(A_2 = 1) - P(A_2 = 1, B_2 = 1),
\end{aligned}
\tag{12.13}
$$

we have that, for a local hidden-variable theory,

$$
\begin{aligned}
P(A_2 = 1, B_1 = 1) - P(A_1 = 1, B_2 = 1) \\
+P(A_1 = 1, B_1 = 1) + P(A_2 = 1, B_2 = 1) \\
\leq P(A_2 = 1) + P(B_1 = 1).
\end{aligned}
\tag{12.14}
$$

This is known as the Clauser–Horne (CH) version of the Bell inequality, and, as we shall see, it can be violated by quantum mechanics. This is a clear demonstration that quantum mechanics is not equivalent to *any* possible local hidden-variable theory.

12.2.3 Experimental demonstrations using down-conversion

Our next step is to consider a situation involving photons produced by parametric down-conversion that will lead to a violation of the CH inequality. The earliest route to this type of experiment was using an atomic cascade. Recent experiments use the method proposed by Reid and Walls, of nondegenerate parametric down-conversion, with two spatial modes a and b each having two possible polarizations H and V. The experiments use a pulsed pump beam focused on a small nonlinear crystal, so that the production of output photons is localized in time and space. An important difference between these and the EPR experiments described above is the use of a much smaller squeezing parameter, so that the photon number per mode is much lower.

This process is described in the simplest case by the effective Hamiltonian explained earlier in Chapter 5:

$$H = i\kappa[a_V^\dagger b_H^\dagger - a_H^\dagger b_V^\dagger + h.c.]. \tag{12.15}$$

Here H and V indicate photons with horizontal and vertical polarization, respectively, in the emitted photon. The corresponding mode functions are localized in space, rather than the monochromatic plane-wave modes treated previously. The experiments are designed so that the two spatial modes propagate in different output directions, and so can be coupled, for example, into different optical fibers.

This generates a correlated squeezed state, with the generic form for $\kappa t \ll 1$ of

$$\begin{aligned} |\Psi'\rangle &= \exp(-iHt)|0\rangle \\ &= |0\rangle + \kappa t(|V\rangle_a|H\rangle_b - |H\rangle_a|V\rangle_b) + O(\kappa t)^2. \end{aligned} \tag{12.16}$$

Often the resulting quantum state is written in a projected form,

$$|\Psi\rangle = \frac{1}{\sqrt{2}}(a_V^\dagger b_H^\dagger - a_H^\dagger b_V^\dagger)|0\rangle. \tag{12.17}$$

Here we have considered the action of the Hamiltonian only to first order in the interaction, and we have dropped the vacuum term. As mentioned above, the first step implies a low-intensity pump with low conversion efficiency, so as to operate in a regime with few photons. This last step is justified because we will only consider events in which at least one photon is detected. This projection turns out not to affect the Bell inequality. Other efficiency issues will be treated in more detail below.

We now define rotated polarization operators for the a and b modes:

$$\begin{aligned} c_+ &= a_H\cos\theta + a_V\sin\theta, & d_+ &= b_H\cos\phi + b_V\sin\phi, \\ c_{\{-\}} &= -a_H\sin\theta + a_V\cos\theta, & d_{\{-\}} &= -b_H\sin\phi + b_V\cos\phi. \end{aligned} \tag{12.18}$$

We would like to calculate two probabilities, $P_+(\theta)$, the probability of measuring the a-mode photon to be in the state $c_+^\dagger|0\rangle$, and $P_{++}(\theta,\phi)$, the probability of measuring the two photons to be in the state $c_+^\dagger d_+^\dagger|0\rangle$. We find that

$$\begin{aligned} P_+(\theta) &= \tfrac{1}{2}, \\ P_{++}(\theta,\phi) &= \tfrac{1}{2}\cos^2(\theta+\phi). \end{aligned} \tag{12.19}$$

We also find that $P_+(\phi)$, the probability of measuring the b-mode photon to be in the state $d_+^\dagger|0\rangle$, is equal to 1/2. Now let us consider two values of θ, which we shall denote by θ and θ', and two values of ϕ, which we shall denote by ϕ and ϕ'. We shall identify A_1 with θ, A_2 with θ', B_1 with ϕ, and B_2 with ϕ'. We will also identify the plus states, $c_+^\dagger|0\rangle$ and $d_+^\dagger|0\rangle$, with the corresponding observable taking the value $+1$, and the minus states, $c_-^\dagger|0\rangle$ and $d_-^\dagger|0\rangle$, with the corresponding observable taking the value -1. With these identifications, the CH inequality can be expressed as

$$P_{++}(\theta',\phi) - P_{++}(\theta,\phi') + P_{++}(\theta,\phi) + P(\theta',\phi') \leq P_+(\theta') + P_+(\phi). \quad (12.20)$$

Let us now assume that $\theta, \theta' > 0$ and $\phi, \phi' < 0$. Furthermore, let us choose $|\phi| - \theta = \theta' - |\phi| = |\phi'| - \theta' = \alpha > 0$ and $|\phi'| - \theta = 3\alpha$. Note that this can be accomplished by choosing $\theta = 0$, $|\phi| = \alpha$, $\theta' = 2\alpha$ and $|\phi'| = 3\alpha$. Then the CH inequality then becomes

$$3\cos^2\alpha - \cos^2(3\alpha) \leq 2. \quad (12.21)$$

It is easy to see that this equality can be violated. For example, the choice $\alpha = \pi/6$ yields the value 9/4 for the left-hand side, which is clearly greater than 2. Therefore, we see that the two-photon state produced by parametric down-conversion can violate the CH inequality, thereby demonstrating that this state cannot be described by a local hidden-variable theory.

To demonstrate this experimentally, the spatial separation of the detection events must be great enough so that the choice of polarization direction is made after the generation of the photons. This has been achieved in experiment, giving results in good agreement with the predictions of quantum theory. While the first Bell experiment was carried out by Clauser, the requirement of causal separation was later realized by Aspect.

Another requirement is that detection must be very efficient. If detector inefficiency is taken into account, it turns out that the reduced correlation caused by random loss of one of the two photons means that a hidden-variable theory explanation is possible, for efficiencies less than 83%. Note that the random loss of both the photons – as occurs in the projection of the initial two-photon state – does not cause problems.

As of the time of writing, there are no experiments yet that eliminate both these 'loopholes'. Ion-trap experiments have demonstrated high efficiency, but not causal separation; while photonic experiments demonstrate causal separation, but with efficiency loopholes. It is essential to combine both causal separation and high efficiency to violate the Bell inequality. While it is presumably just a technical issue, current experiments cannot yet rule out all hidden-variable theories.

12.3 Schrödinger cat paradoxes

It is possible to greatly extend the Bell theorem, into a mesoscopic domain. This regime is of increasing importance, as demonstrated by the award of the 2012 Nobel Prize in Physics to D. J. Wineland and S. Haroche. In these experiments, we are moving into territory where quantum theory is poorly tested at the level of quantum superpositions. As asked by Schrödinger, is it possible that ridiculous objects – like cats that are neither dead nor alive – can actually exist in the real world? Here we focus on the essential question that underlies such investigations, namely: 'What can we measure for a mesoscopic quantum system that uniquely demonstrates its quantum character?'

To understand this, we turn to multipartite measurement strategies originally introduced by Greenberger, Horne and Zeilinger (GHZ), and subsequently extended to the mesoscopic domain by Mermin. We start by considering a set of measurements X_j, Y_j at each of $j = 1, \ldots, M$ distinct measurement locations. We suppose that all the measurements are causally separated – that is, the measurement events all have a space-like separation. We require that all possible results are bounded so that $|X_j| \leq 1$ and $|Y_j| \leq 1$.

Typically, this is applied to experiments with binary measurement outcomes. For example, a vertical polarization could be labeled $+1$, and a horizontal polarization labeled -1. Such outcomes are easily obtained, either in experiments that count photons, or in ion-trap measurements where the label is ± 1 depending on spin, or whether an atomic level is occupied or not. Including cases where $|X_j| < 1$ allows for inefficient detectors that can return a value of 0 to indicate that no particle was detected. This usefully generalizes Bell's original idea.

The original proposal of Greenberger, Horne and Zeilinger was for an 'all-or-nothing' contradiction, in which quantum mechanics could be shown to predict different outcomes from an LHV theory on the basis of a single measurement of three correlated spins. Instead, we will treat a mesoscopic case, in which an arbitrarily large number of measurements at different sites are combined together.

12.3.1 Local hidden-variable prediction

Consider the complex correlation function

$$Z = \prod_j Z_j, \tag{12.22}$$

defined so that

$$Z_j = \frac{1+i}{2}[X_j + iY_j]. \tag{12.23}$$

Although this is not a quantum observable – it involves two incompatible observations – in a hidden-variable theory it would have a well-defined value prior to any observation, since each individual operator can be measured with an appropriate setting. The extremal values attained by this observable are

$$Z_j = \pm 1, \pm i, \tag{12.24}$$

and hence all possible values lie within a square on the complex plane.

Next, we take an average over products of the complex variables, at all possible observation points. Such an average experimentally requires repeated measurements with all possible settings of the observables. Nevertheless, in a hidden-variable theory, this result is obtained from a single ensemble average,

$$\langle Z \rangle = \int d\lambda \, P(\lambda) \prod_{j=1}^{n} Z_j(\lambda), \tag{12.25}$$

where $P(\lambda)$ is a probability that a given hidden variable λ exists, for which the corresponding outcomes are $X_j(\lambda)$ or $Y_j(\lambda)$ depending on local measurement settings at location j. As stated earlier, each measurement takes place with space-like separations from each other, so the quantity Z corresponds to a Bell-like correlation.

Within a local realistic hidden-variable theory one has the constraint that each $Z_j(\lambda)$ lies within a square on the complex plane. It is easily verified that the product of any number of $Z_j(\lambda)$ also is located within this square. The same is true for the ensemble averages, due to convexity properties. Next, we notice that there is a maximum distance from the origin to the boundary of the possible values of $\langle Z + Z^\dagger \rangle/2$. Hence one obtains the inequality in an LHV of

$$|\Re(\langle Z \rangle)| \leq 1. \tag{12.26}$$

12.3.2 Quantum Schrödinger cat prediction

This can be violated by an exponentially large amount in quantum mechanics in the limit of large n. Interestingly, this is a counter-example to the usual correspondence principle, where one traditionally expects quantum mechanics to reduce to classical mechanics for large particle number. However, the quantum state required is a very unusual one, as we will see. It is, in fact, a superposition of a mesoscopic quantum system with all spins up and all spins down – a perfect analogy with the original idea of a cat that is dead and alive simultaneously.

To prove the exponential violation, suppose the corresponding quantum observations are of spin operators represented by the Pauli spin matrices. Hence, at the operator level, $X_j = \sigma_j^x$ and $Y_j = \sigma_j^y$, then

$$Z = [1 + i]^n \prod_{j=1}^{n} \frac{1}{2}[\sigma_j^x + i\sigma_j^y]. \tag{12.27}$$

This complex combination of Pauli spin matrices has the structure

$$\frac{1}{2}[\sigma_j^x + i\sigma_j^y] = \begin{bmatrix} 0 & 1 \\ 0 & 0 \end{bmatrix}, \tag{12.28}$$

so that the only nonzero matrix elements are those that cause a transition from the -1 eigenstate to the $+1$ eigenstate. Since the quantum operator corresponding to Z is a product of all such matrices, the only nonzero matrix elements describe an all-or-nothing transition from an eigenstate of all -1 values to one with all $+1$ values. This can take place in a quantum system that is a superposition of these two distinct eigenstates. In fact, this is a perfect example of a mesoscopic quantum superposition, and can be generated (in principle) in ion-trap experiments, apart from the inevitable decoherence issues.

For the GHZ–Mermin quantum superposition states of the form

$$|\Phi_n\rangle = \frac{1}{\sqrt{2}}(|1, 1, 1, \ldots, 1\rangle + e^{i\phi}|-1, -1, -1, \ldots, -1\rangle), \tag{12.29}$$

one obtains an extremal value of

$$\langle \Phi_n | Z | \Phi_n \rangle = \frac{1}{2}([1 + i]e^{i\phi})^n. \tag{12.30}$$

This is maximized if we choose $\phi = -\pi/4$, so that $e^{i\phi} = [1-i]/\sqrt{2}$. The result obtained is therefore

$$F = \frac{1}{2}\langle\Phi_n|Z + Z^\dagger|\Phi_n\rangle = 2^{n/2-1}, \tag{12.31}$$

which violates the hidden-variable bound for all observations with $n > 2$. The preparation and measurement of this GHZ–Mermin violation, with low enough decoherence and sufficiently large precision to definitively prove the existence of a Schrödinger cat for large n, is still an outstanding challenge in physics. In actually measuring Z, one must carry out 2^n measurements of all possible combinations of the real Pauli matrices. This is extremely difficult as n increases, given that the quantum superposition $|\Phi_n\rangle$ must be generated reproducibly.

12.4 Probabilistic simulations of Bell violations

Bell's theorem proves that probabilistic hidden-variable theories are *not* equivalent to quantum mechanics. Does this mean that it is impossible to simulate quantum mechanics using probabilistic methods? The formulation of hidden-variable theory measurements, as described in Eq. (12.10), looks formally identical to the procedure for obtaining an observable in a phase-space representation, as described in Eq. (7.99). In this section, we treat the question of violating Bell inequalities with probabilistic simulations.

12.4.1 LHV and phase-space methods

We have shown in earlier chapters that these phase-space representations can replicate all the predictions of quantum mechanics. Why is this possible, in view of Bell's theorem? Why are not positive phase-space representations hidden-variable theories? At first sight, phase-space simulation of Bell violations would disprove Bell's theorem and the GHZ–Mermin inequalities described above. Alternatively, since these theorems are rigorous, one may wonder if phase-space methods are applicable to Bell violations. However, there is actually no contradiction.

This is because of an important physical distinction. In phase-space representations, the phase-space variables are *not* necessarily the actual measured values. The only correspondences that are required are on means and correlations. For example, complex rather than integer values are utilized in the positive P-representation, although this is used to calculate the results of photon counting experiments. Hence, the conditions required to derive the Bell inequality do not apply.

As an example, the positive P-representation maps bosonic quantum states into $4M$ real coordinates, $\boldsymbol{\alpha}, \boldsymbol{\beta} = \mathbf{p} + i\mathbf{x}, \mathbf{p}' + i\mathbf{x}'$, which is double the dimension of the corresponding classical phase space. This method leads to exact probabilistic mappings between quantum mechanics and a classical-like phase-space description, even for low occupation numbers.

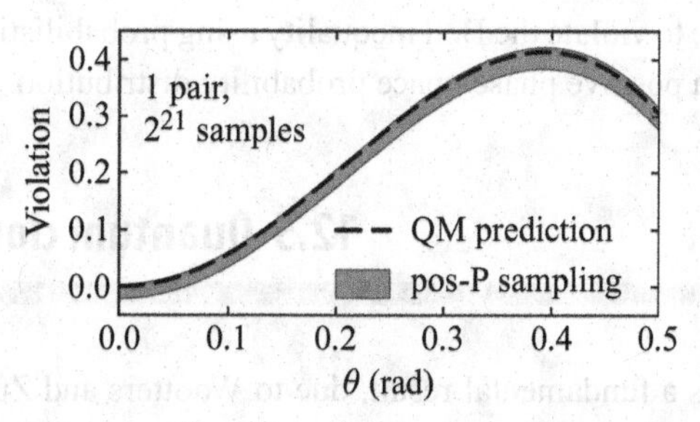

Fig. 12.1 Positive P simulation of a Bell violation.

However, these complex numbers are *not* observable quantities. Instead, they have correlations that are equivalent to certain correlations of the observables.

12.4.2 Phase-space probability for a Bell state

To demonstrate a probabilistic simulation of the bipartite Bell inequalities given above, we use the standard form P_+ of the positive P-representation, given in Eq. (7.102). In terms of this general distribution, the correlations of quantum counts $n_i = a_i^\dagger a_i$ at different locations is

$$\langle n_i \cdots n_j \rangle = \int \alpha_i \beta_i \cdots \alpha_j \beta_j P(\boldsymbol{\alpha}, \boldsymbol{\beta})\, d^{2M}\boldsymbol{\alpha}\, d^{2M}\boldsymbol{\beta}. \tag{12.32}$$

The effects of a polarizer are simply obtained on taking linear combinations of mode amplitudes, just as in classical theory. Here, the relevant operators will be labeled as $\mathbf{a} = (a_{1V}, a_{1H}, a_{2V}, a_{2H})$.

On setting $\lambda = (\boldsymbol{\alpha}, \boldsymbol{\beta})$, there is a remarkable similarity between the hidden-variable theory formula, Eq. (12.10), of Bell, and the positive P formula, Eq. (7.102), for quantum correlations. Yet the positive P theory is fully equivalent to quantum mechanics, and therefore can violate Bell's inequalities. The reason for the difference is due to the different quantities calculated in the correlations. The fundamental observables in Bell's case, of form $X(\lambda)$, are the observed integer-spin or photon count eigenvalues, i.e. $(0, 1, \ldots)$. The corresponding observables in the positive P case, of form $n(\boldsymbol{\alpha}, \boldsymbol{\beta})$, are complex numbers whose mean values and correlations correspond to observable means and correlations. This allows the positive P-distribution to be equivalent to quantum mechanics, even though it looks just like a local hidden-variable theory.

The four-mode state of Eq. (12.17) has the corresponding positive P-distribution, using Eq. (7.102):

$$P_+(\boldsymbol{\alpha}, \boldsymbol{\beta}) = \left\{ \frac{|\bar{\alpha}_{1V}\bar{\alpha}_{2H} - \bar{\alpha}_{1H}\bar{\alpha}_{2V}|^2}{2(2\pi)^8} \right\} e^{-(|\boldsymbol{\alpha}|^2 + |\boldsymbol{\beta}|^2)/2}, \tag{12.33}$$

where $\bar{\alpha}_i \equiv (\alpha_i + \beta_i^*)/2$ is the average coherent amplitude in each projector. This is directly equivalent to the original Bell state, is clearly positive and probabilistic, and violates the Bell inequality. An example of a probabilistic simulation is shown in Figure 12.1. It is similarly possible to violate the Bell inequality using probabilistic sampling of the Q-function, which is also a positive phase-space probability distribution.

12.5 Quantum cloning

There is a fundamental result, due to Wootters and Zurek, that quantum information cannot be cloned. What this means is that there is no device that will transform an input state $|\Psi\rangle \otimes |\Psi_0\rangle$, where $|\Psi\rangle$ is the state we want to copy and $|\Psi_0\rangle$ is a standard state, analogous to a blank piece of paper in a copy machine, to the output state $|\Psi\rangle \otimes |\Psi\rangle$. This is known as the no-cloning theorem. Its proof is quite simple and is based on the linearity of quantum transformations. The idea is that, if $|\Psi\rangle \otimes |\Psi_0\rangle \to |\Psi\rangle \otimes |\Psi\rangle$ is to hold for all states $|\Psi\rangle$, it must hold for an orthonormal basis of the Hilbert space of which $|\Psi\rangle$ is a member. That is, if $|\Psi\rangle \in \mathcal{H}$ and $\{|e_n\rangle\}$ is an orthonormal basis of $\mathcal{H}$, then we must have $|e_n\rangle \otimes |\Psi_0\rangle \to |e_n\rangle \otimes |e_n\rangle$. This, plus the fact that the transformation is linear, completely determines its action, and it is straightforward to see that its action on a superposition of the basis vectors will not give us two copies of that superposition at the output.

12.5.1 Fidelity

While the no-cloning theorem forbids perfect cloning, it is possible to make approximate copies of quantum states. To understand the quality of the copy, we must introduce fidelity, which is the degree to which two quantum states can be called identical. Obviously, if two pure states are identical, their inner product is unity. This provides a motivation for the definition of fidelity for two pure states, $|\Psi\rangle$ and $|\Phi\rangle$, which is that

$$F = |\langle\Psi|\Phi\rangle|^2 \tag{12.34}$$

If one of the two states is not pure, then there is an obvious generalization. We can calculate the mean pure-state fidelity over the diagonal elements of the density matrix, once it is in diagonal form. This leads to the next definition:

$$F = \langle\Psi|\rho|\Psi\rangle. \tag{12.35}$$

If both are impure, then the fidelity measure is more complicated, and there is more than one proposal. These can be found in texts focusing on quantum information. Sometimes the square root of the above quantity is also called the fidelity.

For our purposes, we can measure how good the clones are by their fidelity to a pure initial state, using the second of the above measures. Suppose that the state we want to clone is $|\Psi\rangle$, and at the output we end up with two clones whose reduced density matrices (the approximate clones are generally entangled) are the same and given by ρ. The cloning fidelity is then given by $F = \langle\Psi|\rho|\Psi\rangle$. In the case of qubits, if all states are copied equally well, then the maximum fidelity has been shown to be $5/6$.

12.5.2 Quantum cloning using parametric amplification

Simon *et al.* showed how a parametric amplifier can be used as an approximate cloner, and we will now have a look at their scheme. Our parametric amplifier couples two modes, which we shall call a and b, but each of these modes has two polarization states, horizontal (H) and vertical (V). We therefore have four annihilation operators, a_H, a_V, b_H and b_V, and the corresponding creation operators. If the interaction is of the form

$$H' = \chi(a_V b_H - a_H b_V) + \chi^*(a_V^\dagger b_H^\dagger - a_H^\dagger b_V^\dagger), \tag{12.36}$$

then it is invariant under polarization rotations. In more detail, if

$$\begin{aligned} a_H &\to a_H \cos\theta + i a_V e^{i\phi} \sin\theta, \\ a_V &\to i a_H e^{-i\phi} \sin\theta + a_V \cos\theta, \end{aligned} \tag{12.37}$$

and similarly for b_H and b_V, we find that H_{int} does not change. This makes the performance of this device polarization-independent. If we have an input photon in the a mode in a particular polarization state, we just express H' in terms of operators in the basis defined by the input state, which consists of the input polarization state and the one orthogonal to it, and do the analysis. That means that, without loss of generality, we can consider the case of a horizontally polarized input photon.

Now let us use perturbation theory to see what our state is at time t if the state at time 0 was $a_H^\dagger|0\rangle$. To first order in the interaction, the state at time t will be

$$\begin{aligned} |\Psi\rangle &= a_H^\dagger|0\rangle - \frac{it}{\hbar} H' a_H^\dagger|0\rangle \\ &= a_H^\dagger|0\rangle - \frac{i\chi^* t}{\hbar}(a_V^\dagger a_H^\dagger b_H^\dagger - (a_H^\dagger)^2 b_V^\dagger)|0\rangle. \end{aligned} \tag{12.38}$$

In order to determine how good a cloning process this is, we will not use fidelity but will instead use the fraction of times a photon with the right polarization shows up in the a mode. Given that there are two photons in the a mode, the probability that we get two horizontally polarized photons is $2/3$, and the probability that we get one horizontally polarized photon is $1/3$. This gives an expected number of horizontally polarized photons of $5/3$, and, since the total number of photons in the a mode is 2, the fraction of properly polarized photons (polarization the same as the input photon) in the a mode is $5/6$.

So far, we have paid attention only to the a mode, but let us now have a look at the b mode. In that mode, it is more likely that the photon will come out vertically polarized, that is, with a polarization orthogonal to that of the input photon. Given that there is one photon in this mode, the probability that it is vertically polarized is $2/3$, and the probability that it is horizontally polarized is $1/3$. Since there is one photon in this mode, this gives a fraction of photons whose polarization is orthogonal to that of the input photon of $2/3$. The photons in the b mode have sometimes been referred to as 'anti-clones', because their polarization tends to be orthogonal to that of the input photons.

The b mode output is an example of what is known as a universal NOT, or U-NOT, gate. A perfect U-NOT gate would take a qubit in the state $|\Psi\rangle$ and map it to the state $|\Psi_\perp\rangle$, where $\langle\Psi|\Psi_\perp\rangle = 0$. That is, we would have

$$|\Psi\rangle = \alpha|0\rangle + \beta|1\rangle \quad \longrightarrow \quad |\Psi_\perp\rangle = \beta^*|0\rangle - \alpha^*|1\rangle. \tag{12.39}$$

The complex conjugates are an immediate sign that this mapping cannot be achieved. While quantum mechanical transformations are unitary, the above transformation is anti-unitary. An approximate version of this transformation can be accomplished, however, and the best fidelity that can be achieved is 2/3. In this case, unlike in the case of the cloner, a measurement-based strategy is optimal. One can achieve a fidelity of 2/3 by measuring the qubit along a random axis and then producing a qubit in the direction opposite to the one found by the measurement. For example, if we measure the qubit along the z axis and find the result $+z$, we would produce a qubit in the $-z$ direction. As we see here, however, some cloners also act as U-NOT gates. The cloner has one output that gives clones, and an additional one that gives anti-clones.

The use of a parametric amplifier as both an approximate cloner and an approximate U-NOT gate has been demonstrated in the laboratory. The cloning properties were investigated by De Martini *et al.* and Lamas-Linares *et al.* The U-NOT aspect was studied by De Martini *et al.* The figures of merit for both cloning and the U-NOT processes found in these experiments were very close to the maximum values discussed above.

12.5.3 Cloning coherent states

So far, we have only discussed cloning the polarization degrees of freedom of a photon. Instead, let us investigate cloning the state of the field mode. We cannot clone a general field mode state with any reasonable fidelity, but if the set of states we wish to clone is restricted, then we can approximately clone them. We will focus on cloning coherent states. This will only require a linear phase-independent amplifier and a beam-splitter. The basic idea is to use the amplifier to double the intensity of the coherent state and then send it through a beam-splitter with $T = R = 1/2$. The outputs of the beam-splitter are the clones, and they will not be perfect because both the amplifier and beam-splitter add noise. Let us now start with a coherent state $|\alpha_0\rangle$. This serves as the input to a linear amplifier with an intensity gain of 2. Recall that the action of a linear amplifier is described by the master equation

$$\frac{d\rho}{dt} = g(2a^\dagger\rho a - aa^\dagger\rho - \rho aa^\dagger). \tag{12.40}$$

This yields $\langle a(t)\rangle = e^{gt}\langle a(0)\rangle$, and we are interested in the case $G(t) = e^{gt} = \sqrt{2}$. Now, if the P-representation of the density matrix at time $t = 0$ is $P(\alpha, 0)$, then the P-representation at time t is

$$P(\alpha, t) = \frac{1}{\pi m(t)} \int d^2\beta\, e^{-|\alpha - G(t)\beta|^2/m(t)} P(\beta, 0), \tag{12.41}$$

where $m(t) = G^2(t) - 1$, and, in our case, we will have $m(t) = 1$. This relation can be found by turning the above master equation into a Fokker–Planck equation for the P-representation and then solving it. For the initial coherent state $|\alpha_0\rangle$, we have $P(\alpha, 0) = \delta^{(2)}(\alpha - \alpha_0)$, so that at the time when $G = \sqrt{2}$ and $m = 1$ (we shall denote the P-representation at this time by $P^{out}(\alpha)$), we find

$$P^{out}(\alpha) = \frac{1}{\pi} e^{-|\alpha - \sqrt{2}\alpha_0|^2}. \tag{12.42}$$

We now send this state through a beam-splitter. A beam-splitter is a two-mode device, and we shall denote the modes by a and b. The state from the output of the amplifier goes into the a mode, and the b mode input is the vacuum state. The beam-splitter performs the transformation

$$\int d^2\alpha\, P^{out}(\alpha)|\alpha\rangle_a\langle\alpha| \otimes |0\rangle_b\langle 0| \longrightarrow \int d^2\alpha\, P^{out}(\alpha)|\alpha/\sqrt{2}\rangle_a\langle\alpha/\sqrt{2}| \otimes |\alpha/\sqrt{2}\rangle_b\langle\alpha/\sqrt{2}|. \tag{12.43}$$

We find the reduced density matrix of one of the clones by tracing out one of the modes. These are identical, so let us consider the reduced density matrix of the a mode, ρ_a. The fidelity of the clone is then given by

$$F = \langle\alpha_0|\rho_a|\alpha_0\rangle = \int d^2\alpha\, P^{out}(\alpha)|\langle\alpha_0|\alpha/\sqrt{2}\rangle|^2, \tag{12.44}$$

which we find is equal to 2/3. Consequently, any coherent state, $|\alpha_0\rangle$, can be cloned with a fidelity of 2/3, which is independent of α_0. Though we shall not show this, 2/3 is, in fact, the best fidelity that can be achieved.

12.6 Teleportation

As was mentioned in the introduction to this chapter, what teleportation does is to move a quantum state, which can be viewed as quantum information, from one particle or system to another. The question of whether the terminology is appropriate depends on the quality of the state transfer. If the fidelity – or some other measure of the quality of state transfer – is greater than any classical 'measure and regenerate' strategy, then we are entitled to call this process quantum teleportation.

So far, the idea of beaming material objects from one place to another has no known solution. However, at least we can transport the quantum information. We will first look at this process for qubits and then go on to see how it works with continuous variables. The connection to nonlinear optics is that, in the continuous-variable case, a realistic teleportation process requires squeezed states, and these are produced by nonlinear optical processes.

In principle, if this quantum information is then used to produce a corresponding material object from some local resources, then the method could be used as a type of material object teleportation. However, this is far from easy, even for small objects like atoms.

12.6.1 Teleporting qubits

In the case of qubits, we start with a qubit in the state we wish to teleport, $|\Psi\rangle_a = \alpha|0\rangle_a + \beta|1\rangle_a$, and a singlet state, $|\Phi_-\rangle_{a'b} = (|0\rangle_{a'}|1\rangle_b - |1\rangle_{a'}|0\rangle_b)/\sqrt{2}$. Alice holds both qubits a and a', and Bob holds qubit b. Alice wants to transfer the state of qubit a onto qubit b. In order to do this, she performs a joint measurement on her two qubits. Define the Bell basis for two qubits as

$$\begin{aligned} |\Phi_\pm\rangle &= \frac{1}{\sqrt{2}}(|0\rangle|1\rangle \pm |1\rangle|0\rangle), \\ |\Psi_\pm\rangle &= \frac{1}{\sqrt{2}}(|0\rangle|0\rangle \pm |1\rangle|1\rangle), \end{aligned} \tag{12.45}$$

and notice that the three-qubit state held by Alice and Bob can be expressed as

$$\begin{aligned} |\Psi\rangle_a|\Phi_-\rangle_{a'b} = \frac{1}{2}[&|\Phi_+\rangle_{aa'}(-\alpha|0\rangle_b + \beta|1\rangle_b) - |\Phi_-\rangle_{aa'}(\alpha|0\rangle_b + \beta|1\rangle_b) \\ &+ |\Psi_+\rangle_{aa'}(\alpha|1\rangle_b - \beta|0\rangle_b) + |\Psi_-\rangle_{aa'}(\alpha|1\rangle_b + \beta|0\rangle_b)]. \end{aligned} \tag{12.46}$$

Alice now measures her two particles in the Bell basis, which implies that Bob's qubit will be in one of the four states in the above equation. For example, if Alice gets $|\Phi_-\rangle_{aa'}$ as the result of her measurement, then Bob's particle is in the state $\alpha|1\rangle_b + \beta|0\rangle_b$. Each of the four states that Bob can receive can be transformed into the original state, $|\Psi\rangle$, by an appropriate unitary transformation. Returning to our example, if Bob obtained $\alpha|1\rangle_b + \beta|0\rangle_b$, then applying the transformation σ_x, where $\sigma_x|0\rangle = |1\rangle$ and $\sigma_x|1\rangle = |0\rangle$, to his qubit will leave it in the state $|\Psi\rangle$. Therefore, after performing her measurement, Alice transmits the result of the measurement to Bob by using a classical signal, and he can then apply the appropriate linear transformation to his qubit and obtain Alice's state, $|\Psi\rangle$.

12.6.2 Continuous-variable teleportation

Next, let us look at the continuous-variable case. We will follow a treatment due to Milburn and Braunstein. Before proceeding, however, it is useful to specify the notation. We will be using eigenstates of the quadrature operators defined earlier, $X = a + a^\dagger$ and $Y = i(a^\dagger - a)$. These eigenstates are not normalizable, so they are not proper states, but they are nonetheless quite useful. We shall denote the eigenstates of X by $|x\rangle$, where $X|x\rangle = x|x\rangle$, and those of Y by $|p\rangle$, where $Y|p\rangle = p|p\rangle$. We have that

$$\langle x|x'\rangle = \delta(x - x'), \qquad \langle p|p'\rangle = \delta(p - p'), \tag{12.47}$$

and, since they are complete and orthonormal, both sets of states form resolutions of the identity,

$$I = \int_{-\infty}^{\infty} dx\, |x\rangle\langle x| = \int_{-\infty}^{\infty} dp\, |p\rangle\langle p|. \tag{12.48}$$

The two types of states can be expressed in terms of each other as

$$|x\rangle = \frac{1}{2\sqrt{\pi}} \int_{-\infty}^{\infty} dp\, e^{-ixp/2} |p\rangle, \qquad |p\rangle = \frac{1}{2\sqrt{\pi}} \int_{-\infty}^{\infty} dx\, e^{ixp/2} |x\rangle. \tag{12.49}$$

It is also worth noting that $\exp(ip'X/2)$ is the translation operator for the $|p\rangle$ states, that is

$$\begin{aligned} e^{ip'X/2}|p\rangle &= \int_{-\infty}^{\infty} dx\, e^{ip'x/2} |x\rangle\langle x|p\rangle \\ &= \frac{1}{2\sqrt{\pi}} \int_{-\infty}^{\infty} dx\, e^{ix(p+p')/2} |x\rangle = |p+p'\rangle. \end{aligned} \tag{12.50}$$

In the teleportation protocol, we will have need of two-mode states, where we have denoted the modes by a and b, of the form

$$|\Psi(x,p)\rangle_{ab} = e^{-iY_a X_b/2} |x\rangle_a |p\rangle_b. \tag{12.51}$$

This state is an eigenstate of $X_a - X_b$ with eigenvalue x and of $Y_a + Y_b$ with eigenvalue p. Let us demonstrate the first of these statements. From the Baker–Hausdorff theorem (see Chapter 2), we have that

$$e^{iY_a X_b/2}(X_a - X_b) e^{-iY_a X_b/2} = X_a. \tag{12.52}$$

We then have that

$$\begin{aligned} (X_a - X_b)|\Psi(x,p)\rangle_{ab} &= e^{-iY_a X_b/2} e^{iY_a X_b/2} (X_a - X_b) e^{-iY_a X_b/2} |x\rangle_a |p\rangle_b \\ &= e^{-iY_a X_b/2} X_a |x\rangle_a |p\rangle_b = x|\Psi(x,p)\rangle. \end{aligned} \tag{12.53}$$

Now let us move on to the protocol. We start with the three-mode state $|\Psi\rangle_{a'}|\Psi(x_1,p_1)\rangle_{ab}$, where Alice has modes a and a', Bob has mode b, and $|\Psi\rangle_{a'}$ is the state Alice wishes to teleport. Alice now measures the commuting variables $X_{a'} - X_a$ and $Y_{a'} + Y_a$ on her part of the state. Let us say that she gets x_2 for the first and p_2 for the second. The state that is an eigenstate of these two variables with eigenvalues x_2 and p_2 is

$$|\Phi(x_2,p_2)\rangle_{aa'} = e^{-iX_a Y_{a'}/2} |p_2\rangle_a |x_2\rangle_{a'}. \tag{12.54}$$

Therefore, our three-mode state, up to normalization, after Alice's measurement is

$$|\Phi(x_2,p_2)\rangle_{aa'} Q(x_1,x_2,p_2)|p_1\rangle_b, \tag{12.55}$$

where $Q(x_1,x_2,p_2)$ is an operator that acts on the b mode and is given by

$$Q(x_1,x_2,p_2) = {}_{a'}\langle x_2|\, {}_a\langle p_2| e^{iX_a Y_{a'}/2} e^{-iY_a X_b/2} |\Psi\rangle_{a'} |x_1\rangle_a. \tag{12.56}$$

In order to proceed, we have to evaluate $Q|p_1\rangle_b$. We first use the Baker–Hausdorff theorem to reorder the exponentials:

$$e^{iX_a Y_{a'}/2} e^{-iY_a X_b/2} = e^{-iY_a X_b/2} e^{iX_a Y_{a'}/2} e^{iY_{a'} X_b/2}. \tag{12.57}$$

Inserting this into the expression for Q, we find that

$$Q|p_1\rangle_b = {}_a\langle p_2|x_1\rangle_a\, e^{-ip_2 X_b/2}\, {}_{a'}\langle x_2| e^{i(x_1+X_b)Y_{a'}/2} |\Psi\rangle_{a'}\, |p_1\rangle_b. \tag{12.58}$$

We now have that

$$
\begin{aligned}
&{}_{a'}\langle x_2|e^{i(x_1+X_b)Y_{a'}/2}|\Psi\rangle_{a'}\,|p_1\rangle_b \\
&= \int_{-\infty}^{\infty} dp'\,{}_{a'}\langle p'|\Psi\rangle_{a'}\,{}_{a'}\langle x_2|p'\rangle_{a'}\,e^{i(x_1+X_b)p'/2}|p_1\rangle_b \\
&= \frac{1}{2\sqrt{\pi}}\int_{-\infty}^{\infty} dp'\,{}_{a'}\langle p'|\Psi\rangle_{a'}e^{i(x_1+x_2)p'/2}|p_1+p'\rangle_b \\
&= \frac{1}{2\sqrt{\pi}}e^{ip_1X_b/2}e^{i(x_1+x_2)Y_b/2}|\Psi\rangle_b.
\end{aligned} \tag{12.59}
$$

Finally, putting everything together, we have for the three-mode state after Alice has made her measurements:

$$
|\Psi(x_2, p_2)\rangle_{aa'}\frac{1}{4\pi}\,e^{-ip_2x_1/2}e^{i(p_1-p_2)X_b/2}e^{i(x_1+x_2)Y_b/2}|\Psi\rangle_b. \tag{12.60}
$$

At this point, Bob's state differs from the one Alice is trying to teleport by an inconsequential phase factor and two unitary transformations. If Alice tells Bob the results of her measurements, x_2 and p_2, Bob then knows what the two unitary transformations are, and he can apply their inverses to his state to obtain $|\Psi\rangle_b$.

What we have presented is the idealized procedure. In reality, rather than starting with $|x_1\rangle_a|p_1\rangle_b$, one would start with a product of two one-mode squeezed states that approximate this state, i.e.

$$
\left(\frac{2}{\pi}\right)^{1/2} e^r \int_{-\infty}^{\infty} dx' \int_{-\infty}^{\infty} dp'\, e^{-e^{2r}(x'-x_1)^2}e^{-e^{2r}(p'-p_1)^2}|x'\rangle_a|p'\rangle_b. \tag{12.61}
$$

Using this state instead of the idealized one will result in a fidelity of less than one between the state that is received by Bob and the state that was sent by Alice. The greater the squeezing, that is, the larger the value of r, the closer to one the fidelity will be. The teleportation of coherent states using two-mode squeezed states as a resource was carried out by Furusawa *et al.* in 1998.

Additional reading

Books

- Useful textbooks in quantum measurement and information are:

S. M. Barnett, *Quantum Information* (Oxford University Press, New York, 2009).

J. S. Bell, *Speakable and Unspeakable in Quantum Mechanics* (Cambridge University Press, Cambridge, 1987).

J. Bergou and M. Hillery, *Introduction to the Theory of Quantum Information Processing* (Springer, New York, 2013).

M. Nielsen and I. Chuang, *Quantum Computation and Quantum Information* (Cambridge University Press, Cambridge, 2011).

J. von Neumann, *Mathematical Foundations of Quantum Mechanics* (Princeton University Press, Princeton, NJ, 1955).

Articles

- Six review papers are recommended. The first two on Bell and EPR inequalities, respectively, are:

J. F. Clauser and A. Shimony, *Rep. Prog. Phys.* **41**, 1881 (1978).
M. D. Reid *et al.*, *Rev. Mod. Phys.* **81**, 1727 (2009).

- The next three on quantum cloning are:

V. Buzek and M. Hillery, *Phys. World* **14** (11), 25 (2001).
N. Cerf and J. Fiurasek, *Prog. Opt.* **49**, 455 (2006).
V. Scarani, S. Iblisdir, N. Gisin, and A. Acin, *Rev. Mod. Phys.* **77**, 1225 (2005).

- And the last on continuous-variable quantum information is:

S. L. Braunstein and P. van Loock, *Rev. Mod. Phys.* **77**, 513 (2005).

- For a paper on hidden variables, joint probability and the Bell inequalities, see:

A. Fine, *Phys. Rev. Lett.* **48**, 291 (1982).

- Three papers on the foundations of quantum mechanics are:

A. Aspect, P. Grangier, and G. Roger, *Phys. Rev. Lett.* **49**, 91 (1982).
J. S. Bell, *Physics* **1**, 195 (1965).
A. Einstein, B. Podolsky and N. Rosen, *Phys. Rev.* **47**, 777 (1935).

- And three on experimental investigation of quantum cloning are:

F. De Martini, V. Mussi, and F. Bovino, *Opt. Commun.* **179**, 581 (2000).
F. De Martini, V. Buzek, F. Sciarrino, and C. Sias, *Nature* **419**, 815 (2002).
A. Lamas-Linares, C. Simon, J. Howell, and D. Bouwmeester, *Science* **296**, 712 (2002).

- The paper on which we based our treatment of continuous-variable teleportation is:

G. Milburn and S. Braunstein, *Phys. Rev.* A **60**, 937 (1999).

- The experimental teleportation of coherent states is described in:

A. Furusawa, J. Sorensen, S. Braunstein, C. Fuchs, H. Kimble, and E. Polzik, *Science* **282**, 706 (1998).

Problems

12.1 Using the result in Eq. (5.73), find the state of the field at time t resulting from the interaction given in Eq. (12.36) if the input field is $(1/\sqrt{2})(a_H^\dagger)^2|0\rangle$. Use your result to show how good $2 \to m$ cloning is using the same figure of merit that we did for $1 \to 2$ cloning. That is, given that there are m photons in the output state, find the expected number that have the same polarization as the input photons and then divide by m.

12.2 Find the Fokker–Planck equation for the P-representation that is equivalent to the master equation, Eq. (12.40), and show that the P-representation in Eq. (12.41) is a solution to it.

12.3 It is possible to design state-dependent cloners that work better for a specific class of states than does a cloner that copies all states equally well. An example is the phase-covariant cloner that clones qubit states of the form $|\psi(\phi)\rangle = (|0\rangle + e^{i\phi}|1\rangle)/\sqrt{2}$. It works by taking the input state $|\psi(\phi)\rangle_a|0\rangle_b$ and applying to it the unitary transformation, U, that acts as follows:

$$U(|0\rangle_a|0\rangle_b) = |0\rangle_a|0\rangle_b, \qquad U(|1\rangle_a|0\rangle_b) = \frac{1}{\sqrt{2}}(|1\rangle_a|0\rangle_b + |0\rangle_a|1\rangle_b).$$

(a) Find the reduced density matrix of one of the clones, and calculate its fidelity to $|\psi(\phi)\rangle$.

(b) This transformation can be realized conditionally by a beam-splitter with different transmissivities for different polarizations. The polarization degrees of freedom of the photons are our qubits, with vertical polarization corresponding to $|0\rangle$ and horizontal polarization corresponding to $|1\rangle$. Suppose that the beam-splitter acts on the two modes a and b as

$$a_j^\dagger \to t_j a_j^\dagger + r_j b_j^\dagger, \qquad b^\dagger \to t_j b_j^\dagger - r_j a_j^\dagger,$$

where $j = V, H, r_j$ and t_j are real, and $t_j^2 + r_j^2 = 1$. If we only accept the output from the beam-splitter when there is one photon in each mode, find values of t_V and t_H that realize the cloning transformation given above. What is the probability that there will be only one photon in each mode at the beam-splitter output?

12.4 Using the state in Eq. (12.61) instead of $|x_1\rangle_a|p_1\rangle_b$, with $x_1 = p_1 = 0$, as part of the input state to the teleportation process, find the state that Bob receives. Calling Bob's state $|\psi_B|\rangle$, find $\langle x|\psi_B\rangle$ in terms of $\langle x|\psi\rangle$, where $|\psi\rangle$ is the state Alice wants to teleport.

List of symbols

Quantity	Symbol	SI units and/or typical value
Position	$\mathbf{r} = (x, y, z)$	m
Electric field	$\mathbf{E}$	10^5–10^{10} V m^{-1}
Magnetic field	$\mathbf{H}$	A m^{-1}
Electric displacement	$\mathbf{D}$	C m^{-2}
Magnetic induction	$\mathbf{B}$	10^{-3}–10^2 T
Polarization field	$\mathbf{P}, \mathbf{P}^L, \mathbf{P}^{NL}$	C m^{-2}
Magnetization field	$\mathbf{M}$	A m^{-1}
(Vacuum) permittivity	$(\epsilon_0), \epsilon$	$8.854\,18 \times 10^{-12}$ C V^{-1} m^{-1}
(Vacuum) permeability	$(\mu_0), \mu$	$1.256\,63 \times 10^{-6}$ N A^{-2}
nth-order polarizability tensor	$\boldsymbol{\alpha}^{(n)}$	$(\text{m V}^{-1})^{n-1}$
nth-order susceptibility tensor	$\boldsymbol{\chi}^{(n)}$	$(\text{m V}^{-1})^{n-1}$
nth-order permittivity tensor	$\boldsymbol{\epsilon}^{(n)}$	C m^{-2} $(\text{m V}^{-1})^n$
nth-order inverse permittivity	$\eta^{(n)}$	V m^{-1} $(\text{m}^2\ \text{C}^{-1})^n$
(Vacuum) wavelength	λ	~ 1 μm
Vacuum speed of light	c	$2.997\,924\,58 \times 10^8$ m s^{-1}
Group, phase velocity	v, v_p	$\sim 2 \times 10^8$ m s^{-1}
Refractive index	n_r	~ 1.5
Reduced Planck's constant	$\hbar \equiv h/(2\pi)$	1.05457×10^{-34} J s
Bohr radius	a_0	5.2917×10^{-11} m
Wavevector	$k = 2\pi/\lambda$	$\sim 10^6$ m^{-1}
Slowly varying field envelope	$\mathcal{E}, \mathcal{A}$	10^5–10^{10} V m^{-1}
Intensity (Poynting vector)	I	10^9–10^{20} W m^{-2}
(Peak) power	P	10^{-3}–10^6 W
Waist radius	W_0	10 μm–1 mm
Number density	ρ_n	10^{15}–10^{30} m^{-3}
Linear electric susceptibility tensor	$\chi_{jk}^{(1)}$	0–1
Second-order electric susceptibility tensor	$\chi_{jkl}^{(2)}$	2×10^{-12} m V^{-1}
Third-order electric susceptibility tensor	$\chi_{jklm}^{(3)}$	4×10^{-24} m^2 V^{-2}
Hamiltonian density, Hilbert space	$\mathcal{H}$	J m^{-3}
Oscillation frequency	f	$\sim 10^{14}$ s^{-1}
Atomic resonance (angular) frequency	ν_0, ν_j	rad s^{-1}
Rabi (angular) frequency	Ω	rad s^{-1}
Laser carrier (angular) frequency	$\omega = 2\pi f$	$\sim 10^{15}$ rad s^{-1}
Dipole moment vector	$\mathbf{d}$	10^{-29} C m
Space dimension	d	—
Occupation (photon) number	$\mathbf{n}$	—

Quantity	Symbol	SI units and/or typical value
Total photon number	N	—
Total mode number	M	—
Hamiltonian	H	J
Lagrangian	L	J
Lagrangian density	$\mathcal{L}$	$\mathrm{J\,m^{-3}}$
Raman coupling	α^R	$\mathrm{m^{-1}}$
Damping rate	γ	10^6–$10^9\ \mathrm{s^{-1}}$
Lagrangian action, S-matrix	S	J s
Particle position	$\mathbf{r}_\alpha$	m
Fixed (nuclear) particle position	$\mathbf{R}_\alpha$	m
Transverse coordinate	$\mathbf{r}_\perp = (y, z)$	m
Particle charge	q, q_α	C
Elementary charge	e	$+1.602\,17 \times 10^{-19}$ C
Fixed (nuclear) charge	Q_α	C
Linear input–output matrices	$\mathcal{M}, \mathcal{L}$	—
Coherent displacement	α, β^*	—
Creation, annihilation operators	$a^\dagger, a$	—
Vector potential	$\mathbf{A}$	T m
Dual potential	Λ	$\mathrm{C\,m^{-1}}$
Mode function	$\mathbf{u}_n$	$\mathrm{m^{-3/2}}$
Group velocity	v_k	$\mathrm{m\,s^{-1}}$
Quantization volume	V	$\mathrm{m^3}$
Pauli spin matrix	σ^x	$\begin{bmatrix} 0 & 1 \\ 1 & 0 \end{bmatrix}$
Pauli spin matrix	σ^y	$\begin{bmatrix} 0 & -i \\ i & 0 \end{bmatrix}$
Pauli spin matrix	σ^z	$\begin{bmatrix} 1 & 0 \\ 0 & -1 \end{bmatrix}$
Pauli matrix (raising/lowering)	$\sigma^{(\pm)}$	$[\sigma^x \pm i\sigma^y]/2$
Real quantum field	ϕ	$\mathrm{m}^{-d/2}$
Complex quantum field	ψ	$\mathrm{m}^{-d/2}$
Quantum state	$\lvert\Psi\rangle, \lvert\Phi\rangle$	—
Normalized coherence	$g^{(n)}$	—
Quadrature	X, Y	—
Photon flux amplitude	b^{in}, b^{out}	$\mathrm{s^{-1/2}}$
Characteristic	C_s	—
P-representation	P	—
Gaussian noise field	$\xi(x, t)$	$(\mathrm{m\,s})^{-1/2}$

Index

Printed in the United States
By Bookmasters